WHEN CELLS DIE

WHEN CELLS DIE
A Comprehensive Evaluation of Apoptosis and Programmed Cell Death

Edited by

RICHARD A. LOCKSHIN, Ph.D.
ZAHRA ZAKERI, Ph.D.
and JONATHAN L. TILLY, Ph.D.

 **WILEY-LISS**

A JOHN WILEY & SONS, INC., PUBLICATION

New York • Chichester • Weinheim • Brisbane • Singapore • Toronto

This text is printed on acid-free paper ∞

Copyright @ 1998 Wiley-Liss, Inc.

Published simultaneously in Canada.

While the authors, editor, and publisher believe that drug selection and dosage and the specification and usage of equipment and devices, as set forth in this book, are in accord with current recommendations and practice at the time of publication, they accept no legal responsibility for any errors or omissions, and make no warranty, express or implied, with respect to material contained herein. In view of ongoing research, equipment modifications, changes in governmental regulations and the constant flow of information relating to drug therapy, drug reactions, and the use of equipment and devices, the reader is urged to review and evaluate the information provided in the package insert or instructions for each drug, piece of equipment, or device for, among other things, any changes in the instructions or indication of dosage or usage and for added warnings and precautions.

Library of Congress Cataloging in Publication Data

When Cells Die / edited by Richard A. Lockshin, Zahra Zakeri
   and Jonathan L. Tilly.
      p.   cm.
   "A Wiley-Liss publication."
   Includes bibliographical references and index.
   ISBN 0-471-16569-7 (cloth : alk. paper)
   1. Apoptosis. I. Lockshin, R. A. (Richard A.) II. Tilly,
Jonathan L., 1962-   . III. Zakeri, Zahra, 1952-
   [DNLM: 1. Apoptosis—physiology.   QH 671 P578 1988]
  QH671.P49 1998
  571.9'36—dc21
  DNLM/DLC
  for Library of Congress   97-23884

Printed in the United States of America

10 9 8 7 6 5 4 3 2

CONTENTS

PREFACE

When we started this book, our plans were to help those not yet captive of the field to evaluate the meaning and impact of cell death on what they do. We therefore chose to ask scientists we considered to be interesting, thoughtful, and provocative to reflect on their subjects. Our contributors have risen to this challenge, and have brought forth suggestions that truly enable us to take a larger, more distant, more global view of the field. They have addressed the conceptual and medical significance of cell death. They cover the subject in many tissues and circumstances, as a part of normal development and homeostasis, as a means of defense, and as a pathology. They frequently step back to ask what it all means. This should be a pleasant and enlightening surprise for readers who have wondered what all the fuss was about or who have struggled with rapidly proliferating and hermetic lists of genes, mechanisms, and definitions. In our several sections we address the evolutionary significance and mechanisms of physiological cell death, the means of evaluating it, major areas and situations in which one encounters cell death, and the medical implications of understanding and regulating cell death. Where our authors have touched on several of these aspects, we have attempted to provide sufficient cross-referencing and indexing—printed hypertext—that the reader will be able to locate these ideas easily. These cross-references to other chapters in the book are indicated by [brackets] in the text.

A goal of this effort, and a concern to all of us, is that our readers will not limit themselves to the special type of expertise in which they are interested but will explore all the chapters, for there is much to gain from the cross-fertilization of the several fields that we address. The image of cell death conjured by someone who studies highly elaborated, postmitotic cells with massive cytoplasm is distinctly different from that seen by someone who studies mitotic cells with little cytoplasm and planned or programmed evanescence. These differences of images are extemely provocative, and provide mutual insights as, for instance, one contemplates "partial apoptosis" or "pre-apoptotic changes." It is also gratifying to realize that, in spite of the enlighteningly different viewpoints, there is substantial consensus. The consensus exists both in recognition of more general truths and in the sense of where we must go in the field.

To address these questions we have divided the book into five major sections. In the first section, "The Phenomenon of Cell Death," Trump and

Berezesky address the distinction between necrotic and physiological cell death, emphasizing that most of our assays deal only with *postmortem* events. These themes are picked up elsewhere (by Sikorska and Walker, Papermaster, and Tidball and Albrecht). Ameisen, starting with a consideration of programmed cell death in protozoans and prokaryotes, looks at the origin of cell death as a program, provocatively suggesting that the death–antideath mechanism may be inherited as a module. This hypothesis provides a rationale for localization of the death–survival distinction at the mitochondria–cytoplasm interface as well as the close relationship between cell cycle and cell death. It further suggests that apoptosis genes may perform other important functions—a consideration in contemplated therapies—and sets the stage for later reflections (especially by Mittler, Finkel, and Casella, and Cryns and Yuan) on the efforts by pathogens to prevent cell death and by the host to destroy infected cells. Zakeri outlines the several types of known cell deaths, setting the stage for comparisons between the phenomena described primarily in Section III, "Cell Death Where Mitosis is High and Evanescence is Desirable," and Section IV, "Cell Death in Long-Lived Cells." Gartner and Hengartner continue by reviewing the clarity of the genetics of cell death mechanisms as established in *Caenorhabditis,* which has helped so much to define the field. Finally, Mittler describes cell death in plants, bringing to our attention a branch often neglected but of powerful biological and medical importance. In this well-detailed essay, he again catches the themes of the different appearances of death in different tissues and the struggle between eukaryotes and prokaryotes.

In Section II, "Themes and Approaches to Cell Death," Cryns and Yuan open with one of today's most dominant themes: the role of proteases. They emphasize how higher organisms have elaborated on the simple themes of *Caenorhabditis,* generating redundant and complex proteolytic cascades, often following rapidly changing sequences. In this discussion they also address several themes that echo through the book, including differential vulnerability of cells (discussed later also by Osborne), the importance of the cytoskeleton, and the mechanisms of the formation of classical apoptotic bodies. These latter arguments are picked up by several authors. Finally, they close with discussions of new techniques and very promising therapeutic considerations based on control of proteases. Sikorska and Walter continue the consideration of proteases by emphasizing that proteases initiate the generation of the well-known DNA ladder, bringing in the intellectually obvious but virtually always forgotten point that the appearance of the ladder reflects the failure of final digestion of the DNA, rather than a specific cleavage. Both Cryns and Yuan and Sikorska and Walker are concerned with the question of the specificity of the proteases and the significance of the proteolytic targets; their thoughts are well worthy of consideration by the coming generation of scientists and physicians. Finally, Zakeri addresses more generally the question of how one assesses cell death. This chapter deserves particular attention by students in the field, especially in the light of Tidball and Albrecht's impressive calculations

(Section IV) that a tissue could disappear in 20 days with only four apoptotic cells being visible in any given microscopic section.

Section III, "Cell Death Where Mitosis is High and Evanescence Is Desirable," addresses the overall issue of cell death in the most well-known but to an outsider most inscrutable field, the members of our hematopoietic and lymphatic systems. Osborne takes a global view, emphasizing and elaborating points that are seen elsewhere, such as the dependence of cells on appropriate signals ("cells can die of neglect"), the mechanism by which the interaction of Fas and Fas Ligand can kill cells of the immune system, and the role of proteases. Among these cells and probably others for which inquiry is lacking, proteasomes and ubiquitin-mediated proteolysis are important components of the destruction of the cell. Newell and Vincent follow by pointing out that it is important for an organism to eliminate activated lymphocytes once the crisis has passed, and describe how the interpretation of variant complex, membrane-derived, signals determines whether a cell should proliferate or die. Again, the theme of a highly complex signaling mechanism, used to generate a subtle distinction between a growth response and a death response, is reflected in several other presentations, most prominently those of Budd, Birge *et al.,* Tilly, and, as described in the next paragraph, Papermaster. Budd follows a theme similar to that of Newell and Vincent, but this time pointing out that, in autoimmune disease, the mechanism of inflammation may be that, in chronic infection, FasL lymphocytes are inadvertently brought into long-term contact with Fas-bearing cells, thus provoking an unwanted response and death of bystander cells. Finally in this section, Finkel and Casella examine the situation in AIDS, emphasizing the interest of the virus in protecting its host cell, and of the host in destroying the infected cell. Like Budd, they conclude that the resulting carnage ultimately involves bystander cells. All these authors, recognizing the complexity of their subject, present nicely complementary flow sharts and diagrams as guides to the sequence of events.

Section IV, "Cell Death in Long-Lived cells," turns to a different viewpoint. In looking at cells apart from the immune system—lens and retina, PC-12 cells and differentiated neurons, and muscle—the respective authors consistently recognize the importance of the cytoskeleton and extracellular matrix. While these themes appear elsewhere (as, for instance, in the discussion of proteases by Cryns and Yuan and Sikorska and Walker or in the discussions by Trump and Berezesky, Budd, and Tidball and Albrecht on the mechanisms of bleb formation), the consideration of extracellular matrix is not particularly relevant in the discussion of cells that are, at least for a part of their lives, motile, and are frequently studied in suspension cultures. However, as these authors uniformly conclude, attachment to the extracellular matrix communicates directly through the cytoskeleton to the nuclear matrix and therefore to the chromosomes. They therefore link these apparently different forms of death directly to considerations of the targets of the proteolytic caspases, which have been studied primarily in the lymphatic system. Not only does the issue of extracellular matrix form a theme throughout this section, virtually all authors conclude that the attachment of the cell to the substratum and to specific substratum molecules changes the shape and metabolism of the cell, including its susceptibility to apoptosis. Such an argument has been raised elsewhere (Feldherr and Akin, 1994).

Differential susceptibility to apoptosis is a major theme to these latter authors and returns as a theme in the final section. In this section it appears where Papermaster explores the problem of why only certain cells of the retina die, and why it takes them so long to do so; when Birge, Fajardo, and Hempstead question how it is that minor variants in initial signals divert some cells to growth pathways and others to death pathways; as Cotman *et al.* wrestle with the realization that apoptosis can start in part of a cell (the neurites) and progress to the cell body; and when Cotman *et al.* and Tidball and Albrecht emphasize the futility of examining cell death in *in vitro* models on artificial substrates. Cotman *et al.* also identify what appear to be long-sought-after preapoptotic signals, reflecting Trump's plea for *premortem* analysis, and giving hope that we will be able to identify cells in agony rather than dead. The concept of differential susceptibility generates a further theme, emphasized by Birge *et al.* and based on the general recognition that the pathways of cell death and cell division initially resemble each other and only later diverge. This theme is that the interaction of a membrane receptor molecule with one or more ligands generates an activation that can go in many directions, depending on the nature and timing of appearance of the ligands and the state of metabolism of the cell, including its shape and prior history. Although this theme makes very graphic sense in the context of an extended attached cell, it is surprisingly similar to the concept of activation timing in lymphocytes as expressed by Osborne and by Newell and Vincent, and to Papermaster's cogent observation, based on transgenic mice, that a difference in timing of a few days can turn a pro-proliferation signal into a pro-apoptotic signal.

Our last section, SectionV, "The Clinical Relevance of Apoptosis," implies a division that is in a sense artificial. Most authors address clinical issues and suggest directions that can be taken, and chapters such as those of Budd, Finkel, and Cotman *et al.* directly examine disease states. Both authors in this final section look at mechanisms of death in a manner similar to that used elsewhere in the book. We placed these two chapters here, with extensive cross-referencing, because both authors address problems of major medical importance, fertility and cancer, and because both take a broader overview of the many potential considerations in examining a medical problem. They address, for instance, the possibility that reactive oxygen groups can initiate or provide a signaling pathway for cell death, the recent discoveries of potential roles for the anti-apoptotic gene bcl-2, the interactions of the pro-apoptotic gene p53 and bcl-2, the role of substrate and proteases, and the impact of chemotherapy or environmental toxicants on cell death. Although virtually all authors address the future therapeutic implications—see, especially, the chapters by Budd, Cotman *et al.,* Finkel and Casella, Papermaster, and Cryns and Yuan—these two effectively bring back the general theme that there is great potential for human good in learning to regulate physiological cell death.

Throughout the book, the clearest consensus is that an organism uses cell death in a very positive way—to sculpt its development, to arrange for rapid expansion and subsequent contraction of a cell population in the immune and reproductive systems, and to defend itself by destroying cells that have been infected or attacked. To do so, cells and organisms constantly monitor signals given and received by cells, and respond like secret service agents: An incorrect or confused response means death. This high-security surveillance is a wise choice, for the goal of the attacker is to preserve the cell it is attacking and to use the cell for its own purposes. Thus the bulk of research has looked at death and survival signals, which are many, but which fit a pattern in which infectious pathological agents or potentially malignant cells attempt to stave off death (anti-apoptotic genes) while the organism tries to spot these dangers and invoke cell suicide (pro-apoptotic genes). Occasionally, as in any war, these battles spill out into civilian territory, and, in AIDS and autoimmune inflammatory disease, bystanders die.

A second theme that is easily identifiable is that we should not be confused by occasional differences in morphology or molecular biology. Proteases may differ, but they accomplish the same goals. The morphology may differ, but the differences are perhaps more attributable to biological distinctions—the structure of the cell and the relative urgency of preventing mitosis of a damaged cell, for example. Ultimately, all cells dying by physiological means are more similar than different.

A third theme is a warning related to these differences: Most of our means of identifying and thereby studying dying cells are either measurements of *post mortem* changes or are peculiar concatenations of rate-limiting downstream events. Trump and Berezesky, for instance, argue that the vaunted cell shrinkage that is part of the classical morphology is a physical result of energy available to the cell as calcium enters and disrupts the metabolism of the cell, while Sikorska and Walker note that the other major marker, nucleosomal ladder, probably derives from inhibition of later degradation, and several authors indicate that specific attack on the cytoskeleton produces the third major marker, fragmentation and blebbing. Because of these and similar challenges to our too-facile assumptions, virtually all authors argue that the casualness with which we have addressed our identification and evaluation of cell death will no longer do. We must find means of accurately identifying the threatened or dying cell, rather than the dead cell. It is highly unlikely that a single event will define all types of cell death, or that the phenomena that we use to characterize apoptosis or programmed cell death will exclusively be associated with it. We might better consider the basic biology of the dying cell. For instance, electron microscopy might well reveal putative changes in mitochondria, or the relationship between intracellular levels and kinetics of ceramide and the 500-fold higher levels used to analyze the role of ceramide in cell death might be examined.

This then is the challenge for the student or mature scientist who wants to know what this field is about (and to whom we have addressed this effort): Programmed cell death and apoptosis are very important aspects of a healthy life, and our access to manipulation of it will have vast consequences in many fields of medicine and agriculture. We need to understand and recognize how it comes about. Our authors raise this challenge to you. Our research to present has been good enough for us to recognize that it is superficial. What we desperately need to know is how a cell can integrate a multiplicity of signals, how we can from the exterior read these signals, and how we can use this information to regulate the health of the organism. We hope that this work provides you with the background to provoke new and insightful experiments.

One final note: We considered including here a history of the field but decided that the other subjects took precedence. The history is an interesting subject, especially insofar as much of it occurred before MedLine and is consequently not readily accessible. Some of this story can be found in articles by Majno and Joris (1995), by Clarke and Clarke (1995, 1996), and by Lockshin (1997). Our readers may wish to know the following items: The term "programmed cell death" was, to our knowledge, first used in a doctoral thesis in 1963 and in the publications that resulted (Lockshin, 1963; Lockshin and Williams, 1964, 1965a–d). As in any field, it derived from ideas then extant, most notably those of John Saunders (cf. Saunders, Jr., 1966; Saunders, Jr., and Fallon, 1966). "Apoptosis" was first used by Kerr, Wyllie, and Currie, in 1972 (Kerr *et al.,* 1972). Here, Kerr's immediately previous papers provided the intellectual framework for the more grandiose concept (Kerr, 1965, 1971). The word is most commonly pronounced ăpo′-TO-sis—the "a" as in "apple" and the "to" as in "toe"—or Āpo′-TO-s′is—the "a" as in "ape"—or, especially in North America, ă-pop-TO-sis, the "a" as the first "a" in "abnormal" and the second "p" pronounced on the theory that the "p" of the "pter" root in "helicopter" is pronounced. According to Alexander Antonopoulos (email communication), when "pt" appears at the beginning of the word, the "p" is silent, but when "pt" appears in the middle of the word, the "p" is pronounced.

Richard A. Lockshin
New York, April, 1997

REFERENCES

Clarke PGH, Clarke S (1995): Historic apoptosis. Nature 378: 230.

Clarke PGH, Clarke S (1996): Nineteenth century research on naturally occurring cell death and related phenomena. Anatomy and Embryology 193: 81–99.

Feldherr CM, Akin D (1994): Role of nuclear trafficking in regulating cellular activity. Int Rev Cytol 151: 183–228.

Kerr JFR (1965): A histochemical study of hypertrophy and ischaemic injury of rat liver with special reference to changes in lysosomes. J Path Bact 90: 419–35.

Kerr JFR (1971): Shrinkage necrosis: A distinct mode of cellular death. J Pathol 105: 13–20.

Kerr JFR, Wyllie AH, Currie AR (1972): Apoptosis: a basic biological phenomenon with wide-ranging implications in tissue kinetics. Br J Cancer 26: 239–57.

Lockshin RA (1963): Programmed cell death in an insect. Doctoral dissertation, Harvard University, Department of Biology.

Lockshin RA (1997): The early modern period in cell death. Cell Death Differ 4: 347–351.

Lockshin RA, Williams CM (1964): Programmed cell death. II. Endocrine potentiation of the breakdown of the intersegmental muscles of silkmoths. J Insect Physiol 10: 643–49.

Lockshin RA, Williams CM (1965a): Programmed cell death. I. Cytology of the degeneration of the intersegmental muscles of the Pernyi silkmoth. J Insect Physiol 11: 123–33.

Lockshin RA, Williams CM (1965b): Programmed cell death. III. Neural control of the breakdown of the intersegmental muscles. J Insect Physiol 11: 605–10.

Lockshin RA, Williams CM (1965c): Programmed cell death. IV. The influence of drugs on the breakdown of the intersegmental muscles of silkmoths. J Insect Physiol 11:803–9.

Lockshin RA, Williams CM (1965d): Programmed cell death. V. Cytolytic enzymes in relation to the breakdown of the intersegmental muscles of silkmoths. J Insect Physiol 11: 831–44.

Majno G, Joris I (1995): Apoptosis, oncosis, and necrosis: An overview of cell death. American Journal of Pathology 146: 3–15.

Saunders JW, Jr. (1966): Death in embryonic systems. Science 154: 604–12.

Saunders JW, Jr., Fallon JF (1966): Cell death in morphogenesis. In Locke M. (ed): Major Problems in Developmental Biology, 25th Symp. Soc. Devel. Biol., New York: Academic Press, pp 289–314.

CONTRIBUTORS

RICHARD A. LOCKSHIN, PH.D., Department of Biological Sciences, St. John's University, 8000 Utopia Parkway, Jamaica, NY 11439

ZAHRA ZAKERI, PH.D., Department of Biology, Queens College and Graduate Center of CUNY, 65-30 Kissena Blvd., Flushing, NY 11367

JONATHAN L. TILLY, PH.D., Dept of Obstetrics, Gynecology and Reproductive Biology, Harvard Medical School and The Vincent Center for Reproductive Biology, Massachusetts General Hospital, VBK137E-GYN, 55 Fruit Street, Boston, MA 02114

DOUGLAS E. ALBRECHT, Department of Physiological Science, University of California, Los Angeles, CA 90095-1527

JEAN CLAUDE AMEISEN, Paris University VII, Xavier School of Medicine, Inserm CJF 9707, Hopital Bichat-Claude Bernard, Paris, France

LUIZ BENÍTEZ-BRIBIESCA, M.D., "Leon Weiss" Professor of Oncology, University of Mexico (UNAM), Chief, Oncological Research Unit, Oncology Hospital, National Medical Center, IMSS, Mexico, D. F. Mexico

IRENE K. BEREZESKY, Department of Pathology, University of Maryland School of Medicine, Baltimore, MD 21201

RAYMOND B. BIRGE, Laboratory of Molecular Oncology, The Rockefeller University, 1230 York Avenue, New York, NY 10021

RALPH C. BUDD, M.D., Immunobiology Program, The University of Vermont College of Medicine, Burlington, VT 05405-0068

CAROLYN R. CASELLA, Basic Sciences, National Jewish Center, 1400 Jackson St., Denver, CO 80206

DAVID H. CRIBBS, Institute for Brain Aging and Dementia, University of California Irvine, Irvine, CA 92697-4540

CARL W. COTMAN, Institute for Brain Aging and Dementia, University of California Irvine, Irvine, CA 92697-4540

VINCENT L. CRYNS, Center for Endocrinology, Northwestern University, Tarry 15-755, 303 E. Chicago Ave., Chicago, IL 60611

J. EDUARDO FAJARDO, Laboratory of Molecular Oncology, The Rockefeller University, 1230 York Avenue, New York, NY 10021

TERRI H. FINKEL, Basic Sciences, National Jewish Center, 1400 Jackson St., Denver, CO 80206

ANTON GARTNER, Cold Spring Harbor Laboratory, 1 Bungtown Road, Cold Spring Harbor, NY 11724

BARBARA L. HEMPSTEAD, Department of Medicine and Division of Neuroscience, Cornell University Medical College, New York, NY 10021

MICHAEL O. HENGARTNER, Cold Spring Harbor Laboratory, 1 Bungtown Road, Cold Spring Harbor, NY 11724

KATHRYN J. IVINS, Institute for Brain Aging and Dementia, University of California Irvine, Irvine, CA 92697-4540

RON MITTLER, Department of Plant Sciences, Hebrew University of Jerusalem, Jerusalem 91904, Israel

M. KAREN NEWELL, Division of Immunobiology, Department of Medicine, Given Building C312, University of Vermont College of Medicine, Burlington, VT 05405

BARBARA A. OSBORNE, University of Massachusetts, Department of Veterinary Sciences & Programs in Molecular & Cellular Biology, 304 Paige Laboratory, Amherst, MA 01003

DAVID S. PAPERMASTER, M.D., Neuroscience Program, Department of Pharmacology, University of Connecticut Health Center, 263 Farmington Ave., Farmington, CT 06030-3205

CHRISTIAN J. PIKE, Institute for Brain Aging and Dementia, University of California Irvine, Irvine, CA 92697-4540

MARIANNA SIKORSKA, Apoptosis Research Group, Institute for Biological Sciences, National Research Council of Canada, Ottawa, ON K1A 0R6, Canada

JAMES G. TIDBALL, Department of Physiological Science, University of California, Los Angeles, CA 90095-1527

BENJAMIN F. TRUMP, Department of Pathology, University of Maryland School of Medicine, Baltimore, MD 21201

MICHAEL S. VINCENT, Division of Immunobiology, Department of Medicine, Given Building C312, University of Vermont College of Medicine, Burlington, VT 05405

P. ROY WALKER, Apoptosis Research Group, Institute for Biological Sciences, National Research Council of Canada, Ottawa, ON K1A OR6, Canada

JUNYING YUAN, Departments of Cell Biology and Medicine, Harvard Medical School, Boston, MA 02115

WHEN CELLS DIE

PART I

THE PHENOMENON OF CELL DEATH

CHAPTER 1

THE EVOLUTIONARY ORIGIN AND ROLE OF PROGRAMMED CELL DEATH IN SINGLE-CELLED ORGANISMS: A NEW VIEW OF EXECUTIONERS, MITOCHONDRIA, HOST–PATHOGEN INTERACTIONS, AND THE ROLE OF DEATH IN THE PROCESS OF NATURAL SELECTION

JEAN CLAUDE AMEISEN

Inserm CJF 9707, Hôpital Bichat-Claude Bernard, Paris, and Paris VII University/ Xavier Bichat School of Medicine, Paris, France

Initial questions about the possible existence, mechanisms, and role of physiological programs of cell death emerged at the end of the previous century from the study of animal development, but only in recent years has it been realized that all cells from all multicellular animals are programmed to self-destruct, and that cell survival depends on the repression of this self-destruction program. The identification of the genetic regulation of programmed cell death, of its orderly phenotypes, and of its essential role, not only in development, but also in adult tissue homeostasis, has led to the acceptance of the idea that cells may survive only when signaled by other cells to suppress the induction of a default cell-suicide process.

In this chapter, I will argue that the question of "how" and "why" these programs operate has been obscured to some extent by the spectacular and essential role they perform in multicellular organisms; and that a new view

When Cells Die, Edited by Richard A. Lockshin, Zahra Zakeri, and Jonathan L. Tilly
ISBN 0-471-16569-7   © 1998 Wiley-Liss, Inc.

of programmed cell death emerges when one attempts to ask a different question: Where does programmed cell death come from? In other words, the question of "when" and "how" programmed cell death may have been selected during evolution.

The basic aim I hope this chapter will achieve is to convince the reader that an evolutionary approach may allow new paradigms, new implications, and new questions to emerge on the nature, origin, and role of programmed cell death, and to provide the reader with a change in perspective that will favor the interpretation of previous results in a new way, and the planning and design of new experiments.

ON THE CONCOMITANT EMERGENCE OF PROGRAMMED CELL DEATH AND MULTICELLULAR ORGANISMS: THE PARADIGM AND THE PARADOX

On the Conserved Nature of Programmed Cell Death in Multicellular Animals

Programmed cell death has been found to be an intrinsic part of the development of all invertebrate and vertebrate multicellular animals studied so far, including nematodes, insects, amphibians and mammals (reviewed in Ellis et al., 1991; Raff, 1992; Vaux, 1993; Steller, 1995). It plays an essential role in morphogenesis—the sculpting of the form of embryos and larvae—and in the functional self-organization processes that lead to the maturation of the two most complex regulatory systems of multicellular organisms, the immune system and the nervous system (Glücksmann, 1951; Lockshin and Williams, 1965; Saunders, 1966; Tata, 1966; Cowan et al., 1984; Duvall and Wyllie, 1986; Oppenheim, 1991; Jacobson et al., 1997). In recent years, programmed cell death has also been shown to play an important role in the adult, by allowing tissue homeostasis, regulation of cell numbers, elimination of damaged or abnormal cells, and defense against infections (Raff, 1992; Cohen, 1993; Ameisen, 1994; Green and Scott, 1994; Vaux et al., 1994). Conversely, programmed cell death dysregulation participates in the pathogenesis of several human diseases, ranging from cancer and autoimmunity to AIDS and neurodegenerative disorders (reviewed in Williams, 1991; Ameisen et al., 1994; Nagata and Golstein, 1995; Thompson, 1995).

The evolutionary conservation of programmed cell death in the animal kingdom involves not only its existence and role, but extends to some crucial aspects of its genetic control, and to important aspects of its most frequent phenotype. In all cases that have been studied to date, programmed cell death is regulated by signals provided by other cells, either in the form of cell-lineage information, of soluble mediators, or of cell-to-cell contacts. Induction of programmed cell death may depend essentially on cell-lineage information, such as in the nematode *Caenorhabditis elegans* (Ellis et al., 1991); on the activation of gene transcription, such as in the fruitfly *Drosophila melanogaster* (Steller, 1995); or, in a more stochastic way, on a combination of cell-lineage information, intercellular signaling, transcription factor activation, and cytoplasmic second messengers, such as in mammals (Duvall and Wyllie, 1986; Korsmeyer, 1995; Oppenheim, 1991; Ucker, 1991; Raff, 1992); Schwartz and Osborne, 1993; Vaux, 1993; Williams and Smith, 1993). In all the multicellular organisms mentioned, programmed cell death is a self-destruction process that involves the activation (through proteolytic cleavage) of a family of cysteine proteases (the Ced-3/CPP32/DCP caspase family) that appears to be constitutively expressed in most—if not all—cells from *Caenorhabditis elegans, Drosophila melanogaster,* and mammals and allows the execution of the effector phase of cell death (Martin and Green, 1995; Chinnaiyan and Dixit, 1996; Shaham and Horvitz, 1996; Nagata, 1997). Various phenotypes of programmed cell death have been described, but in all cases, cell suicide appears to occur through an orderly process of morphological disintegration (Lockshin and Williams, 1965; Kerr et al., 1972; Clarke 1990; Schwartz et al., 1993), and the most frequent phenotype constitutes the apoptosis paradigm, with chromatin margination, condensation and fragmentation, and cytoplasmic condensation, vacuolization, and fragmentation (Kerr et al., 1972; Wyllie, 1980; Duvall and Wyllie, 1986). This apoptotic phenotype is also associated with an initial maintenance of cell membrane integrity, extensive cell surface blebbing or pitting, and the expression of specific cell surface molecules that allow a rapid ingestion of the dying cells by phagocytes or neighboring cells, leading to the rapid clearance of dying cells in the absence of any adverse inflammatory reaction, and allowing therefore preservation of the integrity of the tissue, organ, and the whole organism in which cell death is occurring (Morris et al., 1984; Duvall and Wyllie, 1986; Cohen, 1993; Savill et al., 1993).

The concept of cell suicide "by default" (the idea that cells survive only when receiving dominant signals that allow the prevention of the execution of cell suicide) has emerged in recent years from the study of both the intercellular regulation of programmed death in mammalian cells and the genetic regulation of programmed cell death in the nematode *Caenorhabditis elegans* (Ellis et al., 1991; Ucker, 1991; Raff, 1992; Vaux, 1993; Hengartner, 1995). In both models, the activation of the Ced-3/CPP32 executioner caspase family is prevented by the expression of products from the Ced-9/Bcl-2 gene family. According to this view of cell suicide "by default," the basic core of programmed cell death regulation depends on the interplay between the products from two antagonistic gene families. However, this basic core of programmed cell death regulation appears much more complex and sophisticated in mammals than in *Caenorhabditis elegans*. Instead of the single Ced-3 caspase executioner and the single Ced-9 antagonist of Ced-3 yet identified in the nematode (Yuan et al., 1993; Hengartner and Horvitz, 1994), mammals have at least ten members of the caspase family that are homologues of Ced-3 (Martin and Green, 1995; Alnemri et al., 1996; Chinnaiyan and Dixit, 1996), and at least nine members of the Bcl-2 family that are homologues of Ced-9 (Yang and

Korsmeyer, 1996). Additional layers of complexity in mammals derive from the fact that (1) some members of the Bcl-2 family, such as Bcl-2 and BclX$_L$, inhibit apoptosis (as Ced-9), while other members of the Bcl-2 family, such as Bax and Bad, trigger or favor apoptosis induction; (2) the various Bcl-2 family members can hetero- or homodimerize, and programmed cell death regulation appears to depend on the ratio of inhibitors to activators; and (3) these dimerization processes appear to be regulated by phosphorylation processes (reviewed in Yang and Korsmeyer, 1996).

At least in mammalian cells, the ultimate decision to trigger or not to trigger autodestruction appears to be made at the level of the cytoplasm and to result at least from three factors: (1) the proteolytic activation of the inactive caspase precursors into active caspases; (2) the release by mitochondria of molecules required for the optimal proteolytic activation of the caspases; and (3) the interaction of gene products from the Ced-9/Bcl-2 family with the outer mitochondrial membrane (reviewed in Golstein, 1997). In addition, the decision may depend on a still unknown homolog of the Ced-4 *Caenorhabditis elegans* gene product, which appears to act as an activator of the Ced-3/CPP32 caspases and of cell death (when Ced-4 is expressed in the absence of the cell death inhibitor Ced-9/Bcl-2 gene products), and as an adaptor between the caspases and the cell death inhibitor Ced-9/Bcl-2 gene products (when they are expressed), leading to the repression of programmed cell death (Chinnaiyan et al., 1997; Wu et al., 1997).

In mammals, this core effector mechanism of programmed cell death is also regulated upstream by most—if not all—signals involved in cell differentiation and cell proliferation, allowing the whole organism to control the fate of each of its cells in a refined and complex manner (Ucker, 1991; Williams, 1991; Ameisen, 1994; Harrington et al., 1994; Shi et al., 1994). Such a social control of cell survival and cell death has been described as 'altruistic' (from the point of view of each dying cell), but can as well be considered as 'selfish' (from the point of view of the whole organism). The striking conservation, in species of multicellular animals that diverged over several hundred million years ago, of several components of programmed cell death—and of the genetic sequence and functional properties of several of the genes involved in its regulation—has led to the idea that genetic programs of physiologic cell suicide have played an ancient and essential role in the development, functioning and survival of most—if not all—multicellular organisms (Ellis et al., 1991; Raff, 1992; Evans, 1994; Vaux et al., 1994; Steller, 1995).

The Paradigm of a Concomitant Emergence of Programmed Cell Death and Multicellular Organisms

There have been two main reasons that have led to the proposal that programmed cell death emerged with multicellularity (Raff, 1992; Evan, 1994; Vaux et al., 1994). The first pertains to the very nature of multicellular organisms. Cells from multicellular organisms (in contrast to cells from single-celled organisms) are condemned to live together in a given spatial and temporal framework, the body. A multicellular body will survive only if—and as long as—its several differentiated cell families will cooperate in a way that will allow the integrated body to remain adapted to its environment. With the notable exception of the germ cells, no cells from a multicellular organism can outlive the body, nor survive outside the organism. Therefore, the emergence of a genetic program allowing a social form of regulation of cell survival and cell death, through a process of altruistic cell suicide, has been considered as one of the solutions that evolution has provided to the specific problem of social organisms (organisms made of cells condemned to live together (Ucker, 1991; Raff, 1992; Evan, 1994; Vaux et al., 1994; Steller, 1995). This idea—that programmed cell death emerged at the same time as multicellular organisms—has been reinforced by the complementary view that genetic programs allowing regulated cell suicide would have been obligatorily counterselected in single-celled organisms. In contrast to the cells from multicellular organisms, each cell from a single celled organism can be viewed as a germ cell, namely, as an individual that carries an identical theoretical probability to transmit its genome to future generations. Therefore, genetic programs favoring "selfishness"—rather than programs favoring any form of "altruism"—have been postulated to be the only ones that could have become selected in such organisms. According to this view, any mutant gene that might have emerged randomly and allowed regulated cell suicide in a single-celled organism, would have rapidly and obligatorily led to the counterselection of the individual cell expressing such a mutant gene (Evan, 1994; Vaux et al., 1994). Together, these two mutually reinforcing views led to the acceptance of the idea that the evolutionary origin of genetic programs of cell suicide has been concomitant with the evolutionary origin of multicellular organisms.

From the Explicit Paradigm to the Implicit Paradox

The identification of the crucial role of programmed cell death in multicellular organisms, the discovery of the sophisticated nature of its genetic regulation, and the belief that the sole function of the caspase family is to act as executioners of the effector phase of cell suicide (and that the sole function of the cell death inhibitor Ced-9/Bcl-2 family is to act as antagonists of the caspases), has progressively led us to think of programmed cell death in a framework of design, purpose, and intentionality. In other words, programmed cell death has been implicitly viewed as if it had been specifically designed in order to perform its obvious and important function in multicellular organisms. When thinking in evolutionary terms, however, such an approach raises an important paradox. It implies that a complex set of genes allowing the regulation and the phenotype of programmed cell death emerged when—and where—it was required.

Similar explanations have been previously proposed for the origin of complex genetic programs allowing the emergence of sophisticated organs, such as the eye. It was argued that an eye could have become selected only when complex enough to perform its function (allowing an organism to see), and that, therefore, genes required to build an eye appeared—and became selected—when and where the complexity of an organism allowed such a function to be exerted. Such views are intuitively appealing and resort to an evolutionary approach that has been termed "evolution by design." But their appeal is paradoxically related to the fact that they fail to take into account some of the basic and counterintuitive implications of the process of natural selection (Dawkins, 1976; Dawkins, 1982; Sigmund, 1993). It is now believed that the eye (in fact several forms of different eyes) emerged long after the genes that were required for building it, and that the genes required for building an eye became initially selected because they previously allowed other, simpler, phenotypes to be expressed such as using light as a source of energy, chemotaxis, or external clock. Therefore, if one believes that programmed cell death is a ubiquitous, highly conserved, and essential feature of most—or all—multicellular organisms, the question of the evolutionary origin of programmed cell death may benefit from being asked in the following way: "When and how were genes first selected that allowed, upon the emergence of multicellularity, the regulation of programmed cell death?" Was there a period "before" programmed cell death, in which cells died only when they could not do otherwise, and a period "after" programmed cell death, in which regulated self-destruction became one of the possible phenotypes of the cells? Did this frontier arise after the two to three billion years of evolution of the single-celled organism, at the time at which the first multicellular organisms emerged, or did this frontier arise in the kingdom of the single-celled organism, prior to the emergence of multicellular organisms?

There are at least two possible answers to these questions. The first—if one believes that the phenotype of programmed cell death would have been counterselected in single-celled organisms (Evan, 1994; Vaux et al., 1994)—is that the ancestor genes of the executioners were initially ensuring an alternate function (other than self-destruction) in the cell that expressed them. The second possibility is that genes allowing the execution of the cell (and genes allowing the regulated prevention of the executioners) were already present and operational in at least some unicellular organisms; in other terms, that the genes encoding the programmed cell death phenotype emerged and became selected at the level of single-celled organisms.

From the Paradox to the Hypothesis

There were at least two theoretical reasons that led me to speculate that a form of programmed cell death may have emerged in single-celled organisms (Ameisen, 1996a; Ameisen et al., 1995b). Briefly, the first one relates to the fact that single-celled organisms form colonies. It is the survival of the colony, as much as the survival of any of its members, that ensures the survival of the genome of the single-celled organism. Therefore, one may envision that the colony, as a whole, may benefit in some circumstances from the altruistic self-dismissal of part of its cells. The second reason relates to the essential role that has been attributed to programmed cell death in multicellular organisms, namely, the stringent control of cell differentiation, the matching of cell numbers to their environment, and defense against genetic damage and infection, leading to the elimination of abnormal and infected cells. Cell differentiation, regulation of cell numbers, and defense against infection, are not exclusive features of multicellular organisms, but also occur in single-celled organisms (see the following) and may, therefore, be optimized by a form of programmed cell death at the level of single-celled colonies. Finally, a third theoretical reason did not pertain to the potential role of programmed cell death in single-celled organisms, but to the potential nature of the death program and its effector machinery. I will come back later to this last reason.

A PARADIGM SHIFT: PROGRAMMED CELL DEATH AND APOPTOSIS IN SINGLE-CELLED ORGANISMS

On the Choice of an Experimental Model

The experimental model that I chose to test this hypothesis was *Trypanosoma cruzi,* a single-celled eukaryote protist that belongs to one of the most ancient diverging branches of the eukaryote phylogenic tree, the *kinetoplastidae* family (Maslov and Simpson, 1995). *Trypanosoma cruzi* represents one of the first mitochondriated unicellular eukaryotes, is believed to have diverged 1.5 to 2 billion years ago (Sogin, 1991; Knoll, 1992; Maslov and Simpson, 1995; Doolittle et al., 1996), and has several striking features (reviewed in Brener, 1973; De Souza, 1984; Williams, 1985; Mottram, 1994; Schaub, 1994; Maslov and Simpson, 1995): (1) It contains only one giant mitochondrion, the kineto-plast; (2) it is a flagellated protozoan parasite that causes diseases in vertebrates, including Chagas disease in humans, and whose life cycle requires two obligate hosts, a hematophagous insect and a vertebrate; and (3) during its life cycle, *Trypanosoma cruzi* undergoes three developmental changes that involve major modifications in morphology, gene expression, and cell cycle and lead to differentiation into three distinct stages, the epimastigotes, the trypomastigotes, and the amastigotes. In each host, the parasite undergoes a process of differentiation into a cycling cell, the epimastigote in the insect host, and the amastigote in the vertebrate host. Traveling from one host to the other, however, requires its differentiation into a G0/G1-arrested cell, the metacyclic trypomastigote. This differentiation process involves the expression of several sets of genes, which precedes the morphological changes and the cell cycle arrest, and is similar to a process of terminal differentiation in cells from multicellular organisms, with the important difference that the G0/G1 cell cycle arrest is not irreversible. The nonproliferating trypomastigotes are the differentiation states of *Trypanosoma cruzi* that migrate between the two different hosts, and are preadapted for survival in their new host. Therefore, the *Trypanosoma cruzi* life cycle can be viewed as a continuous alternation between proliferative forms in stable environments and a nonproliferative form preadapted to a new and different environment. For example, one of the striking differences between the metacyclic trypomastigote and its epimastigote progenitor in the insect pertains to sensitivity to mammalian complement-mediated immune aggression. Epimastigotes are rapidly killed by complement, whereas metacyclic trypomastigotes express enzymes that inactivate complement (Hall and Joiner, 1993; Nogueira et al., 1975). In the insect vector, *Trypanosoma cruzi* colonizes the digestive tract in which it proliferates in the form of the epimastigotes. Migration along the digestive tract allows the progressive differentiation of epimastigotes in trypomastigotes, which will be excreted on the skin of the vertebrate during the insect's blood meal. In the mammalian host, trypomastigotes infect host cells and differentiate inside these cells into the cycling amastigotes; subsequent differentiation into G0/G1-arrested trypomastigotes, when exiting the host cells, allows the parasite to pursue its life cycle if—and when—the trypomastigote is ingested by an insect vector during a blood meal and will differentiate, in the insect digestive tract, into a cycling epimastigote.

The differentiation of the cycling epimastigotes into the metacyclic trypomastigotes, a crucial step in the life cycle of the *Trypanosoma cruzi,* which takes place in the digestive tract of the insect, can also be achieved *in vitro* in axenic cultures of epimastigotes in conditioned acellular medium, at 27°C, the temperature of the insect vector gut (Camargo, 1964; Castellani et al., 1967; Chiari, 1975; Dusanic, 1980; Sullivan, 1983). In these standard *in vitro* culture conditions, epimastigotes first proliferate exponentially, then, after one to two weeks, in the absence of medium renewal (a condition that is thought to mimic the progressively stressing conditions during the migration along the digestive tract of the insect vector), epimastigotes stop multiplying and differentiate into trypomastigotes, leading to the so-called stationary stage of culture in which the recovery of the G0/G1-arrested trypomastigotes is maximal. These trypomastigotes have all the properties of—and are indistinguishable from—the trypomastigotes that have differentiated *in vivo* in the gut of the insect vector, and are excreted on the skin of a vertebrate host (De Souza, 1984).

The Experiments and the Results: Programmed Cell Death and Apoptosis in *Trypanosoma cruzi*

We investigated whether the stringent process of differentiation of cycling epimastigotes into metacyclic trypomastigotes may involve regulation through programmed cell death. We observed that the stationary stage of epimastigote cultures involved two major events: the differentiation of a minority of the epimastigotes into G0/G1-arrested trypomastigotes and the death of most of the epimastigotes that failed to differentiate into trypomastigotes. Dying and dead epimastigotes appeared as spheroid, condensed cells. Scanning and transmission electron microscopy revealed that the dying spheroid epimastigotes had morphological features typical of apoptosis, including extensive membrane blebbing or deeply pitted surfaces; chromatin condensation and margination along the nuclear membrane; cytoplasmic vacuolization; and in some cases the formation of apoptotic bodies. Strikingly, this process of membrane boiling and blebbing also involved the flagellum of the epimastigotes. Further analyses using the *in situ* terminal deoxytransferase-mediated dUTP nick end labeling (the TUNEL) technique showed extensive DNA fragmentation in the spheroid epimastigotes. Together, these results indicated that a death process, triggered as an alternative response to adequate differentiation, led in the single-celled eukaryote *Trypanosoma cruzi* to a death phenotype that exhibited the typical features of mammalian cell apoptosis (Ameisen et al., 1995b).

One of the crucial features of programmed cell death in multicellular organisms that has led to the concept of the social regulation of cell survival and cell death—and to the view of cell suicide as "default" pathway—has been the observation that all cells from numerous multicellular animals will undergo programmed death (despite the presence of adequate nutrient) if and when they are deprived *in vivo* or *in vitro* from contacts with other cells (Raff, 1992; Raff et al., 1993). In particular, cells cultured *in vitro* can survive only if induction of programmed cell death is actively repressed by signals provided by other cells, and/or by exogenous survival factors that have been added in the test tube. Culturing any cell population below a minimal cell density threshold (and/or in the absence of a minimal concentration of exogenously added survival factors) will obligatorily lead to a rapid induction of programmed cell death. We investigated whether *Trypanosoma cruzi* epimastigote survival may also depend on signals provided by neighboring epimastigote cells. When epimastigotes were suspended in fresh conditioned medium, at the concentration of 10^6/ml, and resuspended every four or five days in fresh conditioned medium, at the same concentration of 10^6/ml, continuous proliferation of epimastigotes could be obtained during several weeks, in the absence of significant epimastigote death and of differentiation into trypomastigotes, indicating that the induction of epimastigote programmed death did not depend on the setting of an internal clock, but on the particular nature of their environmental conditions. When epimastigotes were suspended in fresh conditioned medium below 10^6/ml, at decreasing densities resulting from tenfold dilutions, induction of cell death became massive at a concentration of 10^4/ml, and led, at a concentration of 10^3/ml, to the disappearance of all epimastigotes after two days. When epimastigotes were cultured in poor unconditioned medium, cell death began to be induced at a concentration—10^5/ml—ten times higher than in rich conditioned medium and led, at a concentration of 10^4/ml, to the disappearance of all epimastigotes after two days. Together, these data indicated, first, that epimastigotes continuously require extracellular signals to prevent induction of programmed cell death, and that these survival signals are provided in part by the neighboring epimastigote cells from the colonly: second, that extracellular signals that allow optimal epimastigote self-renewal and prevent trypomastigote differentiation also have a preventive effect on the induction of epimastigote programmed death; finally, that extracellular signals that allow trypomastigote differentiation induce, at the same time, programmed cell death in the epimastigotes that have failed to differentiate properly (Ameisen et al., 1995b). It is interesting to consider, in retrospect, that parasite culture conditions that were empirically selected more than thirty years ago to allow optimal epimastigote growth (Camargo, 1964; Castellani et al., 1967) are in fact conditions that prevent both epimasti-gote death and differentiation, and that parasite culture conditions that were empirically selected more than thirty years ago to allow optimal epimastigote differentiation into trypomastigotes are in fact culture conditions that enhance programmed cell death in the epimastigotes that fail to achieve differentiation into trypomastigotes. In addition, culture conditions that have been empirically selected in recent years to achieve epimastigote cloning and transfection (Kelly et al., 1992) are in fact particular culture conditions that allow, during a limited period of time, the repression of programmed cell death induction in a cycling epimastigote, despite the absence of signals provided by other epimastigote cells. Such *in vitro* culture conditions include solid-phase agar, which provides enhanced adhesion capacities, and hemoglobin (Ameisen et al., 1995b), and may closely mimic the initial *in vivo* conditions in which isolated trypomasti-gotes that have been ingested during a blood meal by an insect vector differen-tiate in the upper insect gut into isolated epimastigotes that will begin cycling until their progeny reach an optimal concentration and provide signals that will substitute for the lack of hemoglobin and may allow survival in the absence of adhesion. Such adhesion-dependent cell regulation of survival is reminiscent of the preventive effect on programmed cell death induction exerted, in several cell populations from mammals, by various adhesion molecules, allowing inter-action with the extracellular matrix, or with other cell populations (Ruoslahti and Reed, 1994).

Together, our findings indicated that in cells from a single-celled organism, as in cells from multicellular organisms, the repression through extracellular signals of a physiological cell death program represents a prerequisite for cell survival, proliferation, and differentiation. We than investigated whether *Trypanosoma cruzi* programmed death may also be regulated by signals pro-vided by their mammalian host, to which the trypomastigote form, but not its epimastigote progenitor, is preadapted. The first host-specific signal that we explored was temperature: Switching the culture temperature from 27°C (the temperature of the epimastigote and tripomastigote insect vector host) to 37°C (the temperature of the trypomastigote mammalian host) led to the rapid induction of epimastigote death by apoptosis in the absence of any apparent differentiation into trypomastigote. The second host signal that we explored was a component of the vertebrate immune defense system that kills epimastigotes, complement. In mammals, complement (as other forms of severe cell membrane injury) has been reported to induce death by necrosis

(Duvall and Wyllie, 1986; Cohen, 1993; Vaux, 1993). Surprisingly, in epimastigotes, complement did not induce necrosis, but a very rapid process of death by apoptosis, including regular DNA fragmentation in multiples of oligonucleosomal length fragments of approximately 200 base pairs. Our finding that epimastigote death in response to complement occurs through apoptosis (and not through necrosis) provided an explanation for the previously reported absence of bystander death of trypomastigotes (which are resistant to complement-induced death) in complement-treated cultures containing both trypomastigotes and epimastigotes (Nogueira et al., 1975), a feature that allows *in vitro* purification of trypomastigotes from complement-treated epimastigote cultures, even when these cultures contain less than 5% trypomastigotes prior to complement treatment.

In contrast to treatment with complement, treatment of epimastigotes with another cell membrane damaging agent (saponin) induced a rapid and catastrophic form of cell death that radically differed from apoptosis, did not induce detectable DNA fragmentation, and resembled necrosis.

The Implications

Our findings indicated that in a unicellular eukaryote organism, an intrinsic cell death program regulated by signals provided by other cells is coupled to a death machinery able to carry out the apoptotic phenotype (Ameisen et al., 1995b; Ameisen, 1996a). This striking conservation of the apoptotic phenotype in *Trypanosoma cruzi* suggests either that the executioners allowing apoptosis in cells from multicellular organisms are present in *Trypanosoma cruzi,* or that different executioners can achieve apoptosis in different organisms. These findings also provided support to the hypothesis that a social form of control of cell survival and cell death could play a role in a single-celled organism. One such possible role of programmed cell death may be to allow the optimal adapation of the single-celled parasite to its different hosts; in particular, regulated limitation of epimastigote numbers in the insect vector gut may optimize the parasite life cycle by minimizing the fitness impairment of the infected vector host. Such a possibility would be consistent with the observation that *Trypanosoma cruzi* has minimal adverse effect on its insect vector (Schaub, 1994). In addition to the possible matching of the number of epimastigote cells to their environment, it is possible that an apoptotic death program in epimastigotes could lead to engulfment of epimastigote corpses by the vector cells present in the gut, thereby preventing an inflammatory response that could be harmful for both the host and the parasite. Although previously published scanning electron microscopy pictures of *in vivo Trypanosoma cruzi* infected insect gut (Böker and Schaub, 1984; Zeledon et al., 1984), show *Trypanosoma cruzi* features that may possibly be reinterpreted in terms of apoptosis the fate of the epimastigote cell corpses remains to be investigated. Programmed cell death may also allow the stringent regulation of cell differentiation of the cycling epimastigotes into their G0/G1-arrested trypomastigote progeny. The process of differentiation of epimastigotes into trypomastigotes is similar to the process of lineage commitment in multicellular organism progenitor cells, with the difference that in trypanosomes such a commitment is reversible, and does not represent a form of terminal differentiation. In multicellular organisms, distinct survival factors have been shown to be required for the survival of the cycling progenitor and for the survival of their differentiated cel progeny (Raff et al., 1993; Linette and Korsmeyer, 1994). The coupling of cell differentiation programs to selective switches in survival factor requirements will ensure that only the fittest differentiated cells, expressing the correctly coordinated set of genes required for their functions, will be allowed to survive. Programmed cell death of proliferating epimastigotes as an obligatory "default" outcome of their failure to differentiate adequately into the G0/G1-arrested infective trypomastigote form in response to appropriate differentiation signals may optimize the spatial distribution of epimastigotes and trypomastigotes in distinct gut compartments of the vector host, and may therefore prevent the infective trypomastigotes in the distal insect gut environment from being diluted by their proliferating progenitors. Published results from scanning electron microscopy of *Trypanosoma cruzi* infected insect vector gut are consistent with this hypothesis, by suggesting the existence of local compartmentalization between trypomastigotes and epimastigotes (Böker and Schaub, 1984). Finally, it is possible that death by apoptosis in response to signals provided by the vertebrate host may have evolved in epimastigotes in order to benefit the invading trypomastigotes. It has been reported that the insect vector occasionally excretes epimastigotes in addition to the trypomastigotes (Zeledon et al., 1984). Complement has been reported to induce necrosis in mammalian cells; the rapid induction of apoptosis in epimastigotes in response to complement, as well as to the temperature shifting upon penetration in the vertebrate host blood, may prevent the triggering of an early inflammatory response that would follow epimastigote necrosis induction and could rapidly target the complement-resistant trypomastigotes for other immune effector mechanisms. An interesting possibility, which remains to be explored, is whether the consequence of apoptotic epimastigote ingestion by mammalian macrophages is the same as that of apoptotic mammalian-cell ingestion, namely, the maintenance of an immunologically silent state of the macrophages (Cohen, 1993; Savill et al., 1993).

In cells from multicellular organisms, including plants,, insects, and mammals, programmed cell death in response to infectious pathogens represents an important defense mechanism against infection (Clem et al., 1991; Levin et al., 1993; Greenberg et al., 1994). Conversely, dysregulation of programmed death in cells from multicellular organisms represents an important strategy selected by infectious pathogens that allows persistence in their multicellular host and escape from immune defense mechanisms (Williams, 1991; Schwartz and Osborne, 1993; Williams and Smith, 1993; Ameisen, 1994; Ameisen et al., 1994; Vaux et al., 1994). Because the evolutionary origin of programmed cell death has been considered concomitant to the emergence of multicellular organisms, the role of programmed cell death in host–pathogen interactions has always been considered from the sole perspective of the infected multicellular host. Our findings suggest, however, that programmed cell death in an infectious single-celled organism may play an important role in its capacity to interact with its infected host in a way that will optimize the fitness of the infectious organism. Therefore, programmed cell death may have played an important bidirectional role in the complex interactions between unicellular and multicellular organisms that have allowed the establishment and persistence of parasitism. If this were the case, the inducible endogenous machinery of single-celled infectious pathogens might represent a new target for the design of therapeutic strategies.

In summary, one of the crucial roles of programmed cell death in both the unicellular and multicellular worlds may be the tight regulation of the processes of growth arrest and differentiation in cells sharing the same genome, as well as the regulation of the coexistence or mutual exclusions of cells in distinct differentiated stages. The potential consequences of the outgrowth of—and competition with—programmed cell death–defective mutants may also share some similarities in unicellular and multicellular organisms. In multicellular organisms, one of the several benefits that have been attributed to the evolutionary conservation of regulated cell death programs is the prevention of the development of cancers. Mutations that allow cells to escape the environmental regulation of their inducible death program also allow these cells to escape the environmental regulation of growth and differentiation, causing the death of the organism to which they belong (Williams, 1991; Vaux, 1993; Harrington et al., 1994). Such mutations represent evolutionary dead ends only if they occur before reproduction of the organism has been achieved, namely, before the *soma* becomes disposable. In the unicellular *Trypanosoma cruzi,* if death of the proliferating epimastigotes is the obligatory "default" outcome of a failure to differentiate adequately into its G0/G1-arrested trypomastigote progeny in response to appropriate inductive signals, the theoretical advantage in terms of outgrowth that could be conferred to an early epimastigote mutant that would escape environmental regulation of its inducible death program may prove to be counterselective in a similar way: The inability of such a proliferating epimastigote mutant to differentiate into the vertebrate pre-adapted trypomastigotes the Trypanosoma cruzi equivalent of a germ cell will lead to an interruption of the parasite life cycle, and therefore to an evolutionary dead end. Once trypomastigotes have been generated and have invaded their vertebrate host, however, the loss of growth control in a late epimastigote PCD-defective mutant may lead to the death of the insect vector, but, at that stage, epimastigotes may be viewed as a form of disposable *soma,* and the possible late occurrence of such mutations may then represent, as in multicellular organisms, an evolutionary neutral accident.

The fact that *Trypanosoma cruzi* is a parasite of multicellular organisms raises another important question regarding the potential evolutionary origin of programmed cell death. The kinetoplastid ancestors of *Trypanosoma cruzi* diverged long before the emergence of the first multicellular organisms, but the present from of the *Trypanosoma cruzi* parasite is adapted to both its vertebrate and insect vector hosts. Therefore, it is possible that the origin of programmed cell death and apoptosis in *Trypanosoma cruzi* does not precede the origin of programmed cell death and apoptosis in its vertebrate and insect hosts, but is a consequence of the evolutionary pressures involved in the establishment of complex interactions between unicellular and multicellular organisms that allowed the emergence of parasitism. Programmed cell death and its apoptotic phenotype may either be a very ancient feature of single-celled eukaryote organisms, or may alternately result from several independent and parallel evolutionary attempts in highly diverging eukaryotes. This important question as to whether the emergence of an apoptotic death program in a single-celled eukaryote is related to its obligate parasite nature may be addressed by investigating the closely related kinetoplastid Bodonids that include both parasites and free-living nonparasite single-celled organisms (Maslov and Simpson, 1995); but whether nonparasite Bodonids directly derive from initial nonparasite ancestors, or from parasite ancestors that subsequently became free living, still remains a matter of speculation.

Other Experimental Findings: Programmed Cell Death in Three Diverging Families of Single-Celled Eukaryotes

Programmed cell death, with features resembling some—or all—aspects of apoptosis, has now been described in five single-celled eukaryote organisms (including three other members of the kinetoplastid family) that belong to three different branches of the eukaryote phylogenic tree that diverged between one and two billion years ago (Cornillon et al., 1994; Ameisen et al., 1995b; Christensen et al., 1995; Ameisen, 1996a; Moreira et al., 1996; Weiburn et al., 1996).

The kinetoplastid protozoan parasite family includes three of the most pathogenic protozoan parasites in humans: *Trypanosoma cruzi,* which causes Chagas disease; *Trypanosoma brucei,* which causes human sleeping sickness, and *Leishmania,* which causes leishmaniasis and visceral Kala Azar (Maslov and Simpson, 1995). An endogenous death program inducing a phenotype identical to apoptosis has been shown to be operational in *Trypanosoma brucei rhodensiense,* in which it is induced by a lectin-binding mechanism (Welburn et al., 1996), and, in *Leishmania amazoniensis,* in which it is induced by heat shock, and prevented by a mammalian cytokine, the monocyte–macrophage colony-stimulating factor (GMCSF: Moreira et al., 1996). Further studies will be required to assess to what extent survival in these parasites depends, as in *Tryanosoma cruzi,* on the prevention of programmed cell death by extracellular signals. At this stage, however, it is already clear that all three pathogenic kinetoplastid parasites share the ability to self-destruct in response to at least some extracellular signals. This feature suggests the possibility of using the intrinsic death machinery of these parasites as a target for the design of new antiparasite drugs, and suggests that some of the drugs already used to kill these organisms may in fact act by inducing apoptosis. It is tempting to speculate that drugs that destroy parasites by inducing necrosis may have several inflammatory side effects that will not be shared by drugs that act by inducing parasite death by apoptosis.

Programmed cell death has also been identified in two diverging families of single-celled eukaryotes that are of more recent evolutionary origin: slime molds and ciliates. The slime mold *Dictyostelium discoideum* represents a particular case of a single-celled eukaryote that can, in certain circumstances, transiently form a multicellular aggregated body and is therefore considered as one of the evolutionary attempts of single-celled organisms at multicellularity. Adverse environmental conditions induce cells from *Dictyostelium discoideum* to aggregate and differentiate, leading to the development of a multicellular organism made of a mixture of viable spores and of dead stalk cells (Wittingham and Raper, 1960; Kaiser, 1986). This developmentally regulated form of programmed cell death involves intercellular signaling, and shares several features with apoptosis (Cornillon ct al., 1994). Programmed cell death has also been reported in a free-living ciliate, *Tetrahymena thermophila* (Christensen et al., 1995), which does not form multicellular bodies, is not a parasite, and is believed to have emerged around one billion years ago. In this single-celled organism, as in *Trypanosoma cruzi* programmed cell death is triggered by default unless certain environmental conditions are provided, and survival depends on the prevention of self-destruction by extracellular signals, including signals provided by neighboring cells.

The Implications

The identification of a regulated cell death program inducing an apoptotic phenotype in five different single-celled eukaryote organisms that belong to three diverging branches of the eukaryote phylogenic tree provides a paradigm for a widespread role for programmed cell death in the control of cell survival that extends beyond the evolutionary constraints that may be specific to multicellular organisms, and raises the question of the origin and nature of the genes that may be involved in the execution and regulation of such a process (Ameisen, 1996a). Evolutionary advantages conferred by such a program of self-destruction could include the constant selection for the survival of the fittest cell in the single-celled eukaryote colony, optimal adaptation of the cell numbers to the environment, and tight regulation of cell cycle and cell differentiation in reponse to environmental changes. Particular usages of programmed cell death may involve intercellular interactions more closely related to multicellularity: in the kinetoplastid parasites, programmed cell death may play an important role in the regulation of the complex interactions between unicellular and multicellular organisms that allow the establishment and persistence of stable host–parasite interactions; in the slime mold, programmed cell death may be crucial for the terminal differentiation of dead stalk cells, which allow the development of a transient multicellular aggregated organism.

Finally, in multicellular organisms, genetic mutations that allow cells to escape environmental regulation of their suicide machinery cause cancer, leading to the death of the organism to which they belong (Williams, 1991; Vaux, 1993; Harrington et al., 1994; Thompson, 1995). The finding that a cell suicide program is operational in the three unicellular eukaryote lineages explored to date suggests that similar genetic escape mutants may also have become counterselected at the level of the colonies of these single-celled eukaryotes.

When and how did unicellular organisms select for the complex genetic programs allowing self-destruction, as well as the coupling of cell survival to the repression of self-destruction? This is the question that I will now try to address. But I would like first to delineate some of the intrinsic problems and limitations that arise when one attempts to travel into the past.

THE ORIGIN OF PROGRAMMED CELL DEATH IN SINGLE-CELLED ORGANISMS: FROM THE QUESTION "WHEN" TO THE QUESTION "HOW"

Limits in Evolutionary Time Traveling: All Contemporaries Are Survivors

There are two possible explanations for the existence of programmed cell death in at least five single-celled eukaryote organisms belonging to different lineages that diverged during a time frame of two billion to one billion years ago. The first possible explanation is that programmed cell death is an ancient and conserved feature in most—if not all—single-celled eukaryote organisms, because programmed cell death emerged at the time—or prior to the time—of the emergence of the kinetoplastids. If this were true, programmed cell death should be found to be present and operational in most—if not all—the branches of the eukaryote phylogenic tree. An alternate explanation, however, as mentioned, is that the ancestors of these five present-day single-celled eukaryotes were devoid of the capacity to self-destruct, and that programmed cell death has been selected in several branches of the phylogenic tree in response to selective pressures that may have involved competition, interaction, and cooperation with the eukaryotes of more recent origin. If this were true, the present-day surviving progeny of the three branches of single-celled organisms would have undergone a process of parallel or convergent evolution, leading to a somewhat recent acquisition of programmed cell death, which may have happened around the emergence of the first multicellular organisms. One way to attempt discriminating between both possibilities is to investigate to what extent most—if not all—single-celled eukaryotes are endowed with the capacity to self-destruct. But the interpretation of such findings may still remain elusive. Indeed, if all single-celled eukaryote organisms are found to have operational physiological self-destruction programs, this will not allow excluding the possibility of strong and ubiquitous selective pressures leading to a recent convergent acquisition of programmed cell death. Conversely, if some single-celled eukaryote organisms are found to be devoid of the capacity to self-destruct, this will not allow excluding the possibility of a form of reductive evolution, namely, a loss of an ancient evolutionary shared feature.

At the present time, the investigation of programmed cell death in single-celled eukaryotes has not led to the identification of its genetic regulation. How many different genetic programs can induce a phenotype of programmed cell death and of apoptosis? Do unicellular eukaryotes share effectors and regulators of self-destruction with multicellular animals? The only way to address this question is to attempt to identify the genes that allow the execution and regulation of programmed cell death in these single-celled eukaryotes. Interestingly, in this regard, it has been shown that the transfection in the budding yeasts of the mammalian Bax gene, which induces programmed death in mammalian cells, also induces death in the yeast and that the cotransfection of the mammalian Bcl-2 gene, which counteracts the effect of the Bax gene product in mammalian cells, also counteracts Bax-induced death in yeasts (Hanada et al., 1995). Although it is still not known whether yeasts can undergo programmed cell death during their normal life cycle, this suggests the possibility that at least some of the downstream regulators and/or effectors of self-destruction are shared in single-celled and multicellular organisms.

A Paradox Concerning the Evolution of Programmed Cell Death: How to Select Genes That Induce a Vanishing Phenotype

Natural selection does not act directly on genes, but on the phenotypes they achieve, in the cell that encodes them, in a given environment. Genetic mutation leads to genetic diversification, and hence to phenotypic diversification. Natural selection sanctioned by death is a process of competition between phenotypes whereby genomes encoding for poorly adapted phenotypes in a given environment do not get transmitted into the next generation, while genomes that allow fitter phenotypes become selected. Simply put, the phenotype represents either the grave of the genome or its vehicle into the next generation (Dawkins, 1982). Genetically regulated programmed cell death is a phenotype that is characterized by the rapid disappearance of the cell that is expressing it. In other words, genes that encode for programmed cell death encode for a phenotype that is the grave of the genome that expresses them. This paradox, as previously mentioned, renders difficult (and counterintuitive) any reasoning on programmed cell death in single-celled organisms in terms of natural selection. But there are other ways to address the question. The first is to think that the genes allowing programmed cell death may not have initially become selected for their capacity to induce programmed cell death, but for their capacity to ensure an additional function that may be essential for the survival of the single-celled organism. An implication of this view is that executioners of self-destruction do not solely function as executioners. An implication of such a view is that programmed cell death is a phenotype that has been initially selected as a price to pay for an additional important phenotype encoded by the same genes. Another way to address the question is to think of the genome that becomes selected, not in terms of single cells, but in terms of a colony of single cells. When thinking in the context of colonies, one can ask the question of the selection of the programmed cell death phenotype from the point of view of the cell that will survive, rather than from the point of view of the cell that will die. In such a context, the question becomes whether the lack of genes allowing programmed cell death may result in the counterselection of the whole colony.

I will now argue that there are ways to look at the nature and role of programmed cell death that are very different from those to which we have been accustomed by thinking in the context of multicellular organisms. But before we begin our time traveling into the past, let us first reflect for a moment on how influential a paradigm (such as that of the obligate emergence of programmed cell death in multicellular organisms) and a paradox (such as that mentioned previously) can be in preventing interpretation of already existing—and sometimes quite ancient—experimental results.

Paradigms, Anomalies, and Retrorecognition Processes

It is interesting to consider that there have been several experimental findings—some reported forty years ago—that could have been interpreted as suggestions that programmed cell death may be operational in single-celled organisms. I believe that the reason these results were not considered in terms of programmed cell death was not related to a lack of the technical expertise allowing to address this question, but to the weight of the existing conceptual framework that linked the origin and role of programmed cell death to the advent of multicellular organisms. Anomalies in science are experimental observations that do not fit existing paradigms, and that should therefore lead to a change in paradigms that renders the new observation no longer an anomaly. It has been argued, however, that anomalies can often remain unrecognized for long periods of time. Retrorecognition has been defined as the phenomenon by which anomalies are not recognized until they have been given an explanation in a new conceptual framework (Lightman and Gingerich, 1991).

There are at least three examples of such unrecognized anomalies that may be relevant to the question of programmed cell death in single-celled organisms. The first concerns developmental programs that lead to the concomitant formation of dead cells and surviving spores, in the context of transient multicellular aggregated bodies. Such developmental programs have long been known to exist in single-celled eukaryotic organisms, such as the slime mold *Dictyostelium discoideum* (Wittingham and Raper, 1960; Kaiser, 1986), but also in prokaryotes, such as *Myxobacteria* (Wireman and Dworkin, 1975; Wireman and Dworkin, 1977; Kaiser, 1986). In other prokaryotes, such as *Streptomyces* or *Bacillus subtilis,* developmental programs leading to the concomitant formation of dead cells and spores have also been known to occur, although in the absence of multicellular bodies. All these programs are triggered by changes in environmental conditions, involve intercellular signaling, and are considered an integral part of the organism life cycle (Kaiser and Losick, 1993). Although the question of the apoptotic phenotype of such developmentally regulated cell death programs could have been raised only in the eukaryote single-celled organisms, the question of the relationship with programmed cell death could have been raised in both the prokaryote and the eukaryote single-celled organisms. During more than three decades, however, questions about the mechanism, role, and genetic control of developmentally regulated cell death programs have remained solely addressed in multicellular organisms. The second example, although much more recent, is also of interest. Five years ago, it was reported that a machinery similar to that carrying out the nuclear components of apoptosis in cells from multicellular organisms (the nuclear chromatin condensation and the DNA fragmentation into multiples of oligonucleosome length fragments) was present and operational in the unicellular ciliated single-celled eukaryote *Tetrahymena* (Davis et al., 1992). This nuclear execution machinery and program was reported, however, not to be involved in the induction of cell death, but in the elimination of supernumerary old macronuclei during *Tetrahymena* conjugation, a situation in which these supernumerary nuclei are destroyed, but the cell is not affected. This finding led to the proposal that programmed cell death, in multicellular organisms, evolved from genetic programs that were originally involved, in single-celled eukaryote organisms, in the elimination of supernumerary macronuclei. In other words, the identification in a single-celled organism of part of the executioners allowing self-destruction did not provide a rationale for the investigation of self-destruction in such organisms, but was used to reinforce the paradigm that self-destruction emerged with multicellularity. In cells from multicellular organisms, as in cells from single-celled organisms, restricted usage of the apoptotic self-destruction machinery can also achieve means other than cell death and allow particular forms of cell differentiation: in mammals, for example, a process of selective induction of apoptotic chromatin and DNA fragmentation, which eliminates the nucleus while sparing the cell, is involved in the formation of the lens cell in the eye. A similar restricted use of the apoptotic effector machinery appear to have evolved in single-celled eukaryotes—such as *Tetrahymena*—in addition to its involvement in programmed cell death.

In a more general way, I believe that retrorecognition phenomena may have occurred at several time points in the history of programmed cell death research, and it is possible that this may be related to the very counterintuitive nature of the idea of self-destruction. If one attempts briefly to recapitulate the history of the successive views of programmed cell death during the past hundred years, self-destruction was first considered an unlikely process; this view was progressively replaced by the idea that programmed cell death may play an essential role in embryonic development, but only in given cell populations, at given time points, and at given locations. In other words, programmed cell death became considered, during a very long period as a price to pay for the complexity of the problems that have to be resolved during embryonic development. More recently, it was proposed that programmed cell death may also be operational—and play an essential role—in adult cells; that dysregulation of physiological cell death programs may play a crucial role in the pathogenesis of several diseases; and finally that most, if not all, cells from the bodies of multicellular animals are constantly programmed to self-destruct unless signaled by other cells to repress induction of self-destruction. Such a negative regulation of cell survival was, however, once again considered an exceptional price to pay for complexity, complexity being in this case the complexity of the multicellular organisms. If one has to consider programmed cell death as a price to pay for the emergence of complexity during evolution, the ultimate—and simplest—level of complexity that we are now addressing is that of single-celled organisms (Ameisen, 1996a).

The third example of retrorecognition, which we will discuss in the following section, concerns various forms of regulated cell death that have been described in various bacterial species for several decades, but that were not considered, until very recently, potential examples of programmed cell death (Jensen and Gerdes, 1995; Yarmolinsky, 1995; Franch and Gerdes, 1996). Such primitive forms of programmed cell death had been described not only, as mentioned, in circumstances that include the terminal differentiation of *Myxobacteria, Streptomyces,* and *Bacillus subtilis,* but also in several circumstances that involve competition between plasmid or viral genome and bacterial genome within a given bacterial colony, as well as competition between bacteria from different species.

TWO HYPOTHESES FOR THE EVOLUTIONARY ORIGIN OF PROGRAMMED CELL DEATH: THE "EVOLUTIONARY ARMS RACE" HYPOTHESIS AND THE "ORIGINAL SIN" HYPOTHESIS

From "Killing Others" to "Killing Self": The "Evolutionary Arms Race" Hypothesis

On Toxins, Antidotes, and Interspecies Bacterial Killing. When competing for the control of environmental resources, several bacterial species use strategies based on the killing of other bacterial species. They do this by secreting toxins (antibiotics) that induce the death of other bacteria. Such toxins include colicin El; colicin E7; microcin (Mcc) B17, and streptomycin, and act either by inserting pores in the bacterial membrane that induce membrane depolarization, or by damaging bacterial DNA through direct mechanisms or through indirect mechanisms that involve the activation of enzymes involved in the modification of DNA topology, such as DNA gyrases (Liu, 1994; Yarmolinsky, 1995). The reason why bacteria can release a toxin that kills other bacteria without getting killed themselves is that they also synthesize an intracellular antidote that protects them against the lethal effect of the toxin (Liu, 1994; Yarmolinsky, 1995). The selection for killer genes encoding toxins used for an offensive evolutionary arms race against other bacterial species, and the concomitant selection for genes encoding antidotes allowing self-protection, is a process that provides the bacteria with both executioner genes and survival genes that can prevent the effect of the executioners. This genetic module bears striking functional resemblance to the basic core of the genetic module that allows, in cells from multicellular organisms, the regulation of programmed cell death: the Ced-3/CPP32 caspase executioners and the Ced-9/Bcl-2 antidotes. In other words, the ability to regulate self-destruction may have simply evolved as a consequence of a capacity to kill others without killing itself. But there are other important and striking aspects of the evolutionary arms race in bacteria that also pertain to the potential evolutionary origin of programmed cell death.

Plasmids, Viruses, and Bacteria: On Toxins, Antidotes, Addiction Modules, and Postsegregational Killing. In the prokaryote world, competition between heterogeneous genomes for environmental resources is not restricted to competition between different bacterial species. Mobile genetic elements, such as plasmids and bacteriophage viruses, also compete with bacteria: in this case, it is the bacterium itself that is the resource, and the evolutionary arms race involves the spreading of the mobile genetic elements in the bacterial colony. Strategies selected by plasmids and bacteriophages in order to propagate in the bacterial colony involves various mechanisms that allow spreading from one bacteria to another (Nordström and Austin, 1989). A striking complementary strategy relies on mechanisms enforcing the bacterial retention of the mobile genetic element, by setting up the obligate death of any bacteria that will inactivate or reject the plasmid or the virus. Several plasmids achieve this by encoding for both a toxin and an antidote. There are various forms of toxins and antidotes, but they all share a similar feature: the toxins are stable and long-lived; the antidotes are unstable and short-lived (Gerdes et al., 1986; Jensen and Gerdes, 1995; Yarmolinsky, 1995). Following infection, this strategy involves the constant expression in the bacteria of the plasmid or viral gene that encodes the toxin and of the plasmid or viral gene that encodes the antidote. If the plasmid becomes disabled or rejected, or if some of the daughter cells of the infected bacteria escape receiving the plasmid from their mother cell, both the toxin and the antidote stop being synthesized in the cured cell. Since the half-life of the toxin extends that of the antidote, this will lead to the obligate and programmed death of the plasmid- or virus-cured cell.

The exogenous genetic modules that encode the toxin and the antidote have been called "addiction modules" because the bacteria, once infected, become addicted to the continuous presence and expression of the infectious genetic element. The death process that is triggered by the inactivation or loss of these addition modules has been called "postsegregational killing" because all plasmid- or virus-free bacterial segregants are condemned to die. Although the addition modules are of foreign origin and are not encoded in the bacterial chromosomes, they endow the bacteria cell with a genetic program allowing a regulated form of self-destruction. What is the nature of the toxins and antidotes encoded by the addition modules, and how is their respective half-life determined? All toxins are long-lived proteins; antidotes are of two kinds. The first kind of antidotes are labile antisense RNAs, of short half-life, that block the translation of the toxin-encoding RNA, as exemplified by the Hok/Sok module, in which the Hok toxin protein induces lethal membrane depolarization and the Sok antisense RNA prevents translation of the Hok encoding RNA (Gerdes et al., 1986; Franch and Gerdes, 1996). Most known antidotes, however, are proteins. Some antidotes are methylases that protect DNA against direct fragmentation by the type II restriction enzyme toxins; other antidotes act by protecting DNA against indirect damage induced by the toxin (through the activation of enzymes such as DNA gyrases), as exemplified by the CcdB/CcdA addiction modules or by the Kid-PemK/Kis-Pem1 a addiction modules (Jensen and Gerdes, 1995; Yarmolinsky, 1995). In all these models in which the interaction of the toxins and the antidotes occur at a protein–protein level (including the ParE/ParD and the Doc/Phd addiction modules), the reason the antidote protein has a shorter half-life than the toxin protein is that the antidote is cleaved by a protease. Surprisingly, in all the known models, the protease is not encoded by the plasmid or the prophage, but the plasmid addiction module relies on constitutively expressed bacterial chromosomal-encoded proteases, which include the Lon- or the ClpP-ATP-dependent serine proteases (Jensen and Gerdes, 1995). Thus the efficiency of the addiction module depends on a form of enforced cooperation between gene products expressed by the infectious agents and gene products constitutively expressed by the chromosome of the bacteria. The reason these bacterial serine proteases are constitutively expressed in the bacterial targets of the plasmid is that they appear to perform essential roles in bacterial survival, which will lead to the counterselection of protease loss of function mutants that may have otherwise escaped plasmid addiction. In some bacterial species, some of these essential roles performed by the Lon and ClpP proteases have been uncovered. For example, they are required for bacterial cell division to proceed, through the temporally regulated proteolysis of methylases required to duplicate chromosomal DNA, as well as for bacterial cell-cycle regulation, through proteolysis of the cell cycle regulator SulA in adverse environmental conditions that lead to DNA damage and/or to the induction of an SOS-stress response (Katayama et al., 1988).

Because it seems obvious—when thinking of programmed cell death in multicellular organisms—to consider self-destruction an "altruistic" cell decision, the question of the nature of the evolutionary constraints that may have led to the selection of the genetic modules allowing programmed cell death has always been equated with the question of the nature of the evolutionary constraint that may have favored the selection of altruistic cell behavior (Raff, 1992; Vaux, 1993; Evan, 1994; Vaux et al., 1994). Through a radical change in perspective, we have now seen that the emergence of programmed cell death can be envisioned as resulting from the selective "selfish" spreading advantage conferred by such genetic modules to the genomes able to express them. This is true in the context of interspecies bacterial competition, in which toxin–antidote modules allow the bacterial genome that encodes them to propagate at the expense of the genome of other bacterial species that do not encode them; this is also true for the host–pathogen competition between plasmid and bacteria, in which the infectious toxin–antidote addiction module favors the propagation of the bacterial cell that expresses it, at the expense of the bacterial cell that succeeds in repressing its expression. But both situations are even closer than it appears: Most—if not all—toxin–antidote modules that allow interspecies bacterial killing are in fact encoded by plasmids (Liu, 1994; Jensen and Gerdes, 1995; Yarmolinsky, 1995). In this context, plasmid-encoded toxin–antidote modules can be seen as genetic modules that allow the propagation of the plasmid that encodes them. Such infections genetic modules may induce death in an autocrine way, as do the addiction modules; or they can induce death in a paracrine way, as do the modules involved in interbacteria species killing. In both instances, they lead to an increase in the numbers of bacterial cells that express them, by inducing the elimination of the bacterial cells that do not express them. In both instances, they are addictive, because "withdrawal" induces death.

Such toxin–antidote modules, by coupling the survival of the bacterial genome to that of the infections genome, may have provided several selective advantages to the infected bacterial cell colony: in addition to an edge in interbacterial species competition, they may have provided bacteria with defense mechanisms against superinfection by other plasmids or bacteriophages. It is easy to see how such strategies can be combined in diverse and complex manners. But overall it is important to realize that whether toxin–antidote modules provide selective advantages to the bacterial cell that expresses them, whether they provide no selective advantage at all, or whether they are detrimental to the cells that carry them, the basic reason for which they propagate is that they are addictive.

Natural selection can favor the propagation of given genes for the sole reason that they are successful at propagating themselves, while being of no advantage—or sometimes while being detrimental—to the fitness of the organisms that carry them. Such genes have been called "selfish genes" (Bull et al., 1992). It is interesting to think that the genetic modules allowing regulated programmed cell death may have initially emerged and become selected as "selfish genetic modules," for the sole reason that encoding for an executioner and for an inhibitor of the executioner just made them good at propagating themselves (Naito et al., 1995). Is there any evidence for such an evolutionary scenario? Is there any indication that such "selfish" toxin–antidote addiction modules may have subsequently become captured and selected by bacterial genomes in order to allow subsequent regulation of an "altruistic" form of programmed cell death?

From Evolutionary Arms Race to Regulated Self-Destruction: The "Self-Addiction"–Programmed Cell Death Module. Very recently, a bacaterial chromosome-encoded "self-addiction" module was discovered in *Escherichia coli* (Aizenman et al., 1996). This genetic module encodes for the long-lived stable toxin MazF, which induces DNA damage, and for the short-lived antidote MazE, which counteracts the toxin. The MazE protein antidote is short-lived because it is constantly cleaved by the constitutively expressed bacterial ClpP-ATP-dependent serine protease.

In appropriate environmental conditions, the MazF toxin, the MazE antidote, and the ClpP proteins are constitutively expressed, leading to a constant *de novo* synthesis of the toxin and to a constant *de novo* sythesis and cleavage of the antidote, a dynamic equilibrium that allows bacterial survival. In adverse environmental conditions, such as nutrient shortage, intracellular signaling—in particular an increase in $3',5'$-bispyrophosphate—leads to the inhibition of the MazE/MazF operon: both the MazE and MazF proteins stop being synthesized. The ClpP protease continues to be expressed, the residual MazE antidote continues to be cleaved, and bacterial self-destruction occurs, as a consequence of MazF toxin-mediated irreversible DNA damage.

This finding is consistent with the multistep scenario for the evolutionary emergence of programmed cell death outlined previously, in which extrachromosomal "selfish" gene modules encoding toxin and antidote, and involved in an evolutionary arms race between heterogeneous genomes, become at some point integrated at the chromosomal level and subsequently used for their capacity to allow the regulation of death and survival in adverse environmental conditions in cells sharing the same genome. Such "altruistic" regulated use of programmed cell death may provide selective advantages to the bacterial colony by allowing, in starving conditions, the survival of a part of the cell population at the expense of the rapid dismissal of others. The "selfish" and infections addiction module has now become an integrated part of the genome of the bacterial colony and has become involved in a form of "altruistic" regulation of cell survival and cell death that strongly resembles programmed cell death (Aizenman et al., 1996).

Interestingly, this form of programmed cell death in single-celled eukaryotes shares another features with programmed death in cells from multicellular organisms: It is a "default" pathway. Because the bacterial cells constitutively express the executioner protein allowing cell suicide, cell survival constantly depends on the expression of a dominant but short-lived antidote protein that prevents activation of the executioner. This programmed cell death module acts in fact as a "self-addiction" module: Once it has begun to be expressed, cells can only survive by continuing to express it. A surprising view of programmed cell death emerges when one fully realizes that it is not the expression of the programmed cell death module that induces cell suicide, but its repression; self-destruction in *Escherichia coli* is a phenotype that results from the regulated repression (in response to environmental signals) of a "self-addiction" genetic module encoding a toxin and an antidote. The coupling of such a repression of the expression of the addiction module to given exogenous signals could have become selected only if it allowed the concomitant survival of at least some members of the colony. In other words, such a program has to be socially regulated at the level of the colony population in order not to lead to the indiscriminate self-destruction of all the bacterial cells in response to adverse environmental conditions. How may such a decision become integrated at the level of the bacterial colony? How may bacteria decide, at a single-cell level, when to die and when to survive?

How to Decide When to Die: On Quorum Sensing and Cell Differentiation in Bacterial Colonies. Although this is often neglected, the ability to differentiate is a feature of most—if not all —single-celled organisms, including prokaryotes. In bacteria, as in single-celled eukaryotes, coordinated changes in gene expression lead to changes in cell cycle regulation, in morphology, and in intercellular signaling (Losick and Stragier, 1992; Hengge-Aronis, 1993; Kaiser and Losick, 1993; Amon, 1996; Kaiser, 1996; Ohta and Newton, 1996). Striking aspects of differentiation in bacteria include fruiting body formation in *Myxobacteria;* light production in luminescent *Vibrios;* spore formation in *Myxobacteria, Streptomyces* and *Bacilli;* and asymmetric cell division in *Bacillus subtilis* and *Caulobacter crescentus.* Several other forms of differentiation have been described in bacteria, including the SOS stress and repair response. One of the most ubiquitous environmental signals that trigger differentiation in bacteria is selective nutrient shortage (Losick and Stragier, 1992; Hengge-Aronis, 1993; Yarmolinsky, 1995; Kaiser, 1996). In appropriate environmental conditions, bacteria undergo exponential vegetative growth. Upon nutrient shortage, a developmental program is triggered in most bacteria species that leads to the concomitant induction of cell differentiation in a part of the population and of cell death in the rest of the colony. In *Myxobacteria, Streptomyces,* and *Bacilli,* nutrient shortage induces the terminal differentiation—followed by the death—of a part of the cells from the colony; these terminally differentiated cells help the other part of the cells from the colony to differentiate into long-lived noncycling and highly resistant spores (Losick and Stragier, 1992; Kaiser, 1996).

Although environmental changes represent the initial and necessary trigger for the complex changes that will lead to this process of alternate and complementary differentiation, the environmental signals by themselves are not sufficient: an additional step of intercellular signaling is required that will lead to a coordinated set of changes in gene expression. In *Myxococcus xanthus,* for example, the decision upon nutrient shortage either to continue to grow (at a reduced rate) or to trigger a developmental program that will lead to concomitant and alternate programmed cell death or sporulation depends on two limiting factors: the density of cells in the bacterial colony and the density of individual cells, in that colony, that decide to respond to the environmental change. Individual cells decide to respond to nutrient shortage by expressing genes that allow the synthesis and release of a given quorum factor (a kind of pheromone) that binds quorum sensors—receptors—that are expressed by each bacterial cell of the colony and are sensitive to the concentration of quorum factor present in the environment. Quorum factor binding induces gene expression only when a threshold concentration of quorum factor is reached that greatly exceeds the quantity of quorum factor that can be synthesized by any given cell (Kaiser, 1996). In other words, differentiation results from a collective decision that depends on the number and proximity of neighbor cells that have taken the individual decision to favor differentiation. Such a process provides an interesting model for understanding how important individual decisions that will affect the future of the whole colony are not taken at the level of any individual cell, but integrated at the level of the colony. But how is this collective step of decision translated in the specification of alternate cell fates that allow restricting terminal differentiation leading to self-destruction to only a part of the bacterial population? I will now argue that there is a striking example that illustrates how such a complex process can be achieved and may have evolved.

A Model for the Resolution of the Paradox of Self-Destruction in Single-Celled Organisms: Asymmetric Cell Division in Bacillus Subtilis.

In order to avoid being counterselected, a cell suicide program has to be regulated in such a way that the sacrifice of some individuals in a unicellular colony will benefit (or at least will not prevent) the survival of other members of the colony. As mentioned, a coupling of programmed cell death regulation to that of cell differentiation and of intercellular signaling represents one of the essential steps towards such solution. But how is this solution achieved? *Bacillus subtilis* provides a spectacular and extreme example of how such major theoretical problem concerning the evolution of programmed cell death in unicellular organisms can be solved (Ameisen, 1996b).

In favorable environmental conditions, *Bacillus subtilis* undergoes vegetative growth through symmetrical cell division. In adverse environmental conditions, in particular upon nutrient shortage, *Bacillus subtilis* undergoes a complex developmental program whose initiation depends, as mentioned, on cell density and on the concentration of released quorum factors (Kaiser, 1996). When initiated, this program begins with a process of asymmetric cell division. The *septum* becomes positioned not at the middle of the cell (as during vegetative growth) but closer to one pole of the developing cell. Cell division, however, is not completed: the polar *septum* separates the cells in two different territories, and the two asymmetric future cells remain attached one to the other (Losick and Stragier, 1992; Shapiro, 1993; Errington, 1996). The biggest part of the cell, called the mother cell, will become terminally differentiated and will subsequently undergo a form of programmed death, after having helped the smaller part of the cell, called the prespore cell, to become a nonproliferating resistant and long-lived spore. In other words, the initiation of a process of asymmetric cell division allows *Bacillus subtilis,* at the level of each single cell, to differentiate into a somatic cell (the mother cell) and into the equivalent of a germ cell (the spore). Because each cell in the colony becomes the coupling unit, differentiation will obligatory lead in the colony to an equal number of self-destructing cells and surviving cells.

An essential aspect in *Bacillus subtilis* of the genetic regulation of cell differentiation that will lead to the coupling of sporulation and programmed cell death is the complex intercellular regulation of the expression and activation of four transcription factors, σ^E to σ^K (Losick and Stragier, 1992). The initial step of asymmetric division will have the important consequence of leading to a different amount of proteins in the prespore cell and in the premother cell. Briefly, the higher amount in the mother cell of proteins such as the SpoIIE phosphatase will induce the selective activation of the σ^F transcription factor in the prespore cell (Arigoni et al., 1995). Activated σ^F from the prespore cell will then lead to the activation of the σ^E factor in the mother cell, which will in turn activate the σ^G factor in the prespore cell; the σ^G factor from the prespore cell will then activate the σ^K factor in the mother cell; finally, the σ^K factor from the mother cell will lead to the final differentiation of the prespore cell into a spore cell, a process that is followed by death of the mother cell. Although the molecular mechanisms involved in the process of mother cell death are unknown and remain to be explored, this crisscross regulation ensures that the σ^K sporulation factor is expressed only in the mother cell and not in the spore, and is expressed in the mother cell at a late stage only of the forespore differentiation, in order to prevent premature mother cell death and subsequent inadequate spore formation.

Such a sophisticated temporal and spatial regulation of gene expression provides a spectacular example of how the coupling of programmed cell death to intercellular communication can avoid the death of the whole colony in adverse environmental conditions, and ensures that the sacrifice of one half of the progeny will provide a selective survival advantage to the other half of the progeny.

Sporulation occurs only in some bacterial species, but as previously mentioned, cell differentiation associated with cell death is a usual response of most bacterial species to adverse environmental conditions. The view that I have proposed is that *Bacillus subtilis* represents an example (rather than an exception) of the intercellular communication that may operate (maybe in a more stochastic manner) in most single-celled organisms, and may allow a coupling of programmed cell death and survival at the level of the colony. Asymmetric cell division is an important and conserved mechanism involved in the specification of cell differentiation in bacteria (Shapiro, 1993), yeasts (Amon, 1996), and multicellular organisms (Horvitz and Herskowitz, 1992). In particular, spatially and temporally crisscross regulation of rate-limiting factors (such as homeotic gene products) is a conserved process in the developmental program allowing the emergence of multicellular organisms from a population of initially identical cells derived from single-celled eggs. Assessing to what extent the mechanisms of social control of cell survival and cell death that operate in single-celled prokaryotes, in single-celled eukaryotes, and in multicellular organisms may be conserved is an important question that remains to be investigated.

In summary, the evolutionary scenario that I have outlined suggests a multistep process for the emergence of programmed cell death that originated in bacteria. The most commonly accepted proposal for the origin of the eukaryote cell—an endosymbiont that arose from the capture of a bacteria by an ancestor of the eukaryote cell—suggests the hypothesis that programmed cell death has undergone a further step of stabilization in the first eukaryote cell, in which an evolutionary arms race between two heterogeneous genomes condemned to live together (the mitochondrial genome and the preeukaryote cell nuclear genome), led to a resolution of these genomic conflicts through a process of enforced cooperation (Ameisen, 1996a). Such a process, as I will discuss later, may provide an explanation for the recently described—and surprizing—role of mitochondria in the execution of programmed cell death.

An interesting aspect of this multistep scenario for the evolutionary origin of programmed cell death is the suggestion that the "altruistic" genetic modules regulating programmed cell death may have initially emerged from the propagation of "selfish" infections genetic modules that were selected through their ability to "addict" the cells that expressed them. This scenario, however, may not be the only possible scenario for the evolutionary origin of programmed cell death. I have proposed an alternative model that has another interesting aspect: it removes the need for a multistep process in the emergence and selection of the genetic modules allowing the regulation of self-destruction (Ameisen, 1996a).

From Self-Organization to Self-Destruction: The "Original Sin" Hypothesis

Effectors and Regulators of Programmed Cell Death in Cells from Multicellular Animals. In cells from multicellular animals, conserved gene families have been identified that are considered to function solely as executioners (the Ced-3/CPP-32/DCP caspase cysteine protease family) or as inhibitors (part of the Ced-9/Bcl-2 family and the IAP family) of programmed cell death. The paradigm that executioners of programmed cell death function only as executioners (and that inhibitors of programmed cell death function only as inhibitors) has emerged from the study of the *Caenorhabditis elegans* nematode model. In *Caenorhabditis elegans,* Ced-9 loss-of-function or gain-of-function mutants, and Ced-3 loss-of-function mutants, show no other phenotypic modifications than those involving developmentally regulated programmed cell death. It is possible, however, that this simple multicellular organism (in which development appears strictly regulated through cell lineage information, and leads to the formation of an organism of fewer than 1000 somatic cells, all somatic cells being terminally differentiated postmitotic cells) may represent an exception to—rather than an example of—the rule. First, Bcl-2, the mammalian homolog of the Ced-9 gene, can partially substitute for Ced-9 in preventing developmental programmed cell death in Ced-9 loss-of-function mutants of *Caenorhabditis elegans;* although no phenotype other than prevention of programmed cell death has been reported in Bcl-2 expressing *Caenorhabditis elegans* (Vaux et al., 1992), Bcl-2 has been shown to exert a regulatory effect on cell cycle in mammalian cells (Linette et al., 1996), including a delay in the G_0/S phase transition, an increase in the levels of the cyclin-dependent kinase inhibitor $P27^{kip1}$, and an impairment in the nuclear translocation of the transcription factor NFAT. These findings suggest that genes that act as inhibitors of the executioners of programmed cell death may also exert additional and important effects on cell activation, differentiation, and cycling. Second, recent results obtained in the fruitfly *Drosophila melanogaster* indicate that the inactivation (through gene deletion experiments) of the *Drosophila* caspase homolog DCP induces several developmental defects affecting imaginal discs and trachea that do not appear related to a prevention of programmed cell death (Song et al., 1997). This finding suggests that the role of caspases in cell function during development may extend beyond the execution of cell suicide. Third, it has been shown in mammalian cells that the proteolytic activation of the CPP32 caspase, which is considered the distal executioner of programmed cell death, requires in some instances the prior activation of the ICE caspase (Nagata, 1997), and appear to require, in all instances, the release by mitochondria of cytochrome C (Kluck et al., 1997; Yang et al., 1997). It is still not known whether CPP32 may have a function other than execution of programmed cell death. But it is already known that the ICE caspase has other roles than the proteolytic activation of the CPP32 executioner (in particular the cleavage of pro-IL-1β, the precursor of the IL-1β cytokine that can occur in the absence of cell death), and it is also known that cytochrome C, when in the mitochondria, plays a vital role in the process that allows cell respiration (Alberts et al., 1994). This suggests that obligate upstream activators of the executioners may have been selected for functions other than the induction of self-destruction.

More generally, most mammalian genes involved in the control of cell cycle and cell differentiation, including protooncogenes, tumor suppressor genes, cyclins, cyclin-dependent kinases, have also been shown to participate in the control of programmed cell death (Ucker, 1991; Raff, 1992; Vaux, 1993; Ameisen, 1994; Harrington et al., 1994; Shi et al., 1994). This has led to the proposal that in cells from multicellular organisms—and in particular in mammalian cells—the cell suicide machinery may have evolved from—and become part of—cell cycle machinery (Ucker, 1991). Such interrelationships between the regulation of the cell cycle and the regulation of self-destruction are, however, not restricted to cells from multicellular organisms, but also exist in single-celled organisms: in yeasts, as in mammalian cells, mitotic catastrophes resulting from the uncoordinated activation of cyclins have a phenotype similar, or identical, to apoptosis (Lundgren et al., 1991; Heald et al., 1993); in some bacteria species, the autolysins that are required for cell division by breaking the peptidoglycan bacterial wall can also induce self-destruction in adverse environmental conditions (Oshida et al., 1995).

The hypothesis that I have proposed is that effectors of the cell cycle machinery may have an intrinsic and unavoidable capacity of inducing the self-destruction of the cell in which they operate. If effectors of the cell cycle machinery can also be effectors of the self-destruction of the cell in which they operate, then the requirement for coupling cell survival to the prevention of self-destruction may be as old as the origin of the first cell (Ameisen et al., 1995b; Ameisen, 1996a).

A "Gedanken Experiment" in the First Living Cells: On the Intrinsic Dangers of Gene Products Required for Survival.

Is it possible to envision the gene products that participate in cell division or cell differentiation as potential executioners? Let us consider, in the ancestor cells of the prokaryotes, the topological manipulations of DNA required for replication, transcription, and recombination; the DNA repair mechanisms required to correct DNA damage; the cell membrane repair mechanisms; and the segmentation process of the cytoplasm required for cell division; and let us consider in a nucleated ancestor of the eukaryotic cell, the processes of rearrangement of chromatin organization, of remodeling or dissolution of the nuclear membrane, and of chromosomal migration required for cell division. All cellular processes have intrinsic error rates; and most—if not all—the processes mentioned involve enzymes that, if not tightly regulated, have the intrinsic potential to lead to cell death. If we attempt to continue this thought experiment in the first living cells, it is tempting to propose that genes allowing cell division and cell differentiation could become selected only if they were associated to genes that encoded for inhibitors able to control their activity by restricting their error rate.

In such a view, any genetic module that encodes a potent enzyme—that is both vital and dangerous—and its inhibitor represents an ancestor of the executioner–antidote modules. An interesting aspect of this scenario is that it does not postulate the existence of any real evolutionary transition between single-celled organisms unable to undergo regulated self-destruction and single-celled organisms able to achieve programmed cell death (Ameisen, 1996a). Another interesting aspect is that such an evolutionary scenario for the origin of programmed cell death accommodates both a phylogenic tree in which prokaryotes are the ancestors of eukaryotes (Sogin, 1991; Knoll, 1992), as most often proposed, and one in which both prokaryotes and eukaryotes diverged from a common extinct ancestor, as recently suggested (Doolittle et al., 1996).

The Evolution of Genetic Diversification as a Model for the Evolution of Programmed Cell Death.

I have argued that such evolution of programmed cell death would share similarities with the evolution of genetic diversification (Ameisen, 1996a). The intrinsic inability of a cell to avoid random genetic mutations and alterations has led to the concomitant and apparently antagonistic selection of both the repression of genetic changes, through various mechanisms of DNA proofreading and repair, and the amplification of DNA diversification through various mechanisms of genetic reassortment including recombination, transformation, conjugation, transduction, retrotransposition, and finally sexuality. The view that an intrinsic inability to avoid random self-destruction is an "original sin" of the cell—an inherent consequence of DNA repair, cell differentiation, and progression through the cell cycle—implies that evolution has led to the concomitant and apparently antagonistic selection of both the repression and the amplification of the self-destruction machinery. In other words, such a model implies that evolution has led to a continuous fine tuning of the regulation of self-destruction, rather than to the emergence, at a given time point, of a cell suicide machinery. Such a model provides a simple mechanism for the selection of upstream inhibitors of self-destruction that enhance the efficiency of cell cycle and cell differentiation; at the same time, such a model provides a simple mechanism for the selection of upstream inducers of programmed cell death that allow enhanced fitness of the colony through the rapid dismissal of cells once a mistake has been made during the cell cycle, and for the altruistic deletion of part of the cells in order to provide a selective advantage to the best adapted cells in adverse environmental conditions.

At this point of the chapter, I have proposed two different scenarios for the potential origin of programmed cell death in single-celled organisms: the "evolutionary arms race" hypothesis and the "original sin" hypothesis. I will now suggest that these two hypotheses are not antagonistic, but can be seen as complementary and even synergistic.

ON THE GENETIC HETEROGENEITY OF "SELF": THE SYNERGISTIC NATURE OF THE "EVOLUTIONARY ARMS RACE" HYPOTHESIS AND THE "ORIGINAL SIN" HYPOTHESIS

"From Killing Others to Killing Self" and "from Killing Self to Killing Others"

The "evolutionary arms race" hypothesis postulates that "selfish" genetic modules encoding for killer genes (toxins) and antagonists of the killer genes (toxin antidotes) were first selected for the advantage they provided to the genetic modules themselves in terms of propagation, and subsequently for the advantages they provided to the genome encoding them in terms of competitions with heterogeneous genomes. When such "selfish" genetic modules became integrated in the chromosome of bacteria, they then became selected for the additional evolutionary advantages provided to the colony by their regulated use in self-destruction processes that enhanced the fitness of the colony. In other words, the "evolutionary arms race" hypothesis postulates that the ancestors of the executioners involved in programmed cell death were genes that were initially involved in murder. On the other hand, the "original sin" hypothesis postulates that the ancestors of the genetic modules allowing regulated programmed cell death were already present at the origin of life, and were constituted of genes that allowed essential cell function to be performed and of genes that prevented dysregulated activation of these effectors to induce the destruction of the cell. An implication of the "original sin" hypothesis is that these genetic modules allowing both the regulation of essential cellular functions and the regulation of self-destruction became subsequently selected as toxin and antidote modules for the advantage they provided in evolutionary battles against heterogeneous genomes. In other words, this model postulates that antibiotic toxins may have evolved from genes that initially allowed essential cell functions (as well as regulated self-destruction).

An extreme implication of the latter view is that bacterial antibiotic toxins such as those involved in membrane depolarization, in DNA fragmentation, or in the activation of enzymes involved in DNA topological changes (such as DNA gyrases) may have important and still undiscovered functions other than killing cells. Whether such toxins could participate (maybe at lower concentrations than those at which they induce killing) in cell signaling, DNA repair, DNA topology, and/or other cell functions is a fascinating possibility that deserves investigation. Another implication of this view is that plasmids and viruses may have acquired the toxin–antidote addiction modules from bacterial chromosomal elements rather than the other way around. Interestingly, when thinking in the framework of the "evolutionary arms race" perspective, it seems likely that toxins used for the killing of other bacterial species could have become selected only if they were already associated with antidotes. If the toxin emerged first, the genome expressing them may have been counterselected. One possible way for toxin–antidote modules to have emerged is through the prior emergence of the antidote; it seems likely, however, that such an antidote (in the absence of an already existing toxin) may have become selected if it participated, by itself, in a particular cellular function. Another possibility is that toxins and antidotes became selected together, because either the toxins were achieving important functions that required regulation by the antidote, or because both toxins and antidotes achieved important cellular functions that allowed cross regulation. In both cases, we end up envisioning toxins and/or antidotes involved in "evolutionary arms race" as initially selected for a role in cellular functions (and allowing regulated self-destruction) prior to any use in the context of weapons. Therefore, I would like propose the "evolutionary arms race" hypothesis and the "original sin" hypothesis not as alternative models, but as complementary models with several shared features able to reinforce each other during evolution. Such a view may become clearer when I now attempt to define further the respective notions of self-destruction and of murder.

Self-Destruction or Murder: On the Heterogeneous Nature of "Self"

In single-celled organisms, the distinction between self-destruction and killing of others does not appear to raise any problem provided that one adopts the following view: If the cell that expresses the executioner genes is the cell that will die, the process can be defined as self-destruction; if the cell that will die is not the cell expressing the executioner gene, the process can be defined as a murder. We will see that the frontier becomes blurred, however, when we begin to attempt more precisely to discriminate between self-destruction and murder by asking the question of the definition of the "self." Although this may come as a surprise, the "self" can be seen as genetically heterogeneous at several levels.

A first level of heterogeneity derives from the moving frontier between infectious "nonself" genes and the "self" genome. Plasmids can infect bacteria as circular extrachromosomal genetic elements, but they can also integrate the bacterial chromosome and become chromosomal episomes; lysogenic bacteriophage viruses do the same (Lewin, 1997). Therefore, addiction modules may have at least two possible origins: (1) foreign mobile genetic elements that have become a permanent part of the genome (the "self"); and (2) foreign mobile genetic elements that have remained extrachromosomal and coexist with the genome in the cell (infectious "nonself"). A second level of heterogeneity is related to the fact that the reproductive success of an addiction module does not depend solely on bacterial cell division (when the module is in the "self"), nor on the infective capacity of the foreign extrachromosomal genetic elements (when the module is encoded by plasmids or bacteriophages). Addiction modules can also spread in a bacterial colony through genetic exchanges between bacteria chromosomes, by the process of conjugation or mating, which involves plasmid episomes (Lewin, 1997). In other words, initially foreign infectious genetic elements (plasmids), once having become integrated in the bacterial chromosome, confer to the bacterial chromosome a capacity to insert part of this "new self" chromosome (as well as part of the "ancient self" chromosome) in other bacteria.

But there are additional layers of genetic heterogeneity of the "self" that occur independently of any horizontal propagation of genes: they are the consequence of genetic mutations. A first level of this heterogeneity can be seen at the level of the cell colony: genetic mutants arise constantly, and the mutant genes express themselves on the background of the "former self" genome. Therefore, at the level of the colony, the "self" will be constantly redefined as the equilibrium between the "former self" and a "new self." A second level of heterogeneity can be seen at the level of each mutant cell itself: The "new self" results from an equilibrium between the mutant genes and the rest of the genome, the "former self." If one considers a regulated cell death program selected for its ability to achieve the destruction of cells harboring mutant genes, this program can be viewed either as achieving the murder of heterogeneous "nonself" genomes, or alternately as achieving the "self-destruction" of the genome because it has been altered. In other words, autodestruction of a cell in response to genetic changes or DNA damage can be seen either as a "selfish" behavior or the genome that induces the murder of cells that have become foreign, or alternately as an "altruistic" sacrificial behavior of the self genome in response to alteration.

But there is still another level of genome heterogeneity. The chromosomal genomes are made of a congregation of genes that propagate if—and when—they achieve sufficient cooperation in order to allow the survival of the cells that harbor them. In principle, the propagation of a given genome—the importance of its progeny—results from selective advantages that such genetic cooperation provides to the cells that harbor it. As we have previously seen, however, "selfish genes" can be successful at propagating themselves without providing any advantage to the cells that express them, and even more surprisingly, for the paradoxical reason that they can be detrimental to the cells that carry them (Bull et al., 1992; Naito et al., 1995). As we have also seen, genetic modules encoding for toxins and antidotes are successful at propagating themselves simply because they counterselect any progeny that will achieve inactivating them. In this context, the destruction of a progeny can be seen either as a process of "self-destruction" (these genetic modules are part of "the self" genome) or as murder (the "selfish genes" are inducing death for the sole reason that the genome is trying to exclude this module from "the self"). In summary, each cell can be viewed as an environment resulting from competition and cooperation between different genetic modules in a given genome, and each colony of single-celled organisms as an environment of competition and cooperation between different cells harboring different mutant genomes.

In such a context of constantly evolving genetic equilibrium, distinctions between the "self" and the "nonself" and between "altruism" and "selfishness" become difficult to make. The propagation of a given genome implies, however, that it has, at some level, evolved a form of stable congregation of genes. Such a situation is best illustrated when initially foreign genetic elements have reached a stage in which they are condemned to live together, have become addicted one to the other and are unable to leave each other. I will now argue that such a situation may have occurred during the evolutionary process that led to the emergence of the eukaryotic cell, and may have acted as an important stabilization factor for the evolution of programmed cell death.

Endosymbiosis As a Stabilization Factor for Programmed Cell Death: The Eukaryote Cell, the Nucleus, and the Mitochondria

All eukaryote cells—from single-celled eukaryote organisms to multicellular animals and plants—harbor at least two genomes: the nuclear genome, which contains most of the cellular genes, and the cytoplasmic organelle genomes, which are small circular DNAs present in the mitochondria (single-celled plants and multicellular plants contain an additional organelle genome, the plastid chloroplast circular DNA). (For a review, see Alberts et al., 1994; Lewin, 1997.) The size of the mitochondrion genome greatly varies depending on the organism, from around 16 Kb in mammalian cells to around 80 Kb in the *S. cerevisiae* yeast and to more than 500 Kb in some multicellular plants. Numbers of mitochondria organelles per cell also greatly vary depending on the organism, ranging from one giant mitochondrion kinetoplast in the ancient kinetoplastid protozoan single-celled eukaryotes, such as the trypanosomes, to several hundred mitochondria per cell in several single-celled eukaryotes and in multicellular animals. Mitochondria play a vital role in all eukaryote cells from single-celled and multicellular organisms: They perform aerobic metabolism, which allows energy production through ATP synthesis by a respiratory process that involves an electron transport chain and a chemiosmotic process. Loss of mitochondrial function forces cells to rely only on anaerobic metabolism, which is impossible—and therefore lethal—in cells from all multicellular animals and, probably, from most single-celled eukaryotes (except some yeast mutants). All the respiratory activity of the cell is performed in the mitochondria. Each mitochondrion is bound by two highly specialized membranes that create two separate compartments, the internal matrix space and the intermembrane space. Mitochondria (as well as chloroplasts) have several striking features: (1) They are not made *de novo* by the cell, but always arise by growth and division of preexisting mitochondria; (2) in organisms endowed with sexual reproduction, such as yeast and multicellular animals, mitochondrial genes show a non-Mendelian (or cytoplasmic) inheritance pattern, with uniparental maternal inheritance in multicellular animals; (3) the mitochondrial DNA organization—as well as the transcription and translation apparatus involved in mitochondrial protein synthesis—are very different from those of the nuclear genes; finally (4) the mitochondrial genome from most organisms is devoid of introns, the fidelity of mitochondrial DNA replication is lower than that of the nuclear genome, and there are even some differences between the mitochondrial genetic code and the universal genetic code.

All these features have strongly suggested that mitochondria are of ancient bacterial origin, and are consistent with the endosymbiotic hypothesis of the origin of eukaryote cells, which postulates that mitochondria arose from bacteria able to perform aerobic metabolism that were ingested by the ancestors of eukaryote cells (Margulis, 1981). According to this hypothesis, eukaryotic cell ancestors were initially anaerobic single-celled organisms without mitochondria. Around two billion years ago, significant amounts of oxygen are believed to have entered the earth atmosphere, as a consequence of the metabolism of oxygen-producing photosynthetic bacteria. The endosymbiont hypothesis postulates that the anaerobic ancestors of eukaryote cells then evolved mitochondria through the capture of bacteria that possessed oxidative phosphorylation systems that became subverted by the eukaryote cell for aerobic metabolism.

The Nucleus and the Mitochondria as Reciprocal Addiction Module

Present-day mitochondria and eukaryotic cells are condemned to live together, and this symbiotic equilibrium is usually viewed as a consequence of the "selfish" enslavement by the ancestors of eukaryotic cells of the aerobic bacteria they captured. But an alternate view, which may be more appropriate, is to consider that such enforced cooperation equilibrium rather results from the stabilization of an initially bidirectional host–pathogen evolutionary arms race. Indeed, there are still several examples, in various organisms, of intracellular bacteria using eukaryote cells as an appropriate environment for their "selfish" propagation. This is true of some of the pathogenic bacteria that invade mammals and replicate in their cells—such as *Ricketsia, Listeria,* and *Shigella*—and subvert host cell signaling processes in order to invade these cells (Ménard et al., 1996). But there are also more complex ways in which bacteria subvert eukaryotic organisms. In several crustacean species and insect Hymenoptera species, bacteria such as *Wolbachia* are able to infect oocysts and to persist in these cells without killing them, resulting in a complex host–parasite equilibrium that leads to non-Mendelian maternal transmission of the bacteria together with the mitochondrion, and that can lead to changes in the phenotypic sex of the infected organism, to male sterility, or to a complex process of pseudo speciation through the prevention of mating between organisms that are infected by different bacterial strains (Rousset et al., 1992; Juchault et al., 1993; Rigaud et al., 1996). Striking and complex interactions with bacteria can also occur in single-celled eukaryote organisms. First, the amoeba *Palomyxa palustris,* one of the rare single-celled eukaryotes that lack mitochondria, is a symbiont that contains aerobic bacteria in its cytoplasm; and it is these bacteria that perform the respiratory activity required for the aerobic metabolism of the amoebae (Alberts et al., 1994). Second, a recent bacterial infection of laboratory cultures of a mitochondriated amoeba has led to the emergence of an endosymbiotic association, in which several thousands of bacteria reproduce in the cytoplasm of each ameoba cell: The amoebae have become dependent on the existence of the bacteria in their cytoplasm, and die if one achieves selective killing of the bacteria (Jeon, 1983). If these situations are to be considered examples of intermediate evolutionary steps towards endosymbiosis, they strongly suggest that the evolution of the present-day eukaryote cell may have resulted from an initial attempt of bacteria to manipulate their host, rather than from the opposite situation. In this context, preliminary unpublished results from Douglas Green (LIAI, La Jolla, CA) on the mechanisms involved in amoeba–bacteria interactions are consistent with the idea that the bacteria behave as an addiction module for the single-celled eukaryote colony.

If the bacterial ancestors of mitochondria initially behaved as addiction modules for the cell they infected, eukaryotic cells themselves appear to have responded by taking countermeasures. Indeed, most of the genes encoding present-day mitochondrial proteins are located in the cell nucleus and seem therefore to have been progressively transferred from the mitochondrial genome to the nuclear genome. In mammalian cells, around a hundred nuclear genes of apparent mitochondrial origin encode around hundred proteins, which include mitochondrial DNA and RNA polymerases, RNA processing enzymes and ribosomal proteins (reviewed in Alberts et al., 1994; Lewin, 1997; Palmer, 1997). These proteins are synthesized on cytosolic ribosomes, but are all imported into the organelle; once they have been synthesized and imported in the mitochondria, these proteins are believed never to leave the mitochondria, at least as long as the cell survives. Interestingly, the strategy of transferring to the nucleus the whole mitochondrial genome may have been rendered impossible at some time point by changes in the mitochondrial genetic code that may have prevented the transcription and translation of these mitochondrial genes from the nucleus. The reason for such a complex arrangement is not obvious unless it is considered in the context of an "evolutionary arms race." On the one hand, eukaryotic cells from most organisms are condemned to retain mitochondria, since they have become dependent on aerobic mitochondrial respiratory metabolism; on the other hand, mitochondria cannot leave the cell, since most of the proteins that constitute them have become encoded by nuclear genes. Thus the endosymbiotic nature of the eukaryote cell may be seen as a result of a complex "evolutionary arms race" between a host and a pathogen that become somehow frozen, at a given point, into a state of enforced cooperation.

The view that I propose is that the actual outcome of such an ancient evolutionary battle—the present-day eukaryote cell—may result from the complex interaction between two reciprocal addiction modules: the mitochondrial addiction module, which has rendered the cell dependent on the presence of mitochondria; and the nuclear addiction module, which has rendered the mitochondria dependent on the cell. This evolutionary view may provide an interesting framework for understanding the recent and surprising findings that an important part of programmed cell death regulation in cells from multicellular animals occurs at the mitochondria–cytoplasm interface; and that this regulation involves gene products that share some unexpected features with bacterial gene products, such as bacterial toxins.

The Use of Bacteria-Related Tools to Control Bacterial-Related Behavior in Eukaryotic Cells: The Control of Programmed Cell Death at the Mitochondrial–Cytoplasmic Interface

Several results have suggested that mitochondria may play an important role in the effector phase of programmed cell death (reviewed in Golstein, 1997; Kroemer et al., 1997): (1) In cell-free systems of nuclear apoptosis induction (in which cytoplasmic extracts from cells undergoing programmed cell death induce the fragmentation of exogenous nuclei extracted from living cells) mitochondrial extracts have been reported to be required in order for the cytoplasmic extracts to induce nuclear fragmentation; (2) cell death has been reported to be correlated with a loss of mitochondrial membrane potential that induces mitochondrial permeability transition and leads to the release, in the cytosol, of some mitochondrial proteins; (3) permeability transition has been reported to induce the release from mitochondria of a protease (AIF, apoptosis inducing factor, which still remains to be characterized) that appears to act as a caspase; finally (4) even in the absence of a mitochondrial permeability transition, mitochondria have been reported to release cytochrome C (an important electronic carrier between the three major enzyme complexes of the mitochondrial respiratory chain) during the effector phase of programmed cell death; cytochrome C (in association with a still unknown component of the cytoplasm) induces the activation of the CPP32 caspase in the cytosol (Kluck et al., 1997; Yang et al., 1997). Thus the intracellular signaling pathways that lead to the executionary phase of programmed cell death appear to require the induction of mitochondrial changes that allow the release of proteins that are normally kept sequestered in these organelles. Very recent findings also suggest that this mitochondrial step is under the control of the Bcl-2 and BclX$_L$ programmed cell death antagonist gene products. Most of the Bcl-2 family gene products are predominantly located on intracellular membranes: the endoplasmic reticulum membrane, the nuclear membrane, and the outer mitochondrial membrane. One of the important mechanisms by which Bcl-2 exerts its suppressive effect on programmed cell death appears to be through the prevention of the release by mitochondria of cytochrome C and AIF in response to signals that favor programmed cell death induction (Golstein, 1997; Kluck et al., 1997; Kroemer et al., 1997; Yang et al., 1997). Bcl-2, when bound to the outer mitochondrial membrane, may also recruit other proteins to this membrane, such as the Raf-1 kinase (which appear to reinforce the suppressive effect of Bcl-2 on programmed cell death) as well as the Ced-4 gene product (whose mammalian homolog remains unknown), which appear to act in mammalian cells either as an activator of members of the Ced-3/CPP32 caspase family (when the Ced-9/Bcl-2/BclX$_L$ gene family products are absent), or as an adaptor between Ced-9/Bcl-2/BclX$_L$ gene family products (when they are present) and the caspases, leading to the prevention of caspase activation. Therefore, the inner side and the outer membrane of the mitochondria appear to be a locus of clustering of the main players in the execution and regulation of the effector phase of programmed cell death. Although some of these players, such as Bcl-2 and the caspases are located in the cytosol of the cell, and the others, such as cytochrome C, are released from the inside of mitochondria, they are all encoded by nuclear genes. This is consistent with previous findings that the mitochondrial DNA contained in the organelle is neither required for the induction of the effector phase of programmed cell death, nor for its prevention by the Bcl-2 gene product, a finding that was initially interpreted as suggesting the absence of any role for mitochondria in the regulation of programmed cell death.

The gene encoding cytochrome C—as most genes encoding mitochondrial proteins—is located in the cell nucleus; it is highly conserved in eukaryotes and encodes a protein synthesized in the cytosol in its precursor form (apocyto-chrome C), and then exported into the mitochondria. Although now located in the cell nucleus, such genes encoding for mitochondrial proteins are pre-sumed to be of ancient bacterial origin. On the other hand, the genes encoding for the Bcl-2 family proteins that participate in the suppression of programmed cell death are presumed to be of ancient eukaryote origin. The recent determi-nation of the nuclear magnetic resonance and crystal structure of the BclX$_L$ gene product (an important member of the Bcl-2 family that also acts as a programmed cell death suppressor) has revealed, however, that the arrange-ment of its alpha helices is reminiscent of the membrane translocation domain of bacterial toxins (Muchmore et al., 1996), in particular the colicins and the diphtheria toxin. In addition, BclX$_L$ and Bcl-2 have been reported to form membrane pores and to induce membrane depolarization, a feature that is similar to that of the colicins and the membrane translocation domain of diphtheria toxin. Thus the gene products of the Bcl-2 family may regulate programmed cell death induction through an ability to form pores in—and change the polarization of—the outer membrane of mitochondria. In other words, the involvement of an organelle of bacterial origin (the mitochondria) in the effector phase of programmed cell death is regulated by a family of genes of apparent eukaryote origin (Bcl-2) through three-dimensional features that they share with bacterial toxins.

I strongly believe that these apparently odd features of programmed cell death regulation in the eukaryote cells only begin to be understandable when replaced in the perspective of an "evolutionary arms race" between bacteria and the ancestors of the eukaryote cell that led to the emergence of the present-day eukaryote cell. The progressive transfer to the nucleus of the essential mitochondrial genes involved in the regulation of programmed cell death may have represented an important step in the regulated use of the mitochondria as a tool for self-destruction. In this context, the mechanisms by which the cell induces an apparently irreversible loss of mitochondrial membrane potential that leads (when not prevented by Bcl-2) to permea-bility transition may be seen as a mechanism by which the cell triggers self-destruction through the deletion of an addiction module (the mitochondria), a process that may mimic—and recapitulate—ancient features of the initial bacteria–eukaryote ancestor cell interactions. Finally, if we attempt to think in terms of toxin–antidote addiction modules, and consider cytochrome C as a toxin and Bcl-2 as an antidote, it is interesting to note that the ability of cytochrome C to activate the CPP32 caspase, and trigger programmed cell death, seems not to involve the cytochrome C domains that are involved in its life-sustaining role in mitochondrial respiration (Kluck et al., 1997).

Therefore, the view of mitochondria and the eukaryote cell as "reciprocal addiction modules" involved in a symbiotic partnership can be integrated both in the framework of the "evolutionary arms race" hypothesis, in which the initial selection of toxin–antidote modules as addiction modules led to their subsequent selection because of the additional functional advantages they provided, and in the framework of the "original sin" hypothesis, in which the initial selection of genetic modules performing vital but dangerous cellular functions led to their subsequent propagation because of the addictive value they provided as toxin–antidote modules.

ON THE NATURE OF PROGRAMMED CELL DEATH EXECUTIONERS

The Requirement for Mitochondria in the Execution of Single-Celled Eukaryote Programmed Cell Death

An implication of the view I have presented is that mitochondria may have played an evolutionary conserved role in the ability of single-celled eukaryote organism to self-destruct. Interestingly, some recent experimental results may provide support to this hypothesis. First, preliminary findings obtained in our laboratory indicate that a loss in mitochondrial membrane potential and a permeability transition similar to those involved in mammalian cell programmed death also occurs in the giant kinetoplastid mitochondrion of the single-celled protozoan *Trypanosoma cruzi* during *Trypanosoma cruzi* programmed cell death (T. Idziorek, A. Ouaissi, J. C. Ameisen, unpublished results). Second, as previously mentioned, there are yeast mutants (such as the petite yeasts) that can survive the loss of mitochondria respiratory function by resorting to anaerobic metabolism as a (reduced) source of energy. Interestingly, transfection of the mammalian Bax gene product (a member of the Bcl-2 family that induces programmed cell death) has been reported to induce death only in yeasts that have functional mitochondria performing respiratory activity (and not in the mitochondrial loss-of-function mutants), suggesting that Bax may induce a form of programmed cell death in yeasts through a mechanism that involves the inhibition of mitochondrial function (Greenhalf et al., 1996). Also in yeasts, death induced by Bax transfection can be antagonized by cotransfection of the Bcl-2 or $BclX_L$ genes (as in mammalian cells), provided that these genes are not deleted in their domains that allow the targeting of the protein to the outer membrane of the mitochondrion (Hanada et al., 1995). Together, these findings imply that an important part of the regulation of Bax-induced death in yeast may occur at the level of the outer mitochondrial membrane. Finally, in the slime mold *Dictyostelium discoideum,* it has been reported on the one hand that the differentiation-inducing factor (DIF)—which is involved in the differentiation of cells into stalk cells that undergo programmed death—is an uncoupler of mitochondrial respiratory activity; and on the other hand, that drugs that act as uncoupling agents of mitochondrial respiratory activity also act as inducers of stalk cell terminal differentiation (Williams, 1995).

Although these data suggest the possibility of an involvement of mitochondria in the regulation of programmed cell death in single-celled eukaryote organisms, it will be important to investigate to what extent this process (and the gene products involved in the regulation of this process) are conserved in single-celled and multicellular organisms. Three experimental models may be expected to be of particular importance for addressing this question: (1) the members of the kinetoplastid family, such as trypanosomes and leishmanias, because they appear to be the most ancient mitochondriated eukaryotes; (2) yeasts, because mitochondria loss-of-function mutants should allow the investigation of potential mechanisms of self-destruction that may not require mitochondria (and that may be the remnants of primitive self-destruction mechanisms that operated in the ancestor of the eukaryote cell); and for the same reason, (3) the rare single-celled eukaryote anaerobic organisms that lack mitochondria, live in oxygen-poor conditions, and include the microsporidia and the diplomonads (such as *Giardia*). These single-celled eukaryotes of ancient diverging origin have neither mitochondrion nor chloroplast, and contain only nuclear DNA. Such single-celled organisms are either the diverging progeny of an ancestor of the eukaryote cell that preceded the period of mitochondria capture, or more probably the progeny of an ancient mitochondriated eukaryote cell that has succeeded, at some point, in losing its mitochondrion (Palmer, 1997), without undergoing self-destruction.

Finally, the previously mentioned striking model of the amoeba *Palomyxa palustris,* which lacks a mitochondrion, but depends on the presence of aerobic bacteria in its cytoplasm to perform oxidative metabolism (Alberts et al., 1994), may represent a valuable model to address the question of the nature of the gene products involved in the regulation of eukaryote cell programmed death in the context of an ancient bacterial (mitochondrion)–eukaryote cell interface.

Programmed Cell Death in Plants

The question of the existence of programmed cell death in plants has long remained neglected. Recently, it has been shown that programmed cell death can occur in plants, but the extent of its role, the nature of the mechanisms involved, and the nature of its genetic regulation still remain to be assessed. Cell death is an important part of plant development, involved in the generation of xylem, and also appear to participate in the senescence of leaves and flowers (Greenberg, 1996). Some striking forms of developmentally regulated processes of programmed cell death that synchronize the release of the gametes and the subsequent death of all the remaining somatic cells (a process similar to the synchronized spore–programmed cell death differentiation models that we have discussed in single-celled prokaryote organisms) have been reported in numerous species of green algae (Hay, 1997). But the best studied forms of genetically regulated programmed cell death in plants is the "hypersensitivity response" to infectious pathogens, a process of induction of self-destruction in infected cells and in neighboring cells that plays a crucial role in defense against infection (Greenberg et al., 1994) and shares some phenotypic features with apoptosis (Levine et al., 1996). In all these instances of programmed cell death, the nature of the executioners and the potential role of mitochondria are unknown. But plant cell programmed death should represent a valuable model in order to address an additional question, that of the potential role in the regulation of programmed cell death of another DNA containing organelle, the plastic–chloroplast. Do chloroplasts behave as addiction modules in the plant cells? Is disruption of the chloroplast an important step of programmed cell death? Does programmed cell death in plants depend on the release of particular proteins from mitochondria, from chloroplasts, or from both?

Interestingly, DNA-containing organelles similar to plant plastid–chloroplasts have been recently uncovered in a family of protozoan parasites, the *apicomplexans,* which include both the human pathogens *Toxoplasma,* which causes toxoplasmosis, and *Plasmodium,* which causes malaria (Köhler et al., 1997). These single-celled eukaryotes may represent an interesting and valuable model to address the question of the respective role of mitochondria and plastids in the regulation of eukaryote cell death.

The Nature of Executioners

We will now go back to the general question of the execution of programmed cell death in both the prokaryote and the eukaryote organisms. We have seen, in this chapter, various examples of different forms of the executioner–suppressor—or toxin–antidote—modules, the basic modules of antagonistic genes that allow the core regulation of programmed cell death. We have seen that executioner caspases, such as DCP in *Drosophila,* may have other functions than that of executioners of cell death. We have seen that suppressers of execution such as Bcl-2 in mammals may have other functions than that of caspase antidote. We have seen that the mitochondrial cytochrome C, an apparent obligate upstream activator of the CPP32 mammalian executioner caspase (when released from the mitochondria into the cytosol) performs an essential function required for cell survival (when present inside the mitochondria). We have also seen that most toxin–antidote addiction (or self-addiction) modules that regulate programmed cell death in bacteria depend on the constitutive expression of a chromosome encoded bacterial protease, most often the Lon or Clp serine protease, which cleaves the antidote; and we have seen that these bacterial proteases, which play an essential role in the execution of cell death, also perform essential functions required for bacterial cell survival.

As previously discussed, all these examples suggest the possibility that there is no such a thing as *bona fide* executioners whose sole function are to execute or to prevent self-destruction. But these examples have additional implications. The first concerns the mechanism by which (a) gene product(s) involved in execution is (are) believed to perform its (their) executioner function. The caspases from multicellular animals are cysteine proteases that cleave substrates at aspartate residues; several of these substrates cleaved by the downstream executioner caspase-3 (CPP32) during programmed cell death in mammals have been identified, such as the DNA repair enzyme poly (ADP-ribose) polymerase (PARP), DNA-PK, SRE/BP, rho-GDI (Nagata, 1997), and it is believed that self-destruction of the cell is the direct consequence of caspase-dependent cleavage of cellular proteins essential for cell survival. The model of the bacterial addiction module-induced programmed cell death, however, suggests the alternate possibility that proteases, such as caspases, in eukaryote cells may not be the terminal effectors of programmed cell death but may act by cleaving of (a) still unknown antagonist protein(s) that counteract(s) the lethal effect of other still unrecognized executioner gene products. If this is true, there are still more downstream effectors of programmed death in eukaryote cells than is now expected.

A second important implication of the models that I have discussed in this chapter is the following: When we think that executioners may have essential functions other than execution, we do not have to obligatorily expect such function to depend on the same molecular domains than those that are involved in execution. There may at least be two such examples. The first pertains, as previously mentioned, to cytochrome C; the molecular domains that appear essential for the role of cytochrome C in the induction of programmed cell death appear different from the domains essential for its vital role in mitochondrial respiratory activity (Kluck et al., 1997). The second example concerns the Lon protease. We have seen that the ATP-dependent serine proteases Lon and Clp are involved in bacteria, in both the induction of programmed cell death and the regulation of cell cycle. Both activities depend on the proteolytic activity of these proteases. Interestingly, Lon (in contrast to Clp) has been conserved in eukaryote cell mitochondria, and as cytochrome C, Lon is now encoded in the nuclear genome of the cell, synthesized in the cytosol, and exported into the mitochondria. An interesting question that deserves to be investigated is whether Lon is released in the cytosol during programmed cell death, and whether it participates (in the eukaryote cell as in bacteria) in the effector phase of programmed cell death (and/or whether it may be part of the still uncharacterized AIF protease). But let us return to the previous question of the nature of the domains involved in execution and in other potential functions of executioners. In eukaryote yeast cells, the Lon protease has recently been shown to have an essential function, unrelated to its proteolytic activity: a chaperonelike function allowing the assembly of the respiratory protein complexes at the inner mitochondrial membrane (Rep et al., 1996). This chaperonelike function is preserved upon inactivation of the proteolytic site of Lon (Rep et al., 1996). Therefore, if caspases perform important functions other than the execution of programmed cell death, such additional functions may not obligatorily involve their protease activity. In this respect, as far as I have heard to date, all known Ced-3 loss-of-function mutants in *Caenorhabditis elegans* are the consequence of small mutations (and not of gross insertion or deletion processes); therefore, in order to assess whether the Ced-3 caspase in *Caenorhabditis elegans* may ensure functions other than execution of programmed cell death—which may be independent of their proteolytic activity—it would be important to perform

Ced-3 knockout experiments and to investigate their phenotype. All these questions may appear to the reader mostly of theoretical interest. But they have potential implications for the therapeutic approach of programmed cell death regulation in disease, in particular for the outcome of therapeutic strategies based on the inhibition of programmed cell death through inhibition of the proteolytic activity of caspases.

There are other important questions that await further investigation. The first is whether at least some of the toxin–antidote modules that regulate programmed cell death in bacteria have been evolutionarily conserved, or whether there has been a convergent evolution of numerous genetic modules that were diversely selected and retained by different organisms. Another question is whether all toxin–antidote modules that became selected also perform other important cellular functions, or whether there are two different kinds of toxin–antidote addiction modules: some that have been solely retained because of their addictive capacity to regulate cell death and survival, and others that have been retained because of their additional capacity to encode for important cell functions.

But I would like to end this chapter with another implication of the single-celled programmed cell death paradigm that does not concern the nature of the possible mechanisms involved in the emergence and selection of programmed cell death, but rather the possible role that self-destruction programs may have played in the evolution of life.

"DEATH FROM WITHIN" AND "DEATH FROM WITHOUT": TOWARDS A NEW VIEW ON THE ROLE OF DEATH IN THE PROCESS OF NATURAL SELECTION

Genetic Mutation and Natural Selection: The Phenotype As the Vehicle or the Grave of the Genome

The current view of the Darwinian theory of evolution may be summarized in the following way: The basis for evolution of living organisms is the existence of random genetic changes; such random genetic changes may or may not lead to phenotypic changes; natural selection does not directly act at the level of the genome, but at the level of the phenotypes they induce, through a process of competition between phenotypes for survival of the fittest in a given environment. Phenotypes best adapted to their environment allow survival and propagation of their genome; phenotypes misadapted to their environment do not allow the propagation of their genome. In brief, the phenotype that results from the expression of a given genome will behave either as a vehicle or as a grave for this genome. In such a view, death represents a sanction that results from an inability of the genome to allow sufficient fitness and survival of the phenotype it encodes. When a genetic mutation provides enhanced fitness to the organism that expresses it in a given environment, the mutant organism becomes selected at the expense of the organism expressing the former nonmutated genome. When a genetic mutation leads to a decreased fitness of the organism that expresses it, this organism may become counterselected by the organism expressing the former nonmutated genome (for a review, see Dawkins, 1976, 1982; Sigmund, 1993).

In multicellular organisms, the genetic mutations that allow the propagation (or the counterselection) of the genome are the genetic mutations that occur at the level of the germ cells; in single-celled organisms, which represented the only form of life during the first two billion years of evolution, any cell from the colony can be considered as a potential germ cell. Therefore, in single-celled organisms (as in multicellular organisms) the processes of random genetic mutations and of natural selection have been considered as the sole major driving forces in evolution. I will now argue that this view may have been too restrictive, because it failed to take into consideration the implications of the capacity of single-celled organisms to differentiate and to self-destruct.

Cell Differentiation in Single-Celled Organisms: Several Vehicles and Graves for a Given Genome

Cell differentiation is a process by which a given genome can achieve different phenotypes through the differential repression of some of its genes. Phenotypic diversity depends on the diversity of the genes that can be repressed in a given cell without being incompatible with a minimal cell fitness. Phenotypic diversity can result from the concomitant presence of different phenotypes in different cells at a given time, from the temporal succession of different phenotypes in the same cells, or from a combination of both processes. As we have seen, most—if not all—single-celled prokaryotes and eukaryotes can achieve various forms of cell differentiation. This feature has several important implications. First, we now see that a given genome has several phenotypes (several ways) through which it can present itself to natural selection. A given phenotype can represent a grave for the genome in a given environment, while another phenotype can represent a vehicle for the same genome in the same environment. Therefore, irrespective of the occurrence of random genetic mutations, cell differentiation endows the genome with a level of phenotypic plasticity that will enhance its fitness in changing (and adverse) environmental conditions. Second, the ability of different phenotypes, representing discrete differentiation stages of the same genome, to interact and cooperate with other phenotypes may result in an enhanced fitness of each phenotype, and therefore contribute to the survival and propagation of the genome.

Programmed Cell Death in Single-Celled Organisms: Using the Grave to Optimize the Vehicle

As we have also seen in this chapter, the coupling of cell differentiation to regulated self-destruction programs may favor the survival and propagation of the genome that can express such self-destruction phenotypes. In adverse environmental conditions, the regulated expression of such programs of terminal differentiation leading to self-destruction may allow, as we have seen, the emergence in the colony of both terminally differentiated somatic cells and of long-lived germlike spore cells, the somatic cells contributing at their own expense to the survival of the germlike spore cells. Such a program may also allow a constant process of optimal adaptation of the numbers of cells sharing the same genome to the environment, allowing self-destruction of part of the cells to preserve survival of the colony. Finally, by coupling the expression of antidote survival genes to the expression of genes allowing precise differentiation phenotypes, programmed cell death may enhance the fitness of the colony by preventing competition of the optimally differentiated fittest cells with flawed offspring that have not achieved an appropriate differentiation stage.

It has been recently argued, in the context of multicellular animal development, that there may be more to evolution than the sole interplay between random genetic mutation and natural selection, and that some intrinsic constraints related to the self-organization processes required for development may have been acting as internal constraints exerted on natural selection (Kauffman, 1993). The view that I propose is that such internal constraints exerted on natural selection may be best exemplified, at the level of the single-celled organisms (and long before the emergence of multicellular organisms), by the coupling of cell differentiation programs to programs of self-destruction. In other words, the ability to regulate phenotypic self-destruction may have represented an important and ancient mechanism by which genomes have resisted to some extent their dismissal by natural selection. Natural selection, sanctioned by death, a central feature of evolution, has been traditionally viewed as operating (1) as a form of "death from without," and (2) at the level of competition between genetically heterogeneous individuals. The existence in both single-celled and multicellular organisms of a physiological cell death program allowing the continuous competition for the survival of the fittest cell in a given individual implies, however, that a form of natural selection sanctioned by death also operates as a process of "death from within" at the level of interactions between cells with the same genome.

On the one hand, the ability of genomes to encode for a genetic module allowing "death from within" represents a barrier against "death from without" (the traditional view of natural selection); on the other hand, the propagation of such genomes will be favored by natural selection (in the traditional sense of "death from without"), not only because such genomes will better resist "death from without," but also because competing genomes that do not encode for "death from within" modules may be more readily counterselected by the process of "death from without." Simply put, programmed cell death allows a process of natural selection "from within" that protects, to some extent, the genome encoding it against natural selection "from without."

Addiction As a Point of No Return: Programmed Cell Death As an Evolutionary Switch between Chance and Necessity

As we have seen, addiction modules that allow programmed cell death regulation may have propagated—and may have favored the propagation of the genomes that encoded them—for the main reason that they are addictive. Whatever advantage they may provide to the genome that encodes them, genetic addiction modules can be seen as a form of "point of no return" in evolution. Once genetic addiction has begun, there may be no way of going back. True, the addicted genome may become counterselected by the process of "death from without" if it does not encode (a) phenotype(s) able to adapt to environmental changes. The addicted genome may therefore disappear, or evolve through the effect of mutations; but if it evolves, it will almost always be as an addicted genome. Although chance must obviously have played an essential major role in the emergence and refinement of the addiction (programmed cell death) modules, once these modules emerged, they may have the striking property of becoming a necessity.

In a sense, any selfish addiction genetic module may be viewed as behaving as an infectious pathogen that spreads at the expense of the host cell and genome that harbors it. This is true for bacterial addiction modules and, as we have seen, might have been important in the propagation of bacteria-eukaryotic cell interactions that have led to the mitochondriated eukaryote cells. But similar forms of "selfish" genes behaving like toxin–antidote addiction modules have also been described in sexually reproducing multicellular animals. They can be involved in competition between different alleles at the same locus, with one allele causing the death of offspring that do not carry them. Examples of such alleles that segregate as a chromosomal gene and spreads through postfertilization killing of diploid progeny that do not express them have been described in flour beetles (the M/m Medea alleles; Bull et al., 1992) and in mice (the scat+/scat alleles; Hurst, 1993). In such a perspective, the regulation of cell survival and cell death may be viewed as an essential component of the process that favors the probability of condemning heterogeneous genes to survive together. As we have previously discussed, one way of looking at living organisms is seeing them as genetically heterogeneous. There are many layers of heterogeneity: infectious genes integrating themselves into chromosomes; mutant genes emerging in the chromosomes; mitochondria and chloroplasts integrated in eukaryote cells; and in each multicellular organism endowed with sexual reproduction, a nuclear genome made of the congregation of half the genome from its father and the other half from its mother, and a mitochondrial genome inherited from its mother. At each of these levels of genetic heterogeneities, a reciprocal sharing in the regulation of cell survival and cell death may be required, in order for reciprocal addiction to allow a state of "no return" leading to the emergence of enforced cooperation.

I propose that this is the essential reason why programmed cell death is so widespread in living organisms. Because genetic addiction modules allowing programmed cell death regulation may have represented an essential driving force in the evolution of life, I believe that our current view of natural selection has to be modified in a way that allows the integration of the process of "death from within" in the framework of "death from without."

ACKNOWLEDGMENTS

JCA thanks Fabienne Ameisen for the initial suggestion, eight years ago, that programmed cell death may have emerged in single-celled organisms, and for constant and insightful discussions thereafter; Doug Green, for insightful and animated discussions, during the past five years, and more recently during a most interesting sabbatical in his laboratory, on the hypothesis of the possible origin, existence, and role of programmed cell death in evolution; Françoise Russo-Marie, for first pointing out to me five years ago that a form of programmed cell death had been described, long ago, in the single-celled eukaryote *Dictyostelium discoideum* and prokaryote *Myxobacteria;* Doug Green and Pierre Golstein for sharing unpublished results; Pierre Sonigo for very stimulating discussions on evolution; and several members of the apoptosis research community, including Andrew Wyllie, Michael Hengartner, Martin Raff, Pat Williamson, and many others for interesting discussions and argumentation during the past three years.

The research programs of JCA are supported by institutional grants from the Institut National de la Santé et de la Recherche Médicale (INSERM), the Paris VII University, and from grants from the Agence Nationale de Recherche sur le Sida (ANRS) and the Fondation pour la Recherche Médicale (FRM).

REFERENCES

Aizenman E, Engelberg-Kulka H, Glaser G (1996): An *Escherichia coli* chromosomal 'addiction module' regulated by 3',5'-bispyrophosphate: A model for programmed bacterial cell death. Proc. Natl Acad Sci USA 93:6059–63.

Alberts B, Bray D, Lewis J, Raff M, Roberts K, Watson JD (1994): Molecular Biology of the Cell. New York. Garland Publishing, Inc.

Alnemri E, Livingston D, Nicholson D, Salvesen G, Thornberry N, Wong W, Yuan J (1996): Human ICE/Ced-3 protease nomenclature. Cell 87:171.

Ameisen JC (1994): Programmed cell death (apoptosis) and cell survival regulation: Relevance to AIDS and cancer. AIDS 8:1197–213.

Ameisen JC (1996a): The origin of programmed cell death. Science 272:1278–79.

Ameisen JC (1996b): Programmed cell death. (Response). Science 274:20–21.

Ameisen JC, Estaquier J, Idziorek T (1994): From AIDS to parasite infection: Pathogen-mediated subversion of programmed cell death as a mechanism for immune dysregulation. Immunol Rev 142:9–51.

Ameisen JC, Estaquier J, Idziorek T, De Bels F (1995a): Programmed cell death and AIDS: Significance, perspectives and unanswered questions. Cell Death and Differentiation 2:9–22.

Ameisen JC, Idziorek T, Billaut-Mulot O, Loyens M, Tissier JP, Potentier A, Ouaissi MA (1995b): Apoptosis in a unicellular eukaryote (*Trypanosoma cruzi*): Implications for the evolutionary origin and role of programmed cell death in the control of cell proliferation, differentiation and survival. Cell Death and Differentiation 2:285–300.

Amon A (1996): Mother and daughter are doing fine: Asymmetric cell division in yeast. Cell 84:651–54.

Arigoni F, Pogliano K, Webb CD, Stragier P, Losick R (1995): Localization of protein implicated in establishment of cell type to sites of asymmetric division. Science 270:637–40.

Böker CA, Schaub GA (1984): Scanning electron microscopic studies of *T. cruzi* in the rectum of its vector *Triatoma infestans*. Z. Parasitenkd 70:459–69.

Brener Z (1973): Biology of *T. cruzi*. Annu Rev Microbiol 27:347–82.

Bull JJ, Molineux IJ, Werren JH (1992): Selfish genes. Science 256:65.

Camargo EP (1964): Growth and differentiation in *T. cruzi*. I. Origin of metacyclic trypanosomes in liquid media. Rev Inst Med Trop Sao Paulo 6:93–100.

Castellani O, Ribeiro LV, Fernandes JF (1967): Differentiation of *T. cruzi* in culture. J Protozool 14:447–51.

Chiari E (1975): Differentiation of *T. cruzi* in culture. In American Trypanosomiasis Research. Proceedings of an International Symposium. Washington, DC: Scientific Publications, PAHO/WHO, pp. 144–45.

Chinnaiyan A, Dixit V (1996): The cell death machine. Curr Biol 6:555–62.

Chinnaiyan A, O'Rourke K, Lane V, Dixit V (1997): Interaction of Ced-4 with Ced-3 and Ced-9: A molecular framework for cell death. Science 275:1122–26.

Christensen ST, Wheatley DN, Rasmussen MI, Rasmussen L (1995): Mechanisms controlling death, survival and proliferation in a model unicellular eukaryote *Tetrahymena thermophila*. Cell Death and Differentiation 2:301–8.

Clarke PG (1990): Developmental cell death: Morphological diversity and multiple mechanisms. Anal Embryol 181:195–213.

Clem RJ, Fechheimer M, Miller LK (1991): Prevention of apoptosis by a baculovirus gene during infection of insect cells. Science 254:1388–90.

Cohen JJ (1993): Apoptosis. Immunol Today 14:126–30.

Cornillon S, Foa C, Davoust J, Buonavista N, Gross JD, Golstein P (1994): Programmed cell death in *Dictyostelium*. Cell Sci 107:2691–704.

Cowan WM, Fawcett JW, O'Leary DDM, Stanfield BB (1984): Regressive events in neurogenesis. Science 225:1258–1265.

Davis MC, Ward JG, Herrick G, Allis CD (1992): Programmed nuclear death: apoptotic-like degradation of specific nuclei in conjugating *Tetrahymena*. Dev Biol 154:419–32.

Dawkins R (1976): The selfish gene. Oxford: Oxford University Press.

Dawkins R (1982): The extended phenotype. Oxford: Oxford University Press.

De Souza W (1984): Cell biology of *T. cruzi.* Ann Rev Cytol 86:197–283.

Doolittle RF, Feng DF, Tsang S, Cho G, Little E (1996): Determining divergence times of the major kingdoms of living organisms with a protein clock. Science 271:470–77.

Dusanic DG (1980): In vitro production of metacyclic trypomastigotes of *T. cruzi.* J Parasitol 66:1046–49.

Duvall E, Wyllie AH (1986): Death and the cell. Immunol Today 7:115–19.

Ellis RE, Yuan J, Horvitz HR (1991): Mechanisms and functions of cell death. Ann. Rev. Cell Biol 7:663–98.

Errington J (1996): Determination of cell fate in Bacillus subtilis. Trends Genet 12:31–34.

Evan GA (1994): Old cells never die, they just apoptose. Trends Cell Biol 4:191–92.

Franch T, Gerdes K (1996): Programmed cell death in bacteria: translational repression by mRNA end-pairing. Mol Microbiol 21:1049–60.

Gerdes K, Rasmunssen PB, Molin S (1986): Unique type of plasmid maintenance function: Postsegregational killing of plasmid-free cells. Proc Natl Acad Sci USA 83:3116–20.

Glücksmann A (1951): Cell deaths in normal vertebrate ontogeny. Biol Rev 26:59–86.

Golstein P (1997): Controlling cell death. Science 275:1081–82.

Green DR, Scott DW (1994): Activation-induced apoptosis in lymphocytes. Curr Op Immunol 6:476–87.

Greenberg J, Guo A, Klessig D, Ausubel F (1994): Programmed cell death in plants: A pathogen-triggered response activated coordinately with multiple defence functions. Cell 77:551.

Greenberg JT (1996): Programmed cell death: A way of life for plants. Proc Natl Acad Sci USA 93:12094–97.

Greenhalf W, Stephan C, Chaudhuri B (1996): FEBS Lett 380:169–75.

Hall BF, Joiner KA (1993): Developmentally-regulated virulence factors of *T. cruzi* and their relationship to evasion of host defences. J Euk Microbiol 40:207–13.

Hanada M, Aimé-Sempé C, Sato T, Reed JC (1995): Structure–function analysis of Bcl-2 protein. Identification of conserved domains important for homodimerization with Bcl-2 and heterodimerization with Bax. J Biol Chem 270:11962–69.

Harrington EA, Fanidi A, Evan GI (1994): Oncogenes and cell death. Curr Op Gen Dev 4:120–29.

Hay M (1997): Synchronous spawning: When timing is everything. Science 275:1080–81.

Heald R, McLoughlin M, McKeon F (1993): Human wee1 maintains mitotic timing by protecting the nucleus from cytoplasmically activated cdc2 kinase. Cell 74:463–74.

Hengartner MO (1995): Life and death decisions: Ced-9 and programmed cell death in *C. elegans.* Science 270:931.

Hengartner MO, Horvitz HR (1994): *C. elegans* cell survival gene *ced-9* encodes a functional homolog of the mammalian proto-oncogene *bcl-2.* Cell 76:665–76.

Hengge-Aronis R (1993): Cell 72:165.

Horvitz HR, Herskowitz I (1992): Mechanisms of asymmetric cell division: Two Bs or not two Bs, that is the question. Cell 68:237.

Hurst LD (1993): *scat+* is a selfish gene analogous to *Medea* of *Tribolium castaneum.* Cell 75:407–8.

Jacobson MD, Weil M, Raff MC (1997): Programmed cell death in animal development. Cell 88:347–54.

Jensen RB, Gerdes K (1995): Programmed cell death in bacteria: Proteic plasmid stabilization systems. Mol Microbiol 17:205–10.

Jeon KW (1983): Int Rev Cytol Suppl 14:29.

Juchault P et al. (1993): J Evol Biol 6:511.

Kaiser D (1986): Control of multicellular development: *Dictyostelium* and *myxococcus.* Ann Rev Genet 20:536–66.

Kaiser D (1996): Bacteria also vote. Science 272:1598–99.

Kaiser D, Losick R (1993): How and why bacteria talk to each other. Cell 73:873–85.

Katayama Y, Gottesman S, Pumphrey J, Rudikkoff S, Clark WP, Maurizi MR (1988): The two-component, ATP-dependent Clp protease of *Escherichi coli.* J Biol Chem 263:15226–36.

Kauffman S (1993): The Origin of Orders. Self-Organization and Selection in Fvolution. Oxford: Oxford University Press.

Kelly JM, Ward HM, Miles MA, Kendall G (1992): A shuttle vector which facilitates the expression of transfected genes in *Trypanosoma cruzi* and *Leishmania.* Nucleic Acids Res 20:3963–69.

Kerr JFR, Willie AH, Currie AR (1972): Apoptosis: A basic biological phenomenon with wide-ranging implications in tissue kinetics. Br J Cancer 26:239–57.

Kluck RM, Bossy-Wetzel E, Green DR, Newmeyer D (1997): The release of cyto-chrome c from mitochondria: A primary site for Bcl-2 regulation of apoptosis. Science 275:1132–36.

Knoll AH (1992): The early evolution of eukaryotes: A geological perspective. Science 256:622–27.

Köhler S, Delwiche CF, Denny CP, Tilney LG, Webster P, Wilson RJM, Palmer JD, Roos DS (1997): A plastid of probable green algal origin in apicomplexan parasites. Science 275:1485–89.

Korsmeyer S (1995): Regulators of cell death. TIBS 11:101–5.

Kroemer G, Zamzami N, Susin SA (1997): Mitochondrial control of apoptosis. Immunol Today 18:44–51.

Levine A, Pennel RI, Alvarez ME, Palmer R, Lamb C (1996): Calcium-mediated apoptosis in plant hypersensitive disease resistance response. Curr Biol 6:427–37.

Levine B, Huang Q, Isaacs J, Reed J, Griffin D, Hardwick J (1993): Conversion of lytic to persistent alphavirus infection by the *bcl-2* oncogene. Nature 361:739–42.

Lewin B (1997): Genes VI. Oxford, New York, Tokyo: Oxford University Press.

Lightman A, Gingerich O (1991): When do anomalies begin? Science 255:690–95.

Linette GP, Korsmeyer SJ (1994): Differentiation and cell death: Lessons from the immune system. Curr Opin Cell Biol 6:809–15.

Linette GP, Li Y, Roth K, Korsmeyer SJ (1996): Cross talk between cell death and cell cycle progression: Bcl-2 regulates NFAT-mediated activation. Proc Natl Acad Sci USA 93:9545–52.

Liu J (1994): Microcin B17: Post-translational modifications and their biological implications. Proc Natl Acad Sci USA 91:4618–20.

Lockshin RA, Williams CM (1965): Programmed cell death: Cytology of degeneration in the intersegmental muscles of the silkmoth. J Insect Physiol 11:123–33.

Losick R, Stragier P (1992): Nature 355:601.

Lundgren K, Walmorth N, Booker R, Dembski M, Kirschner M, Beach D (1991): Cell 64:1111.

Margulis L (1981): Symbiosis in Cell Evolution. New York: W.H. Freeman.

Martin SJ, Green DR (1995): Protease activation during apoptosis: Death by a thousand cuts? Cell 82:349–52.

Maslov DA, Simpson L (1995): Evolution of parasitism in kinetoplastid protozoa. Parasitol Today 11:30–32.

Ménard R, Dehio C, Sansonetti PJ (1996): Bacterial entry into epithelial cells: The paradigm of *Shigella*. Trends Microbiol 4:220–26.

Moreira MEC et al. (1996): Heat shok induction of apoptosis in promastigotes of the unicellular organism *Leishmania amazonensis*. J Cell Physiol 167:305–13.

Morris RG, Hargreaves AD, Duvall E, Wyllie AH (1984): Hormone-induced cell death. 2. Surface changes in thymocytes undergoing apoptosis. Am J Pathol 115:426–36.

Mottram JC (1994): cdc2-related protein kinases and cell cycle control in trypanosomatids. Parasitol Today 10:253–57.

Muchmore SW, Sattlet M, Liang H, Meadows RP, Harlan JE, Yoon HS, Nettesheim D, Chang BS, Thompson CB, Wong SL, Ng SL, Fesik SW (1996): X-ray and NMR structure of human Bcl-xL, an inhibitor of programmed cell death. Nature 381:335–41.

Nagata S (1997): Apoptosis by death factor. Cell 88:355–65.

Nagata S, Golstein P (1995): The Fas death factor. Science 267:1449–56.

Naito T, Kusano K, Kobayashi I (1995): Selfish behavior of restriction-modification systems. Science 267:897–899.

Nogueira N, Bianco C, Cohn Z (1975): Studies on the selective lysis and purification of *T. cruzi*. J Exp Med 142:225–29.

Nordström K, Austin SJ (1989): Mechanisms that contribute to the stable segregation of plasmids. Annu Rev Genet 23:37–69.

Ohta N, Newton A (1996): Signal transduction in the cell cycle regulation of *Caulobacter* differentiation. Trends Microbiol 4:326–32.

Oppenheim RW (1991): Cell death during development of the nervous system. Annu Rev Neurosci 14:453–501.

Oshida T, Sugai M, Komatsuzawa H, Hong YM, Suginaka H, Tomasz A (1995): A *Staphylococcus aureus* autolysin that has an *N*-acetylmuramoyl-L-alanine amidase domain and an endo-β-*N*-acetylglucosaminidase domain: Cloning, sequence analysis, and characterization, Proc Natl Acad Sci USA 92:285–89.

Palmer JD (1997): Organelle genomes: Going, going, gone! Science 275:790–91.

Raff MC (1992): Social controls on cell survival and cell death. Nature 356:397–400.

Raff MC, Barres BA, Burne JF, Coles HS, Ishizaki Y, Jacobson MD (1993): Programmed cell death and the control of cell survival: Lessons from the nervous system. Science 262:695–700.

Rep M, vanDijl JM, Suda K, Schatz G, Grivell LA, Suzuki CK (1996): Promotion of mitochondrial membrane complex assembly by a proteolytically inactive yeast. Lon Sci 274:103–106.

Rigaud T et al. (1996): Genetics 133:246.

Rousset F et al. (1992): Proc R Soc Lond (Biol) 250:91.

Ruoslahti E, Reed JC (1994): Anchorage dependence, integrins, and apoptosis. Cell 77:477–78.

Saunders JWJ (1966): Death in the embryonic systems. Science 154:604–12.

Savill J, Fadok V, Henson P, Haslett C (1993): Phagocyte recognition of cells undergoing apoptosis. Immunol Today 14:131–36.

Schaub GA (1994): Pathogenicity of trypanosomatids on insects. Parasitol Today 10:463–68.

Schwartz LM, Osborne BA (1993): Programmed cell death, apoptosis, and killer genes. Immunol Today 14:582–90.

Schwartz LM, Smith S, Jones MEE, Osborne BA (1993): Do all programmed cell deaths occur via apoptosis? Proc Natl Acad Sci USA 90:980–84.

Shaham S, Horvitz H (1996)· Developing *C. elegans* neurons may contain both cell death protective and killer activities. Genes Dev 10:580–591.

Shapiro L (1993): Protein localization and asymmetry in the bacterial cell. Cell 73:841.

Shi L, Nishioka WK, Th'ng J, Bradbury EM, Litchfield DW, Greenberg AH (1994): Premature p34^{cdc2} activation required for apoptosis. Science 263:1143–45.

Sigmund K (1993): Games of Life. Oxford: Oxford University Press.

Sogin ML (1991): Early evolution and the origin of eukaryotes. Cur Op Gen Dev 1:457–63.

Song Z, McCall K, Steller H (1997): DCP-1, a *Drosophila* cell death protease essential for development. Science 275:536–40.

Steller H (1995): Mechanisms and genes of cellular suicide. Science 267:1445–49.

Sullivan JJ (1983): Metacyclogenesis of *T. cruzi* in vitro: A simplified procedure. Trans R Soc Trop Med Hyg 76:300–2.

Tata JR (1966): Requirement for RNA and protein synthesis for induced regression of the tadpole tail in organ culture. Dev Biol 13:77–94.

Thompson CB (1995): Apoptosis in the pathogenesis and treatment of disease. Science 267:1456–62.

Ucker DS (1991): Death by suicide: One way to go in mammalian cellular development? New Biol 3:103–109.

Vaux DL (1993): Towards an understanding of the molecular mechanisms of physiological cell death. Proc Natl Acad Sci USA 90:786–89.

Vaux DL, Haeker G, Strasser A (1994): An evolutionary perspective on apoptosis. Cell 76:777–79.

Vaux DL, Weissman IL, Kim SK (1992): Prevention of programmed cell death in *c. elegans* by human Bcl-2. Science 258:1955–57.

Welburn SC, Dale C, Ellis D, Beecroft R, Pearson TW (1996): Apoptosis in procyclic *Trypanosoma brucei rhodesiense in vitro*. Cell Death and Differentiation 3:229–36.

Williams GT (1985): Control of differentiation in *T. cruzi*. Curr Top Microbiol Immunol 117:1–22.

Williams GT (1991): Programmed cell death: Apoptosis and oncogenesis. Cell 65:1097–98.

Williams GT, Smith CT (1993): Molecular regulation of apoptosis: Genetic controls on cell death. Cell 74:777–79.

Williams J (1995): Morphogenesis in *Dictyostelium:* New twists to a not-so-old tale. Curr Op Gen Dev 5:426–31.

Wireman J, Dworkin M (1975): Morphogenesis and developmental interaction in Myxobacteria. Science 189:516–23.

Wireman J, Dworkin M (1977): Developmentally-induced autolysis during fruiting body formation by *Myxococcus xanthus.* J Bacteriol 129:796–802.

Wittingham WF, Raper KB (1960): Non-viability of stalk cells in *Dictyostelium discoideum.* Proc Natl Acad Sci USA 46:642–49.

Wu D, Wallen H, Nunez G (1997): Interaction and regulation of subcellular localization of Ced-4 by Ced-9. Science 275:1126–29.

Wyllie AH (1980): Glucocorticoid-induced thymocyte apoptosis is associated with endogenous endonuclease activation. Nature 284:555–56.

Yang E, Korsmeyer SJ (1996): Molecular thanatopsis: A discourse on the Bcl-2 family and cell death. Blood 88:386–401.

Yang J, Liu X, Bhalla K, Kim CN, Ibrado AM, Cai J, Peng TI, Jones DP, Wang X (1997): Prevention of apoptosis by Bcl-2: Release of cytochrome c from mitochondria blocked. Science 275:1129–32.

Yarmolinsky MB (1995): Programmed cell death in bacterial populations. Science 267:836–37.

Yuan J, Shaham S, Ledoux S, Ellis HM, and Horvitz HR (1993): The *C. elegans* cell death gene *ced-3* encodes a protein similar to mammalian interleukin-1β converting enzyme. Cell 75:641–52.

Zeledon R, Bolanos R, Rojas M (1984): Scanning electron microscopy of the final phase of the life cycle of *T. cruzi* in the insect vector. Acta Trop (Basel) 41:39–43.

CHAPTER 2

THE REACTIONS OF CELLS TO LETHAL INJURY: ONCOSIS AND NECROSIS—THE ROLE OF CALCIUM

BENJAMIN F. TRUMP AND IRENE K. BEREZESKY
Department of Pathology
University of Maryland School of Medicine
Baltimore, Maryland 21201

INTRODUCTION

The purpose of this chapter is to review the prelethal cellular changes that follow a lethal injury, including oncosis and apoptosis; to compare oncosis and apoptosis; and to review the morphologic and biochemical changes characteristic of oncosis and necrosis. We will examine existing knowledge concerning events that occur along with putative relevant mechanisms and will develop our working hypothesis for the mechanisms of cell death, emphasizing the role of intracellular ionized calcium ($[Ca^{2+}]_i$) deregulation.

Much of medicine is involved with cell death. Both prevention and therapy of disease are divided between efforts to prevent cell death and necrosis (e.g., trauma, stroke, myocardial infarction, and organ transplantation) and efforts to cause or promote cell death (e.g., treatment of cancer and infectious disease). It is, therefore, essential that we explore and analyze the events that occur during this process, develop hypotheses that lead to critical experiments, and fashion interventions that could modify the course of events.

REACTIONS OF CELLS TO INJURY

Following injury, cells alter their structure and function in a continuum of change that we have divided into a series of three stages: *the prelethal phase;*

When Cells Die, Edited by Richard A. Lockshin, Zahra Zakeri, and Jonathan L. Tilly
ISBN 0-471-16569-7 © 1998 Wiley-Liss, Inc.

the point of cell death; and *the changes characteristic of necrosis* (Fig. 2.1) (for reviews, see Trump and Ginn, 1969; Trump and Arstila, 1971; Trump and Berezesky, 1994).

Although there has been some confusion in the literature, it is now clear that two principal and distinct types of prelethal phases exist: *apoptosis* (considered in detail in Zakeri, Chapter 3, this volume) and *oncosis* (Fig. 2.1). These two types of change differ remarkably in structure and function. Although apoptosis and oncosis are the two principal types of prelethal change that have been described thus far, intermediate forms, difficult or impossible to classify at the present time, also probably exist; additionally, future studies may well reveal additional patterns of change. A major problem exists in interpreting the literature in that standard techniques of morphology for distinguishing between apoptosis and oncosis are often not used, making it difficult to assess the contribution of a given experiment or intervention to one or the

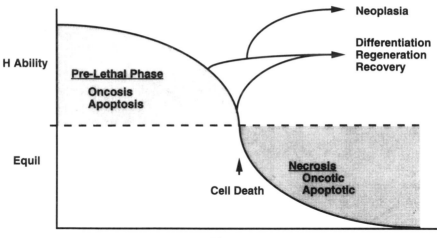

FIG. 2.1. Diagram illustrating the reaction of cells to injury. An injury is applied at time 0. This is followed by a prelethal change that refers to the alterations that occur prior to the "point of no return" or point of cell death. In the case of many injuries such as ischemia or anoxia where the injurious stimulus can be removed, such cells may recover under subsequent transformation to form neoplasms or enter the mitotic cycle leading to regeneration and differentiation. During this phase, the changes can be classified, depending on the morphologic and biochemical findings, into what are termed the changes of oncosis (see text) or the changes of apoptosis. Once the cell passes the "point of no return," the *post mortem* changes in the cell are referred to as *necrosis.* During this phase, as indicated in the diagram, the cells approach physical–chemical equilibrium with the environment. Again, two types of necrosis corresponding to the two types of prelethal changes are observed: oncotic necrosis and apoptotic necrosis (see text). H ability = homeostatic ability; EQUIL = equilibrium. (Reprinted with permission from Trump and Berezesky, 1996a.)

other process. Figure 2.2 depicts schematically the morphological differences as a normal cell passes through either oncosis or apoptosis.

In this context, it is extremely important to differentiate between cellular death and cellular necrosis. Cell death refers to the "point of no return" or point where the cell is unable to maintain homeostasis from the environment, whereas necrosis refers to the *post mortem* changes that occur in cells following death; the prelethal phase may be either described as "apoptosis" or "oncosis."

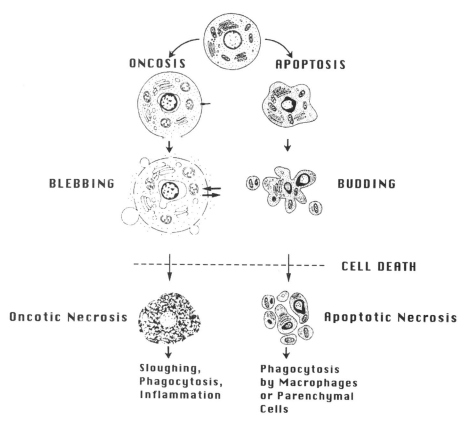

FIG. 2.2. The two pathways of cell death, namely, oncosis and apoptosis, which lead a normal cell (top) to necrosis. The left side of the diagram depicts schematically a cell entering and passing through oncosis while the right side that of apoptosis. Note the differences in morphology, e.g., cell swelling with blebbing and increased permeability of the plasma membrane (arrows) in oncosis, while in apoptosis there is cell shrinkage with budding and karyorrhexis. Note also the marked nuclear chromatin clumping with near-normal cell organelles in apoptosis. Both pathways pass through the "point of no return" or cell death and onto necrosis. In oncotic necrosis, changes include sloughing, phagocytosis, and inflammation, while in apoptotic necrosis, cells break up into clusters of apoptotic bodies with phagocytosis by macrophages or parenchymal cells. (Modified from Majno and Joris, 1995.)

It is thus illogical to refer to "death by apoptosis"or "death by necrosis," since the term *necrosis* represents the *post mortem* changes that occur in cells and tissues following death.

In our view, it is most appropriate to refer to the prelethal phases as "apoptosis" or "oncosis," and the *post mortem* changes as "necrosis." The difference between cell death and necrosis can be readily distinguished by considering the examples of tissue fixation or rapid freezing where the cells are immediately killed and stabilized; however, in these cases, necrosis (*post mortem* changes) does not occur as the cells are dead and the enzymes of the necrotic phase are totally inactivated instantaneously. In contrast, following a lethal injury, such as complete ischemia *in vivo,* the degradation processes take place following the death of the cell.

Prelethal Changes in Cells

Oncosis. The term *oncosis* is derived from the Greek word *oncos,* meaning swelling, and refers to the fact that, in this type of prelethal injury, cell swelling is the predominant early response (Majno and Joris, 1995). Oncosis often follows a variety of injuries, such as toxins and ischemia applied *in vivo* or *in vitro* (Table 2.1), because many of these interfere with ATP synthesis and thus destroy control of the interior environment of the cell by destroying control at the plasma membrane. The lack of ATP and/or loss of cell membrane integrity mean that control of cellular ion content, normally resulting from a balance between Na^+ entry and active Na^+ extrusion, is lost.

In vivo, oncosis typically involves large zones of cells, for example, centrilobular necrosis, which is widespread in the liver following hepatoxins such as carbon tetrachloride; diffuse involvement of particular segments of the proximal tubule in the kidney following nephrotoxins such as $HgCl_2$; or the early stages leading to infarction following loss of blood supply in organs such as the brain, heart, and kidney. As the cells die, an acute inflammatory reaction develops at the periphery of the necrotic zone.

Apoptosis. Apoptosis refers to a type of prelethal change that involves cell shrinkage and fragmentation (Kerr et al., 1972) (Fig. 2.3). *In vivo,* the shed fragments are phogocytized by adjacent parenchymal or mesenchymal cells prior to their death or shed into the extracellular space, for example, a kidney tubule lumen. These fragments, in contrast to cells undergoing oncosis, undergo necrotic change within the phagolysosomal system of the phagocytizing cell. In cell cultures, however, where phagocytizing cells are not present, the cells show swelling as they die as well as other changes typical of necrosis following either oncosis or apoptosis.

Necrosis. Necrosis refers to the changes that occur following the death of the cell. Therefore, these changes refer to the structural and functional events that occur following either apoptosis or oncosis. In this phase following cell

TABLE 2.1. Causative Agents in Oncosis

A. Inhibition of ATP synthesis
  1. Anoxia/ischemia
  2. Metabolic inhibitors
    a. Respiration (e.g., CN, antimycin, CO, etc.)
    b. Oxidative phosphorylation (e.g., 2,4-dinitrophenol, FCCP, etc.)
    c. ATPase–oligomycin
B. Modification of plasma membrane permeability
  1. Mechanical damage
  2. Modification of membrane proteins
    a. Sulfhydral modifiers $HgCl_2$
    b. Oxidant stress
    c. Proteases
    d. Mutations–degenerins
  3. Modification of membrane lipids
    a. Lipid solvents and detergents
    b. Oxidant injury
    c. Phospholipases
  4. Pore and channel formation
    a. Nitric oxide (NO)—NMDA receptors, etc.
    b. Porins and diphtheria toxin, Bcl-2, etc.
    c. Complement C5–9
  5. Ionophores
    a. Calcium (e.g., A23187 ionomycin)
    b. Sodium (e.g., amphotericin B)

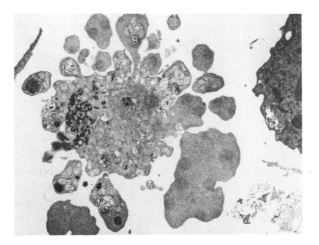

FIG. 2.3. Transmission electron micrograph (TEM) of a JB6 CI41 mouse keratinocyte showing extensive blebbing, many of which contain organelles. Marked nuclear chromatin condensation is also seen, both changes being indicative of apoptosis. $1,000\times$. (Reproduced with permission from Trump and Berezesky, 1996a.)

death, the alterations represent those that accompany the transition of the living cell to physico-chemical equilibrium with the environment; they principally involve degradative responses that result in the decomposition of organelles and other macromolecular systems. This is an important process, as it constitutes a format whereby the organ can result in reconstitution of its elements through the regenerative and remodelling process. These changes are characteristic and can be readily detected structurally by electron microscopy and are mirrored by biochemical and biophysiological changes.

Concept of Programmed Cell Death

The concept of "programmed" (now a computer metaphor) cell death was first utilized in the cell death field by developmental biologists who used the term to refer to a predictable onset of cell death during tissue remodeling in development and metamorphosis. Initially, this referred to changes in hormone levels and growth factor withdrawal, but did not refer at all to the type of prelethal or postlethal (necrotic) change (see Zakeri, Chapter 3, this volume).

Unfortunately, in recent literature, the phrase "programmed cell death" has been equated with apoptosis since apoptosis is usually, but not always, the type of prelethal change observed during such embryologic cell deaths. While it was initially thought that apoptosis required new protein synthesis, this has not been supported by subsequent studies. Moreover, apoptosis, like oncosis, can be readily induced by external stressors (injuries), including environmental toxins or therapeutic agents.

Many studies have clearly established that all cells are "programmed" to die. By this we mean that cells have the enzymatic and signaling mechanisms to trigger a cell death "program" when injured in the appropriate way. This process may or may not require the synthesis of new proteins. In the case of human disease, it most often does not.

Differences between Oncosis and Apoptosis

Cells have apparently developed complex systems to distinguish between types of injurious stimuli. Although each cell has a built-in program to effect death, these mechanisms are differentially activated by different stimuli. For example, withdrawal of a growth factor in cultured cells typically leads to apoptosis, while total ischemia to the myocardium *in vivo* typically leads to oncosis. Therefore, what is the difference in these cellular programs?

The difference seems to involve whether or not the injury primarily or secondarily interferes with ATP synthesis and/or cellular $[Na^+]_i$ regulation. Because in the case of oncosis there is typically no source of ATP, cellular $[Na^+]_i$ increases, and the cells swell with the attendant consequences described in the following. On the other hand, in the case of apoptosis, the cells shrink. Why should this be, why do both have such different characteristics during the prelethal phase, and why should they be so similar after cell death during the necrotic phase? Additionally, what genes are activated toward cell death in these two pathways and what are the respective signaling pathways?

The elucidation of answers to these questions, as well as the refinement of these questions, awaits further characterization. However, at the present time, it would appear that the difference has much to do with the state of energy metabolism and cellular ion regulation.

Oncosis—Stages

Many years ago, we began to investigate and characterize the reversible and irreversible morphological changes observed following prelethal and lethal injury in a variety of cell systems (Trump and Ginn, 1969; Trump and Arstila, 1971). Based on our observations, we classified these changes into stages of cell injury and began to correlate them with structure and function so as to be able to compare differing injuries (for reviews, see Trump et al., 1981, 1982; Trump and Berezesky, 1994). The morphological changes seen during oncosis normally proceed through this series of reproducible stages, which are described in the following and shown in Table 2.2 and Figs. 2.4–2.7). Some of the details of each stage as well as the rate of progression are dependent on the type of injury, the type of cell, and the temperature. However, in general, in a variety of cells and tissues involving vertebrates and invertebrates, these changes are remarkably similar. In the terminology to follow, Stage 1 is defined as the state of the normal cell (Fig. 2.4A, 2.5A).

Stage 2

Morphology. In this stage, which can begin within minutes after application of the injurious stimulus, cells begin to swell with blebbing of the cytoplasm at the cell periphery, swelling of the cytosol, beginning disruption of actin filaments in the region of the plasma membrane, and swelling of the elements of the endoplasmic reticulum (ER) and the Golgi apparatus (Fig. 2.4B). The mitochondria retain virtually a normal appearance, but often show loss of the normal small intramatrical granules. There is no change in appearance of the peroxisomes or lysosomes, but the nuclear chromatin begins to clump very rapidly.

Biochemistry/Function. The functional and biochemical changes in this phase are dramatic. Cellular [ATP] is typically reduced, and, with some types of injury, such as total ischemia of the kidney *in vivo,* ATP becomes nonmeasurable within seconds or minutes after initiation of ischemia.

Among the immediate consequences of the ATP deficiency is loss of control of cellular ion and water regulation. As the ATPases that control Na^+, K^+, and Ca^{2+} become inactive, $[Na^+]_i$, $[Ca^{2+}]_i$, $[Cl^-]_i$, and cellular water begin to increase, while $[K^+]_i$ begins to decrease. The rate of change of these parameters is highly dependent on the specific type of injury. Thus, with types of injury

TABLE 2.2. Organelle Changes in Stages 1–5 of Cell Injury

Organelle	Stage 2	Stage 3	Stage 4	Stage 5
Plasma membrane	Blebs	Blebs and distortion of microvilli	Large blebs begin to detach; marked distortion of microvilli and detachment of blebs	Interruptions in continuity; distortion of microvilli and beginning myelin forms
Cytosol	Blebs; swelling	Blebs; swelling; scattering of polysomes	Same	Same
Cytoskeleton	Irregularities of actin and tubulin filaments	Detachment of actin from plasma membrane	Disappearance of most actin filaments and of tubulin contours	Most elements no longer visible
Mitochondria	Loss of normal intramatrical granules	Condensation	Inner compartment swelling	Inner compartment swelling with flocculent densities and/or calcification
Endoplasmic reticulum/ Golgi	Dilated ER cisternae and Golgi elements	Dilated	Dilated and beginning fragmentation	Fragmentation and formation of vesicles
Lysosomes	Normal	Normal	Swelling; clarification of matrix	Interruptions in membrane continuity
Nucleus	Clumping of chromatin along nuclear envelope	Clumping of chromatin along nuclear envelope	Clumping of chromatin along nuclear envelope	Beginning dissolution of the clumped chromatin

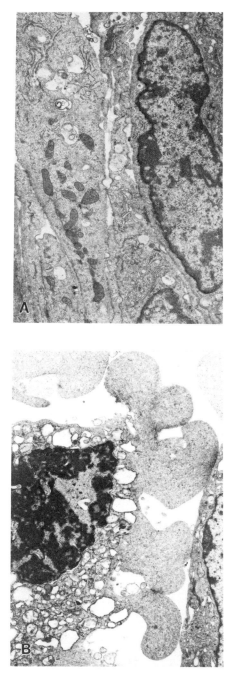

FIG. 2.4. TEMs of promotion-resistant JB6 mouse keratinocytes. (A) Untreated control cells from a monolayer exhibiting well-developed endoplasmic reticulum, normal mitochondria, and prominent nuclei with chromatin clumping adjacent to the nuclear membrane. (B) Following 30 min treatment with benzoyl peroxide (500 μM), some cells display morphologic characteristics of apoptosis including extensive marked chromatin condensation in the nuclei, blebs containing no organelles, dilated ER, and moderately swollen mitochondria. (Reprinted with permission from Jain et al., 1995.)

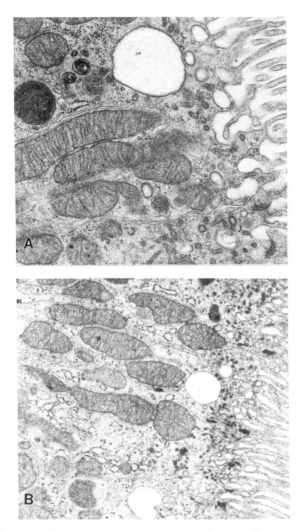

FIG. 2.5. (A) TEM of a portion of Stage 1 control (normal) proximal convoluted tubule cell (PCT) from rat kidney *in vivo.* (B) TEM of Stage 3 rat PCT following 15 min of *in vivo* ischemia. Note condensed mitochondria. 12,000×. (Reproduced with permission from Trump and Berezesky, 1994.)

that directly modify plasmalemmal permeability, such as activation of complement or modification of plasma membrane proteins with mercury salts, the changes occur rapidly with cell swelling, progressing through subsequent stages, and leading to cell death within minutes. On the other hand, with injuries primarily directed toward inhibition of ATP synthesis, such as ischemia, the progression of ion deregulation is much slower as during these initial phases, plasma membrane permeability remains largely unmodified.

Stage 3

Stage 3 is a striking phase morphologically though the changes are reversible (Fig. 2.3B). It can occur within 15–30 min at 37°C depending on the cell type; however, cells may stay viable in this stage for 1–2 days after anoxia if maintained at 0–4°C.

Morphology. In Stage 3, the plasma membrane becomes further distorted with modification of existing specializations such as microvilli and continued presence of blebs, which, by time-lapse cinematography (Trump et al., unpublished observations), alternately protrude and retract. The cytosol is more swollen, and there is more disorganization of actin and tubulin, especially in the region of blebs near the plasma membrane. Elements of the ER and Golgi apparatus become even more dilated, with fragmentation and vesiculation of cisternae. The polysomes, attached to the membrane of the ER, become dissociated.

The mitochondria are extremely condensed (Fig. 2.5B) with dense matrix and loss of intramatrical granules; the intracrystal spaces are dilated. This is characteristic of mitochondria with inhibition of electron transport or oxidative phosphorylation. If the injurious stimulus is removed at this point, for example, reflow after ischemia, the mitochondria resume a normal configuration. The lysosomes are clear, but the membranes are intact, probably reflecting increasing swelling of this compartment as well. The chromatin is markedly condensed, though reversible, and the nuclear envelope is dilated.

Biochemistry/Function

Energy Metabolism. Mitochondria are inactive unless blood flow or injurious stimuli are removed or the mitochondria are isolated and placed in physiologic media. Stimulation of glycolysis may occur in some cells, but in many epithelia, such as the renal or hepatic epithelium, ADP levels are not measurable at this time.

Ion Regulation. In many systems, maximal increases of $[Na^+]_i$, $[Cl^-]_i$, and H_2O, maximal decreases of $[K^+]_i$ and maximal increases of $[Ca^{2+}]_i$ have already occurred. This is reflected by the extreme swelling and the inactivity of ion transport systems in the plasma membrane, the mitochondria, and the ER.

Signalling/Nuclear Function. Much remains to be understood concerning gene levels, stabilization of transcripts, and protein function during this stage. In some systems, damage to DNA may result in increased concentrations of p53 in the nucleus, though these levels are not due to increased transcription but to modification of the turnover of the p53 protein.

Stage 4

Morphology. Stage 4 represents the final of the reversible stages with morphologic and functional changes poised either to recover or the pass the "point of no return" (Fig. 2.4A). Continued blebbing occurs at the cell surface and in highly developed apical plasma membranes such as the microvilli in the proximal tubule. There is formation of unrecongnizable distortion often filling the collapsed tubular lumen with traceries of cell membrane protrusions. Time-lapse cinematography shows withdrawal and formation of extended tendrils of the cell surface. The vesiculation of the distortion of ER in Golgi components can make these difficult to recognize and to separate from invaginations and interdigitations of cell processes. The mitochondria show high-amplitude inner compartment swelling with interruptions of the continuity of the outer membrane (Fig. 2.6A). In appropriate systems, this can be easily recognized by light microscopy. Analyses of mitochondrial phospholipids show loss of cardiolipin and major changes in lipid composition of mitochondrial membranes. The swelling probably corresponds to the so-called "mitochondrial permeability transition" (Pastorino et al., 1996).

Biochemistry/Function. The biochemical features that characterize this stage continue to involve the effects of loss of cell volume regulation with continued swelling of virtually all intracellular compartments. Clearly striking is the transition of mitochondria from the normal or condensed configuration to high-amplitude swelling of the inner compartment. This, of course, is in the face of ATP depletion, increased levels of phospholipase, and increased concentrations of $[Ca^{2+}]_i$. Measurements of mitochondrial lipids during this phase reveal considerable modification with particular loss of cardiolipin. In experiments using fluorescent probes, there is considerable increase of phospholipase activity at this phase following $HgCl_2$ poisoning (Smith et al., 1991).

Also at this phase, the mitochondrial membrane potential ($\Delta \Psi$) disappears; this may represent both the activation of phospholipase A2 as well as the activation of the mitochondrial permeability transition. Niemenen et al., (1996), studying the effects of N-methyl, D-aspartate on cortical neurons, found that cyclosporine A, an inhibitor of the mitochondrial permeability transition, inhibited or delayed depolarization of the mitochondria. Susin et al. (1996) noted that Bcl-2 hyperexpression in the outer mitochondrial membranes not only impeded the mitochondrial permeability transition, but also blocked the result of an apparent mitochondrial protease within the mitochondria.

The change in mitochondrial permeability transition (MPT) may also relate to the cytoskeletal alterations described in this phase. Evtodienko et al. (1996) found that microtubule-disrupting drugs, such as taxol or colchicine, affect the MPT and that modification of microtubules is somehow involved. In the case of oncosis produced by oxidant injury, it appears that oxidation may increase the sensitivity of the MPT to Ca^{2+}. Furthermore, Bernardi (1996) and Petronilli (1996) showed that the MPT, which is a voltage-dependent

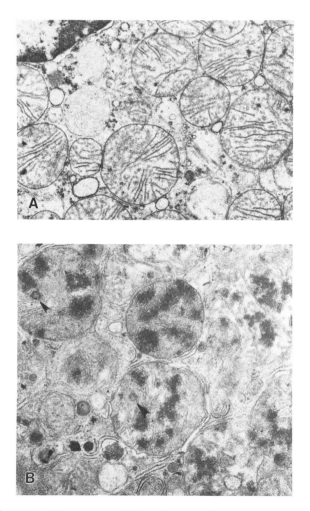

FIG. 2.6. (A) TEM of Stage 4 rat PCT following 60 min of *in vivo* ischemia. Note swollen mitochondria, some of which contain small dense aggregates. (B) TEM of Stage 5 rat PCT following 120 min of *in vivo* ischemia followed by 24 h of reperfusion. Note the swollen mitochondria with large flocculent densities, some of which contain calcifications (arrowheads). 24,000×. (Reproduced with permission from Trump and Berezesky, 1994.)

channel, is activated by matrical Ca^{2+}, inhibited by H^+ and functions in the same manner as a Ca^{2+}-release channel. This probably also relates to the protective effects of reducing cellular pH on the progression of changes that lead to necrosis.

The mechanism of this loss is under considerable study and appears to be clearly related to Ca^{2+} influx and phospholipase A2 activation. Aguilar et al. (1996) recently showed that activation of a mitochondrial calpain-like protease activity is apparently also involved and can function as a cytolytic trigger to initiate mitochondrial MPT. Thus there is increasing evidence that this is a critical step in the transition between reversible and irreversible changes. An example is the TNFα-induced killing of L929 fibroblasts. Pastorino et al. (1996) observed that complete prevention of killing was obtained with cyclosporin A or aristolochic acid (a phospholipase A2 inhibitor). In contrast, agonists of the peripheral benzodiazapine receptor, known to potentiate induction of the MPT, potentiated cytotoxicity. These authors go on to suggest that the MPT occurs as a consequence of the formation of ceramide.

Necrosis

The term *necrosis* refers to the changes that occur following cell death. In this phase, continued degradation of organelles and macromolecules takes place, together with continued protein denaturation. This does not mean, however, that all metabolic activity ceases; indeed, in early phases of necrosis, there may still be electron transport in mitochondria and isolated mitochondria continue to be capable of Ca^{2+} accumulation.

Description. *Necrosis* is an ancient term used to refer to the *post mortem* changes that occur in cells and tissues following death. In this phase, the organelles and macromolecules within the cell are degraded through the action of a series of hydrolytic enzymes including phospholipases, proteinases, and nucleases, ultimately converting the cell to debris in physical–chemical equilibrium with its environment. Necrosis is sometimes termed "autolysis," referring to the autodigestion of cells following death through activation and action of cellular enzymes.

In the pathology literature, however, "autolysis" is often used to refer to changes that occur *post mortem* following the patient's death or somatic death to distinguish it from "necrosis," which occurs following cell death in a living patient. However, many of the cellular changes can be similar after total ischemia either because of cardiac arrest at the time of somatic death or because of ischemia through occlusion of a vessel *in vivo*. Usually, however, differences do exist, probably related to effects of collateral flow at the periphery of necrotic zones *in vivo* and also to the presence of an intense acute inflammatory reaction seen after necrosis *in vivo*. This, of course, contrasts with apoptotic necrosis, which does not incite an inflammatory reaction, occurring as it does usually on a single-cell basis with the apoptotic fragments being phagocytized by adjacent parenchymal or inflammatory cells to their death at the onset of necrosis.

Stage 5

Morphology. One of the hallmarks of this phase is the interruption of continuity of the plasma membrane, which can be readily seen in electron micrographs and which is responsible for uptake of vital dyes, such as propidium iodide, or Trypan blue, as well as for the release of cytosolic enzymes, such as lactic dehydrogenase. In addition, the plasma membrane often forms elaborate myelin forms that are particularly dramatic when seen in time-lapse cinematography as long, threadlike extension of the plasmalemma. The cytosol continues to be swollen, though in the case of apoptotic necrosis, the cytoplasm may remain shrunken for a longer period of time. When studied *in vitro,* apoptotic and necrotic cells also eventually show swelling. Cytoskeletal elements are markedly altered, and little stainable F actin or tubulin can be seen. The ER remains fragmented and ultimately becomes difficult to distinguish from other membrane vesicles found in the cytosol.

Another very valuable hallmark of this phase is the appearance of flocculent densities in the mitochondria (Fig. 2.6B). These electron-dense, fluffy aggregates represent denatured mitochondrial proteins, probably including those from both the matrix and the inner membrane as well (Collan et al., 1981).

Following some etiologic agents, the mitochondria show deposits of calcium phosphate, usually in the form of hydroxyapatites, in addition to the flocculent densities (see Fig. 2.6B). These deposits result from active Ca^{2+} accumulation though at a slow rate, and, therefore, this type of calcification is only seen following injuries that do not primarily inhibit mitochondrial function. Thus, for example, in complete ischemia, the mitochondria in Stage 5 only show flocculent densities, whereas with agents such as $HgCl_2$, which primarily attack the plasma membrane, the mitochondria also show calcifications.

At this stage, the lysosomes show interruption in membrane continuity, and it is probably also at this stage that the hydrolytic enzymes in the lysosomes escape and initiate degradation of intracellular macromolecules. Degradation of chromatin is occurring in the nucleus, which is reflected in electron micrographs by erosion of the clumped chromatin and the appearance of aggregates that form in this material in the nucleoplasm.

Biochemistry/Function. In Stage 5, the predominant reactions are degradative with hydrolysis of proteins, lipids, and nucleic acids. In the early points of this stage, the mitochondria are still capable of electron transport though oxidative phosphorylation is virtually absent. They do, however, appear to retain a capacity for Ca^{2+} uptake, though at a slow rate. Denaturation of proteins is widespread in the cell at this time and accounts for the dense flocculent intramatrical aggregations mentioned above. The clumped chromatin is undergoing degradation, and, over the next several hours, the stainable chromatin will be totally removed. At this stage, there also is greatly increased lysosomal fragility, and it is probable that at some point lysosomal hydrolases find access to their substrates through the fragile membranes.

Stage 6

In this stage, there is acceleration of intracellular digestion measured, reflected by increases of free amino acids, acid-soluble phosphorus, and decreases in total protein, DNA, and RNA. Karyolysis becomes almost complete, and the increased eosinophilia of the cytoplasm correlates with increased binding of anionic dyes to denatured cytoplasmic proteins. Lysosomes become difficult to distinguish morphologically and are not stainable for acid phosphatase, suggesting loss of hydrolases to the cytosol. There are also major rearrangements in intracellular membrane systems that show vesiculation, formation of tubular forms in the mitochondria, and numerous membrane wrappings and whorls.

Stage 7

In this, the final stage we have defined, further fragmentation of organelles is evident (Fig. 2.7) and in addition, there are large dense inclusions, often larger than the nucleus, composed of dense osmiophilic material with internal lamellar regions found in lacunae. These may represent formation of myelin bodies as described by Virchow (1859) and may contain mixtures of lipid, phospholipids, fatty acids, and perhaps even metal soaps of fatty acids. At this stage by light microscopy, the cells have the appearance of an old infarct with intense eosinophilia, total karyolysis, and indistinct cell outlines, though perhaps because of denaturation, cellular outlines remain distinct. However, in some forms of necrosis, such as gummatous or caseous necrosis, total loss of architecture occurs to form more homogeneous eosinophilic areas that result from additional enzymatic activity of microbial products.

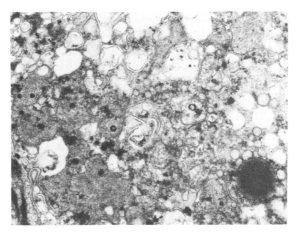

FIG. 2.7. TEM of Stage 7 rat PCT following 60 min of *in vivo* ischemia followed by 24 h of reperfusion. Disrupted organelles are seen with swollen mitochondria containing flocculent densities and some calcifications, all indicative of oncotic necrosis. 10,000×. (Reproduced with permission from Trump and Berezesky, 1994.)

Mechanisms of Oncosis Leading to Cell Death

Altered Cell Volume Regulation. Characteristic of oncosis is the rapid loss of cell volume that ensues. This principally results from inactivation of the Na^+/K^+-ATPase at the plasmalemma, either through direct toxicity, for example, treatment of cells with cardiac glycosides such as ouabian, or through ATP deficiency resulting in activation of ion transport systems as in the case of total ischemia. Stimuli that greatly increase Na^+ permeability, such as activation of the C5–9 components of complement, quickly exceed the capacity of the Na^+/K^+-ATPase even though it is initially not inhibited. Other etiologic factors include mutations in membrane channels that render them continually open, such as mutations in the degenerin series of proteins (Lints and Driscoll, 1996). This work has led to the consideration of targeting membrane channels as a type of modification both of genetic defects and for protection against early changes and diseases such as stroke.

The increased concentrations of $[Na^+]_i$, $[Cl^-]_i$, and H_2O in the cell are accompanied by decreased levels of $[K^+]_i$, all of which lead to cellular swelling because of the Donnan effect. The swelling, however, is not uniform within the cell, and some compartments, such as the ER and Golgi, apparatus, undergo early expansion along with the cytosol, while others in the earlier phases of Stages 2 and 3 (see above) are normal or contracted such as the inner compartment of mitochondria.

In parallel with these changes, there is also increased $[Ca^{2+}]_i$ due to increased entry from the extracellular space and/or redistribution from intracellular compartments such as the ER and the mitochondria (Fig. 2.8). This increased $[Ca^{2+}]_i$ results typically both from reduced ATP concentrations, which drive the Ca^{2+} ATPases in the plasmalemma, the ER, and mitochondria, as well as the effects of diminished Na^+/Ca^{2+} exchange resulting from the increased concentrations of $[Na^+]_i$. During this phase of cell swelling prior to the "point of no return," these altered ion concentrations, particularly $[Ca^{2+}]_i$, appear to affect the organization of actin and other elements of the cytoskeleton and, in turn, to result in the formation of cytoplasmic blebs, which, as seen by time-lapse cinematography, extend and retract continuously at the cell surface. In this regard, Negulyaev et al. (1996) demonstrated that disruption of actin filaments increases the activity of plasmalemmal sodium channels—thus perpetuating the cycle.

The significance of the K^+ loss from cells has been less well studied in this phase. The mechanism of K^+ loss is not well understood; however, Reeves and Shah (1994), studying hypoxic injury to rabbit proximal tubules, observed that the K^+ efflux was inhibited by K^+ channel blockers including Ba^{2+} and tetraethylammonium. That K^+ influx contributed to the injury was indicated by the observation that both TEA and glibenclamide (an inhibitor of ATP-sensitive K^+ channels) reduced hypoxic injury.

Altered $[Ca^{2+}]_i$ Regulation. Deregulation of $[Ca^{2+}]_i$ appears to be a critical process in the events leading from injury to cell death, affecting most of the

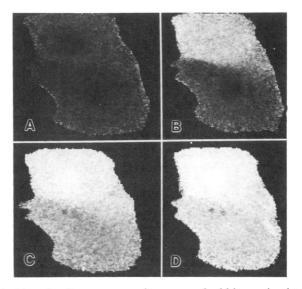

FIG. 2.8. Digital imaging fluorescence microscopy of rabbit proximal tubule epithelial cells cultured in normal (1.37 nM) extracellular Ca^{2+} media, loaded with the fluorophore Fura 2 to measure $[Ca^{2+}]_i$, and treated with 5 mM KCN + 1 mM IAA. Images were generated by pixel-by-pixel ratios from fluorescent image pairs collected at 340 and 380 nm excitation. Increases in ratio values indicate increases in $[Ca^{2+}]_i$ in the upper cell. (C) 40 min. Note $[Ca^{2+}]_i$ has increased also in the lower cell. (D) 60 min. Both cells have similar elevations in $[Ca^{2+}]_i$. 1,447×. (Reproduced with permission from Smith et al., 1992.)

other phenomena that have been described (Trump et al., 1982, 1981; Trump and Berezesky, 1992, 1995, 1996a,b). Indeed, simply treating cells with a Ca^{2+} ionophore can reproduce all the stages of oncosis leading from injury to death. Figure 2.8 shows results from this laboratory using the injury model of anoxia and chemical ischemia (KCN + IAA) in suspended rabbit proximal tubule epithelium (Smith et al., 1992).

Depending on the nature of the initial injurious interaction, increases of $[Ca^{2+}]_i$ can result from increased Ca^{2+} influx from the extracellular space, redistribution from internal stores, or both. The most rapid increases of $[Ca^{2+}]_i$ are seen following damage to, or modification of, the integrity of the plasma membrane, for example, activation of the C5–9 components of complement (Papadimitriou et al., 1994), which lead to explosive cell swelling and increased $[Ca^{2+}]_i$. On the other hand, when the injury is directed at ATP synthesis, as in the case of ischemia, the influx is more gradual, as it is based on ATP deficiency to the $[Ca^{2+}]_i$ and Na^+/K^+ ATPases; however, as membrane integrity is initially intact, $[Ca^{2+}]_i$ increases more gradually until plasma membrane integrity is disrupted.

Following cell injury, $[Ca^{2+}]_i$ can increase from its normal value approximating 100 nM in most cells to millimolar levels during the course of events between cell injury and cell death. In some cases, particularly striking increases in $[Ca^{2+}]_i$ within the nucleoplasm are observed. Subsequent actions of Ca^{2+} involve the activation of a variety of hydrolytic enzymes and binding to Ca^{2+}-binding proteins, which result in many of the other charges discussed below.

Increased $[Ca^{2+}]_i$ also affects volume regulation. As $[Ca^{2+}]_i$ increases, it typically stimulates efflux of K^+ and Cl^- via activation of K^+ and Cl^- regulatory channels (Roman et al., 1996). In many cells, this is the normal method of volume regulation when cells are swollen in hypotonic media or by stimulation of Na^+ influx (Churchwell et al., 1996). Crowe et al. (1995) demonstrated that neuroblastoma cells treated with ionomycin in isosmotic medium showed gradual shrinkage to ~40% of their water volume at a rate of ~1.2%/minute.

Therefore, we suggest that in oncosis this mechanism attempts to regulate volume; however, in the absence of ATP and in the face of damage to the plasma membrane integrity, the process cannot keep pace with the rate of volume increase. On the other hand, in apoptosis, where ATP is present, we suggest that Ca^{2+} influx stimulates the initial and sometimes extreme cellular shrinkage through this mechanism (see Zakeri, Chapter 3, this volume).

Altered Energy Metabolism. Reduction of ATP concentration and inhibition of ATP synthesis are typical of conditions that lead to oncosis. Early maintenance of ATP concentration appears to be one important mechanistic characteristic that distinguishes oncosis from apoptosis. As will be mentioned, this is reflected by the loss of cell volume control and cellular swelling.

Diminished ATP concentration may result from primary inhibition of mitochondrial respiration or oxidative phosphorylation by ischemia, anoxia, or respiratory inhibitors such cyanide or carbon monoxide or uncouplers of oxidative phosphorylation, such as 2,4-dinitrophenol or FCCP. Inhibition of ATP synthesis and uncoupling can also result secondarily from toxins or agents that modify cell membrane permeability, such as activation of complement or insertion of pores, such as bacterial porins into the cell membrane, which, along with Ca^{2+} ionophores, results in rapid increases of $[Ca^{2+}]_i$, and, in turn, uncouple the mitochondria. Thus the series of conditions, toxins, and injuries that result in oncosis typically have early effects on ATP synthesis.

As mentioned, during the reversible Stages 2, 3, and 4, the mitochondria first condense. This condensation is typical of nonfunctional mitochondria that lose ions and H_2O. However, as membrane damage occurs, probably through actions of phospholipases, the inner membrane lipids are modified and the mitochondria pass a transition of permeability called "permeability transition," as discussed. After this, some mitochondria become swollen in Stage 4, though such swollen mitochondria are still reversible and membrane repair can take place.

Mitochondria also provide one of the most reliable hallmarks of cells that have passed the "point of no return," namely, flocculent densities within the matrical compartment (Fig. 2.6B). As mentioned, these flocculent densities represent denatured proteins and apparently irreversible loss of mitochondria membrane function through denaturation of both matrical and probably membrane proteins (Collan et al., 1981). It is probably much more than a correlation that such aggregates of mitochondrial denatured proteins signify the loss of reversibility because without mitochondrial function, most cells are unable to reestablish homeostasis even if the injurious stimulus is removed.

Protein Denaturation. Denaturation begins very early in the oncotic process and as Majno and Joris (1995) have shown, can be detected by light scattering and dark-field microscopy as early as 30 min after anoxia, long before the "point of no return." This progresses as reversibility is lost and accounts, for example, for the intramatrical densities found within mitochondria. This is related to the induction of heat shock proteins, such as the members of the hsp70 family mentioned elsewhere. It also involves ubiquitins, which target presumed irreparable proteins for lysis. Reinduction of heat shock proteins may modify the response and delay or protect against cell death by more effectively restoring denatured proteins to normal function (see Cryns and Yuan, Chapter 6, and Sikorska and Walker, Chapter 7, this volume).

The role of proteosomes during the development of oncosis has not been explored in detail. These, however, could also be involved in remodelling during the prelethal phase. It has been reported that, following some types of injury, proteosomes are removed from the nucleus and accumulate within blebs at the periphery of the cell. This phenomenon, so far, has been mainly linked to apoptosis, but may well also play a role in the oncotic process.

Activation of Hydrolytic Enzymes. Activation of hydrolytic enzymes has been classically associated with necrosis for decades, as they obviously are involved in conversion of dead cells to necrotic debris. The introduction of the lysosome concept by de Duve and his group in the 1950s and 1960s focused even more attention on this group of enzymes (see review by de Duve and Wattiaux, 1966). At that time, de Duve proposed that lysosomal release of hydrolases actually occurred prior to cell death and, in fact, was responsible for cell killing (the so-called suicide bag hypothesis). This hypothesis was never fully confirmed because in fact it often appeared that lysosomes remained intact until after cell death. Unquestionably, lysosomal enzymes, such as cathepsins, lipases, DNases, and RNases, active at the reduced pH that typically occurs during necrosis, are responsible for the necrotic degradation process. More recently, however, the entire subject of the role of hydrolases during the prelethal phase, and their role in the execution step has been revisited, stimulated in large part by genetic studies that have elucidated a number of genes that relate to the killing process (for review, see Chinnaiyan and Dixit, 1996).

Proteases

Cysteine Proteases/ICE Family. These proteases (see also Cryns and Yuan, Chapter 6, and Sikorska and Walker, Chapter 7, this volume) now termed "caspases" (Alnemri et al., 1996), were highlighted by work on *C. elegans* development, where the gene ced-4 was found to be important in the killing process (for review, see Lints and Driscoll, 1996). Although most of the current evidence indicating a role for the ced-4 series of proteins relates to apoptosis, there is a growing body of evidence that it also may be important in oncosis (Yuan et al., 1992, 1993). It consists of the relationships between activation of these enzymes and Bcl-2 expression and from work using inhibitors (Hockenbery et al., 1993). Furthermore, Strack et al. (1996) have shown that other proteases, such as the HIV protease, actually cleave Bcl-2 prior to inducing cell death. Shimizu et al. (1996), however, have shown that the ICE inhibitors act at a different step during inhibition of cell death induced by respiratory inhibitors, such as rotenone and antimycin, than does Bcl-2. These investigators also showed that oncosis and necrosis are also recharted by inhibitors of ICE-like proteases, possibly through multiple pathways. We have shown in our laboratory that Bcl-2 inhibits oncosis induced by oxidant stress (Amstad et al., 1996, 1997). A number of papers have been recently published that suggest that Bcl-2 expression blocks activation of the ICE/ced-3 protease family, as shown by Srinivasan et al. (1996) in the neuronal cell line GT1-7. It was found that overexpression of Bcl-2 does not alter the expression of 5 members of the ICE/ced-3 family, namely, CPP32, ICE, Mch2, Nedd2, and TX, but it did prevent the apoptosis-induced processing of pro-Nedd2 to the cleaved form. This suggests that Bcl-2 is produced upstream from the function of ICE/ced-3 proteases and may inhibit cell death by preventing the post-translational activation of ICE/ced-3 proteases.

The question of the target substrates for this series of cysteine proteases as related to cell death is also uncertain. Candidates include the elements of the cytoskeleton, plasmalemmal proteins, and other critical proteins, such as those of the mitochondrial inner membrane. This is an important area for research as development of appropriate inhibitors could greatly modify the course of cell injury. Other possibilities, as suggested by Morana et al. (1996), include the action of serine/threonine protein phosphatase inhibitors, such as okadaic acid, calycukin A, and cantharidin, in preventing the activation of the ICE/ced-3 proteases. Substrates involved include the retinoblastoma protein Rb and indicate that Rb dephosphorylation has been observed in a number of models of apoptosis, but investigations in oncosis have, thus far, not been performed. This question of an imbalance between protein phosphatase 1 and Rb kinase resulting in activation of ICE/ced-3 proteases remains to be established.

Calpains. Calpains are Ca^{2+}-activated proteases that appear to play a role in a variety of cellular functions, including reactions of cells to lethal injury. A variety of substrates probably exist for these Ca^{2+}-activated enzymes, but the one that has been studied in detail is the cytoskeleton, including keratin, intermediate filaments, and actin. In our studies, we observed, following lethal injury with $HgCl_2$ *in vitro,* oncosis followed by cell death and necrosis (Elliget et al., 1991, 1994). The use of inhibitors of calpain resulted not only in the delay of cell death, but in a pronounced modification of the pattern of bleb formation that resulted. The relative importance of calpains versus the ICE family of proteases is under study in many laboratories, but it would appear that interventions targeted at either class could result in modification of the progression of cell injury.

Lysosomal Proteases. Precise definition of the role of lysosomal proteases in cell injury and cell death has been elusive. Following sublethal injury, there is considerable proteolysis due to autophagy, but this degradation is believed to be entirely within the lysosomal compartment. Within lysosomes, proteins turn over with an apparent half-life of about 8 minutes, through the action of endopeptidases, particularly cathepsins D and L (Bohley and Seglen, 1992). As lysosomes become more fragile during the oncosis phase, there is the possibility that escape to the cytosol might occur and, at the low pH characteristic of this phase, have access to all intracellular membranes and other substrates. Some studies utilizing proteinase inhibitors do, in fact, suggest a role for lysosomal proteinases. Yang and Schnellmann (1996) observed that inhibition of cathepsins B and L, by the cysteine proteinase inhibitor E-64, correlated with cellular protection following injury with antimycin or tetrafluoroethyl-L-cysteine. However, they observed that protection only occurred after some cells had already died, indicating that release of lysosomal proteinases from irreversibly altered cells caused lethal injury in the other prelethal injured epithelium.

Nucleases. Both DNases and RNases are important in determining events that occur in both oncosis and apoptosis (Arends et al., 1990). In oncosis, however, chromatin breakdown begins relatively late, probably about the time of the "point of no return," and results in fragmentation in chromatin lysis rather than the ladder patterns seen in apoptosis. Similarly, hydrolysis of RNA probably proceeds at approximately the same time. Lysosomal hydrolyses/nucleases may be very important at that phase following irreversible injury in removing ribosomes in conjunction with other enzymes, notably proteases.

More recently, however, more attention has been directed toward the role of endonucleases in injuries such as hypoxia or chemical hypoxia, which are normally associated with the induction of oncosis. For example, Hagar et al. (1996) observed that reactive oxygen species played an important role in the injury induced by antimycin; endonuclease-mediated DNA strand breaks were prevented by a number of hydroxyl radical scavengers, by iron chelation, and by inhibitors of endonuclease such as aurintricarboxylic acid (Hagar et al., 1996). In studies resulting in oncosis/necrosis following hypoxia reoxygenation injury, Ueda et al. (1995) observed that the DNA-degrading activity was the result of an apparent 15 kDa endonuclease, which was entirely Ca^{2+} dependent.

Phospholipases (see Zakeri, Chapter 3, Benitez, Chapter 17, this Volume). Activation of phospholipases is critical to the development of oncosis and necrosis at virtually all stages. Some, such as phospholipase C, may be important in early signalling responses. In the early phases following injury, phospholipase C activates the IP3 signalling pathway, resulting in Ca^{2+} release from the ER, while phospholipase A2 may be important both early and late and responsible for first reversible and then irreversible degradation of intracellular membranes such as the mitochondrial inner membrane and ultimately the plasma membrane. The products of phospholipase action, including arachidonic acid and fatty acids, also may play important roles in the pathogenesis of change.

Because of this, phospholipase inhibitors are attracting increasing attention as modifiers of the course of cell injury and preventors of cell death. Activation of phospholipase can be measured with various methods, including the use of fluorescent probes and imaging, and measurement of membrane phospholipids and their metabolites with chromatography, including arachidonic acid, and fatty acids. Phospholipase A2 is both pH and Ca^{2+} dependent. Yang et al. (1994) showed that chelation of $[Ca^{2+}]_i$ prevented release of labeled arachidonic acid, which was blocked by an intracellular Ca^{2+} chelator. Dibucaine and mepacrine, phospholipase A2 inhibitors, also inhibited the KCN-mediated arachidonic acid release. Incubation at pH 6.5 reduced the effect and, at pH 7.9, there was enhancement, similar to findings reported by other investigators. Activation of phospholipase often induces membrane lipid peroxidation, which further contributes to the enhanced membrane degradation. Malis and Bonventre (1986) showed that Ca^{2+} potentiates oxygen-free radical injury to renal mitochondria, which is most likely due to activation of phospholipase A2. In an additional study, Nath et al. (1996) concluded that $HgCl_2$-treated cells generate massive amounts of H_2O_2 and that redox-sensitive genes are expressed, including heme oxygenase and some members of the Bcl-2 family. In this regard, synergism between phospholipase A2 and reactive oxygen species (ROS) has often been suggested and $HgCl_2$ toxicity seems to be an especially appropriate example. Saperstein et al. (1996), however, noted that cytosolic phospholipase A2 may be an important mediator of oxidant damage to renal epithelial cells. Overexpression of phospholipase A2 in LLC-PK1 cell lines was associated with increased susceptibility to injury from H_2O_2 and menadione, which was ameliorated by treatment with a Ca^{2+} chelator.

Signalling Pathways

Immediate–Early Genes (see Osborne, Chapter 8, Papermarker, Chapter 12, Birge et al., Chapter 13, Tilly, Chapter 16, This Volume). A number of immediate–early genes including c-fos, c-jun, and c-myc are rapidly induced within the initial 10–15 min during the early phases of oncosis following lethal injuries such as oxidant stress and $HgCl_2$ (Maki et al., 1992; Gu et al., 1994; Yamamoto et al., 1993, 1994). The exact role of these genes in this process is currently not well understood. In apoptosis, however, induction of both c-jun and c-myc may play a role in some conditions. It is also possible that part of their function relates to cells that survive at the edge of an area, such as at the edge of an infarct, and are related to subsequent regeneration. Early induction of immediate genes, such as c-fos, is protein kinase dependent and highly dependent on the rise in $[Ca^{2+}]_i$ as chelation of $[Ca^{2+}]_i$ with Quin 2 or affecting the injury with Ca^{2+}-free media minimizes the induction.

Activation of p53. p53, when activated during cell injury by DNA damage induced by a variety of agents and hypoxia, is a potent cell death effector gene that acts through mechanisms as yet unknown to induce cell death (see reviews by Harris 1996a,b). p53 overexpression commonly results in apoptosis, though the relationship of p53 overexpression to oncosis is not yet clear. Phosphorylated p53 acts as a transactivator of a number of other genes, including Bax, Gadd45 (waf, p21), and others. At the present time, Gadd 45 and p21 seem more related, however, to cell cycle arrest. Additionally, p53 may have direct protein–protein interactions involving as yet unknown signalling pathways. Mutations in p53 result in modification of this effector death pathway, and, as such, the most common mutation in human cancer may relate to the progression of cancer by eliminating the programmed cell death response. We have recently performed studies by microinjecting p53 genes into NRK cells followed by time-lapse cinematography and electron microscopy of the process (Chang et al., 1997). After a delay, while p53 expression occurs, the cells round up and then exhibit bubbling and blebbing around the cell surface. Following this phase, the cells undergo swelling, often with the formation of large cytoplasmic blebs, and then die, as shown by uptake of propidium iodide.

Some studies indicate that Bax is a primary response gene for p53 involved in a p53 regulated pathway inducing cell death. Bax is part of the bcl-2 family group, which contains both effector and repressor genes for cell death. Bax knockout mice, though viable, exhibit significant aberrations in cell death involving both thymocytes and B-lymphocytes, resulting in hyperplasia and ovaries that contained unusual atretic follicles with excess granulosa cells; male deficient mice were infertile as a result of the accumulation of atypical premeiotic germ cells (Knudson et al., 1995). The mechanism of Bax mediation of cell death is not known, nor is it known how it relates to oncosis as opposed to apoptosis. However, since the mechanism of Bax action is opposed by Bcl-2, and since Bcl-2 is protective against both oncosis and apoptosis, we suggest that p53 induction of Bax might function in both modes of cellular death.

Since hypoxia is a known inducer of both oncosis and apoptosis, the question arises as to how hypoxia can induce increased p53 expression (Graeber et al., 1994). Although oxygen may be necessary to form peroxynitrite and nitric oxide (NO), evidence indicates that NO can be formed during hypoxia if particular cytokines such as $TNF\alpha$ are present. However, it is clear that NO can induce p53, which, in turn, results in down regulation of nitric oxide synthase (NOS2) expression through inhibition of the NOS2 promoter (Forrester et al., 1996). These relationships between NO, hypoxia, and p53 expression demand more investigation. Mei et al. (1996) have also demonstrated a relationship between NO formation and increases in $[Ca^{2+}]_i$ in that, early after initiation of hypoxia, NOS inhibitors can greatly retard the increase. During ischemic stroke, excess release of glutamate causes neurotoxicity by activating the NMDA receptors, resulting in formation of NO, which mediates cell death. Zhang and Steiner (1995) observed that the immunosuppressant drug FK506 inhibits the protein phosphatase calcineurin, resulting in phosphorylation of NOS, which inhibits its activity in forming NO.

Stress Genes. Rapid induction of stress genes, such as hsp70, also accompanies the early stages of oncosis following injury with a variety of agents. This induction is again Ca^{2+} dependent and is presumably due to phosphorylation of a transcription factor. Experiments with inhibitors of ATP synthesis in Ehrlich ascites tumor cells have indicated that the resistance to cell killing that occurs is stationary as opposed to exponentially growing cells and is correlated with expression of hsp68, hsp27, and elevated hsp90. Kabakov and Gabai (1995) also showed that prior induction of hsp70 by heat treatment considerably suppressed the growth-induced actin aggregation and rate of necrosis.

Cytoskeleton. The cytoskeleton actively participates in the process of oncosis and necrosis. Some studies have been performed that show that dynamic alterations in the cytoskeleton are quite evident in the very early stages during the reversible phase. It appears that some of these alterations play important roles in the cell-shape changes and ion transport changes that occur during the swelling phase, and that these changes may have implications for the signalling of immediate and early genes that precedes following most injuries that cause oncosis.

Actin. Changes in the pattern of actin filaments are among the earliest changes following a number of injuries that result in oncosis, such as chemical ischemia. During the early phase, the confirmation of active filaments changes markedly, and as cytoplasmic blebs form, a band of actin is often found at the base of the blebs. We have reported that this correlates with increasing $[Ca^{2+}]_i$ at a threshold of approximately 300–400 nM (Phelps et al., 1989).

$[Ca^{2+}]_i$ regulation is involved in the modification of actin. As the actin changes are delayed following several injuries if the experiments are conducted in Ca^{2+}-free media, this implies the probability of Ca^{2+} and Ca^{2+}-binding proteins involvement in the process. Experiments with $HgCl_2$-induced lethal injury in rat proximal tubule epithelium showed that changes such as blebbing could be significantly delayed through the use of calpain inhibitors and, furthermore, that calpain inhibitors also were protective against cell death (Elliget et al., 1991, 1994). Recent evidence also suggests that actin participates in cell volume regulation through control of Na^+ channels, since modification of actin with cytochalasin results in Na^+-channel opening. Thus degradation of actin, such as through Ca^{2+}-activated proteases, could also contribute to the cellular swelling that occurs in early oncosis.

Modification of actin seems particularly important in bleb formation as the pattern of actin filaments is markedly different in the bleb region. In typical cytoplasmic blebs, the cytosol is clear within the area of the bleb and contains only polysomes and no organelles. There is often a band of actin at the base of the bleb, but filament staining is negative within the bleb itself. One suggestion put forth is that modification of actin-binding proteins to the plasmalemmal proteins results in detachment and in movement of fluid to this blebbed area. This could be aided by the actin-induced opening of Na^+ channels as mentioned.

Tubulin. Destabilization of microtubules also occurs during the stages of oncosis, and, indeed, chemical modification of tubulin within vinblastine induces a somewhat similar type of cytoplasmic blebbing. It thus would appear that both actin and tubulin could be involved in bleb formation. Microtubules, as shown with antibody staining, persist well into the later stages such as Stage 5 when the cell is extremely swollen; however, the microtubule profiles become wavy, possibly due to destabilization and depolymerization of actin.

These changes also correlate with the sustained increases of $[Ca^{2+}]_i$ that occur at the same time. Taxol, which stabilizes microtubules to disaggregation, has been found to protect against calcium ionophore-mediated cell death in PC12 cells, suggesting that the disaggregation does, indeed, play a role in the cell death process (Yang et al., 1994). In this regard, disruption of microtubules in cultured hepatocytes with vinblastine or colchicine induce cell death with cytoplasmic blebbing and DNA fragmentation.

A link between microtubule integrity and the function of Bcl-2 has recently been provided by Haldar et al. (1997), who reported that modification of microtubules by taxol, taxotere, vinblastine, or vincristine treatment of human cancer cell lines induces Bcl-2 phosphorylation, which results in a loss of Bcl-2 function. As this predisposes such cells to death, propagation of cells with damage to the mitotic apparatus is prevented.

Intermediate Filaments. Specific roles for intermediate filaments and changes therein during oncosis and necrosis have not been defined. These obviously resistant polymers persist in cells after severe treatments, but some studies have shown loss of neurofilaments after oxidant stress and neurofilaments may be an important substrate for calpain.

Agents and Gene Products That Protect against Cell Death

Bcl-2 (see Birge et al., Chapter 13, Tilly, Chapter 16, Benitez, Chapter 17, Cryns and Yuan, Chapter 6, This Volume). Beginning with studies on *C. elegans* development, it was observed that the autosomal dominant death repressor gene ced-9 was shown to repress normal cell death in a gain of function form. Investigations of the mammalian homolog, Bcl-2, initiated the elucidation of a family of genes that contain both effector and repressor genes related to Bcl-2. Bcl-2 overexpression can result in protection against a variety of injuries that result in cell death, including those that initiate oncosis as well as those that initiate apoptosis.

The mechanism of action of Bcl-2 is currently under intense investigation. Current evidence indicates that Bcl-2 is localized to the outer mitochondrial membrane, the membranes of the ER, and the outer nuclear membrane. Targeting to these membranes is via a carboxy-terminal hydrophobic domain. Three-dimensional studies have shown that Bcl-x_L in the absence of this domain contains two central hydrophobic helices surrounded by five amphipathic helices (Muchmore et al., 1996). This structure is very similar to the pore-forming domains of some bacterial toxins including diphtheria toxin and colicins A and E1. These domains form membrane channels and also facilitate transmembrane transfer of proteins such as the A fragment of diphtheria toxin. Minn et al. (1997) have shown that Bcl-x_L also forms ion channels in synthetic lipid membranes. The channels induced by Bcl-x_L are increased as pH is reduced and are partially ion selective as follows: $K^+ = Na^+ > Ca^{2+} > Cl^-$.

A relationship of Bcl-2 to $[Ca^{2+}]_i$ regulation has been emphasized in several studies in which Bcl-2 may control Ca^{2+} release from the ER, Ca^{2+} entry through the plasmalemma, and Ca^{2+} regulation by mitochondria-protecting mitochondrial membrane potential (probably through modification of the MPT) in the face of injury. For example, our laboratory has recently reported that cells stably transfected with Bcl-2 have markedly increased protection against cell death induced by oxidant injury induced by H_2O_2 or menadione (Ichimiya et al., 1996). These injuries induce a major rise in $[Ca^{2+}]_i$ during the prelethal phase, which is a major factor in the pathogenesis as elimination of Ca^{2+} from the medium, or buffering of $[Ca^{2+}]_i$ with EGTA-AM are protective against cell killing. Wang et al. (1996) recently proposed another mechanism for protection of mitochondria with their observation that Bcl-2 can target the protein kinase Raf-1 to mitochondria, permitting the phosphorylation and inactivation of the pro-death gene Bad and possibly other protein substrates. When Bcl-2 expression is increased, Murphy et al. (1996) found that mitochondria have enhanced ability to sequester large quantities of calcium without undergoing profound respiratory impairment. It has also been proposed that Bcl-2 might modify release of Ca^{2+} from the ER. However, Distelhorst and McCormick (1996) found that treatment of a lymphoma cell line with thapsigargin, which releases Ca^{2+} from the ER (Thastrup et al., 1990), resulted in an increase in $[Ca^{2+}]_i$ even in cells that were overexpressing Bcl-2. These data suggest that it acts at a more distal location, such as the mitochondria or the plasma membrane.

The range of lethal injuries that are protected by Bcl-2 is expanding, and includes both classic inducers of apoptosis, such as p53 or dexamethasone-induced apoptosis in lymphocytes to cyanide, antimycin, rotenone-induced oncosis, oxidant stress-induced oncosis, and an expanding list of injuries.

Low pH. There have been a number of studies indicating the protective effects of low pH on oncosis and necrosis since its initial discovery. Reduction of extracellular pH to values approximating 6.4 can greatly modify membrane damage and cell death as measured by Trypan blue staining, LDH release, and propidium iodide staining (Pentilla and Trump, 1974; Pentilla et al., 1976). This protection has been observed following injury with many agents including metals, anoxia and ischemia, heat, and others. A somewhat similar effect can be achieved with amiloride, which blocks the Na^+–H^+ exchange system and prevents cytoplasmic alkalinization.

Mechanisms involved in this effect include reduction of increase of $[Ca^{2+}]_i$ as reported by Astma et al. (1996), and also inhibition of the Na^+–Ca^{2+} exchanger. Particularly striking is the so-called "pH paradox" in ischemia-reperfusion injury (Lemasters et al., 1996). In this latter study, it was shown that pH less than 7.0 protected greatly against cell death during ischemia since manipulations that decreased the increase of $[pH]_i$ during reperfusion prevented loss of viability and also included inhibition of the Na^+–H^+ exchange with dimethylamiloride or HOE694. The Na^+–Ca^{2+} exchanger, however, did not affect killing, and thus these authors did not find it to be Ca^{2+} dependent; instead they hypothesized that it was probably due to inactivation of hydrolytic enzymes, such as phospholipases and proteases. Other possibilities we have suggested include antagonism of Ca^{2+} activation of cellular receptors including calmodulin.

There are many studies that indicate that modification of $[Ca^{2+}]_i$ deregulation is markedly protective in preventing cell death. Such protection can come from adding chelators or incubating the cells in low-Ca^{2+} mediators during experimentation by loading with intracellular Ca^{2+} buffers, such as EGTA-AM, by blocking Ca^{2+} entry with Ca^{2+}-entry blockers, such as verapamil, and by the use of calmodulin antagonists. As most Ca^{2+}-mediated effects are accomplished through Ca^{2+}-binding proteins, these represent important targets for the mechanism. For example, inhibitors of calmodulin kinase and calcinorin can protect against cell killing, as can inhibitors of phospholipase A2, which are also regulated by Ca^{2+}. Ca^{2+} binding proteins, the calpain group of proteases, are also involved in cell killing as inhibitors of these enzymes offer protection. Finally, there are growing data that the actions of Bcl-2 in cell protection involve control of $[Ca^{2+}]_i$. Also, cells that have been induced to express higher levels of Ca^{2+}-binding proteins, such as calbindin D-28K, appear more resistant to cell injury and cell death.

Hypothermia. Hypothermia is an ancient method that has been utilized to prevent necrosis and subsequent deterioration and hydrolysis of cellular structures. Indeed, in the "aging" and controlled hydrolysis of beef, the steaks are maintained at body temperature or higher to produce the tenderizing (necrotic) process.

At the same time, it has become obvious that maintenance of organs for transplantation is best performed at low temperature. Certainly, hypothermia retards the process of necrosis. On the other hand, reduction of temperature does not prevent and, indeed, promotes the process of oncosis. In this regard, hypothermia results in a clear demonstration of these stages of cell injury and cell death.

Incubation of mammalian cells at 0–4°C results in a demonstration of the early stages of oncosis leading to cell death that is difficult to reproduce in any other circumstance. Such experiments provide a validation of the reversibility of Stages 2–4 and clearly demonstrate that ion deregulation and cellular swelling are fundamental, but precursory, to the events that ultimately lead to cell killing and subsequent necrosis. For example, when renal cortical slices are incubated at 0–4°C in Ringer's solution for 24–48 hours, Stages 3 and 4 exist for 24–48 hours. It is clear from these studies that the ion deregulation resulting from inhibition of transport systems and the consequent changes in cellular structure and function can be clearly dissociated from the enzymatic events that occur at the "point of no return" and beyond.

Phospholipase Inhibitors. There has been a long history of papers indicating an important role for phospholipases, especially phospholipase A2, in the initiation of oncosis and necrosis. It has been a somewhat difficult subject to study because of the lack of highly specific inhibitors as well as difficulties of assay; however, there now seems to be good evidence on a series of phospholipase inhibitors, which can clearly modify the course of cell injury and delay cell death. Such inhibitors include dibucaine, mepacrin, chlorpromazine, and others. To some extent with certain injuries, this may correlate with the effects of such phopholipase inhibitors on *in vitro* microsomal lipid peroxidation, which they inhibit following a variety of stimuli. This extensive cell injury, due to reactive oxygen species, may be particularly susceptible to inhibitors such as dibucaine. Phospholipase A2 is also pH dependent and inhibited by reduction of $[pH]_i$. Phospholipase A2 inhibition may thus relate to the protection considered by low pH described elsewhere. Recently, Wang et al. (1996) demonstrated that phospholipase A2 mRNA significantly increased in rat hepatocytes shortly after induction of chemical hypoxia and remained high until cell death. In addition, this group showed that pretreatment of hepatocytes with PLA2 specific antisense DNA oligonucleotides abolished its stimulation, decreased the enzymatic activity, and significantly delayed cell death.

Protease Inhibitors. A variety of lysosomal and nonlysosomal protease inhibitors have been used in attempts to modify cell injury and prevent cell death given the putative role that these hydrolases play in oncosis and necrosis. Development of cell permeant protease inhibitors could also result in the development of pharmaceutical agents to modify and prevent the progression of change toward cell death (see Cryns and Yuan, Chapter 6, this volume).

Calpains represent a particularly appealing target because of the increases of $[Ca^{2+}]_i$ that appear to be a common lethal signal following a variety of lethal injuries. We observed that treatment of cells that were lethally injured by $HgCl_2$ were protected by the calpain inhibitors antipain and leupeptin (Elliget et al., 1994) and, furthermore, that calpain inhibitors were much more effective if Ca^{2+} was at normal concentration in the extracellular medium. Moreover, the calpain inhibitors greatly modified the formation of cytoplasmic blebs and prevented disruption of the actin filaments following $HgCl_2$ treatment. Edelstein et al. (1996) also found that calpain inhibitors could protect against both hypoxia and ionomycin-induced cell membrane damage in rat renal proximal tubules. Using the nonpeptide calpain inhibitor PD 150606, this group observed that the cytoprotective effects of low $[pH]_i$ and low $[Ca^{2+}]_i$ are mediated, at least in part by inhibition of calpain activity and that the active form of calpain is the isoenzyme mu-calpain. A possible role of mu-calpain activation in the release of lysosomal enzymes has been suggested by the work of Yamashima et al. (1996), who found that the increased $[Ca^{2+}]_i$ and PIP2 following brain ischemia to the hippocampus activates mu-calpain, which, they suggest, acts at the lysosomal membrane to induce hydrolase release. In addition, activation of calpain may explain the protection to anoxic injury of liver and kidney provided by glycine. Nichols et al. (1994) suggested, for example, that the cytoprotective effect of glycine could be due to calpain inhibition.

The possible utilization of caspase inhibitors to prevent cell death is currently under intense investigation. While clearly implicated in apoptosis, the role of caspases in oncosis is less clear. However, the protective effects of Bcl-2-possible protectors against caspase activity suggest a possible role. The role of inhibitors against lysosomal cysteine proteases is less clear.

Nuclease Inhibitors. There have been several papers studying the protective effects of nuclease inhibitors, such as aurintricarboxylic acid, on apoptosis and thymocytes, but still relatively little is known about effects in oncosis. However, Ueda and Shah (1992) observed that hydrogen peroxide, which induces oncosis in proximal tubule cells, induces DNA damage prior to death in a dose-dependent manner. The nuclease inhibitors aurintricarboxylic acid, Evans blue, and zinc ion prevented the strand breaks and cell death, and similar effects were found with DNA damage induced by treatment with micrococcal endonucleases also prevented by these inhibitors; therefore, it appears that endonuclease activation occurs as an early event leading to DNA damage and that it does not require protein synthesis.

On the other hand, Cantoni et al. (1989) observed in Chinese hamster ovary cells exposed to hydrogen peroxide that the DNA strand breaks could be prevented by the intracellular Ca^{2+} chelator, Quin 2. These authors suggested that the strand breaks resulted from altered Ca^{2+} homeostasis, possibly through activation of Ca^{2+}-dependent endonucleases rather than direct oxidant effects on DNA. Gardner et al. (1997), studying H_2O_2 toxicity, observed that exposure of isolated nuclei in the absence of Ca^{2+} and Mg^{2+} also failed to induce endonucleosomal fragmentation.

Amino Acids. Glycine and other small amino acids have been found to be protective against a variety of lethal injuries that lead to oncosis and necrosis in the renal proximal tubule, endothelium (Weinberg et al., 1992), and liver (Zhong et al., 1996). Miller et al. (1994) demonstrated that strychnine, which binds to glycine receptors in the CNS, mimics the protective properties of glycine in protection of renal proximal tubules from various toxins. At neutral pH, Currin et al. (1996) found that both glycine and strychnine protected against reperfusion injury to sinusoidal endothelial cells of rat livers stored for transplantation. The molecular mechanism of action of glycine protection is still incompletely understood. Recently, Venkatachalam et al. (1994) found that glycine prevented the formation of large water-filled channels in ATP-depleted MDCK cells. These channels permitted the passage of up to 70 kDa fluorescein dextran. Dong et al. (1996) found that glycine permitted survival of ATP-depleted MDCK cells with $[Ca^{2+}]_i$ as high as 100 μM.

Modification of $[Ca^{2+}]_i$. Conditions that lower or buffer $[Ca^{2+}]_i$ can exert marked effects on the kinetics of cell killing through oncosis following a variety of injurious stimuli. This has been accomplished by incubating cells in Ca^{2+}-free media, introduction of intracellular buffers such as Quin 2, EGTA-AM, and, in some cases, by the use of Ca^{2+}-entry blockers (Smith et al., 1991, 1992). The mechanisms of these effects have been further analyzed and found to involve activation of phospholipases, activation of nucleases, and mediation by Ca^{2+}-binding proteins, such as calmodulin. Control of $[Ca^{2+}]_i$ is also important in cellular volume regulation in a variety of cell types. When cells are swollen through the use of Na^+ channel formers, such as veratridine or in hypotonic media, increases of $[Ca^{2+}]_i$ occur, which stimulate efflux of K^+ and Cl^-. The important role of Ca^{2+} in this process is shown through experiments in which removal of extracellular Ca^{2+} or intracellular Ca^{2+} by chelation totally inhibit the recovery of volume regulation.

Hypothesis

Figure 2.9 illustrates our current working hypothesis of the events that occur following cell injury leading to oncosis, cell death, and necrosis, emphasizing the role of deregulation of $[Ca^{2+}]_i$. The initial injurious events leading to oncosis appear to involve primarily injurious agents that interfere with ATP synthesis, modify plasma membrane ion transport systems or plasmalemmal integrity, and redistribute Ca^{2+} from intracellular stores including mitochondria and endoplasmic reticulum.

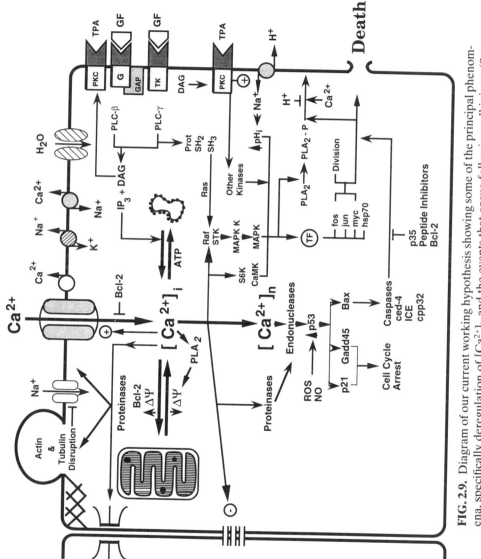

FIG. 2.9. Diagram of our current working hypothesis showing some of the principal phenomena, specifically deregulation of $[Ca^{2+}]_i$, and the events that ensue following cell injury. (See text for details.)

The earliest events involve deregulation of cellular ion regulation, including that of Na^+, K^+, Cl^-, and Ca^{2+}. These are closely interrelated. For example, depletion of ATP results in cessation of Na^+–K^+- and Ca^{2+}-ATPases, resulting in increased $[Na^+]_i$, increased $[Ca^{2+}]_i$, and decreased $[K^+]_i$. Concomitantly, Cl^- moves into the cell. Moreover, the increased $[Na^+]_i$ results in decreased Na^+–Ca^{2+} exchange, further increasing the increase of $[Ca^{2+}]_i$. This increased $[Ca^{2+}]_i$, in turn, increases Ca^{2+} influx from the extracellular space.

These early events result in a loss of cellular volume control and the cells begin to swell through the increased $[Na^+]_i$ and $[Cl^-]_i$. Some oscillatory behavior in cell volume may be seen at this stage as increased $[Ca^{2+}]_i$ activates efflux of K^+ and Cl^-; however, because of the ATP deficiency and the increasing damage to the plasma membrane, volume increase forces prevail. Furthermore, the swelling-induced stretch in the cell membrane may increase these respective ion fluxes.

There are many immediate consequences of a sustained increase of $[Ca^{2+}]_i$, which include activation of proteinases, phospholipases, and nucleases. For example, the early cytoplasmic blebbing appears to result from Ca^{2+}-activated proteinase modification of the cytoskeleton, particularly in the attachment of actin to the plasmalemma. Calpain inhibitors greatly diminish cytoplasmic blebbing. Such inhibitors may also retard the entire process as an important calpain substrate is the Ca^{2+}-ATPase (see Cryns and Yuan, Chapter 6, Sikorska and Walker, Chapter 7, this volume).

As the time after injury increases, the effects of Ca^{2+}-activated phospholipases become increasingly important. In Stage 4, the mitochondria convert from the condensed to the swollen conformation. At this time, control of oxidative phosphorylation is lost, and there is also a loss of the mitochondrial membrane potential, perhaps coincident with activation of the mitochondrial permeability transition. Loss of mitochondrial inner membrane cardiolipin correlates with this phase.

Deregulation of $[Ca^{2+}]_i$ also results in activation of a variety of immediate early genes through activation of their antecedent transcription factors and protein kinases. As illustrated in the upper right quadrant of the diagram, there are common features between the effects of toxic cell injury and those of various "growth factors" including a number of cytokines. These newly transcribed genes include c-fos, c-jun, c-myc, and heat shock proteins, including HSP-70. Their induction appears to depend on protein kinase activation including the MAP kinase pathway. Although the details of these pathways require considerably more study, there seems to be no doubt of their importance in both oncosis and apoptosis.

Activation of p53 also occurs during oncosis following a variety of injuries, including hypoxia and irradiation. Although this does not appear to involve transcription, it nevertheless is effective in resulting in cell death. Through transactivation, increased p53 activates other genes including Bax, Gadd45, and p21(waf). Of these, only Bax appears to be involved in the progression to cell death, while the others are presently only associated with cell-cycle arrest. The effects of Bax are potentially countered by the protective effects of Bcl-2. Among the possible mechanisms of Bcl-2 protection are inhibition of proteinase activity and modification of membrane ion channels, especially of Ca^{2+}. It is of some interest that increased expression of Bcl-2 is protective against both oncosis and apoptosis following an increasing variety of injurious stimuli. Knowledge of the fundamental characteristics of the Bcl-2 protection should lead to improved interventions to promote cell killing or cell survival.

It is our view that the terminal events in cell killing must involve those processes that ultimately destroy the function of the plasmalemma in maintaining homeostasis. We propose that these involve the coordinate function of activated phospholipases and proteinases and that interference with these will constitute a cell death control program.

Conclusion

The pursuit of the elusive mechanisms of cell death is fundamental to preventive and therapeutic medical practice. In this chapter, we have focused on the prelethal changes known as oncosis. This subset of prelethal change is to be contrasted with apoptosis. Both of these patterns of change are represented following many types of cell injury *in vivo;* often they occur in combination, that is, oncosis in the center of a lesion and apoptosis at the periphery. Yet, this is only a beginning; the common factors and those that differ need to be defined in order to elucidate, and manipulate, the factors that lead to cell death.

ACKNOWLEDGMENTS

Supported by NIH Grant DK15440. This is contribution 3908 from the Cellular Pathobiology Laboratory.

REFERENCES

Aguilar HI, Botla R, Arora AS, Bronk SF, Gores GJ (1996): Induction of the mitochondrial permeability transition by protease activity in rats: A mechanism of hepatocyte necrosis. Gastroenterology 110:558–66.

Alnemri ES, Livingston DJ, Nicholson DW, Salvesen G, Thornberry NA, Wong WW, Yuan J (1996): Human ICE/CED-3 protease nomenclature. Cell 87:171.

Amstad PA, Liu H, Ichimiya M, Berezesky IK, Trump BF (1997): Bcl-2 enhances malignant transformation in mouse epidermal cells JB6. Mol Carcinog, in press.

Amstad P, Ichimiya M, Liu H, Chang S, Berezesky I, Gutierrez P, Trump B (1996): The effect of bcl-2 expression on oxidant-induced apoptosis in normal rat kidney (NRK) and mouse epidermal JB6 cells. FASEB J 10:A1425.

Arends MJ, Morris RG, Wyllie AH (1990): Apoptosis. The role of endonucleases. Am J Pathol 136:593–608.

Arora AS, de Groen PC, Croall DE, Emori Y, Gores GJ (1996): Hepatocellular carcinoma cells resist necrosis during anoxia by preventing phospholipase-mediated calpain activation. J Cell Physiol 167:434–42.

Atsma DE, Bastiaanse EM, Van der Valk L, Van der Laarse A (1996): Low external pH limits cell death of energy-depleted cardiomyocytes by attenuation of Ca^{2+} overload. Am J Physiol 270:H2149–56.

Bernardi P, Petronilli V (1996): The permeability transition pore as a mitochondrial calcium release channel: A critical appraisal. J Bioenerg Biomembr 28:131.

Bohley P, Seglen PO (1992): Proteases and proteolysis in the lysosome. Experientia 48:151–57.

Bortner CD, Cidlowski JA (1996): Absence of volume regulatory mechanisms contributes to the rapid activation of apoptosis in thymocytes. Am J Physiol 271:C950–61.

Cantoni O, Sestili P, Cattabeni F, Bellomo G, Pou S, Cohen M, Cerutti P (1989): Calcium chelator Quin 2 prevents hydrogen-peroxide-induced DNA breakage and cytotoxicity. Eur J Biochem 182:209–12.

Chang SH, Phelps PC, Berezesky IK, Wang XW, Elmore LW, Coursen J, Harris CC, Trump BF (1997): Induction of apoptosis by microinjection of wild-type p53 gene into NRK-52E cells. The Toxicologist 36:251.

Chinnaiyan AM, Dixit VM (1996): The cell death machine. Curr Biol 6:555–62.

Churchwell KB, Wright SH, Emma F, Rosenberg PA, Strange K (1996): NMDA receptor activation inhibits neuronal volume regulation after swelling induced by veratridine-stimulated Na^+ influx in rat cortical cultures. J Neurosci 16:7447–57.

Collan Y, McDowell EM, Trump BF (1981): Studies on the pathogenesis of ischemic cell injury. VI. Mitochondrial flocculent densities in autolysis. Virchows Archiv [B], 35:189–99.

Coursen JD, Wang XW, Fornace AJ, Jr, Harris CC (1996): GADD-45-induced aberrant cell-cycle arrest is p53 dependent. Proc Am Assoc Ca Res 37:2.

Cowley RA, Trump BF (1982): Pathophysiology of Shock, Anoxia and Ischemia. Baltimore: Williams and Wilkins.

Crowe WE, Altamirano J, Heurto L, Alvarez-Leefmans FJ (1995): Volume changes in single N1E-115 neuroblastoma cells measured with a fluorescent probe. Neuroscience 69:283–96.

Currin RT, Caldwell-Kenkel JC, Lichtman SN, Bachmann S, Takei Y, Kawano S, Thurman RG, Lemasters JJ (1996): Protection by Carolina rinse solution, acidotic pH, and glycine against lethal reperfusion injury to sinusoidal endothelial cells of rat livers stored for transplantation. Transplantation 62:1549–58.

Davis MA, Smith MW, Chang SH, Trump BF (1994): Characterization of a renal epithelial cell model of apoptosis using okadaic acid and the NRK-52E cell line. Toxicol Pathol 22:595–604.

DeDuve C, Wattiaux R (1996): Functions of lysosomes. Annu Rev Physiol 28:435–92.

Distelhorst C, McCormick T (1996): Bcl-2 acts downstream from Ca^{2+} fluxes to inhibit apoptosis in thapsigargin- and glucocorticoid-treated mouse lymphoma cells. Proc Am Assoc Cancer Res 37:23.

Dong Z, Kaikumar P, Venkatachalam M, Weinberg J (1996): Intracellular calcium levels that determine the survival of glycine protected ATP depleted cells. J Am Soc Nephrol 7:1823.

Edelstein CL, Yaqoob MM, Alkhunaizi AM, Gengaro PE, Nemenoff RA, Wang KK, Schrier RW (1996): Modulation of hypoxia-induced calpain activity in rat renal proximal tubules. Kidney Int 50:1150–57.

El-Deiry WS, Harper JW, O'Connor PM, et al. (1994): WAF1/CIP1 is induced in p53-mediated G1 arrest and apoptosis. Cancer Res 56:1–46.

Elliget KA, Phelps PC, Trump BF (1994): Cytosolic Ca^{2+} elevation and calpain inhibitors in $HgCl_2$ injury to rat kidney proximal tubule epithelial cells. Pathobiology 62:298–310.

Elliget KA, Phelps PC, Trump BF (1991): $HgCl_2$-induced alteration of actin filaments in cultured primary rat proximal tubule epithelial cells labelled with fluorescein phalloidin. Cell Biol Toxicol 7:263–80.

Ellis RE, Yuan Y, Horvitz HR (1991): Mechanisms and functions of cell death. Annu Rev Cell Biol 7:663–98.

Evtodienko YV, Teplova VV, Sidash SS, Ichas F, Mazat JP (1996): Microtubule-active drugs suppress the closure of the permeability transition pore in tumour mitochondria. FEBS Lett 393:86–88.

Forrester K, Ambs S, Lupold SE, Kapust RB, Spillare EA, Weinberg WC, Felley-Bosco E, Wang XW, Geller DA, Tzeng E, Billiar TR, Harris CC (1996): Nitric oxide-induced p53 accumulation and regulation of inducible nitric oxide synthase expression by wild-type p53. Proc Natl Acad Sci USA 93:2442–47.

Fukushima S, Chang SH, Berezesky IK, Trump BF (1995): Apoptosis in the rat kidney following ischemia-reflow. Shock (suppl) 3:58.

Gardner AM, Xu FH, Fady C, Jacoby FJ, Duffey DC, Tu Y, Lichtenstein A (1997): Apoptotic vs. nonapoptotic cytotoxicity induced by hydrogen peroxide. Free Radic Biol Med 22:73–83.

Graeber TG, Peterson JF, Tsai M, Monica K, Fornace, Jr., AJ, Giaccia AJ (1994): Hypoxia induces accumulation of p53 protein, but activation of a G1-phase checkpoint by low-oxygen conditions is independent of p53 status. Mol Cell Biol 14:6264–77.

Gu H, Yamamoto N, Chang SH, Trump BF (1994): Immediate early and stress gene expression in acute renal failure induced by $HgCl_2$. Modern Pathol 7:158a.

Hagar H, Ueda N, Shah SV (1996): Endonuclease induced DNA damage and cell death in chemical hypoxic injury to LLC-PK1 cells. Kidney Int 49:355–61.

Hagar H, Ueda N, Shah SV (1996): Role of reactive oxygen metabolites in DNA damage and cell death in chemical hypoxic injury to LLC-PK1 cells. Am J Physiol 271:F209–15.

Haldar S, Basu A, Croce CM (1997): Bcl2 is the guardian of microtubule integrity. Cancer Res 57:229–33.

Harris CC (1996a): p53 tumor suppressor gene: From the basic research laboratory to the clinic—An abridged historical perspective. Carcinogenesis 17:1187–98.

Harris CC (1996b): Structure and function of the p53 tumor suppressor gene: Clues for rational cancer therapeutic strategies. (Review). JNCI 88:1442–55.

Hengartner MO, Ellis RE, Horvitz HR (1992): Caenorhabditis elegans gene ccd-9 protects cells from programmed cell death. Nature 356:494–99.

Hockenberry DM, Oltvai ZN, Yin XM, et al. (1993): Bcl-2 functions in an antioxidant pathway to prevent apoptosis. Cell 75:241–51.

Ichimiya M, Liu H, Amstad P, Chang S, Berezesky IK, Trump BF (1996): The effect of BCL-2 expression on oxidant-induced cell death and calcium ion mobilization in normal rat kidney (NRK) cells. Shock 5:71.

Jain PT, Chang SH, Berezesky IK, Trump BF (1995): Higher elevation of $[Ca^{2+}]_i$ induced by benzoyl peroxide precedes higher cell death of non-promotable compared to promotable JB6 cells. In Vitro Toxicol 8(3):251–61.

Kabakov AE, Gabai VL (1995): Heat shock-induced accumulation of 70-kDa stress protein (HSP70) can protect ATP-depleted tumor cells from necrosis. Exp Cell Res 217:15–21.

Kerr JFR, Wyllie AH, Currie AR (1972): Apoptosis: A basic biological phenomenon with wide-ranging implications in tissue kinetics. Br J Cancer 26:239–57.

Knudson CM, Tung KS, Tourtellotte WG, Brown GA, Korsmeyer SJ (1995): Bax-deficient mice with lymphoid hyperplasia and male germ cell death. Science 270:96–99.

Lemasters JJ, Bond JM, Chacon E, Harper IS, Kaplan SH, Ohata H, Trollinger DR, Herman B, Cascio WE (1996): The pH paradox in ischemia-reperfusion injury to cardiac myocytes. EXS 76:99–114.

Lemasters JJ, DiGuiseppi J, Nieminen AL, et al. (1987): Blebbing, free Ca^{2+} and mitochondrial membrane potential preceding cell death in hepatocytes. Nature 325:78–81.

Lints R, Driscoll M (1996): Programmed and pathological cell death in Caenorhabditis elegans. In Holbrook NJ, Martin GR, Lockshin RA (eds): Cellular Aging and Cell Death. New York: Wiley–Liss, Inc., pp 235–53.

Lockshin RA and Zakeri Z (1996): The biology of cell death and its relationship to aging. In Holbrook NJ, Martin GR, Lockshin RA (eds): Cellular Aging and Cell Death. New York: Wiley–Liss, Inc., pp 167–80.

Majno G, Joris I (1995): Apoptosis, oncosis, and necrosis. An overview of cell death. Am J Pathol 146:3–15.

Majno G, Joris I (1996): Cells, Tissues, and Disease: Principles of General Pathology. Cambridge, Mass: Blackwell Science.

Maki A, Berezesky IK, Fargnoli J, et al. (1992): Role of $[Ca^{2+}]_i$ in induction of c-fos, c-jun, and c-myc RNA in rat PTE after oxidative stress. FASEB J 6:919–24.

Malis CD, Bonventre JV (1986): Mechanism of calcium potentiation of oxygen free radical injury to renal mitochondria. A model for post-ischemic and toxic mitochondrial damage. J Biol Chem 261:14201–8.

Mei JM, Chi WM, Trump BF, Eccles CU (1996): Involvement of nitric oxide in the deregulation of cytosolic calcium in cerebellar neurons during combined glucose–oxygen deprivation. Mol Chem Neuropathol 27:155–66.

Miller GW, Lock EA, Schnellmann RG (1994): Strychnine and glycine protect renal proximal tubules from various nephrotoxicants and act in the late phase of necrotic cell injury. Toxicol Appl Pharmacol 125:192–97.

Minn AJ, Velez P, Schendel SL, Liang H, Muchmore SW, Fesik SW, Fill M, Thompson CB (1997): Bcl-XL forms an ion channel in synthetic lipid membranes. Nature 385:353–57.

Morana SJ, Wolf CM, Li J, Reynolds JE, Brown MK, Eastman A (1996): The involvement of protein phosphatases in the activation of ICE/CED-3 protease, intracellular acidification, DNA digestion, and apoptosis. J Biol Chem 271:18263–71.

Muchmore SW, Sattler M, Liang H, Meadows RP, Harlan JE, Yoon HS, Nettesheim D, Chang BS, Thompson CB, Wong SL, Ng SL, Fesik SW (1996): X-ray and NMR structure of human Bcl-xL, an inhibitor of programmed cell death. Nature 381:335–41.

Murphy AN, Bredesen DE, Cottopassi G, Wang E, Fiskum G (1996): Bcl-2 potentiates the maximal calcium uptake capacity of neural cell mitochondria. Proc Natl Acad Sci USA 93:9893–9898.

Nath KA, Croatt AJ, Likely S, Behrens TW, Warden D (1996): Renal oxidant injury and oxidant response induced by mercury. Kidney Int 50:1032–43.

Negulyaev YA, Vedernikova EA, Maximov AV (1996): Disruption of actin filaments increases the activity of sodium-conducting channels in human myeloid leukemia cells. Mol Biol Cell 7:1857–64.

Nichols JC, Bronk SF, Mellgren RL, Gores GJ (1994): Inhibition of nonlysosomal calcium-dependent proteolysis by glycine during anoxic injury of rat hepatocytes. Gastroenterology 106:168–76.

Nicholson DW (1996): ICE/CED-3-like proteases as therapeutic targets for the control of inappropriate apoptosis. Nature Biotechnol 14:297–301.

Nicholson DW, All A, Thornberry NA, et al. (1995): Identification and inhibition of the ICE/CED-3 protease necessary for mammalian apoptosis. Nature 376:37–43.

Nieminen AL, Petrie TG, Lemasters JJ, Selman WR (1996): Cyclosporin A delays mitochondrial depolarization induced by N-methyl-D-aspartate in cortical neurons: Evidence of the mitochondrial permeability transition. Neuroscience 75:993–97.

Papadimitriou JC, Phelps PC, Shin ML, Smith MW, Trump BF (1994): Effects of Ca^{2+} deregulation on mitochondrial membrane potential and cell viability in nucleated cells following lytic complement attack. Cell Calcium 15:217–27.

Pastorino JG, Simbula G, Yamamoto K, Glascott PA, Rothman RJ, Farber JL (1996): The cytotoxicity of tumor necrosis factor depends on induction of the mitochondrial permeability transition. J Biol Chem 271:29792–98.

Pentilla A, Glaumann H, Trump BF (1976): Studies on the modification of extracellular acidosis against anoxia, thermal, etc. Life Sci 18:1419–30.

Pentilla A, Trump BF (1974): Extracellular acidosis protects Ehrlich ascites tumor cells and rat renal cortex against anoxic injury. Science 185:277–78.

Phelps PC, Smith MW, Trump BF (1989): Cytosolic ionized calcium and bleb formation after acute cell injury of cultured rabbit renal tubule cells. Lab Invest 60:630–42.

Reeves WB, Shah SV (1994): Activation of potassium channels contributes to hypoxic injury in proximal tubules. J Clin Invest 94:2289–94.

Roman RM, Wang Y, Fitz JG (1996): Regulation of cell volume in a human biliary cell line: Activation of K^+ and Cl^- currents. Am J Physiol 271:G239–48.

Sapirstein A, Spech RA, Witzgall R, Bonventre JV (1996): Cytosolic phospholipase A2 (PLA2), but not secretory PLA2, potentiates hydrogen peroxide cytotoxicity in kidney epithelial cells. J Biol Chem 271:21505–13.

Shimizu S, Eguchi Y, Kamiike W, Waguri S, Uchiyama Y, Matsuda H, Tsujimoto Y (1996): Bcl-2 blocks loss of mitochondrial membrane potential while ICE inhibitors act at a different step during inhibition of death induced by respiratory chain inhibitors. Oncogene 13:21–29.

Smith MW, Phelps PC, Trump BF (1991): Cytosolic Ca^{2+} deregulation and blebbing after $HgCl_2$ injury to cultured rabbit proximal tubule cells as determined by digital imaging microscopy. Proc Natl Acad Sci USA 88:4926–30.

Smith MW, Phelps PC, Trump BF (1992): Injury-induced changes in cytosolic Ca^{2+} in individual rabbit proximal tubule cells. Am J Physiol 262 (Renal Fluid Electrolyte 31):F647–55.

Srinivasan A, Foster LM, Testa MP, Ord T, Keane RW, Bredesen DE, Kayalar C (1996): Bcl-2 expression in neural cells blocks activation of ICE/CED-3 family proteases during apoptosis. J Neurosci 16:5654–60.

Strack PR, Frey MW, Rizzo CJ, Cordova B, George HJ, Meade R, Ho SP, Corman J, Tritch R, Korant BD (1996): Apoptosis mediated by HIV protease is preceded by cleavage of Bcl-2. Proc Natl Acad Sci USA 93:9571–76.

Susin SA, Zamzami N, Castedo M, Hirsch T, Marchetti P, Macho A, Daugas E, Geuskens M, Kroemer G (1996): Bcl-2 inhibits the mitochondrial release of an apoptogenic protease. J Exp Med 184:1331–41.

Thastrup O, Cullen PJ, Drobak BK, et al. (1990): Thapsigargin, a tumor promoter, discharges intracellular Ca^{2+} stores by specific inhibition of the ER Ca^{2+}-ATPase. Proc Natl Acad Sci USA 87:2466–70.

Trump BF, Ginn FL (1969): The pathogenesis of subcellular reaction to lethal injury. In Bajusz E, Jasmin G (eds): Methods and Achievements in Experimental Pathology, Volume IV. Basel: Karger, pp 1–29.

Trump BF, Arstila AU (1971): Cell injury and cell death. In LaVia MF, Hill RB, Jr. (eds): Principles of Pathobiology. New York: Oxford University Press, pp 9–95.

Trump BF, Berezesky IK, Cowley RA (1982). The cellular and subcellular characteristics of acute and chronic injury with emphasis on the role of calcium. In Cowley RA, Trump BF (eds): Pathophysiology of Shock, Anoxia, and Ischemia. Baltimore: Williams & Wilkins, pp 6–46.

Trump BF, Berezesky IK (1995): Calcium, cell death, and tumor promotion. In McClain RM, Slaga TJ, LeBoeuf R, Pitot H (eds): Growth Factors and Tumor Production: Implications for Risk Assessment. New York: Wiley–Liss, Inc., pp 121 31.

Trump BF, Berezesky IK (1995): Calcium-mediated cell injury and cell death. FASEB J 9:219–28.

Trump BF, Berezesky IK (1994): Cellular and molecular pathobiology of reversible and irreversible injury. In Tyson CA, Frazier JM (eds): Methods in Toxicology, Vol. 1B, In Vitro Toxicity Indicators. San Diego: Academic, pp 1–22.

Trump BF, Berezesky IK (1992): The role of cytosolic Ca^{2+} in cell injury, necrosis and apoptosis. Curr Opin Cell Biol 4:227–32.

Trump BF, Berezesky IK, Smith MA (1993): Cellular calcium and mitochondrial dysfunction, Chapter 28. In Jones DP, Lash LH (eds): Methods in Toxicology, Volume 2 (Mitochondrial Dysfunction). V. Mitochondrial Energetics and Transport Processes. London/New York: Academic Press, pp 337–53.

Trump BF, Berezesky IK (1996a): The mechanisms of calcium-mediated cell injury and cell death. New Horiz 4:139–50.

Trump BF, Berezesky IK (1996b): The role of altered $[Ca^{2+}]_i$ regulation in apoptosis, oncosis, and necrosis. Biochim Biophys Acta 1313:173–78.

Trump BF, Berezesky IK (1994): Cellular and molecular pathobiology of reversible and irreversible injury. In Tyson CA, Frazier JM (eds): Methods in Toxicology, Vol. 1B. In Vitro Toxicity Indicators, Chapter 1. New York: Academic Press, pp 1–22.

Trump BF, Berezesky IK, Osornio-Vargas A (1981): Cell death and the disease process. The role of cell calcium. In Bowen ID, Lockshin RA (eds): Cell Death in Biology and Pathology. London: Chapman and Hall, pp 209–42.

Ueda N, Shah SV (1992): Endonuclease-induced DNA damage and cell death in oxidant injury to renal tubular epithelial cells. J Clin Invest 90:2593–97.

Ueda N, Walker PD, Hsu SM, Shah SV (1995): Activation of a 15-kDa endonuclease in hypoxia/reoxygenation injury without morphologic features of apoptosis. Proc Natl Acad Sci USA 92:7202–6.

Venkatachalam M, Weinberg J, Patel Y, Davis J, Saikumar P (1994): Functional characterization of a glycine sensitive porous membrane defect in ATP depleted MDCK cells. J Am Soc Nephrol 5:912.

Virchow R (1859): Cellular pathology as based upon physiological and pathological histology. (Translated from the Second German Edition by Frank Chance.) New York: Dover Publications, Inc., 1971, pp 269–72.

Wang H, Harrison-Shostak DC, Lemasters JJ, Herman B (1996): Contribution of pH-dependent group II phospholipase A2 to chemical hypoxic injury in rat hepatocytes. FASEB J 10:1319–25.

Wang HG, Rapp UR, Reed JC (1996): Bcl-2 targets the protein kinase Raf-1 to mitochondria. Cell 87:629–38.

Weinberg JM, Varani J, Johnson KJ, Roeser NF, Dame MK, Davis JA, Venkatachalam MA (1992): Protection of human umbilical vein endothelial cells by glycine and structurally similar amino acids against calcium and hydrogen peroxide–induced lethal cell injury. Am J Pathol 140:457–71.

Wyllie AH, Kerr JFR, Currie AR (1980): Cell death: The significance of apoptosis. Int Rev Cytol 68:251–306.

Yamamoto N, Maki A, Swann JD, Berezesky IK, Trump BF (1993): Induction of immediate, early and stress genes in rat PTE following injury. Ren Fail 15:163–71.

Yamamoto N, Smith MW, Maki A, Berezesky IK, Trump BF (1994): Role of cytosolic Ca^{2+} and protein kinases in the induction of the hsp70 gene in rat proximal tubular epithelial cells. Kidney Int 45:1093–104.

Yamashima T, Saido TC, Takita M, Miyazawa A, Yamano J, Miyakawa A, Nishijyo H, Yamashita J, Kawashima S, Ono T, Yoshioka T (1996): Transient brain ischaemia provokes Ca^{2+}, PIP2 and calpain responses prior to delayed neuronal death in monkeys. Eur J Neurosci 8:1932–44.

Yang CW, Rathinavelu A, Borowitz JL, Isom GE (1994): Activation of a calcium- and pH-dependent phospholipase A2 by cyanide in PC12 cells. Toxicol Appl Pharmacol 124:262–67.

Yang X, Schnellmann RG (1996): Proteinases in renal cell death. J Toxicol Environ Health 48:319–32.

Yuan J, Shaham S, Ledoux S, et al. (1993): The *C. elegans* cell death gene ced-4 encodes a protein similar to mammalian interleukin-1β-converting enzyme. Cell 75:641–52.

Yuan J, Horvitz H (1992): The C. elegans cell death gene ced-4 encodes a novel protein and is expressed during the period of extensive programmed cell death. Development 116:309–20.

Zhang J, Steiner JP (1995): Nitric oxide synthase, immunophilins and poly(ADP-ribose) synthetase: Novel targets for the development of neuroprotective drugs. Neurol Res 17:285–88.

Zhong Z, Jones S, Thurman RG (1996): Glycine minimizes reperfusion injury in a low-flow, reflow liver perfusion model in the rat. Am J Physiol 270:G332–38.

CHAPTER 3

THE STUDY OF CELL DEATH BY THE USE OF CELLULAR AND DEVELOPMENTAL MODELS

ZAHRA ZAKERI
Department of Biology
Queens College of CUNY
65-30 Kissena Blvd.
Flushing, NY 11367

Cell death, as much a part of normal development as proliferation or differentiation, has been recognized by embryologists for decades, but recently the advent in the field of cellular and molecular biology has made it possible to examine the signaling molecules and the cellular and the molecular responses to the cell-killing signals that shape the embryo. In this chapter we aim to demonstrate some of the different approaches to studying cell death by using two model developmental systems—embryonic mouse limbs and metamorphosis of insect labial gland. These model systems are good examples of different situations that one might encounter in the study of cell death. The use of both *in vivo* and *in vitro* studies to examine mechanisms of cell death and the pitfalls involved will be addressed. Although this chapter uses developmental systems to describe cell death, the logic and the experimental processes can be applied to a number both of normally occurring and abnormally induced cell deaths.

CHARACTERIZATION OF CELL DEATH

It is important for the health of the embryo that the cells that die do so in a very well-controlled manner, producing no adverse effect as the result of their

When Cells Die, Edited by Richard A. Lockshin, Zahra Zakeri,
and Jonathan L. Tilly
ISBN 0-471-16569-7 © 1998 Wiley-Liss, Inc.

demise. This well-controlled death has been referred to by several terms, such as "programmed cell death," "types I and II cell death," "apoptosis," and "active cell death." Programmed cell death (PCD) is often used to describe cell death occurring in developing embryos and in metamorphosing insects. It is characterized by a predictable sequence of steps in target cells that are stimulated by developmental or hormonal stimuli. The term PCD implies a genetic control and differential expression of genes that may either regulate the activation and the progression of cell death or be regulated by these events. Often, a requirement for mRNA and protein synthesis has been identified. The morphology of PCD can be, as will be described, type I—apoptotic—or type II—lysosomal. Both types represent a well-controlled cell death in that there is no damage to the neighboring cells. Although PCD has been equated with the term *apoptosis* or type I cell death, it encompasses both morphologies. This point is illustrated in Fig. 3.1.

Type I Cell Death (Apoptosis)

The classical type I apoptosis is a morphological designation characterized by rapid condensation of the cytoplasm and nuclear chromatin, resulting in DNA fragmentation as well as blebbing of the cell surfaces. In some systems where cells are in contact there appears to be a condensation and breakdown of cell–cell contact. The sequence is followed by the fragmentation of the cells, phagocytosis, and secondary lysosomal degradation of fragments by phagocytic cells derived from the blood supply such as macrophages or neighboring cells. Although there is fragmentation of the apoptotic cells, the integrity of the membrane is conserved. Apoptosis is biochemically defined by dependence on energy and by double-stranded DNA at the linker regions between nucleosomes (see Chapter 2, by Trump and Berezesky, and Chapter 7, by Sikorska and Walker in this volume). This cleavage produces oligonucleosome fragments. An enzyme considered to be a likely candidate for causing this cleavage of DNA is dependent on Ca^{2+} and Mg^{2+}, is active in neutral pH conditions, and is strongly inhibited by zinc (Wyllie, 1980, 1985, 1987; Yamada et al.,

---

FIG. 3.1. Representation of the relationships among programmed cell death, apoptosis, vacuolar cell death, and necrosis. "Programmed cell death" implies a causal or genetic sequence that is not necessarily documented in the physiological cell deaths of differing morphologies. As is suggested in the middle panel, apoptosis or Type I death is best documented in evanescent cells with little cytoplasm, while lysosomal or Type II death is more typical of long-lived cells with massive cytoplasm; there may be intermediaries. The lower panel indicates that programming is best documented in developmental systems, less so for other situations, and presumptively not for toxic pathological situations. Turnover is presumed to fall within the categories of physiological cell death, but is not currently documented.

Type of Death ⇦	Programmed	Apoptosis	Vacuolar	Necrosis
		Active or Physiological Death		
Synonyms ⇩	May include apoptosis and vacuolar	Type I	Type II Lysosomal	--
Description	Documented up- or down-regulation of specific genes; inhibition of protein synthesis blocks cell death	Shrinkage of nucleus & cytoplasm; DNA ladders and positive TUNEL; cell fragmentation and blebbing; may be programmed	Expansion of lysosomes & other vacuoles; late nuclear condensation but eventual positive TUNEL; cell fragmentation	Osmotic swelling and rupture
Occurrence	Developmental; well documented for invertebrates and vertebrates; probably in many physiological situations	Evanescent cells, primarily lymphatic and hematopoietic origin	Longer-lived cells with massive cytoplasm	Any infarcted or severely damaged cell
Examples	*Caenorhabditis*; *Manduca* labial gland, *Antheraea* intersegmental muscle, sympathetic and sensory developing neurons, interdigital regions of developing chick limbs	Lymphocytes and thymocytes under numerous conditions; many other cells	Insect muscle and labial gland; mammary gland; other cells?	[see Trump]

Little cytoplasm, evanescent

Massive cytoplasm, long life

Apoptosis   Vacuolar

Development   Turnover   Pathology

Programming

1981; Duke et al., 1983; Cohen and Duke, 1984; Cohen et al., 1985; Duvall and Wyllie, 1986).

Type II (Lysosomal) Cell Death

Type II or lysosomal cell death is recognized by primary formation of lysosomes and late condensation and fragmentation of nucleus and cytoplasm, as exemplified by the rat mammary gland and *Manduca* labial gland and intersegmental muscle (Lockshin and Beaulaton, 1974a,b, 1979; Beaulaton and Lockshin, 1977; Halaby et al., 1994; Zakeri et al., 1995a,b). In this case, like type I apoptosis, the membrane stays intact. The first morphological signs of this type of death may be expansion of the lysosomal system, while the nucleus appears unperturbed (Zakeri et al., 1993; Halaby et al., 1994). Autophagic vacuoles may selectively remove specific cell organelles, and, in metamorphosing insect tissues or postweaning mammary gland, eventually may consume the bulk of the cytoplasm (Helminen et al., 1968; Helminen and Ericsson, 1970, 1971; Tenniswood et al., 1992). Consequent to the loss of cell mass, the cytoplasm eventually fragments, for instance, in the case of insect muscle, by expansion of the T system (Beaulaton and Lockshin, 1977, 1978, 1982) or in involuting tadpole tail, by fragmentation of the tail muscle into "sarcolytes" (Weber, 1969). Eventually, the nucleus condenses in an apoptotic manner (Zakeri et al., 1993), becomes fragmented and positive to TUNEL (Woo et al., 1994; Jochová et al., 1998), and may manifest a nucleosomal ladder (Zakeri et al., 1993), but this does not occur until the bulk of the cytoplasm is eroded. In insect muscle, nuclear changes are minimal until the muscle depolarizes and becomes flaccid and noncontractile.

In contrast to these forms of physiological cell death, necrotic cell death appears to have as its major marker damage to the cell membrane, leading to loss of membrane integrity. The cell basically appears to be out of control. Necrotic cell death has been characterized by disintegration of the cell without an active role by the lysosomal system or neighboring cells into fragments that are not subsequently detectable (Schweichel and Merker, 1973; Clarke, 1990; Zakeri et al., 1995a). In necrosis, after disintegration of the cellular membranes, chromatin degradation occurs via nucleases as well as lysosomal and other proteases (Kyprianou and Isaacs, 1988; Berges and Isaacs, 1993). Such situations are provoked by exposure of the cells to extreme conditions of temperature, pH, or toxic concentrations of a variety of agents. (See Trump and Berezesky, Chapter 2, this volume.) Necrotic cell death is presumed to be unregulated and not under genetic control. However recent reports on *C. elegans* indicate that mutations that affect particular mechanosensory cells required for touch sensitivity result in swelling and death of neurons. This death is distinctly different from apoptotic cell death. As there is disruption of membrane integrity along with swelling, it is thought that these neurons may be undergoing a necrotic death (Lints and Driscoll, 1996). These results suggest that at least one necrotic cell death may be under genetic control. As is described by Trump and Berezesky (Chapter 2, this volume), this fate may derive from lack of energy available to these cells as they die, since it is likely that these cells ultimately fail from exhaustion engendered by leaky pores.

CELL DEATH IN DEVELOPMENT

Cell death is a major part of development in almost every organ in both vertebrates and invertebrates. The evolutionary conservation of this process suggests that it is as important a part of development as cell differentiation and cell disivion. Cell death in the developing embryo shapes specific organs and perhaps thereby determines their function, and may account for more than half of neurons that are born. Cell death in development is apparently always programmed, in that a series of events takes place once the signal for death is unleashed. An important part of this developmental program is that cell death occurs for the most part during the same window of development in every embryo. Disruption of the timing or the pattern of cell death results in abnormalities. The occurrence and the specificity of cell death in the developing embryo suggests that tight regulatory pathways are involved.

Cell Death in Embryogenesis

Cell death in mammalian embryogenesis occurs as early as inner cell mass differentiation (El-Shershaby and Hinchliffe, 1974; Copp, 1978; Pierce et al., 1989; Hardy et al., 1989). As development progresses, it is found in a variety of tissues such as in the differentiation of the gut mucosa (Harmon, 1984; Pipan, 1979, 1986; Williams and Bell, 1991), the retina (Young, 1984; Penfold, 1986), the immune system (Duvall and Wyllie, 1986; Williams, 1994) and in secondary palate formation (Shapiro and Sweeny, 1969; Chaudhry and Shah, 1973; Pratt et al., 1975; Kerr, 1987; Mori et al., 1994). In the nervous system cell death is involved in the formation of the architectural organization by establishing a definitive pattern of neuronal connections and major axonal pathways (Hamburger and Oppenheim, 1982; Hurle, 1988; Oppenheim 1991; Lo et al., 1995). Cell death is involved in the development of plants, invertebrates, and vertebrates (Mittler, Chapter 5, this volume). In *C. elegans* development cell death is found in neuronal, muscle, epithelial, intestinal and gonadal cells (Gartner and Hengartner, Chapter 4, this volume). During metamorphosis of insects, muscle, gland, and neuronal cells, as well as many other tissues not studied in detail, die (Lockshin, 1981; Schwartz and Truman, 1982; Truman, 1984; Fahrbach et al., 1994; Haas et al., 1995). Cell death in the remodeling of tissues and organs during development in vertebrates is best exemplified by removal of interdigital webs during limb development (Hinchliffe and Johnson, 1980; Hinchliffe, 1981; Hinchliffe and Gumpel-Pinot, 1981; Kerr, 1987; Zakeri et al., 1994; Zakeri and Ahuja, 1994; Coucouvanis et al., 1995). Since the cell death in mouse limbs and the labial gland of the insects provides similarities and differences, we will emphasize the two developmental systems for our discussion.

Cell Death in Developing Limbs

Cell death determines the correct patterning of the limb and is a prominent aspect of the developing limb in all amniotes. The amount of death in the interdigital space of different organisms determines the shape of the hand palette, differentiating a duck limb from that of chick and that of mice and humans. During limb formation, cell death is first observed in the anterior and posterior marginal zones (AMZ and PMZ) of the developing limb bud. As development progresses, there is massive cell death in almost all the interdigital mesenchymal tissue (IMT) located between the chondrifying digits, leading to the formation of free and independent digits (Saunders and Fallon, 1967; Ballard and Holt, 1968; Hinchliffe and Thorogood, 1974; Scott et al., 1977; Milaire and Roze, 1983; Naruse and Kameyama, 1986; Hurle et al., 1996). The domains of dying cells versus the chondrifying cells are first recognized by the separation of cells and the loss of cell–cell contact, leading to the area of interdigital regions in which the cells die. However, in species with webbing between adult digits (ducks), little or no death occurs in the webbed regions of the interdigital areas (Saunders and Fallon, 1967; Fallon and Cameron, 1977; Hurle and Colvee, 1982). The ease of access, the interaction of the live and dying cells, the amount of tissue and the presence of mutant animals with limb deformities that can be linked to cell death render this model valuable for the study of cell death.

The existence of mutant animals with defects in cell death and the reports of the need for RNA and protein synthesis during cell death suggested that the regulation of cell death was genetic. Further evidence that normal cell death in animals is caused by the activation of a suicide program comes from studies of mutants in nematodes (Ellis and Horvitz, 1986), chicks (Hinchliffe et al., 1981; Hurle et al., 1996), and mice (Kochhar, 1977; Zakeri and Ahuja, 1994).

Several limb mutations in birds have been associated with alteration in the pattern of cell death leading to altered limb phenotype. In *ws* (sex-linked wingless mutation), variable expression of the gene results in variable mutant phenotypes, including embryos with no forelimbs and normal hindlimbs to embryos without limbs at all (Hinchliffe and Ede, 1973; Hurle et al., 1996). In contrast, the talpid^3 (*ta^3*) chick mutant is autosomal recessive and displays suppression of cell death in the anterior necrotic zone, the posterior necrotic zone, the opaque patch of the central limb mesenchyme, and the interdigital zone (all of which are zones of cell death in the normal chick limb (Hinchliffe and Ede, 1967; Hinchliffe and Thorogood, 1974). This lethal mutation displays an extended AER, shortened limbs, polydactyly, and failure of bone formation (Kochhar, 1977; Hurle et al., 1996).

There are at least 70 established genetic loci affecting mouse limb development. The phenotypic abnormalities in the mice reflect the variety seen in human limb disorders, making the mouse a good model to study limb morphogenesis in humans. One group of interesting established mouse limb mutants is referred to as the luxoid group. Among these mouse mutations, Dominant hemimelia (Dh) and Hemimelic extra toes (Hx) produce an altered pattern of cell death (Rooze, 1977; Knudsen and Kochhar, 1981). Dh mutant limbs are devoid of cell death, and many will develop a preaxial polydactyly (Rooze, 1977). In the Hx embryos, extended PCD occurs in the opaque patch, resulting in resorption of the tibial precartilage. Hx results in polydactyly of all four feet of heterozygotes (Knudsen and Kochhar, 1981). Another mutation affecting the patterning of cell death in mouse limbs is seen in the Hammertoe mouse, a semidominant mutation that, in homozygous state, shows webbing between digits 2, 3, 4, and 5. The webbing does not reach the distal end of the toes in the heterozygotes. This condition is detectable in homozygote embryos on day 14.5 and in heterozygotes on day 15.5 post coitus (Green, 1981). This phenotype is similar to the human disorder Apert's syndrome, where patients have spoonlike hands due to extensive webbing (Zucker et al., 1991). In these mice there appears to be a direct correlation between the lack of cell death and the defective phenotype (Zakeri et al., 1994).

In addition to the mutations that result in abnormal patterns of cell death, a variety of chemicals can induce cell death in embryos. Several teratogens, such as retinoic acid (RA), affect developing limbs if administered during a specific stage of embryonic development (Kochhar 1975; Sadler and Kochhar, 1976; Sadler et al., 1976; Zakeri and Ahuja, 1994; Ahuja et al., 1997). RA has multiple effects. Although it is necessary for a variety of physiological functions, treatment of pregnant animals with RA results in a variety of facial, neural tube, cardiovascular, and limb abnormalities depending on the stage of treatment (Kochhar et al., 1984; Lammer et al., 1985; Alles and Sulik, 1989; Granstrom et al., 1990; Yasuda et al., 1990; Rizzo et al., 1991; De Luca et al., 1995). The effect of RA is also seen in a variety of species including rats, hamsters, rabbits, and humans (Pinnock and Alderman, 1992; Armstong et al., 1994). There appears to be a temporal and dose-dependent specificity to the effect of RA, ranging from undetectable to embryolethality (Shenefelt, 1972; Armstrong et al., 1994; Ahuja et al., 1997). The action of RA is thought to be regulated by a variety of proteins that bind and transport RA to specific compartments of the cells (Favennac and Cals, 1988).

RA increases the amount of cell death in the interdigital necrotic zone and the mesenchymal core of the embryonic limb (see Schweichel, 1971; Kochhar, 1977; Sulik and Alles, 1991; Zakeri and Ahuja, 1994; Ahuja et al., 1997). These areas represent the areas of normal programmed cell death. While the expansion of death in these zones may lead to the pathogenesis and subsequent malformation, the fact that the pattern of cell death is specific to these preprogrammed areas suggests that RA plays a role in the normal patterning of cell death during development. As will be described, the use of RA tells us more about the relationship of pattern formation to cell death than the role of RA in determining cell death.

Cell Death in Metamorphosing Insect Tissue

Cell death is common in insect embryos and appears to follow an apoptosislike morphology. Overall regulating genes, such as *reaper* and *hid,* are known to control many embryonic deaths (Restifo and White, 1992; Abrams et al., 1993; Steller et al., 1994; White et al., 1994; Grether et al., 1995; White and Steller, 1995). Cell death is also the major means of eliminating larval tissues at metamorphosis. The amount of cell death varies among orders, ranging from relatively little in ametabolic and hemimetabolic insects to essentially all the larval structures in the higher Diptera and Hymenoptera. Among the relatively well-studied Lepidoptera, typical larval organs such as the intersegmental muscles, silk or labial glands, and dermal glands undergo a classic programmed cell death. Although cell death in insect embryos by superficial inspection appears to be apoptotic in form, in most instances of cell death in metamorphosis the morphology is more typically that common to cells with massive cytoplasm—vacuolar or lysosomal, rather than classical apoptosis. These cells of course are very different from lymphocytes or thymocytes: They typically are long since postmitotic, and they are large cells with massive cytoplasm. The death of the cells is presumed to be under endocrine control, since the metamorphosis is controlled by a drop in juvenile hormone while the molting hormone ecdysone rises. However, there are many variants. For instance, in overwintering moths, intersegmental muscles are preserved to aid in the eclosion of the adult. The muscles must therefore survive beyond the period of loss of juvenile hormone and are required for function at a precise moment approximately three weeks after the primary hormonal events. The timing of cell death is therefore regulated by interpolated steps of a drop in molting hormone and ultimately a neural or neurosecretory signal (Lockshin and Williams, 1965a,b; Schwartz and Truman, 1984; Truman, 1992, 1996). Likewise, the death of the motor neurons supporting these muscles as well as those supporting various manifestations of larval behavior is controlled rather precisely by falling ecdysone titers (Schwartz and Truman, 1984; Schwartz, 1992).

IDENTIFICATION OF CELL DEATH

In order to study the mechanisms, one must first be able to identify the dying cell and characterize the type of cell death. Although there are clear definitions of the types of cell death, there is considerable overlap, and a cell may display features common to all three morphologies or a morphology not yet defined. However, having knowledge of some sequential events characteristic of different cell death types, it is possible to develop specific parameters to categorize the dying cells. This categorization will aid in the understanding of the mechanisms involved in regulation of cell death. Markers for cell death may not always be universal and may depend more on the nature of the dying cell; that is, a neuron may die differently from a liver cell or a germ cell. To develop specific markers for identification of cell death, we and others have used the classical definitions as a guide. We used as models cell death in the interdigital regions of the developing mouse embryo as well as the degeneration of the labial gland of *Manduca sexta*. These two models will be used to demonstrate type I and type II PCD. The study of necrotic cell death has been discussed by Trump and Berezesky, Chapter 2, this volume.

One must first detect the dead or dying cell (see Trump and Berezesky, Chapter 2, Tidball and Albrecht, Chapter 15, Cotman et al., Chapter 14, this volume). Given the diversity of the types of cells in a given tissue, any specific marker may or may not be shared. We have therefore developed several markers to use as parameters for cell death. These parameters and markers can be used to characterize cell death in a variety of tissues and organisms and are not unique to these developing systems. It is also possible that one or more of these markers prove not to be useful in a particular system. However, the approach of using a palette of markers remains valid.

Morphological Analysis

Morphological analysis of cell death can be at the level of light or electron microscopy. For light microscopy one can use a variety of stains on either live whole animal or tissue slices. We have used Nile blue sulfate (NBS) to find dead cells in the developing limbs. NBS stains the acidic compartments of the cells such as secondary lysosomes (Schweichel and Merker, 1973; Zakeri and Ahuja, 1994; Zakeri et al., 1995a). The dead cells appear as dark blue dots among the live, transparent or light blue, cells. The staining however depends on a live organism, is time dependent, and increases with time of incubation as more cells die due to the inadequacy of the media. The major advantage of this detection method is that one can look at cell death in a three-dimensional preparation (Fig. 3.2, panel 1). Hematoxylin eosin (H&E) staining of tissue sections can reveal dead cells as darkly stained cell fragments (Fig. 3.2, panel 2), though, unless cell death is relatively massive, apoptotic cells may be uncommon and not easily detected (cf. Tidball and Albrecht, Chapter 15, this volume), and experienced judgment is necessary to avoid misinterpretation. For instance, in our experience we have seen prophase nuclei misread as apoptotic and vice versa.

One of the best methods of detection and characterization of cell death is by electron microscopy. In mouse embryos early as day 10.5 one can detect the dead cells in the developing limbs. By day 14.5 of the limb development there is a great deal of death in the interdigital region of the limb. Like thymocytes, dying cells typically round and condense. The cytoplasm shows little obvious distortion, other than the increasing electron density and the tendency to fragment and bud, but in the nucleus the chromatin coalesces and marginates. The conversion is apparently fairly rapid, as intermediate stages are usually not seen. The prominent nuclear condensation and the changes in the cytoplasm are indicative of type I apoptosis. The apoptotic cell fragments are often inside the larger phagocytic cells (Panel).

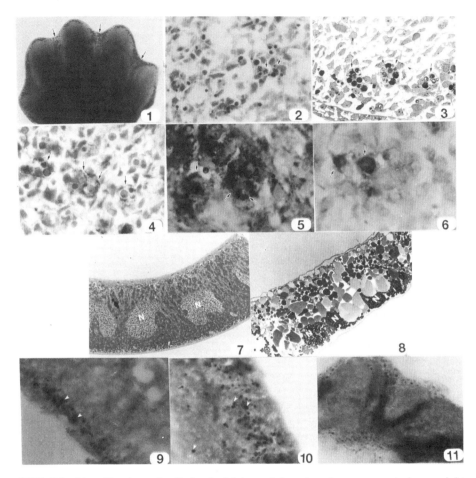

FIG. 3.2. Identification of cell death. Male and female mice were mated overnight. Plugs were checked and the time of plug observation was designated as day 0.5. At day 14.5 of gestation the mothers were sacrificed and the embryos were removed for analysis. Serial sections when possible were used for the different analyses. All the methods used are described in Zakeri et al., 1994. Panel 1: Nile blue sulfate (NBS) staining. The arrows indicate areas where NBS reveals cell death. Panel 2: Hematoxylin-eosin staining; arrows indicate apoptotic cells. Panel 3: Semithin section of interdigital region. Arrows indicate phagocytic cells with engulfed apoptotic cell fragments. Panel 4: 400× magnification of a section of interdigital region stained for the detection of phagocytic cells by the antibody F4/80. The dark brown color indicates peroxidase labeling of the site of the phagocytic cell. Panel 5: 1000× section stained for acid phosphatase. The red focal precipitate indicates the lysosomal enzyme, which is localized to the cytoplasm surrounding the nucleus. Panel 6: Section serial to that in panels 4 and 5, used for detection by *in situ* end labeling of fragmented DNA in the cells (1000×). The nonisotopic ApopTag Peroxidase Kit, Oncor, Gaithersburg, MD was used, and the brown staining indicates fragmented DNA. Panel 7: Cross section of labial gland of *Manduca sexta* just prior to metamorphosis (Day −1). Lumen is up.

In insect tissues as well as postlactational mammary gland and other tissues displaying a vacuolar cell death, changes in cytoplasm well precede obvious nuclear changes. Three examples can be cited: The intersegmental muscles of silkmoths and hawkmoths die over two days, commencing with the eclosion of the adult. The rise in number of lysosomes seen in the muscle becomes apparent after 5 hours; muscle mass, which has begun to decrease slowly as early as 2 days before ecdysis, begins to drop sharply along a curve that extrapolates back to 0 hours. The erosion of the myofilaments is extralysosomal, but by 7 hours autophagic vacuoles are common, consuming successively mitochondria and glycogen granules. By 12 hours, at which time approximately 1/3 of the myofilaments have eroded and the T system is beginning to swell, the muscle depolarizes. Shortly thereafter the myofilaments rapidly disappear, the muscle begins to fragment like an apoptotic cell, and the nucleus condenses.

The labial gland of the hawkmoth is similar (Fig. 3.2, panels 7 and 8). The paired gland (a homologue of the silk gland of *Bombyx mori* and *Drosophila* salivary gland) of *Manduca* is a secretory gland approximately 0.2 mm in diameter, consisting of a single layer of giant (1×10^4 μm^3) polyploid cells. It measures up to 17 cm in length, sinuously extending the length of the larva. The gland secretes a viscous, clear, glycoprotein-containing fluid used presumably to solidify the wall of the pupa's burrow. In the 7-day period during which the larva pupates, the endoplasmic reticulum begins to show unusual alignments on the first day and swirls by the second. Lysosomes, which were confined to the basal border, move centripetally toward the lumen as early as the first day. By the third day (panel 8), the ER is clearly eroding; the cell becomes heavily vacuolated by the fourth day, at which time the previously equivocal changes in the nucleus become marked and the chromatin condenses. By the fourth day, virtually all the cytoplasm consists of lysosomal vacuoles and the nucleus is severely condensed and apoptotic, and by the fifth day most of the cytoplasm has been eliminated (panels 7 and 8) (Halaby et al., 1994; Lockshin and Zakeri, 1994; Zakeri et al., 1995b).

Large, polyploid nuclei (N) are centrally located in the gland. Panel 8: With metamorphosis starting on Day 0, the cytoplasm is completely vacuolated by Day 3 (illustrated here). It is on this day that the nuclei (N) finally clearly condense and, by electrophoresis and end labeling, one begins to detect ladders. Total DNA in the gland also begins to fall on this day. Panels 9–11: The red color and arrows indicate lysosomes in cross sections of the labial gland before metamorphosis begins (Panel 9); on Day 0 (Panel 10), and on Day 4 (Panel 11). Whereas the lysosomes are confined to the basal or adlumenal surface in nonmetamorphosing glands, by Day 0 they begin to spread throughout the cell (Panel 10; here lumen is to the left). Ultimately, the bulk of the cytoplasm is filled with lysosomes. (In panel 11, the nuclei, which still remain, are stained blue by the counterstain.)

The homologous salivary gland of *Drosophila* gives further indication of what is happening. In *Drosophila,* the gland disintegrates during the first 15 hours of metamorphosis. Within the first hour, microfilaments appear less organized than before, and the heretofore centrally located nuclei are pushed to the basal surface. Over the next 8 hours these changes become more progressive, and microtubules appear, by light microscopy, to be thinner but more dense. At the eighth hour, the microfilaments have collapsed to the surface of the nucleus, bringing with them many lysosomes, and vacuoles begin to appear. The vacuolation expands through the tenth hour. At the twelfth hour, microtubules disintegrate rapidly and entirely, the microfilament actin is completely enclosed in autophagic vacuoles, and the cell begins to fragment and to display apoptotic nuclei (Jochová et al., 1997).

In this type of physiological death there is ultimately a rapid apoptotic collapse, in which the chromatin coalesces and marginates, and the cytoplasm condenses and fragments. The pattern in all these synchronous cell deaths is consistent and revealing in that many cytoplasmic and biochemical changes are detected well before the final, rapid, apoptotic collapse. Phagocytosis, or at least cytokine-driven invasion of phagocytes, is uncommon in insects, and the fragments are released into the hemolymph for consumption elsewhere. Except for the role of the phagocytes, this vacuolar type of cell death is similar to that seen in involuting mammary gland (Helminen et al., 1968; Helminen and Ericsson, 1970, 1971; Bursch et al., 1990, 1996; Tenniswood et al., 1992; Guenette et al., 1994; Zakeri et al., 1995a; Lund et al., 1996). Insofar as methodology is concerned, the prominent changes in the position and number of lysosomes are far earlier and more visible markings of the impending death of the cells than are the markers of nuclear change. Potentially at least, the alteration in the cytoskeleton, which may drive the changes in the lysosomes, is another early marker.

Engulfment of Dead Cells

A major element of controlled cell death in vertebrates is the fact that the fragmented cells get cleaned up by the surrounding phagocytic cells. These phagocytic cells can be the neighboring cells or macrophages or macrophage-like cells that come to the area. One can therefore look for phagocytic cells to detect dead cells. To do so one needs specific markers that will recognize the phagocytic cells. These phagocytic cells are presumptively macrophages and therefore would express macrophagelike surface protein. F4/80 antibody recognizes mature mammalian macrophages as well as monocytes. F4/80 by immunohistochemistry recognizes cells that are engulfing the apoptotic cells as found in the interdigital regions of day 14.5 hand plate of the developing mouse embryo (Fig. 3.2, panel 4). These phagocytic cells can be true macrophages recruited to the area or the neighboring mesenchymal cells differentiated into macrophagelike cells and expressing macrophagelike surface antibody. Thus the F4/80 antibody is another reasonable marker for cell death.

In contrast to the dying cells in the limb, the labial gland is not surrounded by other cells, and there are no phagocytic cells that enter the area. Soluble degradation products or cell fragments are released into the hemolymph. The cell fragments are ultimately scavenged by professional phagocytes—some of the prominent changes in coloration that reveal the onset of metamorphosis are in fact phagocytic cells sitting alongside the dorsal heart that are now engorged with cell fragments—but, since they act more as distant sieves or filters than local cleanup squads, in insects they are of little use in evaluating the death of a specific organ. The intersegmental muscle similarly is normally not touched by phagocytes until the very final stages. In mammary gland, prostate, and probably other hormone-dependent tissues in which cell loss can be relatively massive, phagocytes do enter the region to scavenge cell debris. Where cell loss is a more isolated affair, neighboring cells often phagocytose the dying cell. In other instances, such as gut epithelium and epidermis, cells are shed, while in the lens of the eye, the enucleated cells persist (Papermaster, Chapter 12, this volume).

Destruction of the Dead Cell

The destruction of the apoptotic bodies occurs within the phagocytic cells by means of lysosomal enzymes in the phagocyte. Activity of lysosomal enzymes such as acid phosphatase can therefore serve as markers for cell death, as was the common procedure in the 1960s. On sections serial to the ones used for evaluation of macrophages, one can look for the localization of acid phosphatase activity. Although there is a low level of acid phosphatase in all cells of 14.5 day mouse hand plate, this level increases several fold in the area of cell death in correlation with the appearance of phagocytic cells (Fig. 2.5). The lysosomes may be in the phagocytic cells or in the dying cells as they fragment. To distinguish between the two possibilities is difficult when phagocytic cells are present.

In insect tissues and other cells with massive cytoplasm there is less reliance on phagocytes and more on self-destruction of the cytoplasm by autophagy. Typically organelles are surrounded by isolating membranes deriving from the endoplasmic reticulum, these vacuoles are fused with enzyme-containing primary lysosomes, and the resulting autophagic vacuole digests the enclosed organelle. In many situations the autophagic vacuole contains mixed organelles, but in insect metamorphosis glycogen particles or mitochondria may be swept from the cell in synchronous waves of destruction (Beaulaton and Lockshin, 1977). Since lysosomal activity may rise several fold, as measured by acid phosphatase, proteolytic enzymes, or other acid hydrolases (Lockshin and Williams, 1965c; Lockshin, 1969a), the dying tissue is easily recognizable by histochemical staining or direct chemical assay (Lockshin and Beaulaton, 1974a,b; Aidells et al., 1971). In the labial gland, lysosomes both increase in number and relocate. Prior to the beginning of cell death, they are confined to the basal surface of the gland cells (Fig. 3.2, panel 9). The first histochemical sign of metamorphosis, on the first day, is the migration of the lysosomes toward the lumenal or apical regions of the cell. By the third day, the vacuoles stain heavily for lysosomal enzymes, and the bulk of the cytoplasm is seen to be filled with large autophagic vacuoles (Lockshin and Beaulaton, 1974a,b; Halaby et al., 1994). Similarly, in *Drosophila* salivary glands, the vacuolation seen starting at approximately 8 hours is primarily lysosomal; the actin microfilaments are ultimately taken into lysosomes, and the autophagic vacuoles collapse to the surface of the nucleus (Jochová et al., 1997).

The activation of the lysosomes can be also used to examine cellular destruction in cultured cells undergoing cell death. As in the insect gland this can be done both at histochemical and biochemical levels. This particular study may prove very useful since in culture dying cells may release from the surface of the dish or may lyse prior to completing an apoptotic collapse (Chapter 2, this volume).

The activity of the lysosomes has proved to be a very useful tool. Ultimately dead cells are in one manner or another removed by lysosomal activity, whether in phagocytes or by autophagy. The assays are relatively cheap, simple, and universally applicable and are thus useful for screening for regions of cell death. The pattern is easily discernible from that of necrosis, in which large regions of disorganized material are surrounded at the periphery by phagocytes, and lysosomes in the dead cells are typically ruptured. Almost all cells contain lysosomes. The technique is most useful as a screening mechanism, since either the phagocytosing cells (Type I) or the dying cells (Type II) typically contain 2–10 times the number or lysosomes or lysosomal activity of the healthy cells in the vicinity. The screen may also detect cells that are dying but not yet dead (Trump and Berezesky, Chapter 2, Cotman et al., Chapter 14, this volume).

A pitfall for this assay comes when the cells examined natively contain substantial numbers of lysosomes, such as liver or kidney. At the electron microscopic level autophagic or heterophagic vacuoles are relatively characteristic and easily identified. Under conditions of (cellular) starvation, however, such as insulin-dependent diabetes, cells frequently generate autophagic vacuoles, a source of confounding as well as of a provocative question: To date, there is no clear identification of the difference between cell death and atrophy of a damaged or starving cell.

DNA Fragmentation of the Apoptotic Cell

One of the hallmarks of apoptotic cell death has been the fragmentation of DNA at the internucleosomal regions to produce distinct fragments. In most cases the fragmented DNA is shown as separate distinct bands on an agarose gel. This is less complex to detect and use as a marker when a large population of cells dies within a narrow window of time. This scenario is seen in cultured cells or cells marked and separated by a cell sorter. To further amplify the signal detected, radioactive tagging of the isolated DNA can be used to label the free ends of the isolated DNA prior to separation by gel electrophoresis (Zakeri et al., 1993). Even so, in situations in which the numbers of dying cells are few or they are interspersed with live cells, it is difficult to have enough isolated DNA to be able to detect fragmented DNA on a gel. (Also Chapter 15, Tidball and Albrecht, this volume.) For these situations it is more useful to examine DNA fragmentation *in situ* (TUNEL). There are a variety of kits now available for this purpose, most notably from Oncor and Boehringer-Mannheim. Briefly the idea is to add a labeled tag to any free DNA ends by using terminal transferase. The labeled tag is then detected either by fluorescent or nonfluorescent secondary reaction (Zakeri et al., 1994). Using terminal transferase and digoxigenin-11-dUTP, one can identify DNA fragmentation in the dying cells of the limb (Fig. 3.2, panel 6).

The cells of the labial gland are large, and one can obtain the tissue without contamination with other cell types. In this case the DNA can be isolated and analyzed by conventional agarose gel electrophoresis. DNA fragmentation analysis of this tissue showed that the fragmentation is relatively late but is nevertheless detectable as the cell dies. Total DNA does not begin to decline until the fourth day. At this time the nuclei are positive to TUNEL staining and, by using end labeling combined with electrophoresis, one can detect laddered DNA. By the use of the *in situ* TUNEL we also detected labeling at the beginning of the last larval instar, well before metamorphosis. This positive signal correlated with the extensive DNA synthesis and the generation of the free ends of Okazaki fragments, rather than DNA fragmentation at the internucleosomal regions (Jochová et al., 1998). These results suggest caution for the use of these methods in situations in which there might be extensive DNA synthesis, such as in polyploid cells or in circumstances of extensive DNA repair. This latter is of particular concern, since DNA-damaging techniques are often used to induce apoptosis. It is also to be noted that the true initiation of apoptosis may be DNA fragmentation at supranucleosomal level (Sikorska and Walker, Chapter 7, this volume), and that, in Alzheimer brain, far more cells may be TUNEL positive than actually die (Cotman et al., Chapter 14, this volume).

An intriguing means of examining the fragmentation of the DNA has been by the use of pulse-field electrophoresis. (See Chapter 7, Sikorska and Walker, this volume). Pulse-field electrophoresis reveals 50–300 kb DNA at an earlier stage of metamorphosis of the gland, suggesting that DNA damage has begun earlier. As Cotman et al. indicate (Chapter 14, this volume), DNA damage detectable by TUNEL may potentially be reversible. Presumably, by the time that the ladder appears, death is no longer reversible. Finally, even where cell death is relatively high, it may be difficult to identify a ladder. First, extraction techniques, particularly in the face of substantial connective tissue, may be limiting, since the ladder is best seen in situations in which total DNA is extracted and only the low-molecular-weight DNA electrophoresed. Second, the appearance of a ladder is the result of a happy concatenation of kinetics, since the accumulation of ladder-size DNA will occur only if degradation beyond nucleosomal size is slower than degradation to nucleosomal size (Sikorska and Walker, Chapter 7, this volume).

Metabolic Changes in Dying Cells

Particularly in cells dying by type II cell death, predeath changes may be readily detectable. In the labial gland, synthesis of most proteins plummets on the first day, but the gland continues to synthesize and secrete a silklike protein for one more day. The sudden cessation of synthesis of most housekeeping proteins is an early biochemical marker. Mitochondrial function is maintained until the fourth day, though the amount of lysosomal enzymes has begun to rise by the first day, the lysosomes have migrated from a basal position toward the lumen, and autophagic vacuoles have begun to form. With the collapse of the mitochondria on the fourth day, large vacuoles form in the cell, the nucleus begins to collapse in an apoptotic manner, and a DNA ladder is detected for the first time. In insect muscle as well, a drop in synthesis of housekeeping and structural proteins (Lockshin, 1969b; Wadewitz and Lockshin, 1988; Lockshin and Wadewitz, 1990), and a rise of proteolytic enzymes such as cathepsins B and D (Lockshin and Williams 1965c; Lockshin, 1969a, 1975), as well as poly-ubiquitin (Schwartz, 1992; Schwartz et al., 1993; Haas et al., 1995), well precede depolarization and apoptotic collapse (Lockshin and Beaulaton, 1979). Such metabolic markers may be useful in other systems in which the behavior of the cells can be characterized.

ADDRESSING THE MECHANISMS OF CELL DEATH

Comparison of Normal and Abnormal Cell Death in the Limb

Using these markers, we examined cell death in the limb. We first examined the pattern and the nature of cell death in the normal mouse limb. We mapped the areas of cell death and established the cell death to be apoptotic by EM and light microscopy, the latter showing condensing nuclei and the prominent apoptotic features. The characteristic of apoptosis was further confirmed by the detection of the phagocytic cells using immunohistochemistry and the F4/ 80 antibody. In addition, the phagocytic cells had active lysosomes. By the use of *in situ* end labeling, fragmented DNA was detected in the dying cells.

Once the nature of cell death was determined in the normal mouse limb, this information was used to investigate the abnormal pattern of cell death by the use of mutant animals and cell death induced by RA. This approach was used to address the importance of cell death in the correctly patterned developing limb. We determined that the Hammertoe mutant included in its phenotype the failure of normal cell death. The peak of cell death in the wild-type mouse embryo is day 14.5. In homozygous Hammertoe mutant mice, the defect is detectable starting at day 14.5 of gestation. Using the markers presented to characterize cell death, we found that there was a lack of cell death between digits 2 through 5 in the Hm homozygotes. However, dead cells were seen in the anterior and posterior necrotic zones and between digits 1 and 2. This altered pattern of cell death correlated with the abnormal phenotype, that is, webbing between digits 2 and 5 with separation between digits 1 and 2 (Ahuja et al., 1997). These results suggest a genetic regulation of cell death in the limb that is very tightly regulated at a local level, since in this mutant the defect is not in a gene that controls total cell death. On the contrary this defect is local and does not even affect the entire limb, but very specific parts of the developing limb.

Having established this, we asked what regulatory factors might be involved. Several reports had implicated a role for RA in the establishment of correct patterning and cell death in the limb as well as other parts of the embryo. By treating the embryos with RA at different times of development, we established windows of development in which specific doses of RA given to the pregnant mother would increase cell death in sites of naturally occurring cell death. This effect is under tight regulation in that there is a specific temporal sensitivity of the cells to RA. For example, at day 14, cell death is the most dramatic feature involved in shaping the limb. Administration of RA at day 11.5 results in limb malformation, whereas RA administered at day 14 of gestation increases the local number of cell deaths but has little permanent phenotypic effect. Basically by treating the embryos at day 14 of gestation with RA and examining the effect, one can increase the number of dying cells in the interdigital regions as well as in the AMZ and PMZ, all representing regions that normally exhibit cell death. This selective cell killing implies that the sensitivity of the cells in these regions and at specific times that is affected by RA. Hence the RA here acts as the cell killing signal apparently by carrying cells that are already sensitive or primed to die, over the threshold of death. Although the manner of cell killing is apoptotic, whether RA is the cell killing signal in the limb during normal development is still unclear. Furthermore, although it has been suggested that in the breast cancer cell line the cell killing effect of RA can be associated with its receptor RARβ (Liu et al., 1996), this does not appear to be the case in the developing limb (Ahuja et al., 1997).

The fact that RA can increase zones of cell death leads to the question of whether it can induce cell death in situations in which there is a lack of cell death, such as that seen in the Hm mutant mice. The homozygous abnormal limb treated with vehicle continues to exhibit interdigital webbing and a lack of cell death. The RA-treated homozygous abnormal limbs depicted, similar to the +/+ limb, a selective increased cell death in the AMZ, PMZ and between digits 1 and 2 of the Hm/Hm foot. More interestingly, in these embryos, RA induced cell death between digits 2 and 5. In addition, the induction of cell death in the interdigital regions that originally displayed a suppression of cell death resulted in the rescue of the abnormal phenotype. The rescue was more prominent and closer to normal when heterozygous animals were treated with RA (Singh Ahuja et al., 1997). The selective induction of death in mesodermal cells in regions in which the mutant lacks cell death suggests that these cells possess the death machinery and require or are sensitive to a signal such as RA. RA appears to act on its own or in concert with other factors to trigger the sequence of events that lead to cellular demise. It also suggests that the effect of RA is primarily to carry the affected cells, which are for unknown reasons subthreshold in the mutant, to the threshold of programmed cell death. (For another example of rescue of a mutant phenotype by manipulation of cell death, see Papermaster, Chapter 12, this volume.)

ANALYSIS OF CELL DEATH RELATED GENES

The genetic analysis of cell death, most extensive in *C. elegans* (Gartner and Hengartner, Chapter 14, this volume) has led to the identification of several mammalian counterparts. One approach in identifying genes regulating cell death has been to look for differentially expressed genes using a variety of techniques such as examining the expression of known genes during cell death, or looking for differentially expressed cDNA (Cryns and Yuan, Chapter 6, this volume). A number of genes show specific patterns of expression during cell death. Genes that are upregulated during cell death include heat-shock protein 70 (Buttyan et al., 1988), α prothymosine (Berges and Isaacs, 1993), testosterone repressed message-2 (Montpetit et al., 1986), cathepsin B (Narvaez et al., 1996), transforming growth factor-β_1 (Kyprianou and Isaacs, 1989; Selvakumaran et al., 1994), tissue transglutaminase (Piacentini et al., 1991), ubiquitin (Schwartz et al., 1990), and calmodulin (Dowd et al., 1991), all of which have been implicated as cell death markers.

Although these genes may exhibit a specific expression in one instance of cell death, they can be unaffected in another situation. For example, TRPM-2 is upregulated in involuting prostate (Buttyan et al., 1988) and in atretic follicles (Ahuja et al., 1994; see also Tilly, Chapter 16, this volume) but is not found in dying cells of the limb. On the other hand, tissue transglutaminase is found in dying cells of the limb and shows activation with activation of cell death by RA and down regulation with the lack of cell death in the mutant Hm mouse limbs (Ahuja and Zakeri, unpublished observations). Although it is found in the dying cells, it is also found at high levels in blood cells and lower levels in other cells (Ahuja and Zakeri, in preparation).

The gene bcl-2, on the other hand, is often needed for cell survival. The pattern of expression of this gene in the different limb model systems studied does not appear to correlate with cell survival. In fact, we find its highest level of expression in the differentiating cartilage. The fact the bcl-2 knock-out mice give rise to a quasinormal embryo indicated redundancy of function in this gene and has resulted in the discovery of other family members (Birge et al., Chapter 13, Osborne, Chapter 8, Benitez, Chapter 17, Tilly, Chapter 16, this volume). However, the expression of different members of this gene family does not appear to give much information about the role of these genes or their importance in the cell death of the limb (Yong, Ahuja, and Zakeri, in preparation).

The expression during cell death of several immediate early genes including c-fos, c-jun, and c-myc, as well as the tumor suppressor p53, which are associated with cell division, has opened a new area of investigation in the role of the cell cycle genes during cell death. The set of cyclin-dependent kinases (Cdks), related in structure and function to the yeast cell division control kinase Cdc2, allow eukaryotic cells to progress through DNA replication and cell division (Elledge and Spottswood, 1991; Koff et al., 1991; Meyerson et al., 1992). The Cdks are catalytic subunits that require activation by specific regulatory subunits or cyclins. Cell cycling in mammalian cells involves the employment of a family of Cdc2-related kinases (Cdk 1–6) along with a family of cyclins (A–E) that have been implicated in DNA synthesis (Cdks 2, 4, 6 and cyclins A, C–E) and mitosis (Cdk1 and cyclin B, Girard et al., 1991; Pines and Hunter, 1991; Pagano et al., 1992; Zindy et al., 1992; Matsushime et al., 1992; Sherr, 1993; Hartwell and Kastan, 1994). Cdc2 (Cdk1) and Cdk2 have been shown to be activated in *Tat* (HIV-1 transactivator protein) expressing cells that die (Li et al., 1995). Bcl-2 expression has been shown to suppress the levels of Cdk2 as well as apoptosis (Meikrantz et al., 1994). On the other hand, it has been shown that the inactivation rather than the premature activation of Cdc2 also increases the level of apoptosis (Ongkeko et al., 1995). The Cdk5, a homologue of Cdc2, is expressed in the embryonic nervous system, in cells that are not proliferating but rather are differentiating (Tsai et al., 1993, 1994; Ino et al., 1994). Cdk5 is also expressed in adult tissues, with highest levels detected in the brain and testis and lower levels detected in ovary and kidney (Meyerson et al., 1992; Ino et al., 1994). We have shown high levels of expression of Cdk5 in dying cells of developing embryos as well as in adult tissues such as involuting prostate and atretic follicles (Zhang et al., 1997). Not only is the expression of Cdk5 protein increased in the dying cells; its kinase activity is as well. There is also a direct correlation between the expression of Cdk5 and the appearance and disappearance of cell death seen in RA-induced cell death and Hm mutant mice (Singh Ahuja et al., 1998). Unlike tTG, Cdk5 expression is specific to apoptotic cells. These observations provide new insight into the possible function of this novel Cdk during apoptotic cell death (see also Ameisen, Chapter 1, this volume).

EXTRACELLULAR SIGNALS CONTROLLING CELL DEATH

Many approaches are used to induce cell death *in vitro* or *in vivo,* including administration of hormones such as glucocorticoids to lymphocytes or thymocytes, thyroxin to tadpole tissues, TNF to many types of cells, and withdrawal of NGF or other growth factors from PC-12 or other cells, but, surprisingly, the actual mechanism of action of these hormonal signals is not well documented. In general, one assumes that the interaction proceeds through the normal steroid-binding protein–steroid receptor site in the promoter region of a gene for steroids, and otherwise through cell membrane-to-kinase coupling for growth factors, but documentation is actually rather fragile. (See Ameisen, Chapter 1, this volume, for discussion of potential other effects of putative death genes.)

Although α- and β-ecdysterone, the primary molting hormones in insects and generically described as "ecdysone," are steroids and presumed to act like steroid hormones, their role in cell death is not well determined. The steroid regulation of metamorphosis, and cell death in metamorphosis, is complex. In the presence of a terpenoid hormone, juvenile hormone, larval epithelial cells form new cuticles and undergo molts but do not die. At the molt in which the larva pupates, falling titers of juvenile hormone allow ecdysteroids to control the metamorphosis in two steps: first, a small rise in ecdysone commits the metamorphosis (as opposed to a repetition of a larval stage); a day or so later, the beginning of a larger rise actually causes the metamorphic changes. As a steroid hormone, ecdysone binds to receptor proteins, some of which are upregulated in the presence of the hormone, and induces a set of early genes, the products of which induce later genes. In fact, the concept of early and late genes in steroid response was first developed by Ashburner, who recognized the sequential puffing of *Drosophila* salivary gland chromosomes in response to administration of ecdysone. Unfortunately, the role of these hormones in cell death is far less clear. A major problem is that meaningful cell death has not been induced *in vitro,* suggesting either an inadequacy of culture conditions or a possibility that the death of a tissue is a secondary response to the rise in molting hormone, with other tissues adding or withdrawing factors that lead more proximately to the death of the tissue (Halaby et al., in preparation). (It should be noted that the linkages between glucocorticoid induction of cell death in thymocytes and thyroxin induction of cell death in tadpoles are also less clear than is usually assumed.)

SIGNAL TRANSDUCTION PATHWAYS IN CELL DEATH

Although much is known about the genes that may play a role in cell death, less is known about the transmission of the signals that regulate cell death. Most likely in many instances of cell death, specifically in multicellular tissues, the signal to die might be generated by the crosstalk between the cells and detected at the cell surface. The question is then how these signals are transmitted into and throughout the cell leading to the demise of the cell.

Second messengers are parts of signaling cascades defined by the fact that their generation initiates events that translate into cellular responses. Second messengers such as cAMP, fluxes of divalent ions such as Ca^{2+}, and reactive oxygen species (ROS), shown to be critical for the activation and propagation of cellular responses, have also been implicated in cell death but with some degree of controversy (Obeid et al., 1993; Duke et al., 1994; McConkey and Orrenius, 1994; McConkey et al., 1994; Trump and Berezesky, 1995; Whitfield et al., 1995; Tong et al., 1996). More recently ceramide has been added to this list (Pushkareva et al., 1995; Karasavvas et al., 1996a,b; Pronk et al., 1996; Santana et al., 1996; Verheij et al., 1996). Ceramide is a sphingolipid generated (along with phosphocholine) by the hydrolysis of sphingomyelin by neutral and acidic sphingomyelinases (Okazaki et al., 1989; Okazaki et al., 1994; Wiegmann et al., 1994; Heller and Kronke, 1994; Santana et al., 1996). These enzymes are located in the plasma membrane, cytoplasm, or endosomes. The discovery of the inhibitory activity of sphingosine on protein kinase C (Dbaibo et al., 1995) led to the discovery of the sphingomyelin cycle and the observation that sphingolipids such as ceramide, sphingosine 1-phosphate and lysosphingolipids may act as second messengers. Recent reports document the important regulatory function in signal transduction, cell proliferation (Okazaki et al., 1990), cell death (Obeid et al., 1993; Jarvis et al., 1994), cell differentiation (Fishbein et al., 1993), cell growth arrest (Dbaibo et al., 1995), and HIV replication (Rivas et al., 1994). Extracellular agents such as TNF, NGF, interferon, vitamin D_3, IL-1, dexamethasone, and retinoic acid induce the production of ceramide (Hannun, 1994; Dbaibo et al., 1995).

Synthetic cell permeable ceramide analogs have been used to elucidate the biological and biochemical activities of natural ceramide (Hannun, 1994; Jarvis et al., 1994; Karasavvas et al., 1995a,b). Natural ceramide has saturated or unsaturated fatty acid chains 16–24 carbons long and is insoluble in water. Such analogs include short-chain derivatives such as N-acetyl-sphingosine (C_2-ceramide), C_2-dihydroceramide (DHC$_2$-ceramide, C_2-ceramide lacking the trans-4,5-double bond) and N-octanoyl-sphingosine (C_8-ceramide) D-e-C_8-ceramide. While the use of C_2-ceramide has been more popular, our evidence indicates that C_8-ceramide is more likely to mimic the activity of natural ceramide.

One of the most interesting and intriguing properties of ceramide is the induction of apoptosis in cells in culture and *in vivo*. Ceramide generation was measured under various conditions that are known to lead to apoptosis. Indeed, ceramide levels were shown to increase in cells exposed to ionizing radiation (Haimovitz-Friedman et al., 1994), withdrawal of serum (Jayadev et al., 1995), ultraviolet-C radiation, heat shock (Verheij et al., 1996), chemotherapeutic agents such as daunorubicin (Bose et al., 1995), and oxidative stress (Jarvis et al., 1995). Another indication for a role for ceramide as a mediator of cell death comes from the investigation of transgenic mice deficient in acidic sphingomyelinase and lymphoblasts from Niemann–Pick disease. These patients were found to be resistant to apoptosis induced by ionizing radiation (Santana et al., 1996). These studies suggest that the acidic sphingomyelinase is required for apoptosis. However, the mechanism(s) and the signaling pathway(s) of cell death triggered by ceramide have not been elucidated. The activation of the SAPK/JNK (Verheij et al., 1996) and Ras/Raf signal transduction pathway (Yao et al., 1995) by ceramide suggests that these pathways may be involved. It is also unclear whether known ceramide targets, such as CAPK and CAPP, are involved in the signaling leading to cell death (Karasavvas et al., 1996a). It is likely that the regulation of cellular homeostasis by ceramide is complex and multifactorial. This notion is supported by the fact that ceramide seems to be involved in various known signal transduction pathways and activates a cascade of various targets (Hannun et al., 1993; Kolesnick and Golde, 1994; Spiegel et al., 1996). Ceramide is also hypothesized to control the apoptotic pathways of the two members of the TNF superfamily, TNF-α and Fas (Obeid et al., 1993; Cifone et al., 1993; Heller and Kronke, 1994; Jarvis et al., 1994; Cleveland and Ihle, 1995; Pushkareva et al., 1995; Santana et al., 1996).

To investigate the role of ceramide and the sphingolipid cycle in cell death, we used several cell culture and developmental models, including both insects and mice. Different ceramide analogs were constructed to study the kinetics and possible targets for its action. We analyzed the ability of the four synthetic C_8-ceramide stereoisomers, N-octanoyl-DL-*erythro*-dihydrosphingosine (DL-*e*-DH-C_8-ceramide) and a new ceramide derivative N-octyl-D-*erythro*-sphingosine (D-*e*-C_8-ceramine) to induce apoptosis in U937 cells by exposing cells to the compound, and we looked for DNA fragmentation. The *threo* stereoisomers proved to be significantly more potent than the *erythro* stereoisomers in inducing DNA fragmentation. The replacement of a carbonyl group by a methylene group in C_8-ceramine augmented the potency, since this compound was very potent in inducing apoptosis, with maximum DNA fragmentation at 6 h (the maximum level of DNA fragmentation induced by the *threo* stereoisomers is at 12 h). Since the difference between the octyl compound and D-*erythro*-C_8-ceramide is the replacement of the carbonyl group by a methylene group, the carbonyl group emerges as an important component of the ceramide structure for survival and not cell death. CAPK was activated by C_8-ceramide, but was also activated by the other compounds under conditions in which the other compounds did not induce cell death. The postulated CAPP was not a target for ceramide in this system (Karasavvas et al., 1996). Thus the linkage between the phosphatase, the kinase, and cell death cannot be absolute, and remains to be explored. We have also shown that Bcl-2 expression can inhibit cell killing by ceramide; this inhibition is dependent on the potency of the compound (Karasavvas et al., 1996).

Almost all the data linking ceramide to cell killing comes from cultured cells and exposure of these cells to high, unphysiological, levels of ceramide, which usually range in the order of 500-fold higher than the amount commonly measured in cells. We therefore asked if ceramide was also activated in instances of cell death *in vivo* during normal, physiological cell death. We examined the level of endogenous ceramide in different tissues undergoing cell death. In the limbs this provided limited information since in this tissue a small percentage of cells is dying in a limited amount of tissue. However, in the insect labial gland we demonstrated a slight increase in ceramide at the early stages of gland cell death. However, this level is several fold less than that found inside the cultured cells induced to die by treatment with ceramide, and it is difficult to say that the small increase we detected can cause cell death. One can postulate that the small increase can activate the cell death sequence. However, our observation using cell culture and exogenously added ceramide indicates that a several-fold increase in intracellular ceramide will not commit the cell to die within the time frame that other inducers such as TNFα can; one can rescue the cells from death by withdrawing the ceramide and allowing the retained ceramide to degrade. Ceramide has been also implicated in cell killing by TNFα (Kim et al., 1991; Schutze et al., 1992). We have compelling evidence that indicates that one can raise the intracellular level of ceramide well beyond that considered to be effective, for a period longer than considered necessary, and not kill the cell; whereas a brief exposure to TNFα will induce death in a matter of minutes, a several hour exposure to ceramide does not lead to cell death, indicating that the increase in ceramide generated by TNFα is unlikely to be lethal or causative (Karasavvas and Zakeri, submitted). Our findings suggest that increase in the endogenous ceramide does not lead to cell death; it might instead represent a cellular stress response rather than a cell killing factor.

In summary, therefore, we conclude that a palette of detection mechanisms, rather than a single characteristic, is necessary to identify apoptosis, especially since there are several variant forms of physiological cell death. The Hm mutants demonstrate clearly that cells have differing susceptibility to death and that there is a threshold beyond which they become committed to die. Especially in type II death, many biochemical changes well precede the apoptotic collapse, including upregulation of genes that prove on more thorough inspection not to play a causal role in the death. Finally, one must be cautious about extrapolating from *in vitro* studies to presumptive *in vivo* mechanisms, because the quantities of the stimuli and the kinetics of the responses may be very different.

ACKNOWLEDGMENTS

The work presented in this chapter was supported by grants from the National Institute on Aging, K04-AG 0031 to ZZ and AG-R01-10101 to R. Lockshin and ZZ. I would like thank especially the people from my laboratory whose work is reviewed in this chapter: Drs. Harleen Singh Ahuja, Nicos Karasavvas, and Reginald Halaby; Terri Latham, Carol Ko, Chin Kim, Yong Zhu, and Kimmy Woo. I also thank my collaborator Richard Lockshin, and I would like to dedicate this chapter to the memory of Dr. Jana Jochová.

REFERENCES

Abrams JM, White K, Fessler LI, Steller H (1993): Programmed cell death during *Drosophila* embryogenesis. Development 117:29–43.

Aidells B, Lockshin RA, Cullin AM (1971): Breakdown of the silk glands in Galleria mellonella. Acid phosphatase in involuting glands. J Insect Physiol 17:857–69.

Alles AJ, Sulik KK (1989): Retinoic-acid induced limb-reduction defects: Perturbation of zones of programmed cell death as a pathogenetic mechanism. Teratology 40:163–71.

Alles AJ, Sulik KK (1990): Retinoic acid-induced spina bifida: Evidence for a pathogenetic mechanism. Development 108:73–81.

Armstrong RB, Ashenfelter KO, Eckhoff C, Levin AA, Shapiro SS (1994): General and reproductive toxicology of retinoids. In *The Retinoids.* Sporn MB, Roberts AB, Goodman DS (eds): New York: Raven Press, pp 545–72.

Ballard JK, Holt SJ (1968): Cytological and cytochemical studies on cell death and digestion in the foetal rat foot; the role of macrophages and hydrolytic enzymes. J Cell Sci 3:245–61.

Berges R, Isaacs JT (1993): Programming events in the regulation of cell proliferation and death. Clin Chem 39:356–61.

Beaulaton J, Lockshin RA (1977): Ultrastructural study of the normal degeneration of the intersegmental muscles of *Antheraea polyphemus* and *Manduca sexta* (Insecta, Lepidoptera) with particular reference to cellular autophagy. J Morphol 154:39–58.

Beaulaton J, Lockshin RA (1978): Ultrastructural study of neuromuscular relations during degeneration of the intersegmental muscles. Biol Cellulaire 33:169–74.

Beaulaton J, Lockshin RA (1982): The relation of programmed cell death to development and reproduction: Comparative studies and an attempt at classification. Int Rev Cytol 29:215–35.

Bose R, Verheij M, Haimovitz A-F, Scotto K, Fuks Z, Kolesnick R (1995): Ceramide synthase mediates daunorubicin-induced apoptosis: An alternative mechanism for generating death signals. Cell 82:405–14.

Bursch W, Kleine L, Tenniswood M (1990): The biochemistry of cell death by apoptosis. Biochem Cell Biol 68:1071–74.

Bursch W, Ellinger A, Kienzl H, Török L, Pandey S, Sikorska M, Walker R, Hermann RS (1996): Active cell death induced by the anti-estrogens tamoxifen and ICI 164 384 in human mammary carcinoma cells (MCF-7) in culture: The role of autophagy. Carcinogenesis 17:1595–607.

Buttyan R, Olsson CA, Pintar J, Chang C, Bandyk M, Ng PY, Sawczuk IS (1989): Induction of the TRPM-2 gene in cells undergoing programmed cell death. Mol Cell Biol 9:3473–81.

Chaudhry AP, Shah RM (1973): Palatogenesis in the hamster. II. Ultrastructural observations on the closure of the palate. J Morphol 139:329–50.

Cifone MG, De Maria R, Roncaioli P, Rippo MR, Azuma M, Lanier LL, Santoni A, Testi R (1993): Apoptotic signaling through CD95 (Fas/Apo-1) activates an acidic sphingomyelinase. J Exp Med 177:1547–52.

Clarke PGH (1990): Developmental cell death: Morphological diversity and multiple mechanisms. Anat Embryol 181:195–213.

Cleveland JL, Ihle JN (1995): Contenders in FasL/TNF death signaling. Cell 81:479–82.

Cohen JJ (1991): Programmed cell death in the immune system. Adv Immunol 50:55–85.

Cohen JJ, Duke RC (1984): Glucocorticoid activation of a calcium dependent endonuclease in thymocyte nuclei leads to cell death. J Immunol 132:38–42.

Cohen JJ, Duke RC, Chervenak R, Sellins KS, Olson LK (1985): DNA fragmentation in targets of CTL: An example of programmed cell death in the immune system. Adv Exp Med Biol 184:493–508.

Copp AJ (1978): Interaction between inner cell mass and trophoectoderm of the mouse blastocyst. I. A study of cellular proliferation. J Embryol Exp Morphol 48:109–25.

Coucouvanis EC, Martin GR, Nadeau JH (1995): Genetic approaches for studying programmed cell death during development of the laboratory mouse. Methods Cell Biol 46:387–440.

Dbaibo G, Pushkareva MY, Jayadev S, Schwarz JK, Horowitz JM, Obeid LM, Hannun YA (1995): Retinoblastoma gene product as a downstream target for a ceramide-dependent pathway of growth arrest. Proc Natl Acad Sci USA 92:1347–51.

De Luca LM, Darwiche N, Jones CS, Scita G (1995): Retinoids in differentiation and neoplasia. Sci Am 28–37.

Dowd DR, MacDonald PN, Komm BS, Haussler MR, Miesfeld R (1991): Evidence for early induction of calmodulin gene expression in lymphocytes undergoing glucocorticold-mediated apoptosis. J Biol Chem 266:18423–26.

Duke RC, Chervenak R, Cohen JJ (1983): Endogenous endonuclease-induced DNA fragmentation: An early event in cell mediated cytolysis. Proc Natl Acad Sci USA 80:6361–65.

Duke RC, Witter RC, Nash PB, Young JD-E, Ojcius DM (1994): Cytolysis mediated by ionophores and pore-forming agents: Role of intracellular calcium in apoptosis. FASEB J 8:237–46.

Duvall E, Wyllie AH (1986): Death and the cell. Immunol Today 7:115–19.

Elledge SJ, Spottswood MR (1991). A new human p34 protein kinase, CDK2, identified by complementation of a cdc28 mutation in Saccharomyces cerevisiae, is a homolog of Xenopus Eg1. EMBO J 10:2653–59.

Ellis RE, Horvitz HR (1986): Genetic control of programmed cell death in the nematode C. elegans. Cell 44:817–29.

El-Shershaby AM, Hinchliffe JR (1974): Cell redundancy in the zona-intact preimplantation mouse blastocyst: A light and electron microscope study of dead cells and their fate. J Embryol Exp Morphol 31:643–54.

Fahrbach SE, Choi MK, Truman JW (1994): Inhibitory effects of actinomycin D and cycloheximide on neuronal death in adult Manduca sexta. J Neurobiol 25:59–69.

Favennec L, Cals M-J (1988): The biological effects of retinoids on cell differentiation and proliferation. J Clin Chem. Clin Biochem 26:479–489.

Fallon JF, Cameron J (1977): Interdigital cell death during limb development of the turtle and lizard with an interpretation of evolutionary significance. J Embryol Exp Morphol 40:285–89.

Fishbein JD, Dobrowsky RT, Bielawska A, Garrett S, Hannun YA (1993): Ceramide-mediated growth inhibition and CAPP are conserved in *Saccharomyces cerevisiae*. J Biol Chem 268:9255–61.

Girard F, Strausfeld U, Fernandez A, Lamb NJC (1991): Cyclin A is required for the onset of DNA replication in mammalian fibroblasts. Cell 67:1169–79.

Green MC (1981). In *Genetic Variants and Strains of the Laboratory Mouse*. Stuttgart: Gustav Fischer Verlag, p 114.

Grether ME, Abrams JM, Agapite J, White K, Steller H (1995): The *head involution defective* gene of *Drosophila melanogaster* functions in programmed cell death. Genes Dev 9:1694–708.

Guenette RS, Daehlin L, Mooibroek M, Wong K, Tenniswood M (1994): Thanatogen expression during involution of the rat ventral prostate after castration. J Androl 15:200–11.

Haas AL, Baboshina O, Williams B, Schwartz LM (1995): Coordinated induction of the ubiquitin conjugation pathway accompanies the developmentally programmed death of insect skeletal muscle. J Biol Chem 270:9407–12.

Haimovitz-Friedman A, Kan CC, Ehleiter D, Persaud RS, McLoughlin M, Fuks Z, Kolesnick RN (1994): Ionizing radiation acts on cellular membranes to generate ceramide and initiate apoptosis. J Exp Med 180:525–35.

Halaby R, Zakeri Z, Lockshin RA (1994): Metabolic events during programmed cell death in insect labial glands. Biochem Cell Biol 72:597–601.

Hamburger V, Oppenheim RW (1982): Naturally occurring death in vertebrates. Neurosci Comment 1:39–55.

Hannun YA (1994): The sphingomyelin cycle and the second messenger function of ceramide. J Biol Chem 269:3125–28.

Hannun YA, Obeid LM, Wolff RA (1993): The novel second messenger ceramide: Identification, mechanism of action, and cellular activity. Adv Lipid Res 25:43–64.

Hardy K, Handyside AH, Winston RM (1989): The human blastocyst: Cell number, death and allocation during late preimplantation development in vitro. Development 107:597–604.

Harmon B, Bell L, Williams L (1984): An ultrastructural study on the "meconium corpuscles" in rat foetal intestinal epithelium with particular reference to apoptosis. Anat Embryol 169:119–24.

Hartwell LH, Kastan MB (1994): Cell cycle control and cancer. Science 226:1821–27.

Heller RA, Kronke M (1994): Tumor necrosis factor receptor. Mediated signaling pathways. J Cell Biol 126:5–9.

Heller RA, Song K, Fan N, Chang DJ (1992): The p70 tumor necrosis factor receptor mediates cytotoxicity. Cell 70:47–56.

Helminen HJ, Ericsson JL, Orrenius S (1968): Studies on mammary gland involution. IV. Histochemical and biochemical observations on alterations in lysosomes and lysosomal enzymes. J Ultrastruct Res 25:240–52.

Helminen HJ, Ericsson JL (1970): Quantitation of lysosomal enzyme changes during enforced mammary gland involution. Exp Cell Res 60:419–26.

Helminen HJ, Ericsson JL (1971): Effects of enforced milk stasis on mammary gland epithelium, with special reference to changes in lysosomes and lysosomal enzymes. Exp Cell Res 68:411–27.

Hinchliffe JR (1981): Cell Death in Embryogenesis. In Bowen ID and Lockshin RA, eds. Cell Death in Biology and Pathology. Chapman and Hall, London, pp 35–69.

Hinchliffe JR, Ede DA (1967): Limb development in the polydactylous talpid[3] mutant of the fowl. J Embryol Exp Morph 17:385–404.

Hinchliffe JR, Ede DA (1973): Cell death and the development of limb form and skeletal pattern in normal and wingless (ws) chick embryos. J Embryol Exp Morphol 30:753–72.

Hinchliffe JR, Thorogood PV (1974): Genetic inhibition of mesenchymal cell death and the development of form and skeletal pattern in the limbs of talpid³ (Ta³) mutant chick embryos. J Embryol Exp Morph 31:747–60.

Hinchliffe JR, Johnson DR (1980): *The Development of the Vertebrate Limb.* Oxford: Oxford University Press.

Hinchliffe JR, Gumpel-Pinot M (1981): Control and maintenance and anteroposterior skeletal differentiation of the anterior mesenchyme of the chick wing bud by its posterior margin (the ZPA). J Embryol Exp Morphol 62:63–82.

Hurle JM (1988): Cell Death in Developing Systems. In G. (ed) Methods and Achievements in Experimental Pathology. G Karger, Basel 13:55–86.

Hurle JM, Colvee E (1982): Surface changes in the embryonic interdigital epithelium during the formation of free digits: A comparative study in the chick and duck foot. J Embryo Exp Morphol 69:251–63.

Hurle JM, Ros MA, Climent V, Garcia-Martinez V (1996): Morphology and significance of programmed cell death in the developing limb bud of the vertebrate embryo. Micros Res Tech 34:236–46.

Ino H, Ishizuka T, Chiba T, Tatibana M (1994): Expression of CDK5 (PSSALRE kinase), a neural cdc2-related protein kinase, in the mature and developing mouse central and peripheral nervous systems. Brain Res 661:196–206.

Jarvis WD, Kolesnick RN, Fornari FA, Traylor RS, Gewirtz DA, Grant S (1994): Induction of apoptotic DNA damage and cell death by the activation of the sphingomyelin pathway. Proc Natl Acad Sci USA 91:73–77.

Jarvis WD, Grant S, Kolesnick RN (1995): Ceramide and the induction of apoptosis. Clin Cancer Res 2:1–6.

Jayadev S, Liu B, Bielawska AE, Lee JY, Nazaire F, Pushkareva MY, Obeid LM, Hannun YA (1995): Role of ceramide in cell cycle arrest. J Biol Chem 270:2047–52.

Jochová J, Zakeri Z, Lockshin RA (1997): Rearrangement of the tubulin and actin cytoskeleton during programmed cell death in *Drosophila* salivary glands. Cell Death Differ 4:140–49.

Jochová J, Quaglino D, Zakeri Z, Woo K, Sikorska M, Weaver V, Lockshin RA (1998): Protein synthesis, DNA degradation, and morphological changes during programmed cell death in labial glands of *Manduca sexta*. Dev Genetics *in press*.

Karasavvas N, Erukulla RK, Bittman R, Lockshin R, Zakeri Z (1996a): Stereospecific induction of apoptosis in U937 cells by *N*-octanoyl-sphingosine stereoisomers and the novel ceramide analog *N*-octyl-sphingosine. Eur J Biochem 236:729–37.

Karasavvas N, Erukulla RK, Bittman R, Lockshin R, Hockenberry D, Zakeri Z (1996b): *Bcl-2* suppresses ceramide-induced cell killing. Cell Death and Differentiation 3:149–51.

Kerr JFR, Searle J, Harmon BV, Bishop CJ (1987): Apoptosis. In Potten CS (ed): *Perspectives on Mammalian Cell Death.* Oxford: Oxford University Press, pp 93–128.

Kim M-Y, Linardic C, Obeid L, Hannun Y (1991): Identification of sphingomyelin turnover as an effector mechanism for the action of tumor necrosis factor α and γ-interferon. J Biol Chem 266:484–89.

Knudsen TB, Kochhar DM (1981): The role of morphogenetic cell death during abnormal limb-bud outgrowth in mice heterozygous for the dominant mutation Hemimelia extra toe (Hmx). J Embryol Exp Morphol 65:289–307.

Kochhar DM (1975): Assessment of teratogenic response in cultured postimplantation mouse embryos: Effects of hydroxyurea. In Neubert D, Merker HJ (eds): *New Approaches to the Evaluation of Abnormal Embryonic Development.* Stuttgart: Georg Thieme, pp 250–77.

Kochhar DM (1977): Abnormal organogenesis in limbs. In Wilson JG, Fraser FC (eds): Handbook of Teratology. New York: Plenum Press, pp 453–79.

Kochhar DM, Penner JD, Tellone CI (1984): Comparative teratogenic activities of two retinoids: Effects on palate and limb development. Teratog Carcinog Mutagen 4:377–87.

Koff A, Cross F, Fisher A, Schumacher J, Leguellec K, Philippe M, Roberts JM (1991): Human cyclin E, a new cyclin that interacts with two members of the CDC2 gene family. Cell 66:1217–28.

Kolesnick R, Golde DW (1994). The sphingomyelin pathway in tumor necrosis factor and interleukin-1 signaling. Cell 77:325–28.

Kyprianou N, Isaacs JT (1989): Expression of transforming growth factor-β_1 in the rat ventral prostate during castration induced programmed cell death. Mol Endocrinol 3:1515–22.

Kyprianou N, Isaacs JT (1988): Activation of programmed cell death in the rat ventral prostate after castration. Endocrinol 122:552–62.

Lammer E, Chen D, Hoar R, Agnish N, Benke P, Braun J, Curry C, Fernhoff P, Grix AJ, Lott I, Richard J, Sun S (1985): Retinoic acid embryopathy. N E J Med 313:837–41.

Li CJ, Friedman DJ, Wang C, Metelev V, Pardec AB (1995): Induction of apoptosis in uninfected lymphocytes by HIV-1 Tat protein. Science 268:429–31.

Lints R, Driscoll M (1996): Programmed and pathological cell death in *Caenorhabditis elegans.* In Holbrook N, Martin G, and Lockshin RA. *Cellular Aging and Cell Death.* New York: John Wiley and Sons, pp 235–53.

Lo AC, Li L, Oppenheim RW, Prevethe D, Houenou LJ (1995): Ciliary neurotrophic factor promoting the survival of spinal sensory neurons following axotomy but not during the period of programmed cell death. Exp Neurol 134:49–55.

Liu Y, Lee M-O, Wang H-G, Li Y, Hashimoto Y, Klaus M, Reed JC, Zhang X-K (1996): Retinoic acid receptor β mediates the growth-inhibitory effect of retinoic acid by promoting apoptosis in human breast cancer cells. Molec Cell Biol 16:1138–49.

Lockshin RA (1969a). Lysosomes in insects. In Dingle JT, Fell HB (eds): Lysosomes in Biology and Pathology, Amsterdam: North Holland Publishing, pp 363–91.

Lockshin RA (1969b): Programmed cell death. Activation of lysis of a mechanism involving the synthesis of protein. J Insect Physiol 15:1505–16.

Lockshin RA (1975): Failure to prevent degeneration in insect muscle with pepstatin. Life Sci 17:403–10.

Lockshin RA (1981): Cell death in metamorphosis. In Bowen ID, Lockshin RA (eds): *Cell Death in Biology and Pathology.* London: Chapman and Hall, pp 79–122.

Lockshin RA, Beaulaton J (1974a): Programmed cell death. Cytochemical evidence for lysosomes during the normal breakdown of the intersegmental muscles. J Ultrastruct Res 46:43–62.

Lockshin RA, Beaulaton J (1974b): Programmed cell death. Cytochemical appearance of lysosomes when the death of the intersegmental muscles is prevented. J Ultrastruct Res 46:63–78.

Lockshin RA, Beaulaton J (1979): Cytological studies of dying muscle fibers of known physiological parameters. Tissue Cell 11:803–19.

Lockshin RA, Wadewitz AG (1990): Degeneration of myofibrillar proteins during programmed cell death in *Manduca sexta*. In Finch CA, Johnson T (eds): Molecular Biology of Aging, UCLA Symposia in Molecular and Cell Biology, pp 283–97.

Lockshin RA, Williams CM (1965a): Programmed cell death. III. Neural control of the breakdown of the intersegmental muscles. J Insect Physiol 11:605–10.

Lockshin RA, Williams CM (1965b): Programmed cell death. IV. The influence of drugs on the breakdown of the intersegmental muscles of silkmoths. J Insect Physiol 11:803–9.

Lockshin RA, Williams CM (1965c): Programmed cell death. V. Cytolytic enzymes in relation to the breakdown of the intersegmental muscles of silkmoths. J Insect Physiol 11:831–44.

Lockshin RA, Zakeri Z (1994): Programmed cell death: Early changes in metamorphosing cells. Biochem Cell Biol 72:589–96.

Lund LR, Romer J, Thomasset N, Solberg H, Pyke C, Bissell MJ, Dano K, Werb Z (1996): Two distinct phases of apoptosis in mammary gland involution: Proteinase-independent and -dependent pathways. Development 122:181–93.

Matsushime H, Ewen ME, Strom DK, Kato JY, Hanks SK, Roussel MF, Sherr CJ (1992): Identification and properties of an atypical catalytic subunit (p34PSK-J3/ cdk4) for mammalian D type G1 cyclins. Cell 71:323–34.

McConkey DJ, Orrenius S (1994): Signal transduction pathways to apoptosis. Trends Cell Biol 4:370–75.

McConkey DJ, Nicotera DJ, Orrenius S (1994): Signaling and chromatin fragmentation in thymocyte apoptosis. Immunol Rev 142:343–363.

Meikrantz W, Gisselbrecht S, Tam SW, Schlegel R (1994). Activation of cyclin A-dependent protein kinases during apoptosis. Proc Natl Acad Sci USA 91:3754–3758.

Meyerson M, Enders GH, Wu CL, Su LK, Gorka C, Nelson C, Harlow E, Tsai LH (1992): A family of human cdc2-related protein kinases. EMBO J 11:2909–17.

Milaire J, Rooze M (1983): Hereditary and induced modifications of the normal necrotic patterns in the developing limb buds of the rat and mouse: Facts and hypothesis. Arch Biol 94:459–90.

Monpetit ML, Lawless KR, Tenniswood M (1986): Androgen repressed messages in the rat ventral prostate. The Prostate 8:25–30.

Mori C, Nakamura N, Okamoto Y, Osawa M, Shiota K (1994): Cytochemical identification of programmed cell death in the fusing fetal mouse palate by specific labelling of DNA fragmentation. Anat Embryol 190:21–28.

Naruse I, Kameyama Y (1986): Prevention of polydactyly manifestation in polydactyly nagoya (PDN) mice by administration of cytosine arabinoside during pregnancy. Teratology 34:283–289.

Narvaez C, Vanweelden JK, Byrne I, Welsh J (1996): Characterization of vitamin D_3 resistant MCF cell line. Endocrinol 137:400–409.

Nuñez G, London L, Hockenberry D, Alexander M, McKearn JP, Korsmeyer SJ (1990): Deregulated Bcl-2 gene expression selectively prolongs survival of growth-factor-deprived hemopoietic cell lines. J Immunol 144:3602–10.

Obeid LM, Linardic CM, Karolak, LA, Hannun, YA (1993): Programmed cell death induced by ceramide. Science 259:1769–71.

Okazaki T, Bielawska A, Hannun YA (1989): Sphingomyelin turnover induced by vitamin D$_3$ in HL-60 cells. J Biol Chem 264:19076–80.

Okazaki T, Bielawska A, Bell RM, Hannun YA (1990): Role of ceramide as a lipid mediator of 1α,25-dihydroxyvitamin D$_3$-induced HL-60 cell differentiation. J Biol Chem 265:15823–31.

Okazaki T, Bielawska A, Domae N, Bell RM, Hannun YA (1994): Characteristics and partial purification of a novel cytosolic, magnesium-independent, neutral sphingomyelinase activated in the early signal transduction in 1α,25-dihydroxyvitamin D$_3$-induced HL-60 cell differentiation. J Biol Chem 269:4070–77.

Ongkeko W, Ferguson DJP, Harris AL, Norbury C (1995): Inactivation of Cdc2 increases the level of apoptosis induced by DNA damage. J Cell Sci 108:2897–904.

Oppenheim RW (1991): Cell death during development of the nervous system. Annu Rev Neurosci 14:453–501.

Pagano M, Pepperkok R, Verde F, Ansorge W, Draetta G (1992): Cyclin A is required at two points in the human cell cycle. EMBO J 11:961–71.

Penfold PL, Provis JM (1986): Cell death in the development of the human retina: Phagocytosis of pyknotic and apoptotic bodies by retinal cells. Graefe's Arch Clin Exp Opthalmol 224:549–53.

Piacentini M, Autuori F, Dini L, Farrace MG, Ghibelli L, Piredda L, Fesus L (1991): "Tissue" transglutaminase is specifically expressed in neonatal rat liver cells undergoing apoptosis upon epidermal growth factor-stimulation. Cell Tissue Res 263:227–36.

Pierce GB, Gramzinski RA, Parchment RE (1989): Programmed cell death in the blastocyst. Ann NY Acad Sci 567:182–86.

Pines J, Hunter T (1991): Human cell division: The involvement of cyclins A and B1, and multiple cdc2s. Cold Spring Harb Symp Quant Biol 56:449–63.

Pinnock CB, Alderman CP (1992): The potential for teratogenicity of vitamin A and its congeners. Med J Aust 157:804–9.

Pipan N, Sterle M (1979): Cytochemical analysis of organelle degradation in phagosomes and apoptotic cells of the mucoid epithelium of mice. Histochemistry 59:225–32.

Pipan N, Sterle M (1986): Cytochemical and scanning electron-microscopic analysis of apoptotic cells and their phagocytosis in mucoid epithelium of the mouse stomach. Cell Tiss Res 246:647–52.

Pratt RM, Green RM, Hassel JR, Greenberg JH (1975): Epithelial cell differentiation during secondary palate development. In Greulich S (eds): *Extracellular Matrix Influence of Gene Expression.* New York: Academic Press.

Pronk GJ, Ramer K, Amiri P, Williams LT (1996): Requirement of an ICE-like protease for induction of apoptosis and ceramide generation by REAPER. Science 271:808–10.

Pushkareva M, Obeid LM, Hannun YA (1995): Ceramide: An endogenous regulator of apoptosis and growth suppression. Immunol Today 16:294–97.

Restifo LL, White K (1992): Mutations in a steroid hormone-regulated gene disrupt the metamorphosis of internal tissues in *Drosophila:* Salivary glands, muscle, and gut. Roux's Archives of Developmental Biology 201:221–34.

Rivas CI, Golde DW, Vera JC, Kolesnick RN (1994): Involvement of the sphingomyelin pathway in autocrine tumor necrosis factor signaling for human immunodeficiency virus production in chronically infected HL-60 cells. Blood 83:2191–97.

Rizzo R, Lammer EJ, Parano E, Pavone L, Argyle JC (1991): Limb reduction defects in humans associated with prenatal isotretinoin exposure. Teratology 44:599–604.

Rooze MA (1977): The effects of the Dh gene on limb morphogenesis in the mouse. Birth Defects 13:69–95.

Sadler TW, Kochhar DM (1976): Chlorambucil-induced cell death in embryonic mouse limb-buds. Toxicol Appl Pharmacol. 37:237.

Sadler TW, Kochhar DM, Cardell RR (1976): Effects of hydroxyurea on ultrastructure of neuroepithelial cells in mouse embryos. Teratology 13:35A.

Santana P, Peña LA, Friedman A H, Martin S, Green D, McLoughlin M, Cardo C-C, Schuchman ED, Fuks Z, Kolesnick R (1996): Acid sphingomyelinase-deficient human lymphoblasts and mice are defective in radiation-induced apoptosis. Cell 86:189–99.

Saunders JW, Fallon JF (1967): Cell death in morphogenesis. In Locke M, ed. *Major Problems in Developmental Biology.* New York: Academic Press, pp 289–314.

Schutze S, Potthoff K, Machleidt T, Berkovic D, Wiegmann K, Kronke M (1992): TNF activates NF-kB by phosphatidylcholine-specific phospholipase C-induced "acidic" sphingomyelin breakdown. Cell 71:765–76.

Schwartz LM (1992): Insect muscle as a model for programmed cell death. J Neurobiol 23:1312–26.

Schwartz LM, Kosz L, Kay BK (1990): Gene activation is required for developmentally programmed cell death. Proc Natl Acad Sci USA 87:6594–98.

Schwartz LM, Smith SW, Jones MEE, Osborne BA (1993): Do all programmed cell deaths occur via apoptosis? Proc Natl Acad Sci USA 90:980–84.

Schwartz LM, Truman JW (1982): Peptide and steroid regulation of muscle degeneration in an insect. Science 215:1420–21.

Schwartz LM, Truman JW (1984): Hormonal control of muscle atrophy and degeneration in the moth *Antheraea polyphemus.* J Exp Biol 111:13–30.

Schweichel J-U (1971): The influence of oral vitamin A doses on interdigital necrosis in the limb bud of the rat. Teratology 4:501–507.

Schweichel JU, Merker HJ (1973): The morphology of various types of cell death in prenatal tissue. Teratology 7:253–66.

Scott WJ, Ritter EJ, Wilson JG (1977): Delayed appearance of ectodermal cell death as a mechanism of polydactyly induction. J Embryol Exp Morphol 42:93–104.

Selvakumaran M, Lin HK, Sjin RT, Reed JC, Liebermann DA, Hoffman B (1994): The novel primary response to gene MyD118 and the proto-oncogenes myb, myc and bcl-2 modulate transforming growth factor beta1-induced apoptosis of myeloid leukemia cells. Mol Cell Biol 14:2352–60.

Shapiro BL, Sweney LR (1969): Electron microscopical and histochemical examination of oral epithelial–mesenchymal interaction (programmed cell death). J Dent Res 48:652–60.

Sherr CJ (1993): Mammalian G1 cyclins. Cell 73:1059–65.

Singh Ahuja H, James W, Zakeri Z (1997): Rescue of limb deformity in Hammertoe mutant mice by retinoic acid-induced cell death. Devel Dynamics 20:267–275.

Singh Ahuja H, Zhu Y, Zakeri Z (1998): Association of cyclin-dependent kinase 5 and its activator p35 with apoptotic cell death. Develop Genetics *in press*.

Spiegel S, Foster D, Kolesnick R (1996): Signal transduction through lipid second messengers. Curr Opin Cell Biol 8:159–67.

Steller H, Abrams JM, Grether ME, White K (1994): Programmed cell death in *Drosophila*. Philos Trans R Soc Lond (Biol) 345:247–50.

Tenniswood MP, Guenette RS, Lakins J, Mooibroek M, Wong P, Welsh J-E (1992): Active cell death in hormone-dependent tissues. Cancer Metastasis Rev 11:197–220.

Tong JX, Eichler ME, Rich KM (1996): Intracellular calcium levels influence apoptosis in mature sensory neurons after trophic factor deprivation. Exp Neurol 138:45–52.

Truman JW (1984): Cell death in invertebrate nervous systems. Annu Rev Neurosci 7:171–88.

Truman JW (1992): The eclosion hormone system of insects. Prog Brain Res 92:361–74.

Truman JW (1996): Steroid receptors and nervous system metamorphosis in insects. Develop Neurosci 18:87–101.

Trump BF, Berezesky IK (1995): Calcium-mediated cell injury and cell death. FASEB J 9:219–28.

Tsai LH, Delalle I, Caviness VS, Chae T, Harlow E (1994): p35 is a neural-specific regulatory subunit of cyclin-dependent kinase 5. Nature 371:419–23.

Tsai LH, Takahashi T, Caviness VS, Harlow E (1993): Activity and expression pattern of cyclin-dependent kinase 5 in the embryonic mouse nervous system. Development 119:1029–40.

Verheij M, Bose R, Lin XH, Yao B, Jarvis WD, Grant S, Birrer MJ, Szabo E, Zon LI, Kyriakis JM, Friedman A-H, Fuks Z, Kolesnick RN (1996): Requirement for ceramide-initiated SAPK/JNK signalling in stress-induced apoptosis. Nature 380:75–79.

Wadewitz AG, Lockshin RA (1988): Programmed cell death. Dying cells synthesize a coordinated unique set of proteins in two different episodes of cell death. FEBS Lett 241:19–23.

Weber R (1969): Tissue involution and lysosomal enzymes during anuran metamorphosis. In Dingle JT, Fell HB (eds): Lysosomes in Biology and Pathology, Vol. I. Amsterdam: Elsevier North Holland, pp 437–61.

White K, Grether ME, Abrams JM, Young L, Farrell K, Steller H (1994): Genetic control of programmed cell death in *Drosophila*. Science 264:677–83.

White K, Steller H (1995): The control of apoptosis in *Drosophila*. Trends Cell Biol 5:74–78.

Whitfield JF, Bird RP, Chakravarthy BR, Isaacs RJ, Morley P (1995): Calcium-cell cycle regulator, differentiator, killer, chemopreventor, and maybe tumor promoter. J Cell Biochem Suppl 22:74–791.

Wiegmann K, Schutze S, Machleidt T, Witte D, Kronke M (1994): Functional dichotomy of neutral and acidic sphingomyelinases in tumor necrosis factor signaling. Cell 78:1005–15.

Williams GT (1994): Programmed cell death: A fundamental protective response to pathogens. Trends Microbiol 2:463–64.

Williams L, Bell L (1991): Asynchronous development of the rat colon. Anat Embryol 183:573–78.

Woo K, Sikorska M, Weaver VM, Lockshin RA, Zakeri Z (1994): DNA fragmentation and DNA synthesis during insect metamorphosis. 10th Int. Symp. Cellular Endocrin, p 3.

Wyllie AH (1980): Glucocorticoid-induced thymocyte apoptosis is associated with endogenous endonuclease activation. Nature 284:555–56.

Wyllie AH (1985): The biology of cell death in tumors. Anticancer Res 5:131–36.

Wyllie AH (1987): Apoptosis: Cell death under homeostatic control. Arch Toxicol Suppl 11:3–10.

Yamada T, Ohyama H, Kinjo Y, Watanbe M (1981): Evidence for internucleosomal breakage of chromatin in rat thymocytes irradiated in vitro. Radiat Res 85:544–53.

Yao B, Zhang Y, Delikat S, Mathios S, Basu S, Kolesnick R (1995): Phosphorylation of Raf by ceramide-activated protein kinase. Nature 378:307–310.

Young RW (1984): Cell death during the differentiation of the retina in the mouse. J Comp Neurol 229:362–73.

Zakeri ZF (1993): In vitro limb differentiation as an experimental model. In Fallon JF, Goetinck PF, Kelley RO, Stocum DL (eds): *Limb Development and Regeneration.* New York: Wiley–Liss, pp 361–70.

Zakeri Z, Ahuja HS (1994): Cell death in limb bud development. Biochem Cell Biol 72:603–13.

Zakeri Z, Bursch W, Tenniswood M, Lockshin RA (1995a): Cell death. Programmed, apoptosis, necrosis, or other. Cell Death Differ 2:87–96.

Zakeri Z, Quaglino D, Latham T, Woo K, Lockshin RA (1995b): Programmed cell death in the tobacco hornworm, *Manduca sexta:* Alterations in protein synthesis. Microsc Res Tech 34:192–201.

Zakeri ZF, Quaglino D, Latham T, Lockshin RA (1993): Delayed internucleosomal DNA fragmentation in programmed cell death. FASEB J 7:470–78.

Zakeri Z, Quaglino D, Ahuja HS (1994): Apoptotic cell death in the mouse limb and its suppression in the Hammertoe mutant. Dev Biol 165:294–97.

Zhang Q, Ahuja HS, Zakeri ZF, Wolgemuth DJ (1997): Cyclin-dependent kinase 5 is associated with apoptotic cell death during development and tissue remodeling. Devel Biol 183:222–33.

Zindy F, Lamas E, Chenivesse X, Sobczak J, Wang J, Fesquet D, Henglein B, Brechot C (1992): Cyclin A is required in S phase in normal epithelial cells. Biochem Biophys Res Commun 182:1144–54.

Zucker RM, Cleland HJ, Haswell T (1991): Syndactyly correction of the hand in Apert Syndrome. Clinics in Plast Surg 18:357–64.

CHAPTER 4

GENETIC APPROACHES TO PROGRAMMED CELL DEATH IN *C. elegans*

ANTON GARTNER AND MICHAEL O. HENGARTNER
Cold Spring Harbor Laboratory, 1 Bungtown Road, Cold Spring Harbor,
New York 11724

INTRODUCTION

Progammed cell death is a common cell fate in most if not all multicellular organisms. Apoptosis, which will be used as a synonym for programmed cell death throughout this chapter, occurs extensively during development as well as during later life. The development of the nematode worm *Caenorhabditis elegans* provides a good example of the extensive use of programmed cell death. Genetic studies on the apoptotic pathway in *C. elegans* were instrumental in the development of several paradigms of our general understanding of programmed cell death: (1) The execution of programmed cell death during development—which formally may also be considered as a somewhat unusual terminal differentiation program—is subject to precise genetic control. (2) The execution of the apoptotic fate is an active program that can be best described as cellular suicide. In other words, a cell that senses that it is to die actively participates in this process, often induces its own demise, and finally might even induce its proper removal from the organism. (3). Finally, every single cell may have the potential to undergo programmed cell death. Indeed, it has been suggested that cellular suicide is mediated by a default pathway present in every single cell, and only those cells that manage to inactivate this pathway have the privilege to survive (Raff, 1992).

When Cells Die, Edited by Richard A. Lockshin, Zahra Zakeri,
and Jonathan L. Tilly
ISBN 0-471-16569-7   © 1998 Wiley-Liss, Inc.

In this chapter we will focus on the genetic analysis of programmed cell death in the nematode worm *Caenorhabditis elegans* (Fig. 4.1), with a particular emphasis on the genetic methods that are used to study programmed cell death in the worm system. In addition, we will discuss recent biochemical studies showing a direct physical interaction between the key components of the apoptotic pathway. Finally, we will stress some potential future research avenues that might deepen our understanding of programmed cell death.

IDENTITY AND ORIGIN OF CELLS UNDERGOING PROGRAMMED CELL DEATH DURING *C. elegans* DEVELOPMENT

Genetic studies of programmed cell death in *C. elegans* take advantage of its highly reproducible development. This reproducibility allowed the precise elucidation of the pattern of cell division that leads to the generation of all of the 959 somatic cells that make up the adult hermaphrodite worm. Lineage analysis revealed that during somatic development of the hermaphrodite worm, 131 out of the total of 1090 cells born undergo programmed cell death (Sulston and Horvitz, 1977; Kimble and Hirsch, 1979; Sulston et al., 1983). As is true for most other aspects of *C. elegans* development, these deaths show a high degree of uniformity with respect to the identity of dying cells and with respect to the timing of each of these cell deaths during development (Fig. 4.2). This reproducibility of cell death provides unique advantages for genetic analysis: Cell death can be studied on a single-cell level, and even mutations that cause only a very weak defect in programmed cell death or that affect only a small number of cell types might be identified. Programmed cell death of somatic cells is mostly used during embryonic development (113/ 131 deaths).

FIG. 4.1. Adult *C. elegans* hermaphrodite. Freshly hatched larvae, which measure about 100 μm in length, progress through four larval stages (L1–4) to reach adulthood, at which point they measure about 1 mm in length.

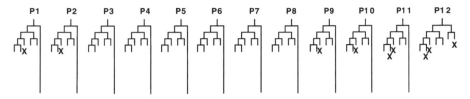

FIG. 4.2. Division patterns of the ventral cord neuroblast cells. The 12 ventral cord blast cells P1–P12 (collectively called Pn) divide in the first larval (L1) stage. Each of the Pn cells thereby generates an anterior daughter (Pn.a) that is a neuroblast cell and a posterior daughter (Pn.p) that is a hypodermal blast cell. In the diagram the lineages of the Pn.a cells descendants are shown (time is shown in the vertical axis, each horizontal line represents a cell division). Each of them follows an identical pattern of division to give birth to 5 neurons. Among these neurons one is involved in the innervation of vulval muscle cells. This particular neuron is only needed in the Pn.a descendants that are close to the vulva (P3-P8); consequently the equivalent cell is eliminated by programmed cell death (X) in the Pn.a 1, 2, 9 10, 11, and 12 descendants. (Diagram adapted from Hengartner and Horvitz, 1992.)

Apoptosis also occurs to a lesser extent during the transition through larval stages. In addition, apoptosis is very prominent in adult hermaphrodites during female germ cell development. The occurrence of apoptosis is neither restricted to particular cell types nor to any particular cell lineage. However, the bulk of apoptotic events affect neuronal cells and to some extent also hypodermal cells (Sulston and Horvitz, 1977; Kimble and Hirsch, 1979; Sulston et al., 1983).

IDENTIFICATION AND QUANTIFICATION OF PROGRAMMED CELL DEATH IN *C. ELEGANS*

Cell death in *C. elegans* can be readily observed in living animals using standard Nomarski optics (Sulston and Horvitz, 1977; Robertson and Thompson, 1982). The first sign of impending death is a decrease in the refractivity of the cytoplasm, which is concomitant with a slight increase in refractivity of the nucleus. Soon thereafter, both nucleus and cytoplasm become increasingly refractile until they resemble a flat round disk. After about 10–30 minutes this flat disk starts to disappear, the nucleus of the dying cell decreases in refractility, begins to appear crumpled, and finally vanishes within less than 1 hour (Sulston and Horvitz, 1977).

Three distinct but related approaches have been used to quantify programmed cell death in *C. elegans*. The first approach takes advantage of the highly reproducible anatomy of the worm, which allows the unambiguous identification of each cell in the body. The absence of programmed cell death can thus be scored indirectly by looking for the presence of extra "undead" cells (cells that should have died but instead survived). Undead cells cannot be distinguished from normal cells by their appearance under the microscope, but sometimes can by their location (e.g., their presence at a position where no cell is normally found). To increase the reliability of the assay, scoring of undead cells is usually performed in the pharynx, the animal's feeding organ, which is separated from the rest of the body by a clearly visible basement membrane. The pharynx is the site of many programmed cell deaths, and thus a large number of deaths can be scored in a single animal, allowing the detection of even very weak effects on cell death (e.g., less than 2% extra cell survival) (Hengartner et al., 1992; Hengartner and Horvitz, 1994a). The drawback of this approach is that it scores the presence of cells that should not be there, rather than deaths *per se*. Consequently, care must be taken to confirm that extra cells are indeed the result of inhibition of death, rather than of extra cell divisions, or of aberrant cell migrations.

The second approach takes advantage of mutations in genes required for the efficient engulfment of apoptotic cells (see the following). In these mutants, cells still die, but many dying cells fail to be engulfed and removed from the animal. These persistent, undegraded cell corpses are very obvious, even to the worm neophyte, and thus can be used as a simple assay for the extent of programmed cell death in the animal (Ellis et al., 1991b; Vaux et al., 1992). Elimination of programmed cell death results in the absence of persistent cell corpses in these mutants (no cells die; so no dead cells fail to be engulfed). The main advantage of this assay is its ease of scoring, as it does not require the experimenter to learn any worm anatomy. However, the number of persistent cell corpses is more variable than the number of surviving cells, and weak effects on cell death cannot be detected with this method. Furthermore, loss of persistent corpses could, at least in principle, also be obtained by bypassing or compensating for the engulfment defect, rather than by preventing programmed cell death.

The third method is to follow the development of the animals under the microscope using differential interference contrast optics, and to identify dying cells by their characteristic morphology (Sulston and Horvitz, 1977). Unfortunately, this approach is very time consuming, as only a handful of animals can be followed at the same time, and only animals of the proper stage of development yield useful information. Because of these constraints, this method is usually only used to confirm results obtained with one of the two alternative approaches that indirectly measure cell death.

MUTANTS DEFINE FOUR DISTINCT STEPS IN THE APOPTOTIC PATHWAY

Genetic analysis has led to the identification of over 100 different mutations that affect programmed cell death. These mutations define 12 genes that affect all programmed cell deaths and a smaller number of genes that are needed to commit specific cells to the apoptotic fate. Since mutants that are defective in all programmed cell deaths are viable and show no obvious defect in their development or adult behavior, it was easy to combine various double mutant combinations to build a genetic pathway for programmed cell death (Sulston, 1976; Hedgecock et al., 1983; Ellis and Horvitz, 1986; Ellis and Horvitz, 1991; Desai et al., 1988; Ellis et al., 1991a,b; Hengartner et al., 1992). This pathway contains four distinct steps, each of them defined by the analysis of various mutants (Fig. 4.3). (1) Initially, specific cell types are committed to the apoptotic fate. Subsequently, (2) the general apoptotic machinery, which is used in all dying cells, is activated. Later, (3) the recognition and engulfment of dying cells by a neighboring cell proceeds, and finally (4) the remnants of engulfed cells are degraded. Genes of the first type generally affect very few cells or cell types, whereas, the genes falling in to the subsequent classes affect all cell deaths. These genes will be described in the following.

THREE GENES, *ced-3*, *ced-4*, AND *ced-9*, DEFINE A GENETIC PATHWAY NEEDED FOR ALL PROGRAMMED CELL DEATHS

(See Cryns and Yuan, Chapter 6, this volume.) The original isolation of mutants defective in all cell deaths during nematode development by Ellis and Horvitz (1986) was successful because all cells that are fated to die use a common pathway to execute programmed cell death. To simplify searching for the desired mutants, screening was originally performed in a genetic background where dead cells are easily recognized because they cannot be degraded (for details see section on mutants defective in the engulfment process). Later screens focused on the identification of mutants with extra undead cells (Ellis and Horvitz, 1986; Ellis et al., 1991a; Hengartner et al., 1992). As a result of many such screens, numerous loss-of-function (lf) alleles of both *ced-3*, *ced-4*, and a single gain-of-function (gf) allele of *ced-9* were identified (Fig. 4.4) (Ellis and Horvitz, 1986; Hengartner et al., 1992; Hengartner and Horvitz, 1994a). The loss of function alleles of *ced-3* and *ced-4* can be classified into allelic series, the strongest alleles of which presumably result in a complete loss of gene function (an assumption supported by molecular and biochemical characterization of these mutations and their corresponding gene products).

   The isolation of *ced-9* loss-of-function (lf) mutants was accomplished by looking for intragenic revertants of a rare *ced-9* gain-of-function allele (Hengartner et al., 1992). The analysis of these revertants revealed extensive apoptosis within the homozygote animal (Fig. 4.4). As a consequence of these extensive deaths, the affected animals die during early embryogenesis (Hengartner et al., 1992). These observations suggest that *ced-9* normally functions to prevent programmed cell death. Genetically *ced-9* acts upstream of both *ced-3* and *ced-4*, as loss-of-function alleles of these genes are able to suppress the extensive cell death phenotype of *ced-9* (lf) animals (Hengartner et al., 1992). Mosaic analysis suggests that *ced-3* and *ced-4* are likely to act cell

The genetic pathway for programmed cell death in *C. elegans*

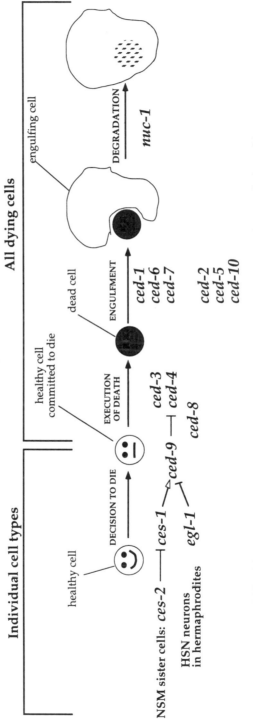

FIG. 4.3. The genetic pathway for programmed cell deaths in *C. elegans*. (→) positive regulatory interaction; (⊣) negative regulatory interaction. (Adapted from Ellis et al., 1991b; Hengartner and Horvitz, 1994c.)

Phenotypes

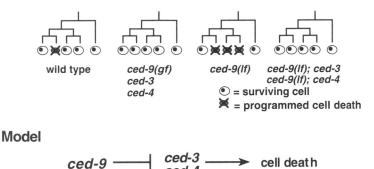

wild type   *ced-9(gf)*   *ced-9(lf)*   *ced-9(lf); ced-3*
           ced-3   *ced-9(lf); ced-4*
           ced-4   ◉ = surviving cell
                                 ✖ = programmed cell death

Model

$$ced\text{-}9 \quad \dashv \quad \begin{array}{c} ced\text{-}3 \\ ced\text{-}4 \end{array} \quad \longrightarrow \quad \text{cell death}$$

FIG. 4.4. Effects of mutations in *ced-3*, *ced-4*, and *ced-9* on the pattern of developmental deaths. Loss of *ced-3* and *ced-4* function, or gain of *ced-9* function prevent all cell deaths. Loss of *ced-9* function results in excessive deaths, which like the normal deaths can be blocked by the inactivation of either *ced-3* or *ced-4*. Genetic model explaining the observed genetic interactions. In this model, *ced-9* acts as a negative regulator of *ced-3* and *ced-4*, and is required to protect cells that should survive from inappropriately activating the cell death program.

autonomously, indicating that they are needed to act within the cells that die (Yuan and Horvitz, 1990). This conclusion is further supported by experiments showing that *ced-3* and *ced-4* overexpression, in neuronal cells that usually do not die, is sufficient to induce programmed cell deaths (Shaham and Horvitz, 1996a). In addition, these studies indicated that *ced-3* is likely to function downstream of *ced-4* because the induction of apoptosis by the overexpression of *ced-3* does not require *ced-4*, whereas in the converse experiment the induction of programmed cell death by *ced-4* overexpression requires the presence of *ced-3* (Shaham and Horvitz, 1996a).

ced-3 AND *ced-9* ARE FUNCTIONALLY CONSERVED THROUGHOUT EVOLUTION

Cloning of the above-mentioned genes revealed that these nematode cell death genes have homologs in mammals that perform similar functions in the control of apoptosis. For example, *ced-9* encodes a product sharing 24% overall identity with the mammalian Bcl-2 oncogene, which like *ced-9* negatively regulates cell death (Hengartner et al., 1994b). Overexpression of *bcl-2* protects cells from death, whereas *bcl-2* (lf) mutations make cells hypersensitive to death inducing signals (for reviews see Korsmeyer et al., 1993; Vaux, 1993; Reed, 1994). The sequence of *ced-3* revealed that it encodes a protein sharing 29% identity with the human interleukin 1β-converting enzyme (ICE), which is a member of a growing family of death-inducing proteases called caspases (Yuan et al., 1993; Alnemri et al., 1992, Cryns and Yuan, Chapter 6, this volume). As is the case for *ced-3* in *C. elegans,* activation of these caspases appears to be necessary and sufficient for apoptotic death of mammalian cells (Miura et al., 1993; Kumar et al., 1994; Wang et al., 1994). Despite the relatively low sequence conservation, the basic mechanism regulating apoptosis seems to be highly conserved since Bcl-2 can substitute for CED-9 in *C. elegans,* and CED-3 can activate apoptosis in mammalian cells (Vaux et al., 1992; Miura et al., 1993; Hengartner and Horvitz, 1994b). A vertebrate homolog of *ced-4* has recently been identified (Zou et al., 1997), confirming previous suggestions that a CED-4–like activity must also exist in mammals (Yuan and Horvitz, 1992; Wu et al., 1997; Chinnaiyan et al., 1997).

CED-3, CED-4, AND *CED-9* ARE PART OF A MOLECULAR FRAMEWORK REGULATING PROGRAMMED CELL DEATH

Recent studies on the biochemical interactions between CED-3, CED-4, and CED-9 indicate that these proteins are part of a molecular complex or network regulating programmed cell death (Chinnaiyan et al., 1997; Spector et al., 1997; Wu et al., 1997). The pattern of interactions between these proteins supports a model according to which the execution of programmed cell death, which is brought about by the active CED-3 protease, is antagonized by the affinity of CED-9 to a putative CED-3/CED-4 complex. The resulting association promotes the formation of an inactive complex that is targeted to mitochondrial and other intracellular membranes via CED-9. According to this model, the execution of the cell death program would be mediated by freeing CED-4 and CED-3 from the inactive CED-9 containing trimer (Fig. 4.5) (Cryns and Yuan, Chapter 6, this volume).

What arguments support this model? Overexpression of *ced-4* in mammalian cells (as is the case in *C. elegans*) is able to induce programmed cell death. This death-promoting activity, however, is inhibited by the concomitant expression of *ced-9* (Chinnaiyan et al., 1997). CED-4 and CED-9 physically interact with each other, as revealed by the two hybrid method and by coimmunoprecipitation experments (Chinnaiyan et al., 1997; Spector et al., 1997; Wu et al., 1997). This interaction appears to be direct, as it can also be reproduced *in vitro* (Spector et al., 1997). Consistent with the physiological importance of this interaction, products of *ced-9* loss-of-function alleles are unable to bind to CED-4 (Chinnaiyan et al., 1997; Spector et al., 1997; Wu et al., 1997). Loss-of-function mutants of CED-4, as well as CED-4L, an alternatively spliced variant of CED-4 that acts as an inhibitor of cell death, still maintain their capacity to bind to CED-9, suggesting that these mutations presumably affect the effector function of CED-4 (Chinnaiyan et al., 1997; Shaham et al., 1996b; Spector et al., 1997; Wu et al., 1997). Consistent with this suggestion, *ced-4L* acts as a dominant negative inhibitor of the *ced-4* death-promoting function

Death-inducing stimulus

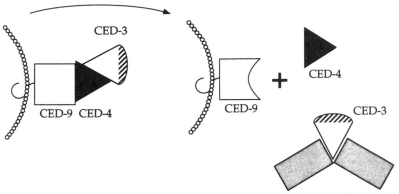

Proteolytic cleavage of CED-3 substrates

FIG. 4.5. A molecular model for CED-9 function. The inactive CED-9/CED-4/CED-3 trimer is bound to the mitochondrial membrane. A death promoting signal causes (by an as yet unknown mechanism) the dissociation of the trimer. This dissociation results into the activation of the CED-3 protease. The cleavage of CED-3 substrates then irreversibly commits the cell to apoptotic death.

if both genes are expressed in mammalian cells (Chinnaiyan et al., 1997). The relevance of the CED-4/CED-9 interaction is further corroborated by immunolocalization studies (in mammalian cells) that indicate that CED-4 is predominantly localized in the cytoplasm but co-localizes with CED-9 to intracellular membranes once CED-9 is expressed (Wu et al., 1997). CED-3 is able to interact directly with CED-4 and also associates to the CED-9 complex indirectly via its interaction with CED-4 (Chinnaiyan et al., 1997).

THE REGULATION OF PROGRAMMED CELL DEATH IN SPECIFIC CELL TYPES

In contrast to our advanced understanding of the molecular nature of the core apoptotic pathway, very little is known about the genes that commit specific subsets of cells to apoptotic death. The *C. elegans* system can also be efficiently used to address this problem. Indeed, mutations affecting specific subsets of cell deaths have been isolated in this species. In principle, there are two different classes of mutants that might affect cell-type-specific deaths. Some mutants might act indirectly, by affecting genes that are needed for specific cells to acquire their proper cell fate (e.g., to know that they have to die or need to survive). Alternatively, genes might directly be involved in the regulation of the cell death machinery in specific target cells. An illustrative example of the first class of mutations are those that affect the homeobox gene *lin-32* (Zhao and Emmons, 1995; Chalfie and Au, 1989). This gene is necessary for neuroblast fate specification, and its inactivation leads to the absence or gross alterations of many neuronal lineages. Thus programmed cell deaths that would normally have occurred cannot happen because cells that should die fail to be generated.

In contrast to the above-mentioned example, some genes are thought to directly affect the decision of specific cell types to undergo programmed cell death. For example, dominant gain-of-function (gf) mutations in the *egl-1* gene (which is not yet cloned) cause the inappropriate induction of programmed cell death of the two seotonergic HSN neurons in hermaphrodites (Desai et al., 1988; Desai and Horvitz, 1989). These two neurons are used in hermaphrodites for the innervation of vulval muscle cells to regulate egg laying. In males, which do not need to lay eggs, the HSNs are also generated, but subsequently eliminated by programmed cell death. The deaths of the HSNs in *egl-1* hermaphrodites elicit an egg-laying defect (hence the name of the gene). The *egl-1* (gf)-induced HSN killing is likely to directly or indirectly affect components of the core apoptotic machinery because both HSN cell death and the egg-laying defect are readily suppressed by mutations in *ced-3* and *ced-4* (Ellis and Horvitz, 1986; Hengartner et al., 1992). One possible explanation of the *egl-1* (gf) phenotype is that these mutations cause the HSN neurons to conclude wrongly that they are in a male, and thus should die.

A further example of genes controlling the induction of apoptosis in a specific subset of cells are the *ces* (cell death specification) genes, which affect the apoptosis of the sister cells of the pharyngeal NSM neurons. Mutations in the two *ces* genes (*ces-1* and *ces-2*) define a liner pathway that leads to the cell-type-specific activation of apoptosis (Ellis and Horvitz, 1991). Epistasis analysis reveals that within this pathway *ces-2* seems to be a negative upstream regulator of *ces-1*. The evidence for the epistatic relationship is derived from the analysis of various single and double mutants. *ces-1* gain-of-function mutants prevent the deaths of the NSM sisters that usually die (as well as preventing the deaths of two additional cells, the sisters of the I2 neurons). NSM sister cell survival is also triggered by loss of function of *ces-2*. Since the cell survival caused by loss of *ces-2* function is suppressed by loss of *ces-1* function, *ces-1* probably acts downstream of *ces-2*. As is the case for *egl-1* mutations, *ces-1* is likely to act upstream of the core apoptotic machinery because the death observed in *ces-1* (lf) is still readily suppressed by *ced-3* mutants (Ellis and Horvitz, 1991). The *ces-2* gene encodes a putative bZIP transcription factor, suggesting that some nematode cell deaths might be regulated at a transcriptional level (Metzstein et al., 1996).

GENES REQUIRED FOR THE ENGULFMENT OF DYING CELLS

We have only a very limited knowledge of the mechanisms required for the engulfment of dying cells in *C. elegans*. Indeed, it is not known what the signals are that lead to the endocytosis of apoptotic cells. Importantly, we do not even know from where these signals emanate. Is the dying cell actively informing its neighbors that it has to be dealt with, or do cells have the capacity to sense that their neighbor is about to pass away and that it has to be engulfed? In any case, the engulfment of dying cells is a highly efficient process that occurs within less than an hour of the first visible morphological changes that announce an apoptotic death. The process of engulfment is a very early event in apoptosis, as it can begin even before a cell that will die has been completely separated from its sister by cell division (Robertson and Thomson, 1982; Ellis et al., 1991a). As engulfment proceeds, the dying cell splits into membrane-bound fragments, and the nuclear membrane degenerates. After dead cells have been engulfed, a nuclease controlled by the gene *nuc-1* digests the DNA of the dead cell (Sulston, 1976; Hedgecock et al., 1983; Hevelone and Hartmann, 1988). Engulfment during embryogenesis is performed mostly by the sisters of the dying cells. In larvae and adults, epithelial cells usually perform this function. There are no professional phagocytes in *C. elegans*.

Six genes (*ced 1, 2, 5, 6, 7, 10*) have been identified, which are required for the engulfment of dying cells (Hedgecock et al., 1983; Ellis et al., 1991a). Unfortunately, cloning of these genes has not yet been reported. In animals mutant for any of the six known engulfment genes, cells death proceeds as usual, but many of the dying cells remain unengulfed and are visible for many hours as flat rounded disks under Nomarski optics (Hedgecock et al., 1983; Ellis et al., 1991a). The fact that cells die in these mutants indicates that the engulfment process is not the cause of most programmed cell deaths (Hedgecock et al., 1983). All the engulfment mutants (even the strongest alleles of a given gene) are only partially defective in the engulfment process, and consequently many of the dead cells are still properly engulfed. The analysis of double mutant combinations led to the division of these six genes into two subgroups, which are proposed to affect two distinct, but partially redundant processes affecting the engulfment of cell corps (Ellis et al., 1991a). In addition to the above-mentioned genes, there might be many other genes involved in engulfment, as a large number of deficiencies results, in the homozygous state, in arrested embryos that contain unengulfed cell corpses (Ahnn and Fire, 1994). Future analysis will have to concentrate on cloning of the genes involved in the engulfment process. One may hope that these studies will help to elucidate the molecular mechanisms needed for the execution of the engulfment program.

POTENTIAL FUNCTIONS OF PROGRAMMED CELL DEATH DURING THE DEVELOPMENT OF *C. elegans*

Genetic analysis of apoptotic cell death in *C. elegans* is facilitated by the fact that mutant animals defective in apoptosis remain viable and seem to have no gross deleterious effects on development, fertility, and behavior (Ellis and

Horvitz, 1986). Only careful analysis reveals that mutant animals develop more slowly, have a slightly reduced fertility, and are affected in complex behavioral tasks such as chemotaxis (Ellis et al., 1991b). Although this lack of an overall phenotype provides advantages for the propagation of mutant animals, it leaves us with the question about the function of cell deaths during *C. elegans* development. The lack of gross alterations in the behavior of animals containing about 12% more cells due to the absence of apoptotic deaths seems especially surprising as many of the "undead cells," although they never further divide, terminally differentiate (Ellis and Horvitz, 1986). Some of them at least partially take over the fate of cells that are closely related to them by lineage (White et al., 1991). An illustrative example of such a process is provided by the "undead" sister of the M4 neuron (Avery and Horvitz, 1987). The M4 neuron is needed for the proper operation of the pharynx in young animals, and its killing (e.g., by laser microsurgery) leads to the starvation of operated animals. However, if the M4 cell is ablated in animals defective for apoptosis, its undead sister cell is able to take over its function and rescues the animals from death by starvation (Avery and Horvitz, 1987).

In general, most of the arguments for the prevalence of apoptosis are in an evolutionary context (for review, see Hengartner, 1997). Apoptosis might, for instance, be a useful evolutionary strategy for morphogenetic modeling as condemning specific cells to undergo apoptosis might be an efficient way to remodel overall morphological structures. Indeed, the death of a single cell can account for major anatomical changes (Sternberg and Horvitz, 1981). For example, the gross difference between the one-armed gonad of *Panagrellus revidus* and the two-armed gonad of *C. elegans* can be explained by the death of the posterior distal tip cell in *P. revidus* (distal tips cells direct the formation of gonadal arms). Another example of morphological modeling might be the refinement of sterotypical lineages that are used repeatedly. Within the ventral cord, the 12 Pn.a neuroblasts follow a stereotypical pattern to generate five cells each (Fig. 4.2). One of these five cells innervates the vulval muscles, and presumably is involved in the control of egg laying. However, only the descendants of the middle six Pn.a blast cells are close enough to the vulval muscles to innervate them efficiently. In the other six lineages, the equivalent cell is thus eliminated by programmed cell death (for review, see Ruvkun, 1997). The use of programmed cell death to eliminate the superfluous cell was presumably an easier evolutionary strategy than the creation of a new cell division pattern in which these cells would not be generated (for review, see Hengartner, 1997).

CONCLUSION

Genetic studies in *C. elegans* have led to the identification of key elements of the core apoptotic pathway. Surprisingly, most components of this core pathway might already be uncovered, because genetic screens aimed at their identification seem to be saturated. Consistently, recent biochemical studies indicate that CED-9, CED-4 and CED-3 physically interact with each other. CED-4 and CED-3 have death-promoting functions that are inhibited once they interact with CED-9. Since vertebrate homologs of *ced-3* and *ced-9* are known to perform similar key roles in the control of mammalian apoptosis, the execution of programmed cell death seems to be an evolutionary conserved program. Indeed, the main difference between the *C. elegans* cell death program and its vertebrate counterpart seems to be the higher complexity and redundancy found in the latter system. Therefore, whereas *ced-9* and *ced-3* are unique genes in *C. elegans,* there are multiple homologs of each in mammals (Cryns and Yuan, Chapter 6, this volume).

After the identification of the core components of the apoptotic pathway, future research will have to focus on two different aspects of apoptosis. One line of research will try ot identify the biologically relevant substrates of the CED-3 protease (see Cryns and Yuan, Chapter 6, and Sikorska and Walker Chapter 7, this volume). Since one expects that there might be several such substrates, future approaches will probably require biochemical methodology. Alternatively, it might also be possibly to define *ced-3* substrates using genetic approaches, such as screening for mutants that are defective only in a distinct subset of apoptotic responses. The second avenue of future research will aim at understanding those mechanisms that shift the death-promoting core machinery into its active state. We know next to nothing about these mechanisms. Activation could, for instance, be accomplished through proteolytic degradation of CED-9 or by preventing it from interacting with CED-3 and CED-4. In addition, apoptosis might also be stimulated by increasing the relative concentrations of CED-4 and/or CED-3. Alternatively, one might also speculate whether any of the components of the core pathway are regulated by posttranslational modifications (see Zakeri, Chapter 3, and Trump and Berezesky, Chapter 2, this volume).

In our search for answers to these questions, studies in *C. elegans* might provide important contributions. Indeed, there are several genes known—and there may be many more to be identified—that affect the deaths of only a specific subset of cells. By carefully analyzing the mechanism leading to the commitment of specific cells to programmed cell death, one might in the long run identify those biologically relevant interactions that impinge on the core apoptotic pathway. In addition to these questions, the mechanisms leading to the engulfment of dead cells are just beginning to be uncovered. Here, one certainly expects a rich harvest of the data that will eventually help to elucidate the mechanisms that signal that a cell has to be engulfed.

ACKNOWLEDGMENTS

Work in the laboratory is supported by U.S. Public Health Service Grant GM52540 and through generous support of the Donaldson Charitable Trust. M. O. Hengartner is a Rita Allen Foundation Scholar. A. Gartner is supported by a Max Kade Post-Doctoral Fellowship.

REFERENCES

Ahnn J, Fire A (1994): A screen for genetic loci required for body-wall muscle development during embryogenesis in *Caenorhabditis elegans*. Genetics 137:483–98.

Alnemri ES, Robertson NM, Fernandez CM, Croce CM, Litwack G (1992): Overexpressed full-length human BCL-2 extends the survival of baculovirus-infected Sf9 insect cells. Proc Natl Acad Sci USA 89:7295–99.

Avery L, Horvitz HR (1987): A cell that dies during wild type *C. elegans* development can function as a neuron in a *ced-3* mutant. Cell 51:1071–78.

Chalfie M, Au M (1989): Genetic control of differentiation of the *Caenorhabditis elegans* touch receptor neurons. Science 243:1027–33.

Chinnaiyan AM, O'Rourke K, Lane BR, Dixit VM (1997): Interaction of CED-4 with CED-3 and CED-9: A molecular framework for cell death. Science 275:1122–26.

Desai C, Garriga G, McIntire SL, Horvitz HR (1988): A genetic pathway for the development of the *Caenorhabditis elegans* HSN motor neurons. Nature 336:638–46.

Desai C, Horvitz HR (1989): *Caenorhabditis elegans* mutants defective in the functioning of the motor neurons responsible for egg laying. Genetics 121:703–21.

Ellis HM, Horvitz HR (1986): Genetic control of programmed cell death in the nematode *C. elegans*. Cell 44:817–29.

Ellis RE, Horvitz HR (1991): Two *C. elegans* genes control the programmed cell deaths of specific cells in the pharynx. Development 112:591–603.

Ellis RE, Jacobson DM, Horvitz HR (1991a): Genes required for the engulfment of cell corpses during programmed cell death in *Caenorhabditis elegans*. Genetics 129:79–94.

Ellis RE, Yuan J, Horvitz HR (1991b): Mechanisms and functions of cell death. Annu Rev Cell Biol 7:663–98.

Hedgecock EM, Sulston JE, Thomson NJ (1983): Mutation affecting programmed cell death in the nematode. *Caenorhabditis elegans*. Science 220:1277–79.

Hengartner MO (1997): Cell death. In *C. elegans II*. Cold Spring Harbor Laboratory Press, pp 383–416.

Hengartner MO, Ellis RE, Horvitz HR (1992): *Caenorhabditis elegans* gene *ced-9* protects cells from programmed cell death. Nature 356:494–99.

Hengartner MO, Horvitz HR (1994a): Activation of *C. elegans* cell death protein CED-9 by an amino acid substitution in a domain conserved in Bcl-2. Nature 369:318–20.

Hengartner MO, Horvitz HR (1994b): *C. elegans* cell survival gene *ced-9* encodes a functional homolog of the mammalian protooncogene *bcl-2*. Cell 76:665–76.

Hengartner MO, Horvitz HR (1994c): The ins and outs of programmed cell death in *C. elegans*. Philos Trans R Soc Lond (Biol) 345:243–46.

Hevelone J, Hartman PS (1988): An endonuclease from *Caenorhabditis elegans:* Partial purification and characterization. Biochem Genet 26:447–61.

Kimble J, Hirsch, D (1979): The postembryonic cell lineages of the hermaphrodite and male gonads in *Caenorhabditis elegans*. Dev Biology 87:396–417.

Korsmeyer, SJ, Shutter DJ, Veis DJ, Merry, DE Oltavi ZN (1993): Bcl-2/Bax: A rheostat that regulates an anti-oxidant pathway and cell death. Semin Cancer Biol 4:327–33.

Kumar SM, Kineshita M, Noda M, Copeland NG, Jenkins NA (1994): Induction of apoptosis by mouse *Nedd2* gene, which encodes a protein similar to the product of the *Caenorhabditis elegans* cel death gene *ced-3,* and the mammalian IL-1β converting enzyme. Genes Dev 8:1613–26.

Metzstein MM, Hengartner MO, Tsung N, Ellis RE, Horvitz HR (1996): Transcriptional regulator of programmed cell death encoded by *Caenorhabditis elegans* gene *ces-2.* Nature 382:545–47.

Miura M, Zhu H, Rotello R, Hartwieg EA, Yuan J (1993): Induction of apoptosis in fibroblasts by IL-1β converting enzyme, a mammailan homolog of the *C. elegans* cell death gene *ced-3.* Cell 75:653–60.

Raff MC (1992): Social controls on cell survival and cell death. Nature 356:397–400.

Reed JC (1994): Bcl-2 and the regulation of programmed cell death. J Cell Biol 124:1–6.

Robertson A, Thomson N (1982): Morphology of programmed cell death in the ventral nerve cord of *Caenorhabditis elegans* larvae. J Embryol Exp Morphol 67:89–100.

Ruvkun G (1997): Patterning the nervous system. In *C. elegans II,* Cold Spring Harbor Laboratory Press, pp 543–82.

Shaham S, Horvitz HR (1996b): An alternatively spliced *C. elegans ced-4* RNA encodes a novel cell death inhibitor. Cell 86:201–8.

Shaham S, Horvitz HR (1996a): Developing *Caenorhabditis elegans* neurons may contain both cell death protective and killer activities. Genes Dev 10:578–91.

Spector MS, Desnoyers S, Hoeppner DL, Hengartner MO (1997): Interaction between the *C. elegans* cell-death regulators CED-9 and CED-4. Nature 385:653–56.

Sternberg PW, Horvitz HR (1981): Gonadal cell lineages of the nematode *Pangrellus redivius* and implications for evolution by the modification of cell lineage. Dev Biol 88:147–66.

Sulston JE (1976): Post-embryonic development in the ventral cord of *Caenorhabditis elegans.* Philos Trans R Soc Lond (Biol) 275:287–97.

Sulston JE, Horvitz HR (1977): Post-embryonic cell lineages of the nematode, *Caenorhabditis elegans.* Dev Biol 56:110–56.

Sulston, JE, Schierenberg E, White JG, Thomson JN (1983): The embryonic cell lineage of the nematode *Caenorhabditis elegans.* Dev Biol 100:64–119.

Vaux DL (1993): Toward an understanding of the molecular mechanisms of physiological cell death. Proc Natl Acad Sci USA 90:786–89.

Vaux DL, Weissman IL, Kim SK (1992): Prevention of programmed cell death in *Caenorhabditis elegans* by human *bcl-2.* Science 258:1955–57.

Wang LM, Miura L, Bergeron L, Zhu H. Yuan J (1994): *Ich-1,* an *Icel/ced-3*-related gene, encodes both positive and negative regulators of programmed cell death. Cell 78:739–50.

White JG, Southgate E, Thomson JN (1991): On the nature of undead cells in the nematode *Caenorhabditis elegans.* Philos Trans R Soc Lond (Biol) 331:263–71.

Wu D, Herschel DW, Nuñez G (1997): Interaction and regulation of subcellular localisation of CED-4 and CED-9. Science 275:1126–1129.

Yuan J, Horvitz HR (1992): The *Caenorhabditis elegans* cell death gene *ced-4* encodes a novel protein and is expressed during the period of extensive cell death. Development 116:309–20.

Yuan J, Horvitz HR (1990): The *Caenorhabditis elegans* genes *ced-3* and *ced-4* act cell autonomously to cause programmed cell death. Dev Biol 138:33–41.

Yuan J, Shaham S, Ledoux S, Ellis HM, Horvitz HR (1993): The *C. elegans* cell death gene *ced-3* encodes a protein similar to mammalian interleukin-1-β converting enzyme. Cell 75:641–52.

Zhao C, Emmons SW (1995): A transcription factor controlling development of peripheral sense organs in *C. elegans*. Nature 377:74–78.

CHAPTER 5

CELL DEATH IN PLANTS

RON MITTLER
Department of Plant Sciences, Hebrew University of Jerusalem,
Jerusalem 91904, Israel

INTRODUCTION

A molecular mechanism for eliminating developmentally misplaced or unwanted cells is essential for the successful development and growth of complex multicellular organisms. Thus, in addition to regulating the rate of cell division, multicellular organisms such as animals and plants contain a biochemical pathway to control cell death. By coordinating the activation of cell division and cell death, animals and plants may direct a variety of developmental processes. These include the generation of developmental patterns and the shaping of cells, tissues, and organs. However, cell death may not be limited to development, and it may also be used in a number of other processes such as control of cell populations and defense against invading microbes (Ellis and Horvitz, 1986; Raff, 1992; Greenberg, 1996; Jones and Dangl, 1996; Mittler and Lam, 1996).

Much as the mechanism of cell division is tightly controlled and complex, so is the mechanism of cell death. Moreover, although the requirements from a cell death pathway appear offhand to be simple, they in fact depend upon the particular organism and the context of cell death. Therefore, cell death may be manifested in many different forms, having different molecular and morphological characteristics. Despite these apparent differences, a general functional term is used to describe all cell deaths that result from the activation of a specific pathway. This term is *programmed cell death* (pcd). The definition of pcd is, therefore, death that results from the activation of a genome-encoded biochemical pathway for cell suicide. This pathway includes not only proteins and enzymes that participate in the regulation and execution of cell death but

When Cells Die, Edited by Richard A. Lockshin, Zahra Zakeri, and Jonathan L. Tilly
ISBN 0-471-16569-7 © 1998 Wiley-Liss, Inc.

also mechanisms involved in the dismantling and recycling of the dead cell's corpse (Schwartzman and Cidlowski, 1993; Vaux, 1939; Martin et al., 1994).

A classic example of pcd is apoptosis. It is characterized by a distinct set of physiological and molecular characteristics. These include the activation of specific proteases and Ca^{2+}-dependent nucleases and an ordered fragmentation of DNA. These biochemical changes are accompanied by drastic morphological alterations, which include cell shrinkage, fragmentation of the nuclei, and the formation of apoptotic bodies that contain cellular debris. The apoptotic bodies that are formed during apoptosis are engulfed and recycled by neighboring cells or specific macrophages, thus resulting in the complete elimination of the cell. However, it should be emphasized that not all cases of pcd take the form of apoptosis (Wyllie et al., 1980, 1984; Schwartz et al., 1993; Kerr, 1995; see also preface and Chapter 3 (Zakeri), this volume.).

Plants serve as an excellent example of how the specific organism and the context of cell death may determine the phenotype but not the function of pcd. For example, unlike animal cells, plant cells have walls that may act as physical barriers preventing the recycling of cellular material from dead cells via apoptotic bodies. Therefore, recycling of cellular content from dead plant cells may occur by degradation of cell debris to low-molecular-weight compounds that are taken up by neighboring cells. Such a process of releasing cellular debris into the intercellular space would have caused an inflammatory response in a mammal. However, plants are also different from animals by not having such an immune response. Therefore, although the function of cell death (i.e., eliminating or killing a cell) may be similar, the mechanisms involved may be very different and specific for the particular organism. The context of cell death is also critical. For example, the cellular differentiation and subsequent formation of certain plant tissues such as the water-conducting system of the plant (i.e., the xylem, Figs. 5.1A, 5.1B, and 5.2), or the cork layer that replaces the epidermis in stems of mature woody plants, requires that cells will die with only the functional cell walls or parts of the cell corpse remaining. These become integral, functional, and essential components of plant anatomy. Therefore, the complete elimination of a cell is not the preferred route in these examples (Greenberg, 1996; Jones and Dangl, 1996; Mittler and Lam, 1996).

In this chapter the occurrence and the particular roles pcd play in plants will be discussed as well as the molecular and physiological mechanisms involved. By studying pcd in organisms that are different from animals in many cellular, subcellular, and anatomical aspects, yet highly ordered and multicellular, we may gain a better overall understanding of this remarkable process and its role in biology.

OCCURRENCE OF PROGRAMMED CELL DEATH IN PLANTS

Like most multicellular organisms, higher plants depend upon pcd for ensuring their proper development and growth. However, unlike animals in which morphogenesis is determined by cell division, cell migration, and cell death, morphogenesis in plants is primarily determined by cell division and cell death (i.e., no cell migration). Another aspect of plant life that involves pcd is the interactions of plants with their environment. Thus the defense of plants against biotic and abiotic agents often involves activation of pcd. The occurrence of pcd in plant development and response to the environment is addressed in the following.

Developmental pcd in Plants: From Reproduction to Senescence

Pcd is activated during many developmental processes in plants. The following is a compendium of some of the known examples of developmental pcd in plants starting from the formation of sexual organs and ending in senescense.

Pcd During Reproduction in Plants

Floral Organ Abortion. Many plant species contain unisexual flowers. However, during the formation of these, primordia for both androecium (male) and gynoecium (female) organs are initiated. In order to generate a unisexual flower, one of these primordia types undergoes developmental arrest, followed by pcd. Recently, a gene with sequence similarity to hydroxy steroid reductase (*TS2*) was shown to be involved in this process in maize (DeLong et al., 1993).

Megaspore Abortion. Pcd is involved in the formation of the female gametes in seed plants. Thus a single meiotic event results in the formation of four haploid cells, three of which undergo pcd, leaving the remaining one to produce, following two additional mitotic divisions, the egg and associated cells of the embryo (Bell, 1996).

Tapetum Degeneration. Pcd is also involved in the formation of the male sexual organs and during the maturation of pollen the surrounding tapetum layer undergoes pcd. This process is believed to supply the developing pollen with nutrients (Greenberg, 1996).

Pcd Preventing Self-Pollination. Plants evolved several mechanisms to prevent self-pollination and increase their genetic variability. One of these involves inhibiting or killing of germinating self-pollen grains upon recognition by the transmitting tissue of the female flower (pistil). This process is mediated by a class of proteins called S proteins that were shown to have an RNase activity, essential for their function (Kao and McCubbin, 1996). At least two more independent types of pcd are known to occur before the successful fertilization of the egg:

1. Pcd in the transmitting tissue: The growth of the pollen tube through the pistil is accompanied by selective cell death. Thus pistil cells along

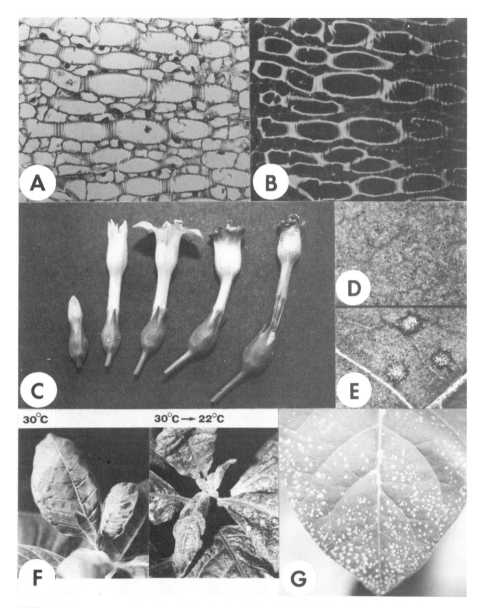

FIG. 5.1. Programmed cell death in plants. (A) and (B) Cell death during development: Pcd of tracheary elements (TEs): (A) a cross-section through a tobacco petiole showing mature TEs with their characteristic thick and highly patterned walls. These TEs are dead cells that function as tubes that conduct water from the stem to the leaf. The process of TE pcd is described in more detail in Fig. 5.2. (B) The same cross-section shown in (A), however, observed with a fluorescent microscope. The autofluorescence of TEs walls is the result of excessive lignification that accompanies the development and death of TEs and results in the formation of sturdy cell walls that provide the

the growth track of the pollen tube undergo pcd while the rest of the pistil tissue remains intact (Wang et al., 1996b).

2. *Synergid death:* Two synergid cells are present at the entry to the egg sack; one of them must undergo pcd in order for the arriving pollen tube to enter and release its sperm cells.

Pcd During Embryogenesis and Germination in Plants

Pcd of the Embryo Suspensor. Two cell lineages are formed by the first division of the zygote. One leads to the formation of the embryo, and the other forms the suspensor, which is the functional equivalent of the trophoblast in mammals. When the developing embryo reaches the heart stage, the role of the suspensor is complete, and it undergoes pcd (Yeung and Meinke, 1993; Jones and Dangl, 1996).

Cell Death in Storage Tissue. The development and germination of the embryo depends upon an ample supply of nutrients, which is found in the seed. This supply originates from endosperm cells that undergo pcd. In barley this process is assisted by lytic enzymes such as DNases that are secreted from the aleurone layer, which surrounds the endosperm (Brown and Ho, 1987).

---

plant with mechanical strength. (C) Organ senescence in plants: The development and senescence of tobacco flowers. Following pollination and fertilization of eggs petals as well as other organs of the flower undergo pcd. This process of organ senescence, as well as the process of whole plant senescence, is regulated by certain plant hormones such as ethylene and cytokinin and involves the activation of many cellular mechanisms. (D) and (E) Tobacco mosaic virus (TMV) induced hypersensitive response (HR) lesions: The antimicrobial defense of certain plants often involves the activation of pcd at the invasion site of the pathogen. This results in the formation of a lesion and the blocking of pathogen spread. (D) A mock-inoculated leaf and (E) a TMV-inoculated leaf with typical HR lesions. TMV–tobacco interactions are described in more detail in Fig. 5.3. (F) Inhibition of TMV-induced pcd in tobacco at 30°C and activation of cell death in systemically infected cells upon a temperature shift to 22°C. The process of TMV-induced pcd in tobacco is inhibited at 30°C resulting in the systemic spread of TMV (left panel). However, upon shifting of systemically infected plants to 22°C pcd is activated (right panel). Since TMV spreads to almost all parts of the plant, the activation of pcd is not localized to the site of invasion [forming a lesion as shown in (E)] but results in the complete death of the plant. (G) Activation of pcd in transgenic tobacco plants expressing a bacterial proton pump: Transgenic tobacco plants that express the bacterio-opsin (bO) gene develop spontaneous lesions, activate multiple defense mechanisms, and become resistant to infection by viruses and bacteria. This transgenically activated pcd is very similar to the pathogen-induced HR. It is thought that the function of the bO gene as a proton channel results in the activation of the HR pcd pathway.

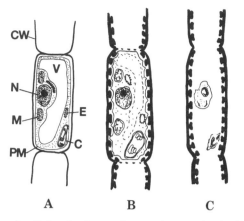

FIG. 5.2. Programmed cell death of a tracheary element (TE). (A) A TE progenitor cell that originated from the procambium. (B) Following elongation and extensive cell wall synthesis the TE undergoes pcd. It is thought that pcd of TEs is initiated by the rupture of the vacuole, which is followed by the destruction of cellular content. No apoptotic bodies are formed and the nuclei does not fragment. Rupture of the plasma membrane occurs late during this pcd process. (C) A mature TE which functions, by end-to-end association with other TEs, as a water-conducting tube. The remains of the nuclei and other cellular debris may be found in mature TEs. Abbreviations: C, chloroplast; CW, cell wall; E, endoplasmic reticulum; M, mitochondria; N, nuclei; PM, plasma membrane; V, vacuole.

Cell Death in Maturing Plant Organs

Formation of Tracheary Elements. Tracheary elements (TE; Figs. 5.1A and 5.1B) are the characteristic, essential part of the zylem tissue of plants. They are dead hollow cells that by end-to-end association form the water- and ion-conducting system of the plant. Therefore, the final stage in their differentiation is pcd (Fukuda, 1996). The process of TE death is described in more detail in the following. During the growth of the mature plant pcd also occurs in roots, leaves, and stems.

Death of Root Cap Cells. Plant cells that cover the root cap function to protect the root tip during its growth into soil by secreting a polysaccharide slime and by providing the root tip with a protective layer of dead cells. It was recently suggested that death of root cap cells may be the result of an active process of cell death (i.e., pcd) (Wang et al., 1996a).

Morphogenesis of Leaf Structure. Although the structure of most leaves is determined by differential cell and tissue growth, at least in the genus *Montsera* patches of cells die at early stges of leaf development, resulting in the formation of slits and holes in the mature leaf (Kaplan, 1984; Greenberg, 1996).

Cell Death at the Stem. The continuous growth of the stem is also thought to result in cell death. Thus cell division in the cambium layer causes cell death in the cork layer, which replaces the ruptured epidermis, and in parenchyma cells at the stem's pith (Fahn, 1982).

Cell Death During Senescence. Two types of senescence are known to occur in plants: (1) organ senescence, a type of cell death that occurs during the senescence of leaves, petals, fruits, branches, and roots (Fig. 5.1C); (2) whole plant senescence, a type of death often found in monocapric plants following fertilization and seed development. Senescence in plants is an active process that follows a defined genetic program and thus should be referred to as pcd. It is accompanied by many biochemical and structural changes, which include among others the induction of cysteine proteases, RNases, and lipoxygenases. In addition, the process of senescence was shown to be modulated by the plant hormones ethylene and cytokinin (Smart, 1994; Gan and Amasino, 1995; Valpuesta et al., 1995; Greenberg, 1996; Orzaez and Granell, 1996); see also Crynsand Yuan (Chapter 6) and Sckorska and Walker (Chapter 7) this volume.

Environmental pcd in Plants: Biotic and Abiotic

In animals pcd is activated in response to many environmental insults such as toxins, various chemicals, and high-energy radiation, and in response to attack by pathogens. Plants also utilize pcd in similar manners.

Activation of pcd in Response to Abiotic Stress

Pcd of Hair Cells. Adaptations of plants to environmental conditions such as high light and low humidity often involves covering their surfaces with a thick layer of dead unicellular hairs. These cells are thought to undergo pcd, resulting in the formation of a protective layer that functions to block damaging high irradiance and trap humidity (Greenberg, 1996).

Autolysis of the Aerenchyma. Submerged roots, stems, and petioles often suffer from a condition of low oxygen tension. In order to facilitate the transfer of gas through these tissues, patches of cells undergo pcd to form pockets and channels of air. This tissue is called *aerenchyma*. Interestingly, unlike most other cases of pcd in plants, this pcd process results in the complete elimination of the cell. Recently, the plant hormone ethylene was implicated in regulating this cell death process. It should, however, be indicated that aerenchyma may also be formed by separation of cells and growth. This process may accompany the normal development of some plant species and may not be the result of anoxic conditions (He et al., 1996a,b).

Activation of pcd in Response to Pathogens (Biotic Stress). At least two types of cell deaths may occur following infection of a plant with a pathogen: (1) The hypersensitive response (HR): a rapid pcd process, which is activated in some plants in order to inhibit the spread of an invading pathogen. During this response the infected plant is sacrificing some of its cells in order to circle the invading pathogen with a layer or a ring of dead cells (Figs. 5.1D and 5.1E). This type of pcd will be discussed in detail in the following (Goodman and Novacky, 1994; Dangl et al., 1996). (2) Disease symptoms: This type of cell death, which appears relatively late during the development of some diseases, is considered to result from toxins produced by the invading pathogen. However, certain mutants were shown to develop cell death associated-disease symptoms in the absence of pathogens. In addition, toxins such as the AAL toxin, which is produced by the fungus *Alternaria alternata,* are thought to induce pcd in plants. Therefore, cell death that occurs during the development of a disease may result from the activation of a cell death program in the infected plant (Dangl et al., 1996; Jones and Dangl, 1996; Walton, 1996).

PCD DURING THE DIFFERENTIATION OF TRACHEARY ELEMENTS

The terminal differentiation of a tracheary element (TE), which functions at maturity as a water-conducting tube, is perhaps the best studied and the most common form of developmental cell death in plants. TEs can differentiate from progenitor cambium and procambium cells during the normal growth and development of the plant, or they may differentiate *in culture* from leaf mesophyll cells following injury. At maturity TEs are dead hollow cells with a highly patterned cell wall structure (Figs. 5.1A and 5.1B). The process of TE differentiation would, therefore, require that a plant cell will synthesize secondary cell walls and die (Fig. 5.2). The study of TE differentiation was greatly facilitated by the development of an *in vitro* system in which leaf mesophyll cells treated with indolacetic acid (IAA) and cytokinin differentiated *in culture* into mature dead TEs. Microscopic observation, video time-lapse microscopy, and molecular and biochemical analysis of TE differentiation revealed that although TE differentiation is a *bona fide* pcd process, it is distinct from apoptosis in animals (Fukuda, 1996).

The differentiation of TEs can be divided into four ontogenic processes: cell origination, cell elongation, secondary cell wall deposition and lignification, and pcd. Some of the molecular events that accompany the early stages of TE differentiation include rearrangement of actin filaments, increase in tubulin and DNA synthesis, increase in endoplasmic reticulum, vesicles, and mitochondria, and high transcription and translation activities. The involvement of the Ca^{2+}/calmodulin system in regulating early stages of TE differentiation was also suggested. During late stages of TE differentiation the majority of cellular activities are directed at secondary cell wall synthesis and thickening;

thus the syntheses of xylan, cellulose, and lignin are greatly facilitated. Peroxidase activity is also induced (Northcote, 1995; Fukuda, 1996; Trump and Berezcoky, Chapter 2; Lockshin, Preface, and Sckorska and Walker, Chapter 7, this volume).

Pcd of TEs is accompanied by an increase in nuclease activities and an increase in the activity of cysteine proteases. However, the phenotype of cell death is unique and may represent a distinct pathway for pcd. Most of the degenerative processes that accompany TEs death initiate with the onset of vacuole rupture (Fig. 5.2). The large plant vacuole, which is a structure relatively unique to plant cells, contains many lytic enzymes that may be released upon its rupture. In addition, the pH of the vacuole is acidic and may cause a drastic acidification of the cytosol. Although the nuclei of TE do not fragment during pcd, TUNEL staining (i.e., staining for the presence of 3' OH ends of degraded nuclear DNA) indicated that TE differentiation is accompanied by an active process of DNA degradation (see also Sckorska and Walker, Chapter 7, this volume). However, no nucleosomal ladders were reported to occur during this process. Therefore, despite some similarities between this process and apoptosis (i.e., increase in nuclease and protease activities, positive TUNEL staining, and acidification of the cytosol), the sequence of morphological events and the ultrastructural characteristics of TEs death are very different from apoptosis in animals. Pcd of maturing TEs may therefore represent a unique pathway for pcd (Thelen and Northcote, 1989; Minami and Fukuda, 1995; Mittler and Lam, 1995b; Fukuda, 1996).

PATHOGEN-INDUCED PROGRAMMED CELL DEATH IN PLANTS

Plant–Pathogen Interactions and the Hypersensitive Response

Plants contain many preexisting structural and chemical barriers that prevent infection by most pathogens (Osbourn, 1996). However, certain pathogens evolved mechanisms that enable them to penetrate these barriers. Invasion of such pathogens into the plant tissue often activates an array of secondary defense mechanisms. These include the synthesis of antimicrobial compounds called phytoalexins, the induction of pathogenesis-related (PR) proteins, and the accumulation of salicylic acid (SA), a key mediator of defense response activation (Malamy et al., 1990; Metraux et al., 1990; Ward et al., 1991; Glazebrook and Ausubel, 1994; Shulaev et al., 1995). Many of the PR proteins studied exhibit *in vitro* antimicrobial activity. These most often encode for lytic enzymes such as chitinases and glucanases, which attack the pathogen cell wall. The PR proteins that are induced during the response of plants to pathogens are secreted into the intercellular spaces of the plant, where they are most likely to encounter the invading bacteria or fungi. In addition, PR proteins accumulate in the plant vacuole and may be released upon cell death

(Bowles, 1990; Linthorst, 1991). Other proteins that may function to attack the invading pathogen together with chitenases and glucanases are RNases, DNases, and phospholipases (Mittler and Lam 1995a; Moiseyev et al., 1994). In addition, signals such as SA that are produced at the site of infection are often transported to other parts of the plant, where they cause the activation of similar defense mechanisms (i.e., PR proteins; Fig. 5.3; Shulaev et al., 1995). The induction of PR proteins in uninfected parts of the plant results in a heightened state of resistance to pathogen attack, which is termed *systemic-acquired resistance* (SAR; Ryals et al., 1996).

The activation of the plant defense mechanisms upon pathogen infection is often accompanied by a rapid process of cell death at and around the site of infection (Figs. 5.1D, 5.1E, and 5.1F). This process, also termed the *hypersensitive response* (HR), results in the formation of a layer or a ring of dead cells around the infection site. It is believed that this response serves to inhibit the growth of the invading pathogen by killing of infected and uninfected cells and producing a physical barrier composed of dead plant cells. The rapid dehydration that follows the death of plant tissue may also have deleterious effects on pathogen growth by limiting the availability of nutrients. In addition, it is thought that during the HR dying plant cells strengthen their cell walls and accumulate certain toxic compounds such as different phenolics and phytoalexins (Goodman and Novacky, 1994; Dangl et al., 1996). It should, however, be noted that different pathogens such as bacteria, fungi, and viruses induce different types of cell death with different morphological and physiological characteristics, and a different rate of cell death. Nevertheless, it appears as if the sacrifice of cells by the plant in order to inhibit pathogen growth and prevent systemic infection is the main function of the HR. This function may be similar to the activation of apoptosis in response to infection with viruses or bacteria in animal cells (Zychlinsky et al., 1992; Mittler and Lam, 1996; Finkel and Casella, Chapter 11, this volume). Thus activation of cell death as a means of preventing further infection by an invading pathogen appears to be a general theme in the biology of multicellular organisms.

Several lines of evidence suggest that the HR results from the activation of a pcd process (Greenberg, 1996; Jones and Dangl, 1996; Mittler and Lam, 1996). These include the activation of cell death in the absence of pathogens by mutations in certain genes that are thought to be involved in the cell death pathway (Walbot et al., 1983; Wolter et al., 1993; Greenberg et al., 1994; Dietrich et al., 1994), the activation of the HR by expression of different foreign genes in plants (Takahashi et al., 1989; Becker et al., 1993; Mittler et al., 1995), and the activation of cell death upon recognition of elicitors, which are compounds produced by the pathogen (He et al., 1993; Hammond-Kosack et al., 1994; Levine et al., 1994). That cell death that resembles the HR can be activated in the absence of a pathogen strongly suggests that this type of cell death is not directly caused by the invading pathogen but rather results from the activation of a host-encoded pathway for cell death (i.e., pcd). Moreover, activation of cell death during the HR was shown to require the activity

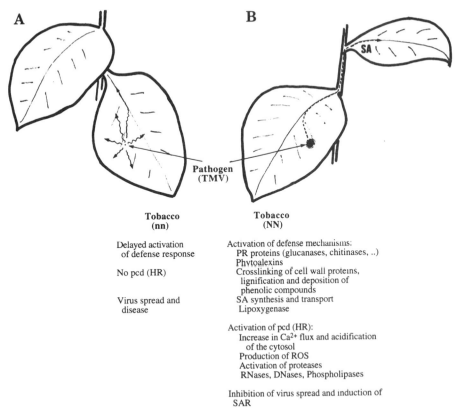

FIG. 5.3. Gene-dependent resistance and plant–pathogen interactions. The interaction of tobacco mosaic virus (TMV) with tobacco plants is outlined: (A) The interaction of TMV with a susceptible tobacco cultivar that does not contain the resistance gene *N* (nn) is shown. During this interaction the plant is incapable of recognizing the pathogen and therefore does not mount a rapid and efficient defense response. Thus pcd and some defense mechanisms are not activated while the activation of other defense mechanisms is delayed. This results in the successful inoculation of the plant, and the growth and spread of the pathogen. (B) The infection of a TMV-resistant tobacco cultivar that contains the resistance gene *N* (NN) results in the activation of multiple defense mechanisms and pcd. Thus an HR lesion is formed and the spread of the TMV pathogen is inhibited. Moreover, salicylic acid (SA), which is synthesized at the site of infection, is transported to other parts of the plant, where it signals the activation of defense mechanisms and the induction of systemic resistance (SAR). The distinction between activation of defense mechanisms and activation of pcd (HR) is not always clear, and some of the molecular mechanisms that are activated during one of these may also be activated during the other (the different mechanisms activated upon pathogen recognition are indicated in the figure). Abbreviations: HR, hypersensitive response; PR, pathogenesis related; ROS, reactive oxygen species, SAR, systemic acquired resistance.

of the plant transcription and translation machinery (He et al., 1993, 1994). These findings further suggest that genes that are encoded by the plant are actively involved in the death process of plant cells during the HR.

Despite the impressive array of antimicrobial defenses, not all plant–pathogen interactions result in the successful defense of the plant (Fig. 5.3). Thus in some instances the invading pathogen is not recognized by the plant and the defense mechanisms, including the HR, are not activated or activated at a slow rate that is insufficient to block the spread of the pathogen (Godiard et al., 1994). Therefore, the recognition of the invading pathogen by the plant is a crucial event. This recognition is traditionally described by the "gene-for-gene" or "corresponding gene" model (Flor, 1956; Keen, 1990; de Wit, 1992; Bent et al., 1994; Hammond-Kosack et al., 1994; Jones et al., 1994; Mindrinos et al., 1994; Whitham et al., 1994; Boyes et al., 1996). In this model, an avirulence factor produced by an *Avr* gene in the pathogen can be specifically recognized by a resistance (*R*) gene encoded by the plant. This recognition triggers the rapid activation of defense mechanisms and inhibits the proliferation of the microbe within the plant. This is termed an "incompatible" interaction. In the absence of either the *Avr* or its corresponding *R* gene, the pathogen is not recognized and disease symptoms may develop. This is termed a "compatible" interaction and usually signifies susceptibility of a plant to the pathogen (Fig. 5.3).

The Genetics of the Hypersensitive Response

Perhaps the most compelling evidence that the HR is a pcd process is the existence of mutants that spontaneously activate the HR in the absence of a pathogen (Dangl et al., 1996). These mutants are often referred to as *disease lesion mimics* since they develop lesions that resemble lesions that occur during the activation of the HR. However, the term *disease lesion mimics* also includes mutants that develop lesions that resemble disease symptoms that are distinct from the lesions formed during the HR. Two additional terms are used to describe mutants that develop HR lesions in the absence of a pathogen. These are *accelerated cell death* (*acd;* Greenberg and Ausubel, 1993) and *lesion simulating disease* (*lsd;* Deitrich et al., 1994). The mutations that cause the appearance of HR lesions in the absence of a pathogen are thought to occur in plant genes that control pcd. Thus mutations in these genes would result in the abnormal activation or suppression of pathogen-induced pcd (Greenberg and Ausubel, 1993; Dietrich et al., 1994; Greenberg et al., 1994). Disease lesion mimic mutants were isolated from tomato, maize, barley, rice, and *Arabidopsis.* They were classified according to their appearance into two groups: initiation and feedback or propagation mutants (Walbot et al., 1983; Dietrich et al., 1994). This classification is based upon the assumption that two mechanisms are involved in coordinating the HR: a pathway for the initiation of pcd that may include genes with functional similarity to the cell death genes *ced-3* and *ced-4* of *Caenorhabditis elegans* (Ellis and Horvitz,

1986) and a mechanism for the inhibition of cell death that may include genes that are functionally similar to the *ced-9* gene of *Caenorhabditis elegans* (Hengartner and Horvitz, 1994). Initiation mutants develop spontaneous lesions with a defined border; they are thought to be defective in regulating the activation of cell death but not its inhibition at the border of the lesion. Thus lesions will spontaneously initiate due to a mutation in the cell death activation pathway, but the cell death process will be inhibited, resulting in a lesion with a defined border. The abnormal activation of cell death in these mutants may result from the lack a negative regulator of cell death activation (a recessive initiation mutant) or from the constitutive activation a cell death signal (a dominant initiation mutant). Propagation or feedback mutants form spontaneous or induced lesions that spread indeterminately and are therefore presumed to be defective in downregulating the process of cell death in cells surrounding a developing lesion (recessive mutations). Thus in these mutants cell death, which is initiated randomly or following an infection with a pathogen or a mechanical injury, will uncontrollably propagate, due to a mutation in the cell death suppression pathway, eventually resulting in the complete death of the leaf.

Several cell death mutants express molecular and biochemical markers that are associated with the antimicrobial defense response of plants. These include the spontaneous synthesis of PR proteins, the accumulation of SA, the deposition of callose or other cell wall strengthening compounds, and the synthesis of phytoalexins (Deitrich et al., 1994; Greenberg et al., 1994). The activation of these antimicrobial defenses in the absence of a pathogen further indicate that the cell death pathway that is activated in these mutants is the same cell death mechanism that is activated during the response of plants to invading pathogens. However, in contrast to the activation of cell death at and around the site of infection during a real HR, the activation of cell death in some of the cell death mutants occurs sporadically throughout the leaf. What then is the cause of cell death initiation in these mutants? Since the mutated gene is present in all the leaf cells, a particular cellular parameter may distinguish certain cells from others and make them more prone to the activation of pcd. This factor may be the level of a particular hormone, the cell cycle stage, or any other cellular signal. A dependence of pcd activation upon a cellular parameter such as the cell cycle is known to occur in animals. Alternatively, cell death may be randomly initiated in different cells of the leaf. The activation of cell death in those cells may cause the induction of a particular cell death repressor mechanism in neighboring cells, thus resulting in the appearance of lesions only around the cells that randomly initiated cell death.

Cell death mutants are powerful tools for the study of pcd in plants. Cloning of the genes responsible for the lesion mimic phenotypes may reveal whether plants use molecular mechanisms that are similar to those used by animals, that is, BCL-2 and ICE-like proteins. In addition, by crossing of these mutants for complementation studies, the order of the cell death genes along the pcd pathway may be determined (Dangl et al., 1996). Cell death mutants may also be used for biochemical and physiological studies of pathogen-induced pcd. They are unique in that they reflect only the mechanisms activated by the plant. Thus one can study biochemical and physiological aspects of pcd without the often interfering presence of the pathogen.

The cloning of two plant genes that regulate hypersensitive response (HR) cell death (i.e., pathogen-induced programmed cell death) was recently reported; Dietrich et al. (1997) reported the cloning of the *LSD1* gene from *Arabidopsis,* and Buschges et al. (1997) reported the cloning of the *Mlo* gene from barley. The *LSD1* gene, which is a negative regulator of cell death, was found to encode a novel zinc finger protein. It was suggested that *LSD1* regulates transcription and either represses a prodeath pathway or activates an antideath pathway, in response to signals such as superoxide that are produced by cells undergoing HR cell death. The *Mlo* gene, which is also thought to be a negative regulator of HR cell death, was found to encode a novel protein that contains six membrane-spanning helices. The cloning of the *LSD1* and *Mlo* genes is likely to provide researchers with a first entry point into the cell death pathway of plants. It also present the first solid evidence for the existence of genes that regulate cell death in plants.

Additional tools that may prove beneficial for the study of pcd in plants are transgenic plants that spontaneously activate the HR in the absence of a pathogen. For example, transgenic tobacco plants that express a bacterial gene encoding the proton pump bacterio-opsin (bO) were found to spontaneously activate the HR and defense mechanisms in a manner similar to some of the dominant initiation cell death mutants (Fig. 5.1G). Moreover, the activation of cell death and defense mechanisms in these plants was affected by the same environmental conditions that effect the pathogen-induced HR (Mittler et al., 1995; Mittler and Lam, 1996). Thus cell death and defense mechanisms may also be studied in these plants in the absence of a pathogen. The current advantage of the transgenic system over the lesion mimic mutants is that the gene responsible for the cell death phenomena is known. In the case of the transgenic tobacco plants expressing the bO gene it is thought that the function of bO as a proton pump mimics some of the early events that accompany the HR (to be discussed in more detail in the following). The expression of other transgenes was also reported to cause the activation of the HR (Takahashi et al., 1989; Becker et al., 1993). The expression of some of these genes, such as a mutant of ubiquitin, is thought to alter drastically the metabolism of the cell. These alterations may result in the activation of pcd in a manner that is similar to the activation of pcd by alterations in cellular homeostasis in animals (Becker et al., 1993; Dangl et al., 1996; Mittler and Lam, 1996). In support of this hypothesis the majority of disease lesion mimic mutants appear to result from dominant mutations. It was, therefore, suggested that some of these mutations alter the homeostasis of the cell, thus generating a cell death signal (Walbot et al., 1983). The infection of a plant cell by an invading pathogen such as a virus may also cause alteration in cellular metabolism and activate pcd.

An additional approach to study genes involved in pcd in plants is to study the effect of overexpressing particular animal genes that are known to regulate pcd. Transgenic plants can be made to express these genes, and the effect of their expression can be studied on several developmental processes such as the differentiation of TEs or their response to pathogens. Of interest are genes that belong to three different families. The first are genes that belong to the *bcl-2* family that are key regulators of pcd in many animal cells (Boise et al., 1993; Korsmeyer, 1995). However, ectopic expression of *bcl-X$_L$* or *bcl-2* in tobacco plants revealed that they do not inhibit pcd (Mittler et al., 1996; Chen and Klessig, personal communication; Papermaster, Chapter 12, this volume). The second group of genes that suppress pcd encode proteins that inhibit ICE-like proteases. This family includes the p35 protein from baculovirus and the *CrmA* gene from cowpox virus (Bump et al., 1995; Ray and Pickup, 1996). Studies on the response of plants, expressing these genes to pathogen attack are underway in several laboratories. The third group of animal genes that inhibit pcd includes homologs of the Defender Against Death 1 (*Dad1*) gene (Nakashima et al., 1993; Sugimoto et al., 1995). This gene was initially identified as encoding a protein that rescues hamster cells from apoptosis. Homologs of *Dad1* were found in human, mice, yeast, *Xenopus, Caenorhabditis elegans,* and plants. *Dad1* is thought to block cell death at a step downstream from or independent of *Bcl-2*. However, recently it was reported that *Dad1* may encode for one of the subunits of oligosaccharyltransferase (Silberstein et al., 1995), complicating the interpretation of results obtained with transgenic plants that overexpress this gene. The study of the effect of overexpressing animal genes such as p35 and *bcl-2* may reveal whether plants utilize pcd mechanisms that are similar to those used by animal cells.

Since, as explained, it is becoming apparent that plants use pcd to block pathogen proliferation, it would be interesting to examine whether some plant viruses or other plant pathogens evolved molecular mechanisms that are similar to those used by animal viruses to suppress apoptosis, such as the E1B-19kD protein of adenovirus (Lakshmi et al., 1992; see also Finkel and Casella, Chapter 11, this volume) or the p35 protein from baculovirus (Bump et al., 1995) that are inhibitors of apoptosis.

Biochemistry, Morphology, and Detection of Pathogen-Induced pcd

Detection of pcd in Plants. Several methods are used to detect pcd in plant cells. Cell death in plants may be observed by staining of single plant cells *in culture* with a dye indicating their viability (Levine et al., 1994, 1996). Recently, revised methods for viability staining *in planta* were introduced (Weymann et al., 1995). The use of these methods enables the detection of dead cells in leaves. Pcd in plants can also be followed by the use of the classic TUNEL staining, which detects the products of DNA degradation (Gavrieli et al., 1992; Zakeri Chapter 3, this volume). However, some modifications may be required in order to convert the commercial kits available for TUNEL staining in animals to staining of pcd in plants (Wang et al., 1996a). It is also useful to include a positive control of differentiating TEs (Mittler et al., 1995). Microscopic observation of cell death in plants is often easier compared to the observation of cell death in animals. The reason for this is that the corpse of a dead plant cell or at least the plant cell walls are typically not eliminated. Nonmicroscopic detection of pcd in plants include the isolation of DNA and the detection of DNA fragmentation via conventional and field-inversion agarose gel electrophoresis (described in more detail in the following and in Sckorska and Walker, Chapter 7, this volume) and the measuring of ion leakage from leaf discs (Mittler and Lam, 1995a; Levine et al., 1996; Mittler et al., 1997a; Wang et al., 1996a; Ryerson and Heath, 1996). The detection of ion leakage from cells is a relatively good measure of cell death in plants, since, as explained, plant cells rarely form apoptotic bodies and the content of the dead cell is typically released into the intercellular fluid.

Cleavage of DNA and Activation of Proteases. Pcd in animals is accompanied by an active process of nuclear DNA fragmentation. This process is considered an irreversible and often late stage of pcd (Peitsch et al., 1994; Bortner et al., 1995; Zakeri, Chapter 3; and Schorska and Walker, Chapter 7, this volume). The degradation of nuclear DNA during pcd in animals occurs via two independent mechanisms; cleavage of chromatin into large DNA fragments of about 300 or 50 kb, and digestion of DNA to smaller fragments. Digestion of DNA into small fragments may occur via specific cleavage between nucleosomes that yield fragments that are multimers of about 180 bp (detected as a "DNA ladder" following agarose gel electrophoresis), or by a nonspecific degradation, which results in a random pattern of DNA fragments (detected as a "smear" following agarose gel electrophoresis). This difference may result from the timing of the proteolytic cleavage of nucleosomal proteins. If these proteins are digested before the nucleases are activated, then the pattern of DNA degradation will be random. However, if the nculeases are activated before the proteolytic digestion of nucleosomes, then they will cleave the DNA between nucleosomes, and the resulting DNA fragments will be multimers of about 180 bp. In plants both random and ordered DNA degradation were reported. The degradation of DNA during virus-induced pcd in tobacco and bacteria-induced pcd in soybean resulted in the formation of large DNA fragments of about 50 kb (Levine et al., 1996; Mittler et al., 1997a). However, further degradation of DNA resulted in a random pattern of DNA degradation (Mittler and Lam, 1995a; Levine et al., 1996). In contrast, toxin- and fungal–induced pcd in tomato and cowpea resulted in the formation of DNA ladders (Ryerson and Heath, 1996; Wang et al., 1996a).

Pcd in tobacco is accompanied by induction of Ca^{2+}-dependent nucleases (Mittler and Lam, 1995a). These may represent the plant counterparts of the animal apoptotic nucleases. Ca^{2+}-dependent nucleases were found in the intercellular fluid of leaves that contain HR lesions as well as in nuclei isolated from cells undergoing pcd (Mittler and Lam, 1997). Although the relationship between the increase in nuclease activities and the degradation of nuclear DNA was not directly demonstrated, it is possible that nuclear localized nucleases participate in the initial degradation of nuclear DNA during early stages of the HR (detected by TUNEL staining; Mittler et al., 1995), while the secreted nucleases found in the intercellular fluid of leaves may participate in the more extensive degradation of DNA that occurs late during pcd when the integrity of the plasma membrane is compromised (Mittler et al., 1997a).

During pcd in animals, a cascade of proteases is activated (Fraser and Evan, 1996; Cryns and Yuan, Chapter 3; and Sckorska and Walker, Chapter 7, this volume). These are thought to be involved in mediating the activation of pcd. Thus a signal for cell death causes the activation of a specific initiator protease that cleaves and activates a set of amplifier proteases that cleave and activate another set of machinery execution proteases. The possible involvement of protease activity in pathogen-induced pcd in plants was suggested by Levine et al. (1996), who demonstrated that treatment of soybean cells with a serine proteinase inhibitor suppressed pcd that is induced by H_2O_2 or a calcium ionophore. The activity of serine proteases was also found to be required for bacteria-induced pcd in *Arabidopsis* (Mittler et al., 1997b). These findings suggest that the activation of proteases is an important step in the induction of pcd in plants.

Morphology of pcd in Plants. Ultrastructural studies of plant tissues undergoing pathogen-induced cell death revealed a variety of different phenotypes. For example, rapid death of cells that came in direct contact with an invading fungus resulted in the rupture of the plasma membrane and the leakage of cellular content (similar to necrosis in animals). On the other hand, cells that died in a relatively slower rate such as virus-infected cells exhibited many ultrastructural changes. These include the condensation of the cytoplasm and nuclear material, shrinkage of cells, and increased vacuolization. Often rupture of the vacuole membrane preceded the disruption of the plasma membrane (Mittler et al., 1997a). However, in almost all cases of pathogen-induced pcd in plants no apoptotic bodies were detected (Bestwick et al., 1995; Mittler and Lam, 1996). In addition, although the nuclear material of plant cells condenses during pcd, the nucleus does not fragment (Mittler et al., 1997a; Mittler and Lam 1995a; Trump and Berezesky Chapter 2; and Sckorska and Walker, Chapter 7, this volume). An exception is a study by Wang et al. (1996a) in which the nuclei of protoplasts undergoing toxin-induced pcd were shown to fragment in a manner that may be similar to the fragmentation of nuclei during apoptosis. In addition, Wang et al., (1996a), as well as Levine et al. (1996), reported the detection of apoptotic bodies in plant cells undergoing pcd. However, the role of these bodies is not clear at present, since they were not shown to be engulfed by neighboring cells. It therefore appears that different plant cells may undergo different ultrastructural changes in response to different pathogens. These may reflect the different requirements from the cell death process that may depend upon the type of pathogen and plant.

Signal Transduction during Pathogen-Induced Cell Death

The recognition of an invading pathogen by a plant cell, that is, the recognition of an avirulance (*Avr*) factor by the resistance (*R*) gene, triggers a signal transduction pathway, which results in the induction of the HR (Godiard et al., 1994; Mittler and Lam, 1996). Some of the early events associated with the activation of the HR are the rapid generation of reactive oxygen species (ROS) in the form of superoxide and H_2O_2, the so-called "oxidative burst" (Baker and Orlandi, 1995; Goodman and Novacky, 1994; Jabs et al., 1996; Levine et al., 1994, 1996; Tenhaken et al., 1995), and a rapid flux of ions across the plasma membrane (Atkinson and Baker, 1989; He et al., 1994; Nurnberger et al., 1994). The rapid production of ROS by plant cells upon the recognition of pathogens may be analogous to the production of ROS by macrophages. Moreover, it is thought that a membrane-bound NADPH–oxidase complex is involved in this process (Levine et al., 1994; Auh and Murphy, 1995). Thus, much like macrophages, which produce ROS to kill pathogens, plant cells may also generate high extracellular levels of ROS. These may function to kill the pathogen, the plant cell (i.e., pcd), or both. It was also suggested that ROS such as H_2O_2 function as signals that activate other defense mechanisms such as the synthesis of SA and induction of PR proteins, and as oxidizing agents that cause the strengthening of cell walls (Bradley et al., 1992; Chen et al., 1993; Bi et al., 1995; Green and Fluhr, 1995; Leon et al., 1995; Neuenschwander et al., 1995). ROS have also been implicated during late stages of the HR, and the HR was associated with an increase in lipid peroxidation (Goodman and Novacky, 1994). The possible involvement of ROS as mediators of pcd in animals is controversial (Jacobson and Raff, 1995; Shimizu et al., 1995; McLaughlin et al., 1996). In plants strong evidence exists for the involvement of ROS in mediating cell death (Hammond-Kosack and Jones, 1996; Jabs et al., 1996; Levine et al., 1994, 1996; May et al., 1996; Mittler et al., 1996). However, the oxidative burst alone may not be sufficient to trigger cell death in all plant–pathogen systems (Glazener et al., 1996).

Recent studies indicated that under conditions of low oxygen pressure or high humidity cell death may be inhibited without affecting the activation of other defense mechanisms such as the synthesis of SA and the induction of PR proteins (Mittler et al., 1996; Hammond-Kosack and Jones, 1996). These results suggest that pcd can be uncoupled from the activation of other defense mechanisms during the HR. This hypothesis is supported by previous studies in which activation of defense mechanisms was shown to occur in the absence of pcd (Gross et al., 1993; Jakobek and Lindgren, 1993; Lawton et al., 1993; Bowling et al., 1994; Hammond-Kosack et al., 1996). Therefore, although the signal transduction events that lead to the activation of pcd and defense mechanisms during the HR are controlled by a single gene-for-gene interaction (*R* in the plant and *Avr* in the pathogen), it appears as if two independent pathways may be set into motion by this recognition event, a pathway for pcd and a pathway for the induction of PR proteins and perhaps other defense mechanisms.

The signal transduction pathway that leads to the activation of pcd was shown to involve increases in proton and Ca^{2+} flux and protein phosphorylation (Felix et al., 1991; Chandra and Low, 1995; Levine et al., 1996; Trump and Berezesky, Chapter 2; and Newell and Vincent, Chapter 9, this volume). The increased intracellular concentration of Ca^{2+}, which is accompanied by the acidification of the cytoplasm (inward increase in proton flux), is thought to signal the activation of kinases and the phosphorylation of several cellular targets. These may include the phosphorylation and activation of the NADPH–oxidase complex (Dangl et al., 1996). The involvement of kinase activity in the transduction of pathogen-related signals is supported by the finding that some of the cloned R genes contain kinase domains (Bent, 1996). Several studies performed with cultured cells indicated the involvement of a variety of other signaling molecules. These include G proteins, phosphatases, and phospholipases. However, care should be exercised when extrapolating from studies performed in tissue culture to the whole plant (He et al., 1994). The role of Ca^{2+} flux in signaling pcd was examined with cultured cells as well as with whole plants. It was found that the Ca^{2+} flux inhibitor lanthanum chloride inhibited pathogen-induced pcd. Moreover, lanthanum chloride was found to specifically inhibit cell death without affecting the activation of some defense mechanisms (Levine et al., 1996; Mittler et al., 1997b). In addition, it was found that treatment of plant cells with a Ca^{2+} ionophore in the presence of Ca^{2+} ions (i.e., inducing an inward Ca^{2+} flux) resulted in the activation of pcd (Levine et al., 1996). Thus it is likely that an increase in Ca^{2+} flux is involved in the signaling of pcd in plants. That expression of a proton pump (bO) in transgenic tobacco plants resulted in the activation of the HR and defense mechanisms suggested that the flux of protons is also important. This suggestion was strengthened by the finding that expression of a mutant version of the proton pump (D85A), which is incapable of proton pumping, did not result in a similar phenotype (Mittler et al., 1995; see also Trump and Berezesky, Chapter 2, this volume).

What Is the Role of pcd in Plant Protection?

Although it is becoming apparent that the HR is a type of pcd (Dangl et al., 1996; Mittler and Lam, 1996; Levine et al., 1996; Ryerson and Heath, 1996; Wang et al., 1996a), the role of this response in the antimicrobial defense of plants is not entirely clear. Cell death during the HR may cause the inhibition of pathogen growth by killing of infected cells, by depriving the invading pathogen from nutrients, by presenting the pathogen with a physical barrier, and/or by causing rapid dehydration. However, it may as well be that plant cell death is the consequence of a mechanism aimed at killing the pathogen, and the death of plant cells that came in close contact with the pathogen is merely a detrimental side effect of this mechanism. Moreover, the activation of other defense mechanisms along with cell death raises the question of whether these mechanism are sufficient to block pathogen growth in the absence of pcd. Thus it is not known whether pathogen-induced activation of defense mechanisms in the absence of pcd (i.e., when pcd is suppressed) will result in the inhibition of pathogen growth. In addition, it is not known whether a reverse situation in which pcd is activated in the absence of defense mechanisms is sufficient for the inhibition of pathogen growth. Two mutants of *Arabidopsis* that may correspond to this reverse situation are *acd1* and *NDR1* (Greenberg and Ausubel, 1993; Century et al., 1995). In these mutants cell death is activated in response to pathogen infection. However, this activation is not sufficient to block the proliferation of the pathogen. In addition, in a limited number of examples invading pathogens were detected beyond the ring of dead cells that surrounds the site of infection, again suggesting that cell death may not be the only factor required for inhibiting pathogen growth. On the other hand, the artificial triggering of cell death upon infection with a pathogen was found to block the spread of the pathogen and to increase the resistance of potato plants (Strittmatter et al., 1995). As is the case with many questions in biology, the answer to "what is the role of pcd in plant protection?" is probably not simple, and it is likely that the coordinated activation of pcd and defense mechanisms is required in order to provide the plant with an effective defense response.

SUMMARY

The many examples of pcd in plants may suggest that plants contain different pathways for cell death. Cell death in plants may be activated by different developmental signals, environmental insults, hormones, toxins, or pathogens. Moreover, pcd caused by different agents may require different rates of cell death. For example, the death of an infected plant cell should occur in a relatively rapid rate in order to prevent the spread of the invading pathogen, while the death of a senescing cell should occur at a relatively slow rate in order to allow the efficient transfer of nutrients from the dying cell to younger cells. In addition, the phenotype of cell death is often very different. For example, the death of an aerenchyma cell results in the almost complete removal of the cell corpse, while the death of a plant cell during the HR does not. It is therefore likely that plants contain different programs for cell death (i.e., different cell death pathways). However, it is also possible that all cell death processes in the plant are controlled by the same set of genes as may be the case with the *reaper* gene in *Drosophila* (White et al., 1994). Such an assumption would imply that the same cell death machinery that is activated during development is also activated during pathogen-induced pcd. Perhaps the cloning of genes that suppress or activate pcd (i.e., functional homologs of *bcl-2* or *ced-3*) will help in determining whether a single core pathway or multiple pathways are utilized for the different examples of cell death in plants.

Plants contain many unique cellular and subcellular structures and mechanisms. These include the plant cell wall, the vacuole, and the chloroplast with its highly complex apparatus for photosynthesis. Despite these and other differences between animals and plants, they appear to utilize pcd in a similar manner. Thus pcd is used during development and in response to different biotic and abiotic stresses (Greenberg, 1996; Jones and Dangl, 1996; Mittler and Lam, 1996). In addition, it appears as if certain mechanisms of cell death such as the involvement of protease and DNase activity is common to plants and animals (Sikorska and Walker, Chapter 7, this volume). However, many differences exist between pcd in plants and pcd in animals. These include among others the different mechanisms for the removal of the cell's corpse. It is likely that these differences result from the different anatomical and physiological characteristics that exist between plants and animals. Recently it was suggested that pcd may occur in unicellular organisms such as *Dictyostelium* (Cornillon et al., 1994) and even bacteria (Yarmolinsky, 1995). Therefore, the presence of a genetic program for a cell suicide pathway may be universal to all organisms. The involvement of pcd in almost all aspects of the life cycle of the plant suggest that pcd is a ubiquitous process in plants.

ACKNOWLEDGMENTS

I Thank Prof. Tsvi Sachs for his critical comments on the chapter and Dr. Eric Lam for his help, advice, and sharing of unpublished data.

REFERENCES

Atkinson MM, Baker CJ (1989): Role of the plasmalemma H^+-ATPase in *Pseudomonas syringae*-induced K^+/H^+ exchange in suspension-cultured tobacco cells. Plant Physiol 91:298–303.

Auh CK, Murphy TM (1995): Plasma membrane redox enzyme is involved in the synthesis of O_2 and H_2O_2 by *Phytophtora* elicitor-stimulated rose cells. Plant Physiol 107:1241–47.

Baker CJ, Orlandi EW (1995): Active oxygen in plant pathogenesis. Annu Rev Phytopatol 33:299–321.

Becker F, Buschfeld E, Schell J, Bachmair A (1993): Altered response to viral infection by tobacco plants perturbed in ubiquitin system. Plant J 3:875–81.

Bell PR (1996): Megaspore abortion: a consequence of selective apoptosis? Int J Plant Sci 157:1–7.

Bent AF, Kunkel BN, Dahlbeck D, Brown KL, Schmidt R, Giraudat J, Leung J, Staskawicz BJ (1994): RPS2 of Arabidopsis thaliana: A leucine-rich repeat class of plant disease resistance genes. Science 265:1856–60.

Bent AF (1996): Plant disease resistance genes: Function meets structure. Plant Cell 8:1757–71.

Bestwick CS, Bennet MH, Mansfield JW (1995): Hrp mutant of *Pseudomonas syringae* pv *phaseolicola* induces cell wall alterations but not membrane damage leading to the hypersensitive response in lettuce. Plant Physiol 108:503–16.

Bi YM, Kenton P, Mur L, Darby R, Draper J (1995): Hydrogen peroxide does not function downstream of salicylic acid in the induction of PR protein expression. Plant J 8:235–45.

Boise LH, Gonzalez-Gracia M, Postema CE, Ding L, Lindsten T, Turka LA, Mao X, Nunez G, Thompson CB (1993): *bcl-x*, a *bcl-2*-related gene that functions as a dominant regulator of apoptotic cell death. Cell 74:597–608.

Bortner CD, Oldenburg NBE, Cidlowski JA (1995): The role of DNA fragmentation in apoptosis. Trends Cell Biol 5:21–25.

Bowles DJ (1990): Defense-related proteins in higher plants. Annu Rev Biochem 59:873–907.

Bowling CA, Guo A, Cao H, Gordon SA, Klessig DF, Dong X (1994): A mutation in Arabidopsis that leads to constitutive expression of systemic acquired resistance. Plant Cell 6:1845–57.

Boyes DC, McDowell JM, Dangl JL (1996): Many roads to resistance. Curr Biol 6:634–37.

Bradley DJ, Kjellbom P, Lamb CJ (1992): Elicitor- and wound-induced oxidative cross-linking of proline-rich plant cell wall protein: a novel, rapid defense response. Cell 70:21–30.

Brown PH, Ho TD (1987): Biochemical properties and hormonal regulation of barley nuclease. Eur J Biochem 168:357–64.

Bump NJ, Hackett M, Hunginin M, Sheshagiri S, Brady K, Patrick C, Ferenz C, Franklin S, Ghayur T, Li P, Licari P, Mankovich J, Shi L, Greenberg AH, Miller LK, Wong WW (1995): Inhibition of ICE family proteases by Baculovirus antiapoptotic protein p35. Science 269:1885–88.

Buschges R, Hollricher K, Panstruga R, Simons G, Wolter M, Frijters A, Van Daelen R, Van der Lee T, Diergaade P, Groenendijk J, Topsch S, Vos P, Salamini F, Schulze-Lefert P (1997): The barley *Mlo* gene: A novel control element of plant pathogen resistance. Cell 88:695–705.

Century KS, Holub EB, Staskawicz BJ (1995): *NDR1*, a locus of *Arabidopsis thaliana* that is required for disease resistance to both a bacterial and a fungal pathogen. Proc Natl Acad Sci USA 92:6597–6601.

Chandra S, Low P (1995): Role of phosphorylation in elicitation of the oxidative burst in cultured soybean cells. Proc Natl Acad Sci USA 92:4120–23.

Chen Z, Silva H, Klessig DF (1993): Active oxygen species in the induction of plant systemic acquired resistance by salicylic acid. Science 262:1883–86.

Cornillon S, Foa C, Davoust J, Buonavista N, Gross JD, Golstein P (1994): Programmed cell death in *Dictyostelium*. J Cell Sci 107:2691–704.

Dangl JL, Dietrich RA, Richberg MH (1996): Death don't have no mercy: Cell death programs in plant–microbe interactions. Plant Cell 8:1793–807.

DeLong A, Calderon-Urrea A, Dellaporta SL (1993): Sex determination gene *TASSLESEED2* of maize encodes a short-chain alcohol dehydrogenase required for stage-specific floral organ abortion. Cell 74:757–68.

Dietrich RA, Delaney TP, Uknes SJ, Ward ER, Ryals JA, Dangl JL (1994): Arabidopsis mutants stimulating disease resistance response. Cell 77:565–77.

Dietrich RA, Richberg MH, Schmidt R, Dean C, Dangl JL (1997): A novel zinc finger protein is encoded by the *Arabidopsis LSD1* gene and functions as a negative regulator of plant cell death. Cell 88:685–94.

Ellis HM, Horvitz RH (1986): Genetic control of programmed cell death in the nematode C. elegans. Cell 44:817–29.

Fahn A (1982): Plant Anatomy. Oxford: Pergamon Press.

Felix G, Grosskopf DG, Regenass M, Boller T (1991): Rapid changes of the protein phosphorylation are involved in transduction of the elicitor signal in plant cells. Proc Natl Acad Sci USA 88:8831–34.

Fraser A, Evan G (1996): A license to kill. Cell 85:781–84.

Flor HH (1956): The complementary genic systems in flax and flax rust. Adv Genet 8:29–54.

Fukuda H (1996): Xylogenesis: Initiation, progression, and cell death. Annu Rev Plant Physiol Plant Mol Biol 47:299–325.

Gan S, Amasino RM (1995): Inhibition of leaf senescence by autoregulated production of cytokinin. Science 270:1986–88.

Gavrieli Y, Sherman Y, Ben-Sasson SA (1992): Identification of programmed cell death *in situ* via specific labeling of nuclear DNA fragmentation. J Cell Biol 119:493–501.

Glazebrook J, Ausubel FM (1994): Isolation of phytoalexin-deficient mutants of *Arabidopsis thaliana* and characterization of their interactions with bacterial pathogens. Proc Natl Acad Sci USA 91:8955–59.

Glazener JA, Orlandi EW, Baker CJ (1996): The active oxygen response of cell suspensions to incompatible bacteria is not sufficient to cause hypersensitive cell death. Plant Physiol 110:759–63.

Godiard L, Grant MR, Dietrich RA, Kiedrowski S, Dangl JL (1994): Perception and response in plant disease resistance. Curr Biol 4:662–71.

Goodman RN, Novacky AJ (1994): The Hypersensitive Response Reaction in Plants to Pathogens, A Resistance Phenomenon. St. Paul, Minnesota: American Phytopathological Society Press.

Green R, Fluhr R (1995): UV-B-induced PR-1 accumulation is mediated by active oxygen species. Plant Cell 7:203–12.

Greenberg JT (1996): Programmed cell death: A way of life for plants. Proc Natl Acad Sci USA 93:12094–97.

Greenberg JT, Ausubel FM (1993): Arabidopsis mutants compromised for the control of cellular damage during pathogenesis and aging. Plant J 4:327–42.

Greenberg JT, Ailan G, Klessig DF, Ausubel FM (1994): Programmed cell death in plants: A pathogen-triggered response activated coordinately with multiple defense functions. Cell 77:551–63.

Gross P, Julius C, Schmelzer E, Hahlbrock K (1993): Translocation of cytoplasm and nucleus to fungal penetration sites is associated with depolymerization of microtubules and defense gene activation in infected, cultured parsley cells. EMBO J 12:1735–44.

Hammond-Kosack KE, Harrison K, Jones JDG (1994): Developmentally regulated cell death on expression of the fungal avirulence gene *Avr9* in tomato seedlings carrying the disease-resistance gene *Cf-9*. Proc Natl Acad Sci USA 91:10445–49.

Hammond-Kosack KE, Silverman P, Raskin I, Jones JDG (1996): Race-specific elicitors of cladosporium fulvum induce changes in cell morphology and the synthesis of ethylene and salicylic acid in tomato plants carrying the corresponding Cf disease resistance gene. Plant Physiol 110:1381–94.

Hammond-Kosack KE, Jones JDG (1996): Resistance gene-dependent plant defense responses. Plant Cell 8:1773–91.

He CJ, Finlayson SA, Drew MC, Jordam WR, Morgan PW (1996a): Ethylene biosynthesis during aerenchyma formation in roots of maize subjected to mechanical impedance and hypoxia. Plant Physiol 112:1679–85.

He CJ, Morgan PW, Drew MC (1996b): Transduction of an ethylene signal is required for cell death and lysis in the root cortex of maize during aerenchyma formation induced by hypoxia. Plant Physiol 112:463–72.

He SY, Huang HC, Collmer A (1993): Pseudomonas syringae pv. syringae Harpin: A protein that is secreted via the hrp pathway and elicits the hypersensitive response in plants. Cell 73:1255–66.

He SY, Bauer DW, Collmer A, Beer SV (1994): Hypersensitive response elicited by *Erwinia amylovora* harpin requires active plant metabolism. Mol Plant Micro Inter 7:289–92.

Hengartner MO, Horvitz HR (1994): Programmed cell death in *Caenorhabditis elegans*. Curr Opin Genet Dev 4:581–86.

Jabs T, Dietrich RA, Dangl JL (1996): Initiation of runaway cell death in an *Arabidopsis* mutant by extracellular superoxide. Science 273:1853–56.

Jacobson MD, Raff M (1995): Programmed cell death and Bcl-2 protection in very low oxygen. Nature 374:814–16.

Jakobek JL, Lindgren PB (1993): Generalized induction of defense response in bean is not correlated with the induction of the hypersensitive reaction. Plant Cell 5:49–56.

Jones DA, Thomas CM, Hammond-Kosack KE, Balint-Kurti P, Jones JDG (1994): Isolation of the tomato Cf-9 gene for resistance to cladosporium fluvum by transposon tagging. Science 266:789–92.

Jones AM, Dangl JL (1996): Logjam at the Styx: programmed cell death in plants. Trends Plant Sci 1:114–19.

Kao TH, McCubbin AG (1996): How flowering plants discriminate between self and non-self pollen to prevent inbreeding. Proc Natl Acad Sci USA 93:12059–65.

Kaplan DR (1984): Alternative modes of organogenesis in higher plants. In White RA, Dickson WC (eds): Contemporary problems in plant anatomy. Academic Press, pp 261–300.

Keen NT (1990): Gene-for-gene complementarity in plant–pathogen interactions. Annu Rev Genet 24:447–63.

Kerr JFR (1995): Neglected opportunities in apoptosis research. Trends Cell Biol 5:55–57.

Korsmeyer SG (1995): Regulators of cell death. Trends Genet 11:101–5.

Lakshmi R, Debbas M, Sabbatini P, Hockenberry D, Korsmeyer S, White E (1992): The adenovirus E1A proteins induce apoptosis, which is inhibited by the E1B 19kDa and Bcl-2 proteins. Proc Natl Acad Sci USA 89:7742–46.

Lawton K, Uknes S, Friedrich L, Gaffney T, Alexander D, Goodman R, Metraux JP, Kessmann H, Ahl-Goy P, Gutrella M, Ward E, Ryals J (1993): The molecular biology of systemic acquired resistance. In Fritig B, Legrand M (eds): Mechanisms of Plant Defense Responses. Dordrecht, Netherlands: Kluwer Academic Publishers, pp 422–32.

Leon J, Lawton MA, Raskin I (1995): Hydrogen peroxide stimulates salicylic acid biosynthesis in tobacco. Plant Physiol 108:1673–78.

Levine A, Tenhaken R, Dixon R, Lamb C (1994): H_2O_2 from the oxidative burst orchestrates the plant hypersensitive disease resistance response. Cell 79:583–93.

Levine A, Pennell RI, Alvarez ME, Palmer R, Lamb C (1996): Calcium-mediated apoptosis in plant hypersensitive disease resistance response. Curr Biol 6:427–37.

Linthorst JM (1991): Pathogenesis-related proteins of plants. Crit Rev Plant Sci 10:123–50.

Malamy J, Carr JP, Klessig DF, Raskin I (1990): Salicylic acid: A likely endogenous signal in the resistance response of tobacco to viral infection. Science 250:1002–4.

Martin JS, Green RD, Cotter TG (1994): Dicing with death: Dissecting the components of the apoptosis machinery. Trends Biol Sci 19:26–30.

May MJ, Hammond-Kosack KE, Jones DG (1996): Involvement of reactive oxygen species, glutathione metabolism, and lipid peroxidation in the *Cf* gene-dependent defense response of tomato cotyledons induced by race-specific elicitors of *Cladosporium fulvum*. Plant Physiol 110:1367–79.

McLaughlin KA, Osborne BA, Goldsby RA (1996): The role of oxygen in thymocyte apoptosis. Eur J Immunol 26:1170–74.

Metraux J-P, Singer H, Rayals J, Ward E, Wyss-Benz M, Gaudin J, Raschdorf K, Schmid E, Blum W, Inverardi B (1990): Increase in salicylic acid at the onset of systemic acquired resistance in cucumber. Science 250:1004–6.

Minami A, Fukuda H (1995): Transient and specific expression of a cysteine endopeptidase associated with autolysis during differentiation of *Zinnia* mesophyll cells into tracheary elements. Plant Cell Physiol 36:1599–606.

Mindrinos M, Katagiri F, Yu G, Ausubel FM (1994): The A. thaliana disease resistance gene RPS2 encodes a protein containing a nucleotide-binding site and a leucine-rich repeats. Cell 78:1089–99.

Mittler R, Lam E (1995a): Identification, characterization, and purification of a tobacco endonuclease activity induced upon hypersensitive response cell death. Plant Cell 7:1951–62.

Mittler R, Lam E (1995b): *In situ* detection of nDNA fragmentation during the differentiation of tracheary elements in higher plants. Plant Physiol 108:489–93.

Mittler R, Shulaev V, Lam E (1995): Coordinated activation of programmed cell death and defense mechanisms in transgenic tobacco plants expressing a bacterial proton pump. Plant Cell 7:29–42.

Mittler R, Lam E (1996): Sacrifice in the face of foes: Pathogen-induced programmed cell death in higher plants. Trends Microbiol 4:10–15.

Mittler R, Lam E (1997): Characterization of nuclease activities and DNA fragmentation induced upon hypersensitive response cell death and mechanical stress. Plant Mol. Biol. 34:209–221.

Mittler R, Shulaev V, Seskar M, Lam E (1996): Inhibition of programmed cell death in tobacco plants during a pathogen-induced hypersensitive response at low oxygen pressure. Plant Cell 8:1991–2001.

Mittler R, Simon L, Lam E (1997a): Pathogen-induced programmed cell death in tobacco. Science, 110:1333–1344.

Mittler R, Del-Pozo O, Meisel L, Lam E (1997b): Pathogen-induced programmed cell death in plant: a possible defense mechanism. Developmental genetics *in press.*

Moiseyev GP, Beintema JJ, Fedoreyeva LI, Yakovlev GI (1994): High sequence similarity between ribonuclease from ginseng calluses and fungus-elicited proteins from parsley indicates that intracellular pathogenesis-related proteins are ribonucleases. Planta 193:470–72.

Nakashima T, Sekiguch T, Kuraoka A, Fukushima K, Shibata Y, Komiyama S, Nishimoto T (1993): Molecular cloning of a human cDNA encoding a novel protein, DAD1, whose defect causes apoptotic cell death in Hamster BHK21 cells. Mol Cell Biol 13:6367–74.

Neuenschwander U, Vernooij B, Friedrich L, Uknes S, Kessmann H, Ryals J (1995): Is hydrogen peroxide a second messenger of salicylic acid in systemic acquired resistance? Plant J 8:227–33.

Northcote DH (1995): Aspects of vascular tissue differentiation in plants: Parameters that may be used to monitor the process. Int J Plant Sci 156:245–56.

Nurnberger T, Nennstiel D, Jabs T, Sacks WR, Hahlbrock K, Scheel D (1994): High affinity binding of a fungal oligopeptide elicitor to parsley plasma membranes triggers multiple defense responses. Cell 78:449–60.

Orzaez D, Granell A (1996): DNA fragmentation is regulated by ethylene during carpel senescence in Pisum sativum. Plant J 11:137–44.

Osbourn AE (1996): Preformed antimicrobial compounds and plant defense against fungal attack. Plant Cell 8:1821–31.

Peitsch MC, Georg M, Tschopp J (1994): The apoptosis endonuclease: Cleaning up after cell death. Trends Cell Biol 4:37–41.

Raff CM (1992): Social controls on cell survival and cell death. Nature 356:397–400.

Ray CA, Pickup DJ (1996): The mode of death of pig kidney cells infected with cowpox virus is governed by the expression of the *crmA* gene. Virology 217:384–91.

Ryals JA, Neuenschwander UH, Willits MG, Molina A, Steiner HY, Hunt MD (1996): Systemic acquired resistance. Plant Cell 8:1809–19.

Ryerson DE, Heath ME (1996): Cleavage of nuclear DNA into oligonucleosomal fragments during cell death induced by fungal infection or by abiotic treatments. Plant Cell 8:393–402.

Schwartz LM, Smith SW, Jones MEE, Osborne BA (1993): Do all programmed cell death occur via apoptosis? Proc Natl Acid Sci USA 90:980–84.

Schwartzman RA, Cidlowski JA (1993): Apoptosis: The biochemistry and molecular biology of programmed cell death. Endocrine Rev 14:133–51.

Shimizu S, Eguchl Y, Kosaka H, Kamiike W, Matsuda H, Tsujimoto Y (1995): Prevention of hypoxia-induced cell death by Bcl-2 and Bcl-xL. Nature 374:811–13.

Shulaev V, Leon J, Radkin I (1995.): Is salicylic acid a translocated signal of sysatemic acquired resistance in tobacco? Plant Cell 7:1691–701.

Silberstein S, Collins PG, Kelleher DJ, Gilmore R (1995): The essential *OST2* gene encodes the 16-kD subunit of the yeast oligosaccharyltransferase, a highly conserved protein expressed in diverse eukaryotic organisms. J Cell Biol 131:371–83.

Smart CM (1994): Gene expression during leaf senescence. New Phytol 126:419–48.

Strittmatter G, Janssens J, Opsomer C, Botterman J (1995): Inhibition of fungal disease development in plants by engineering controlled cell death. 13:1085–89.

Sugimoto A, Hozak RR, Nakashima T, Nishimoto T, Rothman H (1995): *dad1,* an endogenous programmed cell death suppressor in *Caenorhabditis elegans* and vertebrates. EMBO J 14:4434–41.

Takahashi H, Shimamoto K, Ehara Y (1989): Cauliflower mosaic virus gene IV causes growth suppression, development of necrotic spots and expression of defense-related genes in transgenic tobacco plants. Mol Gen Genet 216:188–94.

Tenhaken R, Levine A, Brisson LF, Dixon RA, Lamb C (1995): Function of the oxidative burst in hypersensitive disease resistance. Proc Natl Acad Sci USA 92:4158–63.

Thelen MP, Northcote DH (1989): Identification and purification of a nuclease from *Zinnia elegans* L.: A potential molecular marker for xylogenesis. Planta 179:181–95.

Valpuesta V, Lange NE, Guerrero C, Reid MS (1995): Up-regulation of a cysteine protease accompanies the ethylene-insensitive senescence of daylily (*Hemerocallis*) flowers. Plant Mol Biol 28:575–82.

Vaux DL (1993): Toward an understanding of the molecular mechanisms of physiological cell death. Proc Natl Acad Sci USA 90:786–89.

Walbot V, Hoisington DA, Neuffer MG (1983): Disease lesion mimic mutations. In T. Kosuge and C. Meredith (eds.): Genetic Engineering of Plants. New York: Plenum Publishing Company, pp 431–42.

Walton JD (1996): Host-selective toxins: Agents of compatibility. Plant Cell 8:1723–33.

Wang H, Li J, Bostock RM, Gilchrist DG (1996a): Apoptosis: A functional paradigm for programmed plant cell death induced by a host-selective phytotoxin and invoked during development. Plant Cell 8:375–91.

Wang H, Wu H, Cheung AY (1996b): Pollination induces mRNA poly(A) tail-shortening and cell deterioration in flower transmitting tissue. Plant J 9:715–27.

Ward ER, Uknes SJ, Williams SC, Dincher SS, Wiederhold DL. Alexander DC, Ahl-Goy P, Metraux J-P, Ryals JA (1991): Coordinate gene activity in response to agents that induce systemic acquired resistance. Plant Cell 3:1085–94.

Weymann K, Hunt M, Uknes S, Neuenschwander U, Lawton K, Steiner HY, Ryals J (1995): Suppression and restoration of lesion formation in Arabidopsis *lsd* mutants. Plant Cell 7:2013–22.

Whitham S, Dinesh-Kumar SP, Choi D, Hehl R, Corr C, Baker B (1994): The product of the tobacco mosaic virus resistance gene N: Similarity to toll and the interleukin-1 receptor. Cell 78:1101–15.

White K, Grether ME, Abrams JM, Young L, Farrell K, Steller H (1994): Genetic control of programmed cell death in Drosophila. Science 264:677–83.

de Wit PJGM (1992): Molecular characterization of gene-for-gene systems in plant–fungus interactions and application of avirulence genes in control of plant pathogens. Ann Rev Phytopathol 30:391–418.

Wolter M, Hollricher K, Salamini F, Schulze-Lefert P (1993): The *mlo* resistance alleles to powdery mildew infection in barley trigger a developmentally controlled defense mimic phenotype. Mol Gen Genet 239:122–28.

Wyllie AH, Kerry JFR, Currie AR (1980): Cell death: The significance of apoptosis. Int Rev Cytol 68:251–306.

Wyllie AH, Morris RG, Smith AL, Dunlop D (1984): Chromatin cleavage in apoptosis: Association with condensed chromatin morphology and dependence on macromolecular synthesis. J Pathol 142:67–77.

Yarmolinsky MB (1995): Programmed cell death in bacterial populations. Science 267:836–37.

Yeung EC, Meinke DW (1993): Embryogenesis in angiosperms: Development of the suspensor. Plant Cell 5:1371–81.

Zychlinsky A, Prevost MC, Sansonetti PJ (1992): Shigella flexneri induces apoptosis in infected macrophages. Nature 358:167–69.

PART II

THEMES AND APPROACHES TO
CELL DEATH

CHAPTER 6

THE CUTTING EDGE: CASPASES IN APOPTOSIS AND DISEASE

VINCENT L. CRYNS and JUNYING YUAN
Harvard Medical School Departments of Cell Biology and Medicine
Boston, MA 02115

INTRODUCTION

Many insights into the genetic control of programmed cell death (PCD) have come from studying the nematode *Caenorhabditis elegans*. During hermaphrodite development, 131 of its cells are selectively eliminated by programmed cell death at well defined stages (Sulston and Horvitz, 1977; Sulston et al., 1983; Ellis et al., 1991). Two genes that are essential for the execution of cell death are *ced-3* (cell death abnormal) and *ced-4*. (see Gartner and Hengartner, Chapter 4, this volume) Recessive loss-of-function mutations in either of these genes disrupt this carefully regulated developmental program by suppressing the programmed cell deaths that normally occur (Ellis et al., 1991; Yuan, 1995). *ced-4* encodes a protein of 549 amino acids (Yuan and Horvitz, 1992) which binds to and activates CED-3 (Chinnaiyan et al., 1997a and b; Irmler et al., 1997a). Recently, a mammalian homologue of CED-4, Apaf-1, has been identified (Zou et al., 1997). In contrast, *ced-3* encodes a protein of 503 amino acids with a serine-rich middle region and flanking non-serine-rich regions that are highly conserved among related nematodes (Yuan et al., 1993).

An important clue as to the function of *ced-3* came from the observation that it encodes a protein homologous to the mammalian interleukin-1β-converting enzyme (ICE), a novel cysteine protease that cleaves pro-interleukin-1β (pro-IL-1β) between Asp 116 and Ala 117 to generate the mature cytokine (Cerretti et al., 1992; Thornberry et al., 1992; Yuan et al., 1993). Overall, CED-3 and human ICE are 29% identical. The most highly conserved region (43% identity)

When Cells Die, Edited by Richard A. Lockshin, Zahra Zakeri, and Jonathan L. Tilly
ISBN 0-471-16569-7 © 1998 Wiley-Liss, Inc.

between these proteins is found in their C-terminal domains: amino acids 246–360 of CED-3 and 166–287 of ICE (Yuan et al., 1993). Importantly, both proteins contain a QACRG pentapeptide in this region that includes the active site Cys-285 in ICE essential for its proteolytic activity (Thornberry et al., 1992; Yuan et al., 1993). Moreover, ICE is activated by proteolytic processing (detailed in the following) at four Asp-X cleavage sites (Thornberry et al., 1992); two of these sites are conserved in CED-3 (Yuan et al., 1993). In addition, the vast majority of *ced-3* missense mutations alter amino acid residues that are conserved in human ICE (Yuan et al., 1993).

The similarities between CED-3 and ICE implicated a novel family of cysteine proteases in the execution of programmed cell death in organisms as diverse as worms and humans and strongly suggested that the cell death apparatus is remarkably conserved during evolution (reviewed in Steller, 1995). This rapidly growing family of proteases has been given the name "caspase" (cysteine protease with aspartic acid substrate specificity, see the following) (Alnemri et al., 1996). The remainder of this chapter, then, will examine (1) the structural and enzymatic characteristics of the mammalian members of this family of proteases, (2) the evidence that caspases play a critical role in mammalian apoptosis, and (3) the clinical implications of the above for human diseases.

MULTIPLE MAMMALIAN HOMOLOGUES OF CED-3

Since the identification that ICE (caspase-1) is a mammalian homologue of Ced-3, several additional members of this family have been subsequently cloned: Nedd2/Ich-1 (hereafter referred to as caspase-2) (Kumar et al., 1992, 1994; Wang et al., 1994), CPP32/Yama/Apopain (hereafter referred to as caspase-3) (Fernandes-Alnemri et al., 1994; Nicholson et al., 1995; Tewari et al., 1995), ICH-2/TX/ICE rel-II (hereafter referred to as caspase-4) (Faucheu et al., 1995; Kamens et al., 1995; Munday et al., 1995), ICE rel-III/TY (hereafter referred to as caspase-5) (Munday et al., 1995; Faucheau et al., 1996); Mch2 (hereafter referred to as caspase-6) (Fernandes-Alnemri et al., 1995), Mch3/ICE-LAP3/ CMH-1 (hereafter referred to as caspase-7) (Fernandes-Alnemri et al., 1995; Duan et al., 1996a; Lippkc ct al., 1996), MACH/FLICE/Mch5 (hereafter referred to as caspase-8) (Boldin et al., 1996; Fernandes-Alnemri et al., 1996; Muzio et al., 1996), Mch4/FLICE2 (hereafter referred to as caspase-10) (Fernandes-Alnemri et al., 1996; Vincenz and Dixit, 1997), ICE-LAP6/Mch6 (hereafter referred to as caspase-9) (Duan et al., 1996; Srinavasula et al., 1996) and Ich-3 (hereafter referred to as caspase-11) (Wang et al., 1996). Based on their amino acid sequences, mammalian caspases can be classified into three subfamilies (Fig. 6.1): (1) a caspase-1 subfamily including caspase-1, caspase-4, caspase-5, and caspase-11; (2) a caspase-3 subfamily including caspase-3, caspase-6, caspase-7, caspase-8, and caspase-10; and (3) a caspase-2 subfamily including caspase-2 and caspase-9 (Duan et al., 1996a,b; Alnemri et al., 1996).

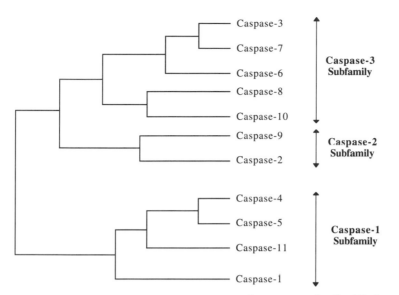

FIG. 6.1. Phylogenetic relationship of the mammalian caspase family. Phylogenetic analysis of the mammalian caspase family was performed using Megalign software (DNASTAR). Based on their amino acid sequences, three subfamilies (indicated on the right) can be discerned: a caspase-3 subfamily, a caspase-2 subfamily, and a caspase-1 subfamily.

Members of the caspase family are characterized by a number of key structural and functional properties (reviewed in Nicholson, 1996). Caspases are normally present in the cell as inactive proenzymes that are activated by proteolytic cleavage at Asp residues. As schematized in Fig. 6.2, proteolytic processing at Asp residues removes an N-terminal prodomain and generates the active protease, a tetramer composed of two self-associating large/small subunit heterodimers (Walker et al., 1994; Wilson et al., 1994). Moreover, caspases all share an unusual substrate specificity: They cleave target proteins between an N-terminal Asp residue (P1 site) and a C-terminal small hydrophobic amino acid (P1' site) (Cerretti et al., 1992; Thornberry et al., 1992; Thornberry and Molineaux, 1995). This substrate Asp specificity has been reported in only one other protease: the structurally unrelated serine protease, granzyme B, implicated in Fas-independent cytotoxic T lymphocyte (CTL) killing of target cells (Shi et al., 1992 and reviewed in Berke, 1995). Taken together, these characteristics strongly suggest that *in vivo* proteolytic activation of a given caspase can be accomplished by autoproteolysis and/or proteolytic processing by other caspases (and/or granzyme B); both mechanisms have been demonstrated *in vitro* (reviewed in Fraser and Evan, 1996).

Important insights into the catalytic mechanism of these proteases have come from the X-ray crystal structure of caspase-1 (Walker et al., 1994; Wilson et al., 1994; Nicholson, 1996). In addition to the active site Cys-285 of this

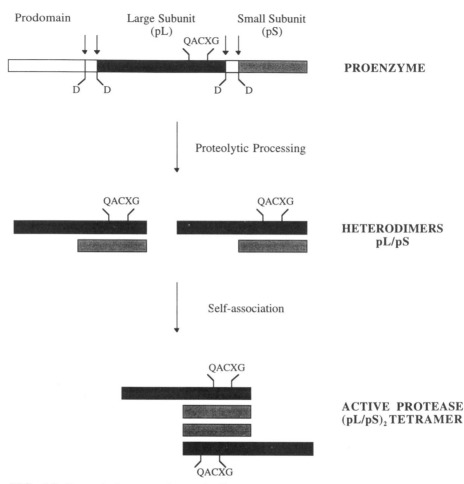

FIG. 6.2. Proteolytic processing and oligomerization of caspases. Caspases are expressed as inactive proenzymes that are activated by proteolytic cleavage at Asp (D) residues, resulting in removal of the regulatory N-terminal prodomain and generation of the large and small subunits. The active protease is a tetramer composed of two self-associating large and small subunit heterodimers. The conserved pentapeptide QACXG including the active-site Cys residue is indicated.

cysteine protease, two other residues in the large subunit (p20) are essential for catalytic activity: His-237 and Gly-238 (Walker et al., 1994; Wilson et al., 1994). These three residues are strictly conserved in all known caspases (Fig. 6.3). In addition, two amino acid residues from the large subunit (p20) (Arg-179 and Gln-283) and two from the small subunit (p10) (Arg-341 and Ser-347) form the binding pocket for the substrate P1 Asp residue (Walker et al., 1994; Wilson et al., 1994). These amino acids are also remarkably conserved

Large Subunit

	P1	P2 ✶ ✶	P1 P2 ✶
CED-3	P T R N G	L S H G E	Q A C R G
Caspase-3	T S R S G	L S H G E	Q A C R G
Caspase-7	G V R N G	L S H G E	Q A C R G
Caspase-6	P E R R R	L S H G E	Q A C R G
Caspase-8	R D R N G	L S H G D	Q A C Q G
Caspase-10	K D R Q G	L T H G R	Q A C Q G
Caspase-9	R T R T G	L S H G C	Q A C G G
Caspase-2	E F R S G	L S H G V	Q A C R G
Caspase-4	P P R N G	M S H G I	Q A C R G
Caspase-5	P A R N G	M S H G I	Q A C R G
Caspase-11	S L R Y G	M S H G T	Q A C R G
Caspase-1	P R R T G	M S H G I	Q A C R G

Arg-179   His-237 Gly-238   Gln-283 Cys-285

Small Subunit

	P2 P2 P2 P1 P2 P2	P2	P1 P2
CED-3	A Q Y V S W R N	S A R G	S W F I Q A V
Caspase-3	P G Y Y S W R N	S K D G	S W F I Q S L
Caspase-7	P G Y Y S W R S	P G R G	S W F V Q A L
Caspase-6	E G Y Y S H R E	T V N G	S W Y I Q D L
Caspase-8	N N C V S Y R N	P A E G	T W Y I Q S L
Caspase-10	P G Y V S F R H	V E E G	S W Y I Q S L
Caspase-9	P G F V S W R D	P K S G	S W Y V E T L
Caspase-2	K G T A A M R N	T K R G	S W Y I E A L
Caspase-4	P H N V S W R D	S T M G	S I F I T Q L
Caspase-5	P H N V S W R D	R T R G	S I F I T E L
Caspase-11	P H H L S Y R D	K T G G	S Y F I T R L
Caspase-1	P D N V S W R H	P T M G	S V F I G R L

Arg-341   Ser-347

FIG. 6.3. Conserved active-site residues in caspases. The amino acid residues that are essential for catalytic activity (indicated at the top of the figure by an asterisk), form the substrate P1 binding pocket (indicated by the designation P1), and are adjacent to the substrate P2-P4 residues (indicated by the designation P2) in caspase-1 are shown (Walker et al., 1994; Wilson et al., 1994). Protein sequences of the caspase family members were aligned using Megalign software (DNASTAR). The residues participating in catalysis and P1 binding are remarkably conserved. In contrast, the residues adjacent to the substrate P2-P4 residues are generally less well conserved and are likely to play a role in the divergent substrate specificities of various caspases. (Adapted with permission from Fernandes-Alnemri et al., 1996)

in all caspases: Other than a conservative Ser to Thr change in caspase-8 at the position corresponding to Ser-347 in caspase-1, they are identical in all family members (Fig. 6.3). An identical pentapeptide beginning with Gln-283 and including the catalytic cysteine-285 (QACRG) is found in all caspases except caspase-8 (QACQG), caspase-10 (QACQG), and caspase-9 (QACGG). Finally, the residues that are adjacent to the substrate P2-P4 in caspase-1 (indicated by the designation P2 in Fig. 6.3) (Walker et al., 1994; Wilson et al., 1994) are generally less well conserved across the family; these residues are likely to contribute to the divergent substrate specificities of various caspases (see the following) (Nicholson, 1996; Rotonda et al., 1996). Importantly, the active site is composed of amino acid residues from both the large and small subunits, thereby accounting for the requirement of both subunits for catalysis. Moreover, the striking conservation of many of these residues argues strongly that caspases have similar catalytic mechanisms (Walker et al., 1994; Wilson et al., 1994; Nicholson, 1996; Rotonda et al., 1996).

Despite these similarities, caspases have several distinguishing characteristics (reviewed in Nicholson, 1996). Although they all cleave substrates at an Asp residue in the P1 position, they differ substantially in their preference for amino acid residues at the P4 position (four amino acids N-terminal to the cleavage site) in their substrates. Interestingly, the P2 position (two amino acids N-terminal to the cleavage site) seems to play little role in substrate recognition (Thornberry et al., 1992). Caspase-1 cleaves pro-IL-1β at a site Tyr Val His Asp^{116}-Ala^{117} to generate mature IL-1β and shows a strong preference for substrates with a hydrophobic residue, such as Tyr, at the P4 position (Thornberry et al., 1992; Thornberry and Molineaux, 1995). In contrast, caspase-3 cleaves the DNA repair enzyme poly(ADP-ribose) polymerase (PARP) at a site Asp Glu Val Asp^{216}-Gly^{217} and prefers Asp residues at the P4 position (Lazebnik et al., 1994; Nicholson et al., 1995). In an effort to develop specific inhibitors of these proteases, the reversible peptide inhibitors acetyl-Tyr-Val-Ala-Asp aldehyde (YVAD-CHO) (Thornberry et al., 1992) and acetyl-Asp-Glu-Val-Asp aldehyde (DEVD-CHO) (Nicholson et al., 1995) have been modeled after the pro-IL-1β and PARP cleavage sequences, respectively. Caspase-1 is preferentially inhibited by YVAD-CHO (K_i 0.76 nM) (Thornberry et al., 1992) while caspase-3 is preferentially inhibited by DEVD-CHO (K_i < 1 nM) (Nicholson et al., 1995). Several of the other caspase-3-like proteases, including caspase-7 and caspase-10, are also inhibited by DEVD-CHO (Fernandes-Alnemri et al., 1995, 1996). The recently deduced crystal structure of caspase-3 provides an explanation for the observed differences in substrate specificity between caspase-1 and caspase-3 (Nicholson, 1996; Rotonda et al., 1996). Although the structure of the active sites of caspase-1 and caspase-3 are quite similar in many respects, they differ markedly in the region that interacts specifically with the P4 residue (the enzyme's S4 subsite). The S4 subsite in caspase-1 is a cavernous pocket capable of handling large P4 residues such as Tyr, while the analogous pocket in caspase-3 is dramatically smaller due to an inserted loop of ten amino acids at a position corresponding to residue 381 of caspase-1 and the side chains of Trp-348 and Phe-381B; several of these amino acids are conserved in other members of the caspase-3 subfamily (Rotonda et al., 1996). The small size of caspase-3's S4 site, together with the spatial configuration of its hydrogen bond donors and acceptors, are ideally suited for an Asp residue at P4 and likely account for caspase-3's observed predilection for cleaving substrates C-terminal to an Asp-X-X-Asp motif (Rotonda et al., 1996). Although less is known about the substrate specificities of the other caspase-1 and caspase-3 subfamily members, they are likely to resemble their subfamily namesake (caspase-1 or caspase-3), at least to some extent, in this regard. For instance, the substrate preferences of CED-3 are more similar to those of caspase-3 than caspase-1 or caspase-2 (Xue et al., 1996).

In addition to their distinct substrate specificities, caspases differ significantly in the length and composition of their N-terminal prodomains. Although this domain is not present in the active enzyme, it is likely to play an important regulatory function (Criekinge et al., 1996; Fraser and Evan, 1996). Recessive inactivating CED-3 mutations have been mapped to this region (Yuan ct al., 1993; Shaham and Horvitz, 1996; Fraser and Evan, 1996). Recent evidence suggests that the prodomain also likely mediates the binding of the CED-4 homologue Apaf-1 to the caspases; this interaction is required for caspase activation (Irmler et al., 1997a; Zou et al., 1997). Moreover, the prodomain can deliver the inactive proenzyme to the activating apparatus via protein–protein interaction motifs. As detailed later in this chapter, both caspase-8 and caspase-10 have two such motifs in their prodomains that target them to the ligand-activated Fas and tumor necrosis factor (TNF) receptor complexes (Boldin et al., 1996; Fernandes-Alnemri et al., 1996; Muzio et al., 1996; Vincenz and Dixit, 1997). In this way, the prodomains may selectively determine which caspases get activated by a given stimulus. Another factor potentially influencing this differential activation is the tissue-specific distribution of the various caspases. Although the published information on this subject is incomplete, caspases seem to be expressed in most, if not all tissues, although the level of their RNA expression varies considerably from tissue to tissue. Clearly, the relative abundance of caspases in a given tissue could be an important determinant of their differential specificity.

ROLE OF CASPASES IN APOPTOSIS

Overexpression of Caspases Induces PCD

Beyond their structural and functional similarities to the *C. elegans ced-3* cell death gene, several lines of evidence indicate that the mammalian homologues of *ced-3* play an analogous role in the execution of apoptosis in higher organisms. To begin with, transient expression of caspase cDNAs induces morphological and biochemical features characteristic of apoptosis (e.g., cellular/nuclear condensation and internucleosomal DNA fragmentation) in mammalian

(Miura et al., 1993; Kumar et al., 1994; Wang et al., 1994; Faucheu et al., 1995; Munday et al., 1995; Boldin et al., 1996; Duan et al., 1996a,b; Lippke et al., 1996; Muzio et al., 1996; Wang et al., 1996) or insect cells (Fernandes-Alnemri et al., 1994, 1995a,b). Induction of apoptosis in these systems requires a functionally active cysteine protease: Constructs with point mutations in the catalytic Cys residue or in the neighboring conserved Gly residue, or which do not contain both the large and small subunits, are unable to induce cell death (Miura et al., 1993; Fernandes-Alnemri et al., 1994, 1995a,b; Kumar et al., 1994; Wang et al., 1994; Faucheu et al., 1995; Kamens et al., 1995; Boldin et al., 1996; Duan et al., 1996a,b; Lippke et al., 1996). Interestingly, overexpression of *ced-3* in mammalian cell lines also results in programmed cell death, albeit somewhat less efficiently than its mammalian homologues (Miura et al., 1993), again underscoring the concept that the cell death machinery has been remarkably well conserved during evolution (see Ameisen, Chapter 1, Sikorska and Walker, Chapter 7, and Preface, this volume). Importantly, apoptosis induced by overexpression of these proteases can be inhibited by bcl-2, the antiapototic mammalian homologue of *ced-9* that likely acts upstream of caspases in the cell death pathway (Chinnaiyan et al., 1996a), and by specific viral and peptide inhibitors of caspases (see the following) (see Birge et al., Chapter 13, Cotman et al., Chapter 14, Benitz-Bribiesca, Chapter 17, Finkel, Amusen, and Casella, Chapter 11, Chapter 1, this volume). Nevertheless, the physiological relevance of these studies is clouded by the finding that overexpression of a variety of nonspecific proteases (e.g., proteinase K, trypsin and chymotrypsin) induces apoptosis (Williams and Henkart, 1994; Fraser and Evan, 1996).

Caspases Are Activated during PCD

Further evidence that caspases play a role in apoptosis comes from the observation that at least several of these proteases are proteolytically processed and activated during cell death. The p45 caspase-1 proenzyme (pro-caspase-1) is cleaved into its active p20 and p10 subunits during apoptosis induced by removal of extracellular matrix (Boudreau et al., 1995) and by granzyme B (Shi et al., 1996). Since proteolytic processing of pro-IL-1β is a specific function of caspase-1, detection of mature IL-1β can be used as a indicator of caspase-1 activity. In HeLa cells, apoptosis induced by TNF-α in the presence of cycloheximide or by granzyme B is associated with the secretion of mature IL-1β in the supernatant, strongly suggesting that caspase-1 is activated by these apoptotic stimuli (Miura et al., 1995; Shi et al., 1996). Pro-caspase-3 (p32) is proteolytically processed into its active p20/17 and p12 subunits during apoptosis induced by a variety of agents, including anti-Fas mAb, TNF-α, the broad-spectrum protein kinase inhibitor staurosporine, phosphatidylinositol 3-kinase inhibitors, the DNA damaging agent etoposide, and by granzyme B (Armstrong et al., 1996; Chinnaiyan et al., 1995; Erhardt and Cooper, 1996; Jacobsen et al., 1996; Martin et al., 1996; Schlegel et al., 1996; Shi et al., 1996; Slee et al., 1996; Wang et al., 1996). Similarly, caspase-2 (Li et al., 1997), caspase-6 (Orth et al., 1996), and caspase-7 (Chinnaiyan et al., 1996; Duan et al., 1996) have all been shown to be processed into their active subunits during the induction of cell death by a variety of stimuli. Importantly, processing of these proenzymes precedes or coincides with the cleavage of apoptotic substrates such as PARP and the appearance of morphological features of apoptosis; hence, protease activation occurs during the onset of apoptosis and could therefore play a role in its systematic execution.

Inhibitors of Caspases Prevent PCD

Viral Inhibitors. The observation that several caspases are activated during apoptosis only *implicates* them in the execution of this process. Conclusive evidence that they play an *essential* role in cell death comes from a multitude of studies demonstrating that specific inhibitors of these proteases prevent or delay a broad spectrum of programmed cell deaths. Several viral gene products have been identified that inhibit key steps in the programmed cell death machinery. These antiapoptotic proteins allow enhanced viral replication and survival by preventing apoptosis in virus-infected host cells (reviewed in Thompson, 1995). Two such viral gene products are inhibitors of caspases: the cowpox cytokine response modifier gene *crmA* (Ray et al., 1992) and the baculovirus p35 gene (Clem et al., 1991). The *crmA* gene encodes a serpin that is a potent active site-directed inhibitor of caspase-1 ($K_i < 4$ pM) (Ray et al., 1992; Komiyama et al., 1994) and a weaker inhibitor of other caspases such as caspase-3 ($K_i < 1\mu$M) (Nicholson, 1995, 1996). CrmA also inhibits granzyme B (Quan et al., 1995; Martin et al., 1996), a serine protease in cytotoxic lymphocyte (CTL) granules (Shi et al., 1992) that mediates the Fas-independent component of CTL-induced apoptosis of target cells (reviewed in Berke, 1995). CrmA is cleaved by caspase-1 (Komiyama et al., 1995) and granzyme B (Quan et al., 1995; Martin et al., 1996), suggesting that it might act as a competitive inhibitor of these proteases. Microinjection of a crmA expression vector into dorsal root ganglion neurons dramatically protects them from apoptosis induced by nerve growth factor withdrawal (Gagliardini et al., 1994). Moreover, overexpression of crmA in mammalian cells delays or prevents apoptosis induced by tumor necrosis factor-α (TNF-α) (Enari et al., 1995; Miura et al., 1995; Tewari and Dixit, 1995), anti-Fas antibody (Enari et al., 1995; Los et al., 1995; Tewari and Dixit, 1995), serum withdrawal (Wang et al., 1994), overexpression of caspase-1 (Miura et al., 1993), and to a lesser extent caspase-2 (Wang et al., 1994), disruption of the extracellular matrix (Boudreau et al., 1995), and CTL-mediated apoptosis of target cells (particularly the Fas-mediated component) (Tewari et al., 1995b), but does not prevent staurosporine-induced cell death (Chinnaiyan et al., 1996; Jacobsen et al., 1996). In addition, thymocytes from transgenic mice overexpressing crmA in T lymphocytes are resistant to Fas-induced apoptosis *in vitro* but are sensitive to γ-radiation and glucocorticoid-induced cell death (Smith et al., 1996). Inhibition of apoptosis requires functional CrmA: Overexpression of a mutant crmA

(Thr to Arg substitution at position 291) that fails to bind/inhibit caspase-1 does not prevent Fas-, TNF-, or CTL-induced cell death (Tewari et al., 1995a,b).

In contrast to CrmA, the baculovirus p35 gene product is a potent inhibitor of at least several caspases (CED-3, caspase-1, caspase-2, caspase-4, and caspase-3) but not granzyme B (Bump et al., 1995; Xue and Horvitz, 1995; Bertin et al., 1996); hence, p35 is a specific, broad-spectrum inhibitor of caspases. Protease inhibition requires cleavage of p35 by these caspases at a site Asp Gly Met Asp^{87}-Gly^{88} (Bump et al., 1995; Xue and Horvitz, 1995; Bertin et al., 1996) that resembles the Asp-X-X-Asp cleavage motif of caspase-3. Substitution of the P1 Asp with a Glu or Ala residue prevents cleavage of the p35 protein and abolishes its ability to inhibit caspases (Bertin et al., 1996; Xue and Horvitz, 1995). Expression of p35 prevents baculovirus-induced apoptosis of insect cells (Clem et al., 1991) and developmental programmed cell death in *Drosophila* (Hay et al., 1994) and *C. elegans* (Sugimoto et al., 1994; Xue and Horvitz, 1995). In addition, overexpression of p35 in mammalian cells inhibits cell death mediated by TNF, anti-Fas antibody (Beidler et al., 1995) and nerve growth factor withdrawal (Rabizadeh et al., 1993; Martinou et al., 1995). In all models examined to date, p35's antiapoptotic activity is dependent on its ability to inhibit caspases. Expression of mutant p35 constructs with a substitution at the P1 Asp of the caspase cleavage site (thereby abolishing its ability to inhibit these proteases) fails to prevent developmental programmed cell death in *C. elegans* (Xue and Horvitz, 1995) and baculovirus-induced apoptosis in insect cells (Bertin et al., 1996). Unlike p35, CrmA is not a substrate/inhibitor of CED-3 and does not prevent developmental cell death in *C. elegans* (Xue and Horvitz, 1995). However, replacement of the caspase-1 cleavage site in CrmA with the CED-3 cleavage site in p35 renders CrmA susceptible to CED-3 cleavage and protective against cell death in *C. elegans*, (Xue and Horvitz, 1995).

Recently, a family of viral proteins which inhibit cell death induced by TNF family members has been identified (Bertin et al., 1997; Thome et al., 1997; Hu et al., 1997). Unlike CrmA and p35, these so-called v-FLIPs (Thome et al., 1997) are not active site-directed inhibitors of caspases. Rather, by virtue of sequence homology with regions in the prodomains of caspases 8 and 10, they prevent the delivery of these caspases to ligand-bound death receptors (see the following), thereby inhibiting apoptosis.

Peptide Inhibitors. Specific peptide inhibitors of caspases have also demonstrated that these proteases play a critical role in mammalian apoptosis. As noted earlier, peptide inhibitors have been modeled after the P1–P4 residues in the cleavage sites of pro-IL-1β (YVHD) and PARP (DEVD), thereby providing some selectivity against caspase-1-like and caspase-3-like proteases, respectively (Thornberry et al., 1992; Nicholson et al., 1995). However, this selectivity is diminished at μM concentrations, an important consideration when interpreting *in vivo* studies in which these inhibitors are used in concentrations often exceeding 10–100 μM (due to their limited penetration of cell membranes).

Peptide aldehydes based on these sequences are reversible, competitive inhibitors of caspases, while the corresponding methyl ketones are irreversible inhibitors that covalently modify the active site cysteine and penetrate cells more readily (Nicholson et al., 1995; Thornberry et al., 1992; Thornberry et al., 1994; Thornberry and Molineaux, 1995; Nicholson, 1996). Pretreatment of a variety of transformed cell lines or primary cells with these specific caspase inhibitors (generally in 10–100 μM concentrations) prevents cell death induced by anti-Fas antibody (Chow et al., 1995; Enari et al., 1995; Los et al., 1995; Armstrong et al., 1996; Schlegel et al., 1996), neurotrophic factor deprivation (Milligan et al., 1995), etoposide and other DNA damaging agents (Zhivotovsky et al., 1995; An and Dou, 1996), glucocorticoids (Zhivotovsky et al., 1995), and staurosporine (Zhu et al., 1995; Jacobsen et al., 1996). These peptide inhibitors also delay interdigital and motor neuron cell death *in vivo* in embryonic chicks (Milligan et al., 1995) and interdigital cell death in explant cultures of embryonic mouse paws (Jacobsen et al., 1996) (see Zakeri, Chapter 3, this volume).

Dominant Negative and Antisense Inhibitors. A final line of evidence indicating that caspases play a pivotal role in apoptosis comes from studies using dominant negative and antisense constructs to inhibit these proteases. Alternative splicing of the caspase-2 gene results in the expression of two isoforms: the caspase-2$_L$ mRNA encodes a functional protease composed of p20 and p10 subunits, while the caspase-2$_S$ mRNA encodes a nonfunctional, truncated protease (Wang et al., 1994). Overexpression of caspase-2$_S$ prevents cell death induced by serum withdrawal (Wang et al., 1994), most likely by interfering with the activity of endogenous caspase-2 or other caspases by acting in a dominant negative fashion by forming proteolytically inactive tetramers. Since subunits of different caspase family members can combine to form novel proapoptotic tetramers (Fernandes-Alnemri et al., 1995; Kamens et al., 1995), a given dominant negative caspase family member could interfere with the activity of multiple additional caspases. Similarly, overexpression of catalytically inactive caspase-8 constructs (a truncated isoform or a point mutant replacing the catalytic Cys with Ser) protects cells from Fas- and TNF-induced apoptosis (Boldin et al., 1996). Microinjection of an expression vector encoding a dominant negative caspase-1 mutant (Cys-285 to Gly) inhibits cell death in dorsal root ganglia neurons induced by nerve growth factor deprivation (Li et al., 1996; Friedlander et al., 1997). Furthermore, ectopic expression of antisense caspase-1 or caspase-2 inhibits apoptosis induced by Fas-ligation (Los et al., 1995) and growth factor withdrawal (Kumar, 1995), respectively. Although the specificity of these interventions is unclear (i.e., overexpression of a dominant negative or antisense caspase-1 construct may also inhibit other caspases), they unambiguously implicate caspases in the execution of apoptotic cell death.

Taken together, the data presented so far indicate that one or more caspases are activated during the induction of apoptosis induced by a wide variety of stimuli and play an essential role in its execution. In contrast to the single

CED-3 protease in *C. elegans,* the large number of mammalian homologues suggests that these proteases may be performing partly redundant functions and/or be sequentially activated in a proteolytic cascade (Nicholson, 1996; Fraser and Evan, 1996). Recent studies suggest that both of these possibilities are likely to be true. Caspase-1, itself, is likely to play an indispensable role in only certain types of apoptotic cell death. Mice with targeted disruption of both caspase-1 alleles develop normally and manifest strikingly few apoptotic defects (Kuida et al., 1995; Li et al., 1995). Thymocytes from caspase-1-deficient mice show partial resistance to Fas-induced but not granzyme B-induced cell death, while B lymphoblasts and embryonic fibroblasts from these mice are resistant to granzyme B-mediated apoptosis (Kuida et al., 1995; Shi et al., 1996). Further evidence that caspase-1 or a caspase-1-like protease (i.e., another caspase-1 subfamily member) plays a critical role in Fas-induced cell death comes from the following observations: (1) 10 nM YVAD-CHO, a selective inhibitor of caspase-1 at this concentration, prevents apoptosis induced by Fas-ligation (Los et al., 1995); and (2) cytosolic extracts from cells treated with anti-Fas monoclonal antibody have a caspase-1-like proteolytic activity as measured by specific cleavage of a YVAD fluorogenic substrate (Enari et al., 1996). In contrast, thymocytes from caspase-3-deficient mice show normal sensitivity to apoptosis induced by a variety of stimuli, including Fas-ligation (Kuida et al., 1996). However, caspase-3-deficient mice have hyperplastic cell masses in their central nervous system and die prematurely at one to three weeks of age, a phenotype that may reflect diminished apoptosis in the developing brain (Kuida et al., 1996). While these findings indicate that caspase-1 and caspase-3 play essential (i.e., nonredundant) roles in only very specific types of apoptotic cell deaths, they do not exclude the possibility that these proteases contribute more broadly (i.e., in a redundant fashion) to other types of cell death. Evaluation of the indispensability of other caspases *in vivo* will have to await their targeted deletion.

Additional insights into the contribution of various caspases to the execution of apoptosis have come from studies of their proteolytic substrates (detailed in a a later section). The DNA repair enzyme poly(ADP-ribose) polymerase (PARP) is cleaved during apoptosis induced by a variety of stimuli (Kaufmann et al.,1993; Lazebnik et al., 1993; Tewari and Dixit, 1995). Studies in cell-free extracts derived from apoptotic cells indicated that the PARP protease is a caspase family member other than caspase-1 that is inhibited by μM concentrations of YVAD-CMK (Lazebnik et al., 1994) or by nM concentrations of DEVD-CHO (Nicholson et al., 1995). These apoptotic extracts reproduce many of the characteristic features of apoptosis seen *in vivo* (see Trump and Berezesky, Chapter 2, this volume). Addition of nuclei to these extracts results in chromatin condensation/margination followed by nuclear distintegration and internucleosomal fragmentation of DNA (Lazebnik et al., 1993; Newmeyer et al., 1994; Enari et al., 1995b; Martin et al., 1995). Importantly, the inhibitor profile for preventing apoptosis *in vitro* (i.e., fragmentation of added nuclei) (Lazebnik et al., 1994; Nicholson et al., 1995) is identical to that of the PARP protease, and selective removal of PARP proteolytic activity from these apoptotic extracts prevents apoptosis *in vitro* (Nicholson et al., 1995). These findings strongly suggest that the caspase(s) responsible for PARP proteolysis is also a key component of the apoptotic cell death machinery. Caspase-3 is one component of the PARP protease (Fernandes-Alnemri et al., 1994; Nicholson et al., 1995; Tewari et al., 1995), although other caspases also likely cleave PARP during cell death (Fernandes-Alnemri et al., 1995a,b; Gu et al., 1995; Duan et al., 1996; Lippke et al., 1996; Muzio et al., 1996). The latter point is underscored by the recent finding that PARP cleavage is not blocked in apoptotic thymocytes from caspase-3-deficient mice (Kuida et al., 1996). Further evidence that caspase-3 and/ or related subfamily members play an important role in apoptosis comes from the observations that caspase-3 is proteolytically processed and activated by a variety of apoptotic stimuli (Armstrong et al., 1996; Chinnaiyan et al., 1996; Erhardt and Cooper, 1996; Jacobsen et al., 1996; Martin et al., 1996; Schlegel et al., 1996; Slee et al., 1996; Wang et al., 1996) and by other caspases, including caspases 6, 8, and 10 (Fernandes-Alnemri et al., 1996; Liu et al., 1996; Muzio et al., 1996), and the apoptotic serine protease granzyme B *in vitro* (Darmon ct al., 1996; Darmon et al., 1995, Martin et al., 1996; Quan et al., 1996). In addition, many of the proteins that are known to undergo proteolysis during cell death are substrates of one or more of these proteases (see the following). Finally, addition of caspase-3 to nonapoptotic cytosolic extracts induces apoptosis *in vitro* (Nicholson et al., 1995; Enari et al., 1996). Taken together, these findings suggest that caspase-3 and/or caspase-3-like proteases are activated by a proteolytic cascade and act as distal effectors to cleave specific proteins during apoptotic cell death (see Sikorska and Walker, Chapter 7, this volume).

ACTIVATION OF A CASPASE PROTEOLYTIC CASCADE DURING PCD

Direct evidence for such a proteolytic cascade involving caspases came from an elegant study by Enari and colleagues who used a cell-free system to study the proteolytic events in Fas-mediated apoptosis (Enari et al., 1996). Using YVAD and DEVD tetrapeptide fluorogenic substrates to detect caspase-1-like and caspase-3-like activity, respectively, they demonstrated that a caspase-1-like protease was activated rapidly (maximal activity within 10 minutes) after treatment with anti-Fas mAb, while a caspase-3-like maximal activity much more slowly (protease was activated after 2.5 hours) and required prior activation of the caspase-1-like protease. Addition of caspase-1 to nonapoptotic extracts induced caspase-3-like activity and fragmentation of nuclei, both of which were prevented by nM concentrations of YVAD-CHO or DEVD-CHO. In contrast, addition of caspase-3 to nonapoptotic extracts did not induce caspase-1-like activity, but did result in disintegration of nuclei; induction of apoptosis by caspase-3 was prevented by DEVD-CHO but not by YVAD-CHO. These results indicate that caspase-1 (or a caspase-1-like protease) is an upstream activator of caspase-3 (or a caspase-3-like protease) during the execution of Fas-induced apoptosis. Once caspase-3 has been activated, caspase-1 is no longer necessary for cell death: Addition of caspase-3 to cytosolic extracts from caspase-1-deficient mice induces apoptotic nuclear changes (Enari et al., 1996). Furthermore, these studies suggest that the resistance of thymocytes from caspase-1-deficient mice to Fas-induced killing stems from a defect in their ability to activate caspase-3. Similarly, the protection conferred by crmA overexpression against a wide variety of apoptotic agents may reflect inhibition of caspase-1 or a caspase-1-like protease necessary for activation of caspase-3 (Enari et al., 1996).

Recently, the most upstream caspases in the proteolytic cascade initiated by Fas ligand and TNF-α have been identified. Their respective receptors for apoptotic signaling, Fas/APO-1/CD95 and TNFR1/p55, are members of the TNF/NGF receptor family and contain a conserved cytoplasmic region ("death domain") that is required for induction of cell death (reviewed in Nagata and Golstein, 1995). Activation of Fas and TNFR1 by ligand binding recruits the death domain-containing proteins FADD/MORT1 (hereafter referred to as FADD) (Boldin et al., 1995; Chinnaiyan et al., 1995) and TRADD (Hsu et al., 1995), respectively, to these receptors via specific protein–protein interactions mediated by the death domain. TRADD acts as an adaptor protein between FADD and TNFR1 (Hsu et al., 1996). Thus, FADD interacts with both ligand-activated Fas and TNFR1 complexes and plays a critical role in apoptotic signaling mediated by these receptors (Chinnaiyan et al., 1996b; Hsu et al., 1996). FADD, in turn, recruits caspase-8 and caspase-10, which each contain two FADD-like regions in their prodomains, to the activated Fas and TNFR1 complexes, thereby leading to their proteolytic activation and release from the receptor (Boldin et al., 1996; Fernandes-Alnemri et al., 1996; Muzio et al., 1996; Vincenz and Dixit, 1997). By virtue of their intimate association with these receptor complexes, caspase-8 and caspase-10 must be positioned at the apex of the caspase proteolytic cascade initiated by Fas ligand and TNF-α. Although the substrates of these caspases are largely unknown, they are likely to include one or more other caspase family members. Caspase-10, for instance, can proteolytically process caspase-3 and caspase-7 *in vitro* (Fernandes-Alnemri et al., 1996). Caspase-8 also likely activates caspase-3 and/or caspase-7, either directly or indirectly by activating caspase-6, an upstream regulator of caspase-3 and caspase-7 (Orth et al., 1996). Clearly, there are many gaps in our understanding of the events in the proteolytic cascade beginning with caspase-8/caspase-10 activation and culminating with the activation of caspase-3 and other caspases that act as the downstream effectors to proteolyze key intracellular protein targets. Given its marked preference for a DEVD fluorogenic substrate (Boldin et al., 1996), caspase-8 is not the caspase-1-like protease detected by Enari and co-workers (Enari et al., 1996) that activates caspase-3 during Fas-induced cell death, but rather, a likely upstream activator of this caspase-1-like protease. Moreover, it is unclear whether caspase-8 or caspase-10 play a role in other apoptotic signaling pathways. Alternatively, novel caspases may be similarly recruited to activated death receptors by virtue of protein–protein interaction motifs in their prodomains.

GRANZYME B MEDIATES CTL-INDUCED APOPTOSIS OF TARGET CELLS BY ACTIVATING CASPASES

As alluded to earlier, cytotoxic T lymphocytes (CTL) are a critical component of the cellular immune defense against tumor and virus-infected cells. CTLs induce apoptosis in target cells by two mechanisms: the interaction of Fas ligand on activated CTLs with Fas (receptor) on target cell membranes (Fas-mediated pathway), and the exocytosis of CTL cytoplasmic granules containing the serine protease granzyme B (and other related granzymes) and the pore-forming protein perforin, the latter facilitating delivery of granzyme B into target cells (granzyme–perforin pathway) (reviewed in Berke, 1995). These pathways are nonredundant: CTLs from mice deficient in Fas, Fas ligand, perforin, or Granzyme B are all severely impaired in their ability to induce apoptosis in target cells (reviewed in Berke, 1995; Nagata and Golstein, 1995). As we have seen, the Fas-mediated signaling pathway results in the activation of a cascade of caspases that play a pivotal role in Fas-induced killing. Since granzyme B is the only other known protease that shares the unusual Asp-substrate specificity of caspases (Shi et al., 1992), it seemed plausible that the delivery of this serine protease into target cells might also trigger a proteolytic cascade involving members of the caspase family. Recently, a number of studies have provided direct experimental support for this hypothesis. To begin with, several caspases are proteolytically processed and activated by granzyme B *in vitro,* including caspase-3 (Darmon et al., 1995; Fernandes-Alnemri et al., 1996; Martin et al., 1996; Quan et al., 1996), caspase-6 (Orth et al., 1996), caspase-7 (Fernandes-Alnemri et al., 1996; Gu et al., 1996), caspase-10 (Fernandes-Alnemri et al., 1996), caspase-8 (Muzio et al., 1996), caspase-9 (Duan et al., 1996b) and caspase-11 (Wang et al., 1996). Moreover, granzyme B processes caspase-3 *in vivo* during the induction of DNA fragmentation in target cells (Darmon et al., 1995). In contrast, neither caspase-1 (Darmon et al., 1994; Martin et al., 1996) nor caspase-2 (Darmon et al., 1994; Martin et al., 1996; Li et al., 1997) are direct substrates of granzyme B *in vitro.* However, proteolytic cleavages of caspase-1 and caspase-2 have been observed in apoptosis induced by granzyme B in the presence of perforin (Shi et al., 1996) or CTL (Li et al., 1997). Proteolytic activation of caspase-3 and/or related subfamily members is essential for granzyme B–mediated apoptosis of target cells. CTLs from mice deficient in granzyme B, in contrast to those derived from wild-type mice, do not cleave caspase-3 in target cells and are severely impaired in their ability to induce DNA fragmentation in these cells (Darmon et al., 1996). Furthermore, granzyme B–dependent DNA fragmentation of target cells is inhibited by μM concentrations of DEVD-CHO but not YVAD-CHO (Darmon et al., 1996). Finally, addition of granzyme B to nonapoptotic cytosolic extracts induces their ability to fragment nuclei and cleave a variety of apoptotic substrates including PARP; granzyme B's ability to induce apoptosis *in vitro* was markedly inhibited by 0.5 μM DEVD-CHO but not by 5 μM YVAD-CHO (Martin et al., 1996). Taken together, these findings indicate that granzyme B's cytolytic actions are mediated by caspase-3 or related proteases. Hence, both pathways (Fas and granzyme-perforin) by which CTLs induce apoptosis in target cells are executed by caspases. However, it has yet to be determined whether granzyme B cleaves substrates other than caspases, and whether such proteolysis contributes to its cytolytic activity. Interestingly, the cowpox viral gene product CrmA, an inhibitor of both caspase-1 and granzyme B, provides the cowpox virus with a strategy to resist both CTL pathways, although the Fas-mediated pathway may be its primary target (Tewari et al., 1995b).

APOPTOTIC SUBSTRATES OF CASPASES

Since caspases are proteases, it is imperative to identify and characterize their cellular substrates. Given the tightly regulated nature of programmed cell death and the inherent specificity of caspases, these targets are likely to be a relatively small subset of the total cellular proteins whose selective proteolysis likely contributes to the execution of apoptotic cell death (see Sikorska and Walker, Chapter 7, this volume). Hence, the identification of these so called "death substrates" can point to key molecular events/pathways in this ordered process. During apoptosis, proteolytic cleavage of a substrate may be activating (e.g., latent death-promoting proteins) or inactivating (e.g., survival-promoting or structural proteins). Finally, substrates are endogenous "indicators" of protease activity during apoptosis, and can thereby help to identify which caspases are acting when during programmed cell death (Rosen, 1996).

Many (if not all) of the proteins that are known to undergo proteolysis during cell death are substrates of one or more caspases. Recently, several such apoptotic substrates have been identified. As proposed elsewhere (Rosen, 1996), these substrates can be divided into three broad categories: (1) proteins that play a role in the homeostatic response to stressful stimuli; (2) structural proteins; and (3) proteins of unknown function. Huntingtin (Goldberg et al., 1996), the Huntington's disease (HD) gene product, and the presenilins (implicated in Familial Alzheimer's disease) (Kim et al., 1997; Loetscher et al., 1997) fall into the third category and will be discussed in the section on clinical implications. Stress-response proteins degraded during apoptosis include the DNA repair enzymes PARP (Kaufmann et al., 1993; Lazebnik et al., 1994) and the catalytic subunit of the DNA-dependent protein kinase (DNA-PK$_{CS}$) (Casciola-Rosen et al., 1995, 1996; Han et al., 1996; Song et al., 1996), several protein components of the RNA splicing complex including U1-70kDa (Casciola-Rosen et al., 1994, 1996) and the heteronuclear ribonucleoproteins C1 and C2 (Waterhouse et al., 1996), and the cell cycle regulator, the retinoblastoma (RB) tumor suppressor gene product (An and Dou, 1996; Jänicke et al., 1996). Cleavage of these substrates likely impairs their ability to repair DNA strand breaks and splice messenger RNA, thereby potentially tipping the balance in favor of cell death over cell repair (Casciola-Rosen et al., 1996). Moreover, the RB gene product is a critical regulator of cell cycle progression and inhibits programmed cell death during development and in response to irradiation and p53 overexpression (Clarke et al., 1992; Jacks et al., 1992; Lee et al., 1992; Haas-Kogan et al., 1995; Haupt et al., 1995). As such, its proteolytic inactivation would be expected to facilitate cell death. In addition to these nuclear proteins, a variety of signal transduction molecules that also likely play a role in the stress response are cleaved during apoptosis. These signal transduction proteins include protein kinase Cδ (PKCδ) (Emoto et al., 1995), the GDP dissociation inhibitor of Rho family GTPases D4-GDI (Na et al., 1996), and the sterol regulatory element binding proteins SREBP1 and SREBP2 (Wang et al., 1995, 1996). The cleavage of D4-GDI, a negative regulator of Rho family GTPases, is particularly intriguing because its proteolysis would likely activate these GTPases. Activated Rho family members have been implicated in cytoskeletal rearrangements and stress-activated protein kinase (SAPK)/jun kinase activation (JNK), two events associated with apoptotic cell death (Hall, 1994; Coso et al., 1995; Minden et al., 1995; Xia et al., 1995; Chen et al., 1996; Verheij et al., 1996) (see Zakeri, Chapter 3, Trump and Berezesky, Chapter 2, Cotman et al., Chapter 14, Tidball and Albrecht, Chapter 15, this volume). In this way, the disruption of homeostatic repair processes and simultaneous activation of proapoptotic signaling pathways (e.g., SAPK/JNK pathway) is likely to commit the cell to an apoptotic fate (Casciola-Rosen et al., 1996).

Other apoptotic substrates of caspases include a number of structural proteins that maintain the integrity of the cytoskeleton or nuclear matrix. Several components of the actin cytoskeleton, including β-actin (Mashima et al., 1995; Kayalar et al., 1996) (see Sikorska and Walker, Chapter 7, this volume), the α-subunit of the actin-binding protein fodrin (nonerythroid spectrin) (Martin et al., 1995; Cryns et al., 1996; Greidinger et al., 1996), and Gas2 (Brancolini et al., 1995), are cleaved during certain types of apoptosis. These cytoskeletal proteins appear to be selective targets of caspases: Other related proteins such as vimentin, tubulin, and talin do not undergo apoptotic proteolysis (Brancolini et al., 1995). Since disruption of actin filaments by cytochalasin B and E, inhibitors of actin polymerization, results in apoptosis (Kolber et al., 1990), it seems plausible that proteolytic cleavage of actin (and other key components of the actin cytoskeleton) might play an integral role in this process. Like the cytochalasins, actin proteolysis by caspase-1 inhibits its polymerization *in vitro* (Kayalar et al., 1996). Furthermore, overexpression of a C terminal deletion of gas2 resembling one of its cleavage fragments induces cytoplasmic shrinkage and the "rouding up" of cells, characteristic morphological features of apoptotic cell death (Brancolini et al., 1995; Trump and Berezesky, Chapter 2, this volume). In addition to their structural roles, these proteins also modify the activity of neighboring enzymes. For instance, fodrin inhibits several signal transducing phospholipases (A2, C, and D) (Lukowski et al., 1996), some of which have been implicated in programmed cell death (Jäätelä et al., 1995). Perhaps, fodrin cleavage by caspases activates these phospholipases by liberating them from fodrin's inhibitory influence. Finally, two nuclear matrix proteins have been demonstrated to undergo proteolytic cleavage by caspases during cell death: the nuclear mitotic apparatus protein (NuMA) (Weaver et al., 1996) and lamins A, B, and C (Lazebnik et al., 1995; Orth et al., 1996; Takahashi et al., 1996; Weaver et al., 1996). Selective proteolysis of these key nuclear matrix proteins likely contributes to the disintegration of the nuclear envelope during apoptosis (Lazebnik et al., 1995; Orth et al., 1996; Takahashi et al., 1996; Weaver et al., 1996; Sikorska and Walker, Chapter 7, this volume).

In addition to providing clues about apoptotic signaling pathways, the identification of these intracellular substrates has enabled us to better characterize the caspases that act to execute the cell systematically. Based on *in vitro* cleavage analyses and their sensitivity to inhibition by nM concentrations of DEVD-CHO, several of the proteins known to be cleaved during apoptosis (PARP, DNA-PK$_{CS}$, U1-70kDa, D4-GDI, huntingtin, and the SREBPs) are likely to be substrates of caspase-3 and/or caspase-3-like proteases (Kaufmann et al., 1993; Casciola-Rosen et al., 1994, 1995, 1996; Lazebnik et al., 1994; Wang et al., 1995; Goldberg et al., 1996; Han et al., 1996; Na et al., 1996; Song et al., 1996; Wang et al., 1996). All these proteins except SREBP1 are cleaved C-terminal to a DXXD motif corresponding to a consensus substrate recognition sequence of caspase-3 (Nicholson et al., 1995); SREBP1 is cleaved after an SEPD sequence (Wang et al., 1995). Since the cleavage sites of PKCδ and RB also contain DXXD motifs, they are likely to be proteolyzed by caspase-3 (Emoto et al., 1995; Jänicke et al., 1996). In contrast, the lamin protease is not caspase-3, but rather the closely related caspase-6 protease (Orth et al., 1996; Takahashi et al., 1996). Lamin proteolysis during apoptosis occurs later than PARP proteolysis, and the two proteases have distinct inhibitor profiles (Lazebnik et al., 1995; Greidinger et al., 1996; Takahashi et al., 1996). Moreover, lamin A is cleaved *in vitro* by caspase-6 but not by caspase-3 (Orth et al., 1996; Takahashi et al., 1996). Fodrin proteolysis is carried out by yet another caspase, which is insensitive to inhibition by DEVD-CHO (unlike caspase-3 and caspase-6) and is one of the earliest proteolytic events in apoptosis (Cryns et al., 1996; Greidinger et al., 1996). Although the specific caspase(s) responsible for cleaving fodrin and the other apoptotic substrates have not been identified, these studies indicate that programmed cell death is likely to require the proteolytic activity of multiple caspases acting on multiple cellular targets. Given the distinct cleavage time course for some of these apoptotic substrates, these studies also provide additional support for the concept of a caspase proteolytic cascade during cell death.

Clearly, these substrate-directed studies raise a number of important questions that have yet to be answered. For instance, what functional role does cleavage of a given substrate play in apoptosis? Although we can speculate that proteolysis of an important cytoskeletal protein might play a role in membrane blebbing or cellular condensation, definitive evidence linking cleavage of any substrate to the morphological events of apoptosis is limited (see Budd, Chapter 10, this volume). One way to address these issues might be to express a cleavage-resistant substrate (selectively mutated at the cleavage site Asp residue) in cells, and then assess its impact on apoptotic cell death. This approach was recently used to demonstrate that lamin proteolysis by caspases does, indeed, play an important role in the condensation/fragmentation of the nucleus during apoptosis (Rao et al., 1996). Nevertheless, it seems unlikely that proteolytic cleavage of any one substrate would be *sufficient* for cell death; rather, cell death is likely to require the proteolysis of many intracellular targets. Another fundamental question is how do we go about systematically identifying additional caspase substrates? Several of these substrates are antigens targeted in autoimmune diseases and have been identified by using specific autoreactive antisera from patients (Casciola-Rosen et al., 1994, 1995). Recently, we have used a novel, small pool expression cloning approach to systematically identify apoptotic caspase substrates (Lustig et al., 1997; Cryns et al., 1997). Briefly, a cDNA library is divided into small pools. Each pool is transcribed/translated *in vitro* and subsequently incubated with cytosolic extracts derived from nonapoptotic and apoptotic cells. Pools containing proteins that are proteolyzed by apoptotic extracts (but not by nonapoptotic extracts) are further divided and re-analyzed as above until an individual cDNA encoding a potential caspase substrate is isolated. These methodologies (and others) should facilitate the identification of the intracellular targets of caspases and thereby shed important insights into molecular events in apoptotic cell death.

ROLE OF CASPASES IN DISEASE

Given the critical role that apoptosis plays in development and in the response to homeostatic challenges, it is not surprising that derangements in this tightly regulated process have been implicated in the pathogenesis of many diseases. Broadly speaking, dysregulated apoptosis can take two forms: (1) increased sensitivity to apoptosis, leading to excessive cell death; and (2) resistance to apoptosis, resulting in deficient cell death (reviewed in Thompson, 1995; Nicholson, 1996) (Finkel and Casella, Chapter 11, and Benitz- Bribiesca, Chapter 17, this volume). Disorders likely to be characterized by increased cell death include neurodegenerative diseases (e.g., Alzheimer's disease, Huntington's disease, amyotrophic lateral sclerosis, spinal muscular atrophy, and others), vascular diseases, (Cotman et al., Chapter 14, and Tidball and Albrecht, Chapter 15, this volume), and AIDS. In contrast, diseases characterized by deficient or inadequate apoptosis include cancer (and benign neoplasia), autoimmune disorders (e.g., lupus, rheumatoid arthritis, and others), and chronic/latent viral infections (reviewed in Thompson, 1995; Nicholson, 1996).

Because caspases are essential effectors of apoptotic cell death, they are likely to play a pivotal role in the pathogenesis of these disorders.

In diseases associated with enhanced sensitivity to apoptosis, activation of caspases is likely to play a central role. Apoptosis is a major pathogenetic mechanism of cell death caused by vascular insufficiency/ischemia. Oligonucleosomal DNA fragmentation, a characteristic feature of apoptotic cell death, has been demonstrated in animal models of cerebral and myocardial ischemia (Hill et al., 1995; Li et al., 1995a,b; Nitatori et al., 1995; Kajstura et al., 1996; Fliss and Gattinger, 1996) and in post mortem studies of human myocardial infarctions (Itoh et al., 1995; Bardales et al., 1996). Apoptotic myocytes can be detected within two to three hours of the onset of ischemia (Bardales et al., 1996; Fliss and Gattinger, 1996; Kajstura et al., 1996). Reperfusion of ischemic regions may accelerate the apoptotic process, presumably by the introduction of free radicals, inflammatory cytokines and calcium (Schumer et al., 1992; Gottlieb et al., 1994; Thompson, 1995; Fliss and Gattinger, 1996). Several lines of evidence indicate that caspases are important mediators of the apoptosis during ischemic/reperfusion cell death. First, cerebral ischemia and subsequent reperfusion are associated with the rapid production of mature IL-1β (Hara et al., 1997). Since caspase-1 is the only known protease that can convert pro-IL-1β to its mature cytokine, these findings strongly suggest that caspase-1 is activated during ischemia/reperfusion. Second, caspase-1 and/or other caspases are proteolytically processed during reperfusion of ischemic brain (Hara et al., 1997). Finally, overexpression of crmA protects cells from hypoxic cell death in a tissue culture model of ischemia (Friedlander et al., 1996).

Like ischemia, neurodegenerative diseases are also characterized by increased apoptosis, leading to the demise of discrete populations of neurons (reviewed in Nicholson, 1996; Thompson, 1995). Numerous *post mortem* studies from patients with Huntington's or Alzheimer's disease have demonstrated oligonucleosomal DNA fragmentation in striatal and hippocampal neurons, respectively (Dragunow et al., 1995; Portera-Cailliau et al., 1995; Smale et al., 1995; Thomas et al., 1995). In addition, spinal muscular atrophy, a heterogeneous autosomal recessive neurodegenerative disease, is caused by deletions in one of two adjacent genes encoding antiapoptotic proteins: the survival motor neuron (SMN) (Lefebvre et al., 1995) and neuronal apoptosis inhibitory protein (NAIP) genes (Roy ct al., 1995). The SMN protein binds to and potentiates the antiapoptotic activity of bcl-2 (Iwahashi et al., 1997), a negative regulator of caspases (Chinnaiyan et al., 1996) (see the following). NAIP is a member of a conserved family related to the baculovirus inhibitor of apoptosis (iap) proteins which prevent apoptosis by directly inhibiting one or more caspases (Liston et al., 1996; Deveraux et al., 1997). In this way, functional disruption of these genes may lead to aberrant activation of caspases. Although there are currently no published studies that have examined whether caspases are activated in apoptotic neurons from these patients, the recent observations that huntingtin (the Hungtinton's disease gene product) and the presenilins (implicated in familial Alzheimer's disease) are specifically cleaved by caspase-3 suggest that these proteases may, indeed, be important pathogenetically in neurodegenerative disorders (Goldberg et al., 1996; Kim et al., 1997; Loetscher et al., 1997). Huntington's disease is caused by expansion of a CAG repeat in the HD gene that results in a variable-length polyglutamine tract in huntingtin (Huntington's Disease Collaborative Research Group, 1993). The size of this polyglutamine tract determines how readily huntingtin is proteolyzed by caspase-3, thereby suggesting a potential mechanism by which triplet expansion may contribute to neuronal apoptosis (Goldberg et al., 1996; reviewed in Rosen, 1996). Similarly, a mutant presenilin protein associated with familial Alzheimer disease was shown to be more readily cleaved by caspase-3 than its wild-type counterpart, perhaps accounting for at least part of its pro-apoptotic effect in neurons (Kim et al., 1997).

In contrast, diseases associated with deficient apoptosis may be characterized by inhibition or decreased activity of caspases. As we have seen, a number of viruses resist the host immune response by specifically inhibiting caspases (e.g., the cowpox CrmA and baculovirus p35 gene products, Chapter 1), thereby preventing apoptosis of virally infected cells (see Ameisen, Chapter 1, and Finkel Casella, this volume). Similarly, autoimmune disorders are characterized by diminished apoptosis of autoreactive and activated lymphocytes (reviewed by Thompson, 1995). Inherited defects in Fas or Fas ligand in mice and humans results in a fatal autoimmune syndrome characterized by profound lymphadenopathy and spenomegaly (reviewed in Nagata and Golstein, 1995; Thompson, 1995). Given the central role that caspases play in Fas-mediated apoptosis, these protcases are likely to be key pathogenetic mediators of autoimmunity. The observation that mice deficient in caspase-1 or caspase-3 do not develop these symptoms and have subtle or no defects in Fas-mediated killing, respectively, suggests that inhibition of additional caspases is necessary to effectively block Fas signaling and thereby give rise to overt autoimmune disease (Kuida et al., 1995; Kuida et al., 1996). Finally, cancer cells acquire a diminished sensitivity to apoptosis (and an enhanced proliferative capacity) as a result of accumulated genetic damage. A dramatic illustration of the importance of resistance to apoptosis in the pathogenesis of cancer comes from studies of the bcl-2 gene, an antiapoptotic mammalian homologue of the nematode ced-9 survival gene (Vaux et al., 1992; Hengartner and Horvitz, 1994). bcl-2 was cloned by virtue of its translocation from chromosome 18 to 14 in follicular lymphomas, an event that leads to its overexpression and results in resistance to apoptotic cell death (Tsujimoto et al., 1984; Bakhshi et al., 1985; Vaux et al., 1988). Importantly, bcl-2 is a negative upstream regulator of caspases: overexpression of bcl-2 in tumor cell lines protects them from apoptosis induced by a wide variety of stimuli (reviewed in Reed, 1995; Yang and Korsmeyer, 1996) and prevents activation of the caspase proteolytic cascade (Chinnaiyan et al., 1996a). More recently, overexpression of an endogenous inhibitor of certain caspases, the so called FLIP protein, has been demonstrated in malignant melanomas (Irmler et al., 1997b). Similarly, survivin, a novel human member of the anti-apoptotic iap family, is overexpressed in a number of human cancers (Ambrosini et al., 1997); iap proteins likely suppress apoptosis by inhibiting caspases (Deveraux et al., 1997). Hence, inhibition of caspases may be a general theme in neoplastic transformation and/or tumor progression. It is tempting to speculate that caspases, like the proapoptotic p53 gene, may function as tumor suppressor genes whose inactivation by somatic or germline mutations renders tumor cells resistant to apoptosis.

Given the likely role that caspases play in the dysregulated apoptosis underlying these diseases, these proteases are appealing candidates for therapeutic manipulation (reviewed in Thompson, 1995 and Nicholson, 1996). In the future, diseases characterized by deficient apoptosis might become amenable to gene therapy, whereby caspases are selectively delivered to specific targets, such as tumor cells. Although there are numerous obstacles to surmount, a recent study demonstrating regression of brain tumors *in vivo* by induction of caspase-1 expression suggests that gene therapy using these proteases may become an important therapeutic strategy (Yu et al., 1996). In addition, the identification of apoptotic substrates of caspases may point to novel targets for cancer chemotherapy. In contrast, diseases characterized by increased apoptosis might respond to treatment with specific inhibitors of caspases. Indeed, two recent reports indicate that peptide inhibitors of these proteases reduce cerebral infarct size by 50% or more in animal models of ischemia (Loddick et al., 1996) or ischemia followed by reperfusion (Hara et al., 1997). Importantly, reduction of infarct size was accompanied by an improved neurological outcome (Hara et al., 1997). Similar dramatic protection from ischemic brain injury and its neurological consequences was observed in a transgenic mouse expressing a dominant negative mutant of caspase-1 (catalytic Cys-285 replaced with Gly) targeted to neural tissue (Friedlander et al., 1997b). This same caspase-1 dominant negative construct also delays disease progression in a mouse model of the neurodegenerative disease amyotrophic lateral sclerosis (Friedlander et al., 1997a). These striking findings suggest that caspase inhibitors may become important adjuncts to current thrombolytic and acute revascularizations strategies aimed at limiting brain or myocardial infarct size, and may be useful in the treatment of neurodegenerative disorders. Whether such strategies will work for chronic and systemic disorders associated with excessive cell death, such as AIDS, remains to be seen. For instance, global inhibition of apoptosis by systemic administration of caspase inhibitors may have its own untoward consequences. Nevertheless, the fundamental role that these proteases play in apoptosis and disease suggest that the ability to selectively regulate their activity may open a myriad of therapeutic possibilities.

ACKNOWLEDGMENTS

This work was supported in part by a Mentored Clinical Scientist Development Award K08-CA01752-03 (to V.L.C.) and grant AG12859-01 (to J.Y.) from the National Institutes of Health, and by an Established Investigatorship from the American Heart Association (to J.Y.). We thank Dr. David Fisher and Dr. Paul Anderson for critical review of this manuscript.

REFERENCES

Alnemri ES, Livingston DJ, Nicholson DW, Salvesen G, Thornberry NA, Wong WW, Yuan J (1996): Human ICE/CED-3 protease nomenclature. Cell 87:171.

Ambrosini G, Adida C, Altieri DC (1997). A novel anti-apoptosis gene, survivin, expressed in cancer and lymphoma. Nat Med 3:917–21.

An B, Dou QP (1996): Cleavage of the retinoblastoma protein during apoptosis: An interleukin 1β-converting enzyme-like protease as candidate. Cancer Res 56:438–42.

Armstrong RC, Aja T, Xiang J, Gaur S, Krebs JF, Hoang K, Bai X, Korsmeyer SJ, Karanewsky DS, Fritz LC, Tomaselli KJ (1996): Fas-induced activation of the cell death-related protease CPP32 is inhibited by bcl-2 and by ICE family protease inhibitors. J Biol Chem 271:16850–55.

Bakhshi A, Jensen JP, Goldman P, Wright JJ, McBride OW, Epstein AL, Korsmeyer SJ (1985): Cloning the chromosomal breakpoint of t(14;18) human lymphomas: Clustering around JH on chromosome 14 and near a transcriptional unit on 18. Cell 41:899–906.

Bardales RH, Hailey LS, Xie SS, Schaefer RF, Hsu SM (1996): In situ apoptosis assay for the detection of early acute myocardial infarction. Am J Pathol 149:821–29.

Beidler DR, Tewari M, Friesen PD, Poirier G, and Dixit VM (1995): The baculvirus p35 protein inhibits Fas- and tumor necrosis factor-induced apoptosis. J Biol Chem 270:16526–28.

Berke G (1995): The CLT's kiss of death. Cell 81:9–12.

Bertin J, Armstrong RC, Ottilie S, Martin DA, Wang Y, Banks S, Wang G-H, Senkevich TG, Alnemri ES, Moss B, Lenardo MJ, Tomaselli KJ, Cohen JI (1997): Death effector domain-containing herpesvirus and poxvirus proteins inhibit both Fas- and TNFRI-induced apoptosis. Proc Natl Acad Sci USA 94:1172–76.

Bertin J, Mendrysa SM, LaCount DJ, Gaur S, Krebs JF, Armstrong RC, Tomaselli KJ, Friesen PD (1996): Apoptotic suppression by baculovirus P35 involves cleavage by an inhibition of a virus-induced CED-3/ICE-like protease. J Virol 70:6251–59.

Boldin MP, Goncharov TM, Goltsev YV, Wallach D (1996): Involvement of MACH, novel MORT1/FADD-interacting protease, in Fas/APO-1- and TNF receptor-induced cell death. Cell 85:803–15.

Boldin MP, Varfolomeev, EE, Pancer Z, Mett IL, Camonis JH, Wallach D (1995): A novel protein that interacts with the death domain of Fas/APO-1 contains a sequence motif related to the death domain. J Biol Chem 270:7795–98.

Boudreau N, Sympson CJ, Werb Z, Bissell MJ (1995): Suppression of ICE and apoptosis in mammary epithelial cells by extracellular matrix. Science 267:891–93.

Brancolini C, Benedetti M, Schneider C (1995): Microfilament reorganization during apoptosis: The role of Gas2, a possible substrate for ICE-like proteases. EMBO J 14:5179–90.

Bump NJ, Hackett M, Hugunin M, Seshagiri S, Brady K, Chen P, Ferenz C, Franklin S, Ghayur, T, Li P, Licari P, Mankovich J, Shi L, Greenberg AH, Miller LK, Wong WW (1995): Inhibition of ICE family proteases by baculovirus antiapoptotic protein p35. Science 269:1885–88.

Casciola-Rosen L, Nicholson DW, Chong T, Rowan KR, Thornberry NA, Miller DK, Rosen A (1996): Apopain/CPP32 cleaves proteins that are essential for cellular repair: A fundamental principle of apoptotic death. J Exp Med 183:1957–64.

Casciola-Rosen LA, Anhalt GJ, Rosen A (1994a): Autoantigens targeted in systemic lupus erythematosus are clustered in two populations of surface structures on apoptotic keratinocytes. J Exp Med 179:1317–30.

Casciola-Rosen LA, Anhalt GJ, Rosen A (1995): DNA-dependent protein kinase is one of a subset of autoantigens specifically cleaved early during apoptosis. J Exp Med 182:1625–34.

Casciola-Rosen LA, Miller DK, Anhalt GJ, Rosen A (1994b): Specific cleavage of the 70-kDa protein component of the U1 small nuclear ribonucleoprotein is a characteristic biochemical feature of apoptotic cell death. J Biol Chem 269:30757–60.

Cerretti DP, Kozlosky CJ, Mosley B, Nelson N, Van Ness K, Greenstreet TA, March CJ, Kronheim SR, Druck T, Cannizzaro LA, Huebner K, Black RA (1992): Molecular cloning of the interleukin-1β converting enzyme. Science 256:97–100.

Chen YR, Meyer CF, Tan TH (1996): Persistent activation of c-Jun N-terminal kinase 1 (JNK1) in gamma radiation-induced apoptosis. J Biol Chem 271:631–34.

Chinnaiyan AM, O'Rourke K, Tewari M, Dixit VM (1995): FADD, a novel death domain-containing protein, interacts with the death domain of Fas and initiates apoptosis. Cell 81:505–12.

Chinnaiyan AM, Orth K, O'Rourke K, Duan H, Poirier GG, Dixit VM (1996a): Molecular ordering of the cell death pathway. Bcl-2 and Bcl-xL function upstream of the CED-3-like apoptotic proteases. J Biol Chem 271:4573–76.

Chinnaiyan AM, Tepper CG, Seldin MF, O'Rourke K, Kischkel FC, Hellbardt S, Krammer PH, Peter ME, Dixit VM (1996b): FADD/MORT1 is a common mediator of CD95 (Fas/APO-1) and tumor necrosis factor receptor-induced apoptosis. J Biol Chem 271:4961–65.

Chinnaiyan AM, O'Rourke K, Lane BR, Dixit VM (1997a): Interaction of CED-4 with CED-3 and CED-9: A molecular framework for cell death. Science 275:1122–26.

Chinnaiyan AM, Chaudhary D, O'Rourke K, Koonin EV, Dixit VM (1997b): Role of CED-4 in the activation of CED-3. Nature 388:728–29.

Chow SC, Weis M, Kass GEN, Holmström TH, Eriksson JE, Orrenius S (1995): Involvement of multiple proteases during Fas-mediated apoptosis in T lymphocytes. FEBS Lett 364:134–38.

Clarke AR, Maandag ER, van Roon, M, van der Lugt NMT, van der Valk M, Hooper ML, Berns A, te Riele H (1992): requirement for a functional RB-1 gene in murine development. Nature 359:328–30.

Clem RJ, Fechheimer M, Miller LK (1991): Prevention of apoptosis by a baculovirus gene during infection of insect cells. Science 254:1388–90.

Coso OA, Chiariello M, Yu JC, Teramoto H, Crespo P, Xu N, Miki T, Gutkind JS (1995): The small GTP-binding proteins Rac1 and Cdc42 regulate the activity of the JNK/SAPK signaling pathway. Cell 81:1137–46.

Criekinge WV, Beyaert R, Van de Craen M, Vandenabeele P, Schotte P, De Valck D, Fiers W (1996): Functional characterization of the prodomain of interleukin-1β-converting enzyme. J Biol Chem 271:27245–48.

Cryns VL, Bergeron L, Zhu H, Li H, Yuan J (1996): Specific cleavage of α-fodrin during Fas- and tumor necrosis factor-induced apoptosis is mediated by an interleukin-1β-converting enzyme/Ced-3 protease distinct from the poly(ADP-ribose) polymerase protease. J Biol Chem 271:31277–82.

Cryns VL, Byun Y, Rana A, Mellor H, Lustig KD, Ghanem L, Parker PJ, Kirschner MW, Yuan J (1997): Specific proteolysis of the kinase protein kinase C-related kinase 2 by caspase-3 during apoptosis: Identification of a novel, small pool expression cloning strategy. J Biol Chem 272:29449–53.

Darmon AJ, Ehrman N, Caputo A, Fujinaga J, Bleackley RC (1994): The cytotoxic T cell proteinase granzyme B does not activate interleukin-1 β-converting enzyme. J Biol Chem 269:32043–46.

Darmon AJ, Ley TJ, Nicholson DW, Bleackley RC (1996): Cleavage of CPP32 by granzyme B represents a critical role for granzyme B in the induction of target cell DNA fragmentation. J Biol Chem 271:21709–12.

Darmon AJ, Nicholson DW, Bleackley RC (1995): Activation of the apoptotic protease CPP32 by the cytotoxic T-cell-derived granzyme B. Nature 377:446–48.

Dragunow M, Faull RL, Lawlor P, Beilharz EJ, Singleton K, Walker EB, Mee E (1995): In situ evidence for DNA fragmentation in Huntington' disease striatum and Alzheimer's disease temporal lobes. NeuroReport 6:1053–57.

Deveraux QL, Takahashi R, Salvesen GS, Reed JC (1997): X-linked IAP is a direct inhibitor of cell-death proteases. Nature 388:300–4.

Duan H, Chinnaiyan AM, Hudson PL, Wing JP, He W-W, Dixit, VM (1996a): ICE-LAP3, a novel mammalian homologue of the *Caenorhabditis elegans* cell death protein Ced-3 is activated during Fas- and tumor necrosis factor-induced apoptosis. J Biol Chem 271:1621–25.

Duan H, Orth K, Chinnaiyan AM, Poirier GG, Froelich CJ, He WW Dixit, VM (1996b): ICE-LAP6, a novel member of the ICE/Ced-3 gene family, is activated by the cytotoxic T cell protease granzyme B. J Biol Chem 271:16720–24.

Ellis RE, Yuan J, Horvitz HR (1991): Mechanisms and functions of cell death. Ann Rev Cell Biol 7:663–98.

Emoto Y, Manome Y, Meinhardt G, Kisaki H, Kharbanda S, Robertson M, Ghayur T, Wong WW, Kamen R, Weichselbaum R, Kufe D (1995): Proteolytic activation of protein kinase C δ by an ICE-like protease in apoptotic cells. EMBO J 14:6148–56.

Enari M, Hase A, Nagata S (1995): Apoptosis by a cytosolic extract from Fas-activated cells. EMBO J 14:5201–8.

Enari M, Hug H, Nagata S (1995): Involvement of an ICE-like protease in Fas-mediated apoptosis. Nature 375:78–81.

Enari M, Talanian RV, Wong WW, Nagata S (1996): Sequential activation of ICE-like and CPP32-like proteases during Fas-mediated apoptosis. Nature 380:723–26.

Erhardt, P, Cooper GM (1996): Activation of the CPP32 apoptotic protease by distinct signalling pathways with differential sensitivity of bcl-x_L. J Biol Chem 271:17601–4.

Faucheau C, Blanchet AM, Collard-Dutilleul V, Lalanne JL, Diu-Hercend A (1996): Identification of a cysteine protease closely related to interleukin-1β coverting enzyme. Eur J Biochem 236:207–13.

Faucheu C, Diu A, Chan AWE, Blanchet A-M, Miossec C, Hervé F, Collard-Dutilleul V, Gu Y, Aldape RA, Lippke JA, Rocher C, Su MS-S, Livingston DJ, Hercend T, Lalanne J-L (1995): A novel human protease similar to the interleukin-1β converting enzyme induces apoptosis in transfected cells. EMBO J 14:1914–22.

Fernandes-Alnemri T, Armstrong RC, Krebs J, Srinivasula SM, Wang L, Bullrich F, Fritz LC, Trapani JA, Tomaselli KJ, Litwack G, Alnemri ES (1996): *In vitro* activation of CPP32 and Mch3 by Mch4, a novel apoptotic cysteine protease containing two FADD-like domains. Proc Natl Acad Sci USA 93:7464–69.

Fernandes-Alnemri T, Litwack G, Alnemri ES (1994): CPP32, a novel human apoptotic protein with homology to *Caenorhabditis elegans* cell death protein ced-3 and mammalian interleukin-1B-converting enzyme. J Biol Chem 269:30761–64.

Fernandes-Alnemri T, Litwack G, Alnemri ES (1994): *Mch2,* a new member of the apoptotic *ced-3/Ice* cysteine protease family. Cancer Res 55:2737–42.

Fernandes-Alnemri T, Takahashi A, Armstrong R, Krebs J, Fritz L, Tomaselli KJ, Wang L, Yu Z, Croce CM, Salvesen G, (1995): Mch3, a novel human apoptotic cysteine protease highly related to CPP32. Cancer Res 55:6045–52.

Fliss H, Gattinger D (1996): Apoptosis in ischemic and reperfused rat myocardium. Circ Res 79:949–56.

Fraser A, Evan G (1996): A license to kill. Cell 85:781–84.

Friedlander RM, Brown RH, Gagliardini V, Wang J, Yuan J (1997a): Inhibition of ICE slows ALS in mice. Nature 388:31.

Friedlander RM, Gagliardini V, Hara H, Fink KB, Li W, MacDonald G, Fishman MC, Greenberg AH, Moskowitz MA, Yuan J (1997b): Expression of a dominant negative mutant of Interleukin-1β converting enzyme in transgenic mice prevents neuronal cell death induced by trophic factor withdrawal and ichemic brain injury. J-Exp. Med. 185:933–940.

Friedlander RM, Gagliardini V, Rotello RJ, Yuan J (1996): Functional role of interleukin 1β (IL-1β) in IL-1β-converting enzyme-mediated apoptosis. J Exp Med 184:717–24.

Gagliardini V, Fernandez PA, Lee RKK, Drexler HCA, Rotello RJ, Fishman MC, Yuan J (1994): Prevention of vertebrate neuronal death by the *crmA* gene. Science 263:826–28.

Goldberg YP, Nicholson DW, Rasper DM, Kalchman MA, Koide HB, Graham RK Bromm M, Kazemi-Esfarjani P, Thornberry NA, Vaillancourt JP, Hayden MR (1996): Cleavage of huntingtin by apopain, a proapoptotic cysteine protease, is modulated by the polyglutamine tract. Nature Genet 13:442–49.

Gottlieb RA, Burleson KO, Kloner RA, Babior BM, Engler RL (1994): Reperfusion injury induces apoptosis in rabbit cardiomyocytes. J Clin Invest 94:1621–28.

Greidinger EL, Miller DK, Yamin T-T, Casciola-Rosen L, Rosen A (1996): Sequential activation of three distinct ICE-like activities in Fas-ligated Jurkat cells. FEBS Lett 390:299–303.

Gu Y, Sarnecki C, Aldape RA, Livingston DJ, Su MS-S (1995): Cleavage of poly(ADP-ribose) polymerase by interleukin-1B converting enzyme and its homologs TX and Nedd-2. J Biol Chem 270:18715–18.

Gu Y, Sarnecki C, Fleming MA, Lippke JA, Bleackley RC, Su MS (1996): Processing and activation of CMH-1 by granzyme B. J Biol Chem 271:10816–20.

Haas-Kogan DA, Kogan SC, Levi D, Dazin P, T'Ang A, Fung YKT Israel MA (1995): Inhibition of apoptosis by the retinoblastoma gene product. EMBO J 14:461–72.

Hall A (1994): Small GTP-binding proteins and the regulation of the actin cytoskeleton. Annu Rev Cell Biol 10:31–54.

Han Z, Malik N, Carter T, Reeves WH, Wyche JH, Hendrickson EA (1996): DNA-dependent protein kinase is a target for a CPP32-like apoptotic protease. J Biol Chem 271:25035–40.

Hara H, Friedlander RM, Gagliardini V, Ayata C, Fink K, Huang Z, Shimizu-Sasamata M, Yuan J, Moskowitz MA (1997): In hibition of interleukin-1β converting enzyme family protein reduce ischemic and cytotoxic neuronal damage. Pro. Nat. Acad. Sci. USA 94:2007–12.

Haupt Y, Rowan S, Oren M (1995): p53-mediated apoptosis in HeLa cells can be overcome by excess pRB. Oncogene 10:1563–71.

Hay BA, Wolff T, Rubin GM (1994): Expression of baculovirus p35 prevents cell death in *Drosophila*. Development 120:2121–29.

Hengartner MO, Horvitz HR (1994): C. elegans cell survival gene *ced-9* encodes a functional homolog of the mammalian proto-oncogene *bcl-2*. Cell 76:665–76.

Hill IE, MacManus JP, Rasquinha I, Tuor UI (1995): DNA fragmentation indicative of apoptosis following unilateral cerebral hypoxia-ischemia in the neonatal rat. Brain Res 676:398–403.

Hsu H, Shu H-B, Pan M-P, Goeddel DV (1996): TRADD-TRAF2 and TRADD-FADD interactions define two distinct TNF receptor-1 signal transduction pathways. Cell 84:299–308.

Hsu H, Xiong J, Goeddel DV (1995): The TNF receptor1-associated protein TRADD signals cell death and NF-κB activation. Cell 81:495–504.

Hu S, Vincenz C, Buller M, Dixit VM (1997): A novel family of viral death effector domain-containing molecules that inhibit both CD-95- and tumor necrosis factor receptor-1-induced apoptosis. J Biol Chem 272:9621–24.

Huntington's Disease Collaborative Research Group (1993): A novel gene containing a trinucleotide repeat that is expanded and unstable on Huntington's Disease Chromosomes. Cell 72:971–979.

Itoh G, Tamura J, Suzuki M, Suzuki Y, Ikeda H, Koike M, Nomura M, Jie T, Iton K (1995): DNA fragmentation of human infarcted myocardial cells demonstrated by the nick end labeling method and DNA agarose gel electrophoresis. Am J Pathol 146:1325–31.

Irmler M, Hofmann K, Vaux D, Tschopp J (1997a): Direct physical interaction between the Caenorhabditis elegans 'death proteins' CED-3 and CED-4. FEBS Lett 406:189–90.

Irmler M, Thome M, Hahne M, Schneider P, Hofmann K, Steiner V, Bodmer J-L, Schröter M, Burns K, Mattmann C, Rimoldi O, French L, Tschopp J (1997b): Inhibition of death receptor signals by cellular FLIP. Nature 388:190–95.

Iwahashi H, Eguchi Y, Yasuhara N, Hanafusa T, Matsuzawa Y, Tsujimoto Y (1997): Synergistic anti-apoptotic activity between Bcl-2 and SMN implicated in spinal muscular atrophy. Nature 390:413–17.

Jäätelä M, Benedict M, Tewari M, Shayman JA, Dixit VM (1995): Bcl-x and Bcl-2 inhibit TNF and Fas-induced apoptosis and activation of phospholipase A_2 in breast carcinoma cells. Oncogene 10:2297–305.

Jacks T, Fazeli A, Schmitt EM, Bronson RT, Goodell MA, Weinberg RA (1992): Effects of an Rb mutation in the mouse. Nature 359:295–300.

Jacobsen MD, Weil M, Raff MC (1996): Role of Ced-3/ICE-family proteases in staurosporine-induced programmed cell death. J Cell Biol 133:1041–51.

Jänicke RU, Walker PA, Lin XY, Porter AG (1996): Specific cleavage of the retinoblastoma protein by an ICE-like protease in apoptosis. EMBO J 15:6969–78.

Kajstura J, Cheng W, Reiss K, Clark WA, Sonnenblick EH, Krajewski S, Reed JC, Olivetti G, Anversa P (1996): Apoptotic and necrotic myocyte cell deaths are independent contributing variables of infarct size in rats. Lab Invest 74:86–107.

Kamens J, Paskind M, Hugunin M, Talanian RV, Allen H, Banach D, Bump N, Hackett M, Johnston CG, Li P, Mankovich JA, Terranova M, Ghayur T (1995): Identification and characterization of ICH-2, a novel member of the interleukin-1β-converting enzyme family of cysteine proteases. J Biol Chem 270:15250–56.

Kaufmann SH, Desnoyers S, Ottaviano Y, Davidson NE, Poirier GG (1993): Specific proteolytic cleavage of poly(ADP-ribose) polymerase: An early marker of chemotherapy-induced apoptosis. Cancer Res 53:3976–85.

Kayalar C, Örd T, Testa MP, Zhong L-T, Bredesen DE (1996): Cleavage of actin by interleukin 1β-converting enzyme to reverse DNase I inhibition. Proc Natl Acad Sci USA 93:2234–38.

Kim T-W, Pettingell WH, Jung Y-K, Kovacs DM, Tanzi RE (1997): Alternative cleavage of Alzheimer-associated presenilins during apoptosis by a caspase-3 family protease. Science 277:373–76.

Kolber MA, Broschat KO, Landa-Gonzalez B (1990): Cytochalasin B induces cellular DNA fragmentation. FASEB J 4:3021–27.

Komiyama T, Ray CA, Pickup DJ, Howard AD, Thornberry NA, Peterson EP, Salvesen G (1994): Inhibition of the interleukin-1β converting enzyme by the cowpox virus serpin CrmA. An example of cross-class inhibition. J Biol Chem 269:19331–37.

Kuida K, Lippke JA, Ku G, Harding MW, Livingston DJ, Su MS-S, Flavell RA (1995): Altered cytokine export and apoptosis in mice deficient in interleukin-1β converting enzyme. Science 267:2000–2.

Kuida K, Zheng TS, Na S, Kuan C-y, Yang D, Karasuyama H, Rakic P, Flavell RA (1996): Decreased apoptosis in the brain and premature lethality in CPP32-deficient mice. Nature 384:368–72.

Kumar S (1995): Inhibition of apoptosis by the expression of antisense *Nedd2*. FEBS Lett 368:69–72.

Kumar S, Kinoshita M, Noda M, Copeland NG, Jenkins NA (1994): Induction of apoptosis by the mouse *Nedd2* gene, which encodes a protein similar to the product of the *Caenorhabditis elegans* cell death gene *ced-3* and the mammalian IL-1B-converting enzyme. Genes Dev 8:1613–26.

Kumar S, Tomooka Y, Noda M (1992): Identification of a set of genes with developmentally down-regulated expression in the mouse brain. Biochem Biophys Res Commun 185:1155–61.

Lazebnik YA, Cole S, Cooke CA, Nelson WG, Earnshaw WC (1993): Nuclear events of apoptosis in vitro in cell-free mitotic extracts: A model system for analysis of the active phase of apoptosis. J Cell Biol 123:7–22.

Lazebnik YA, Kaufmann SH, Desnoyers S, Poirier GG, Earnshaw WC (1994): Cleavage of poly(ADP-ribose) polymerase by a proteinase with properties like ICE. Nature 371:346–47.

Lazebnik YA, Takahashi A, Moir RD, Goldman RD, Poirier GG, Kaufmann SH, Earnshaw WC (1995): Studies of the lamin proteinase reveal multiple parallel biochemical pathways during apoptotic execution. Proc Natl Acad Sci USA 92:9042–46.

Lee EY-HP, Chang C-Y, Hu N, Wang Y-CJ, Lai C-C, Herrup K, Lee W-H, Bradley A (1992): Mice deficient for Rb are nonviable and show defects in neurogenesis and haematopoiesis. Nature 359:288–94.

Lefebvre S, Bürglen L, Reboullets S, Clermont O, Burlet P, Viollet L, Benichou B, Cruaud C, Millasseau P, Zeviani M, LePaslier D, Frézal J, Cohen D, Weissenbach J, Munnich A, Melki J (1995): Identification and characterization of a spinal muscular atrophy-determining gene. Cell 80:155–65.

Li H, Bergeron L, Cryns V, Pasternack MS, Zhu H, Shi L, Greenberg A, Yuan J (1997): Activation of caspase-2 in apoptosis. J Biol Chem 272:21010–14.

Li P, Allen H, Banerjee S, Franklin S, Herzog L, Johnston C, McDowell J, Paskind M, Rodman L, Salfeld J, Towne E, Tracey D, Wardwell S, Wei F-Y, Wong W, Kamen

R, Seshadri T (1995): Mice deficient in IL-1B-converting enzyme are defective in production of mature IL-1B and resistant to endotoxic shock. Cell 80:401–11.

Li W, Fishman MC, Yuan J (1996): Prevention of apoptosis in CNTF-dependent neurons by a mutant ICE and by viral protein CrmA but not by proto-oncogene product Bcl-2. Cell Death Diff 3:105–12.

Li Y, Chopp M, Jiang N, Zaloga C (1995a): In situ detection of DNA fragmentation after focal cerebral ischemia in mice. Mol Brain Res 28:164–68.

Li Y, Chopp M, Jiang N, Zhang ZG, Zaloga C (1995b): Induction of DNA fragmentation after 10 to 120 minutes of focal cerebral ischemia in rats. Stroke 26:1252–57.

Lippke JA, Gu Y, Sarnecki C, Caron PR, Su MS-S (1996): Identification and characterization of CPP32/*Mch2* homolog 1, a novel cysteine protease similar to CPP32. J Biol Chem 271:1825–28.

Liston P, Roy N, Tamai K, Lefebvre C, Baird S, Cherton-Horvat G, Farahani R, McLean M, Ikeda J-E, MacKenzie A, Korneluk RG (1996): Suppression of apoptosis in mammalian cells by NAIP and a related family of IAP genes. Nature 379:349–53.

Liu X, Kim CN, Pohl J, Wang X (1996): Purification and characterization of an interleukin 1B converting enzyme family protease that activates cysteine protease P32 (CPP32). J Biol Chem 271:13371–76.

Loddick SA, MacKenzie A, Rothwell NJ (1996): An ICE inhibitor, z-VAD-DCB attenuates ischaemic brain damage in the rat. NeuroReport 7:1465–68.

Loetscher H, Deuschle U, Brockhaus M, Reinhardt D, Nelboeck P, Mous J, Grünberg J, Haass C, Jacobsen H (1997): Presenilins are processed by caspase-type proteases. J Biol Chem 272:20655–59.

Los M, Van de Craen M, Penning LC, Schenk H, Westendorp M, Baeuerle PA, Dröge W, Krammer PH, Fiers W, Schulze-Osthoff K (1995): Requirement of an ICE/CED-3 protease for Fas/APO-1-mediated apoptosis. Nature 375:81–83.

Lukowski S, Lecomte M-C, Mira J-P, Marin P, Gautero H, Russo-Marie F, Geny B (1996): Inhibition of phospholipase D activity by fodrin. An active role for the cytoskeleton. J Biol Chem 271:24164–71.

Lustig KD, Stukenberg T, McGarry T, King RW, Cryns VL, Mead P, Zon L, Yuan J, Kirschner MW (1997): Small pool expression screening: A novel strategy for the identification of genes involved in cell cycle control, apoptosis and early development. Methods Enzymol 283:83–99.

Martin SJ, Amarante-Mendes GP, Shi L, Chuang TH, Casiano CA, O'Brien GA, Fitzgerald P, Tan EM, Bokoch GM, Greenberg AH, Green DM (1996): The cytotoxic cell protease granzyme B initiates apoptosis in a cell-free system by proteolytic processing and activation of the ICE/CED-3 family protease, CPP32, via a novel two-step mechanism. EMBO J 15:2407–16.

Martin SJ, Newmeyer DD, Mathias S, Farschon DM, Wang HG, Reed JC, Kolesnick RN, Green DR (1995): Cell-free reconstitution of Fas-, UV radiation- and ceramide-induced apoptosis. EMBO J 14:5191–200.

Martin SJ, O'Brien GA, Nishioka WK, McGahon AJ, Mahboubi A, Saido TC, Green DR (1995): Proteolysis of fodrin (non-erythroid spectrin) during apoptosis. J Biol Chem 270:6425–28.

Martinou I, Fernandez PA, Missotten M, White E, Allet B, Sadoul R, Martinou JC (1995): Viral proteins E1B 19K and p35 protect sympathetic neurons from cell death induced by NGF deprivation. J Cell Biol 128:201–8.

Mashima T, Naito M, Fujita N, Noguchi K, Tsuruo T (1995): Identification of actin as a substrate of ICE and an ICE-like protease and involvement of an ICE-like protease but not ICE in VP-16-induced U937 apoptosis. Biochem Biophys Res Comm 217:1185–92.

Milligan CE, Prevette D, Yaginuma H, Homma S, Cardwell C, Fritz LC, Tomaselli KJ, Oppenheim RW, Schwartz LM (1995): Peptide inhibitors of the ICE protease family arrest programmed cell death of motoneurons in vivo and in vitro. Neuron 15:385–93.

Minden A, Lin A, Claret FX, Abo A, Karin M (1995): Selective activation of the JNK signalling cascade and c-Jun transcriptional activity by the small GTPases Rac and Cdc42Hs. Cell 81:1147–57.

Miura M, Friedlander RM, Yuan J (1995): Tumor necrosis factor-induced apoptosis is mediated by a CrmA-sensitive cell death pathway. Proc Natl Acad Sci USA 92:8318–8322.

Miura M, Zhu H, Rotello R, Hartwieg EA, Yuan J (1993): Induction of apoptosis in fibroblasts by IL-1B-converting enzyme, a mammalian homolog of the C. elegans cell death gene *ced-3*. Cell 75:653–60.

Munday NA, Vaillancourt JP, Ali A, Casano FJ, Miller DK, Molineaux SM, Yamin T-T, Yu VL, Nicholson DW (1995): Molecular cloning and pro-apoptotic activity of $ICE_{rel}II$ and $ICE_{rel}III$, members of the ICE/CED-3 family of cysteine proteases. J Biol Chem 270:15870–76.

Muzio M, Chinnaiyan AM, Kischkel FC, O'Rourke K, Shevchenko A, Ni J, Scaffidi C, Bretz JD, Zhang M, Gentz R, Mann M, Kreammer PH, Peter ME, Dixit VM (1996): FLICE, a novel FADD-homologous ICE/CED-3-like protease, is recruited to the CD95 (Fas/APO-1) death-inducing signaling complex. Cell 85:817–27.

Na S, Chuang TH, Cunningham A, Turi TG, Hanke JH, Bokoch GM, Danley DE (1996): D4-GDI, a substrate of CPP32, is proteolyzed during Fas-induced apoptosis. J Biol Chem 271:11209–13.

Nagata S, Golstein P (1995): The Fas death factor. Science 267:1449–56.

Newmeyer DD, Farschon DM, Reed JC (1994): Cell-free apoptosis in Xenopus egg extracts: Inhibition by bcl-2 and requirement for an organelle fraction enriched in mitochondria. Cell 79:353–64.

Nicholson DW (1996): ICE/CED-3-like proteases as therapeutic targets for the control of inappropriate apoptosis. Nature Biotech 14:297–301.

Nicholson DW, Ali A, Thornberry NA, Vaillancourt JP, Ding CK, Gallant M, Gareau Y, Griffin PR, Labelle M, Lazebnick YA, Munday NA, Raju SM, Smulson ME, Yamin T-T, Yu VL, Miller DK (1995): Identification and inhibition of the ICE/CED-3 protease necessary for mammalian apoptosis. Nature 376:37–43.

Nitatori T, Sato N, Waguri S, Karasawa Y, Araki H, Shibanai K, Kominami E, Uchiyama Y (1995): Delayed neuronal death in the CA1 pyramidal cell layer of the gerbil hippocampus following transient ischemia is apoptosis. J Neurosci 15:1001–11.

Orth K, Chinnaiyan AM, Garg M, Froelich CJ, Dixit VM (1996): The CED-3/ICE-like protease Mch2 is activated during apoptosis and cleaves the death substrate lamin A. J Biol Chem 271:16443–46.

Orth K, O'Rourke K, Salvesen GS, Dixit VM (1996): Molecular ordering of apoptotic mammalian CED-3/ICE-like proteases. J Biol Chem 271:20977–80.

Portera-Cailliau C, Hedreen JC, Price DL, Koliatsos VE (1995): Evidence for apoptotic cell death in Huntington disease and excitotoxic animal models. J Neurosci 15:3775–87.

Quan LT, Caputo A, Bleackley RC, Pickup DJ, Salvesen GS (1995): Granzyme B is inhibited by the cowpox virus serpin cytokine response modifier A. J Biol Chem 270:10377–79.

Quan LT, Tewari M, O'Rourke K, Dixit V, Snipas SJ, Poirier GG, Ray C, Pickup DJ, Salvesen GS (1996): Proteolytic activation of the cell death protease Yama/CPP32 by granzyme B. Proc Natl Acad Sci USA 93:1972–76.

Rabizadeh S, LaCount DJ, Friesen PD, Bredesen DE (1993): Expression of the baculovirsu p35 gene inhibits mammalian neural cell death. J Neurochem 61:2318–21.

Rao L, Perez D, White E (1996): Lamin proteolysis facilitates nuclear events during apoptosis. J Cell Biol 135:1441–55.

Ray CA, Black RA, Kronheim SR, Greenstreet TA, Sleath PR, Salvesen GS, Pickup DJ (1992): Viral inhibition of inflammation: Cowpox virus encodes an inhibitor of the interleukin-1β converting enzyme. Cell 69:597–604.

Reed JC (1995): Regulation of apoptosis by bcl-2 family proteins and its role in cancer and chemoresistance. Curr Opin Oncol 7:541–46.

Rosen A (1996): Huntington: New marker along the road to death? Nature Genet 13:380–82.

Rotonda J, Nicholson DW, Fazil KM, Gallant M, Gareau Y, Labelle M, Peterson EP, Rasper DM, Ruel R, Vaillancourt JP, Thornberry NA, Becker JW (1996): The three-dimensional structure of apopain/CPP32, a key mediator of apoptosis. Nature Struct Biol 3:619–25.

Roy N, Mahadevan MS, McLean M, Shutler G, Yaraghi Z, Farahani R, Baird S, Besner-Johnston A, Lefebvre C, Kang X, Salih M, Aubry H, Tamai K, Guan X, Ioannou P, Crawford TO, de Jong PJ, Surh L, Ikeda J-E, Korneluk RG, MacKenzie A (1995): The gene for neuronal apoptosis inhibitory protein is partially deleted in individuals with spinal muscular atrophy. Cell 80:167–78.

Schlegel J, Peters I, Orrenius S, Miller DK, Thornberry NA, Yamin T-T, Nicholson DW (1996): CPP32/apopain is a key interleukin 1B converting enzyme-like protease involved in Fas-mediated apoptosis. J Biol Chem 271:1841–44.

Schumer M, Colombel MC, Sawczuk IS, Gobe G, Connor J, O'Toole KM, Olsson CA, Wise GJ, Buttyan R (1992): Morphologic, biochemical, and molecular evidence of apoptosis during the reperfusion phase after brief periods of renal ischemia. Am J Pathol 140:831–38.

Shaham S, Horvitz H (1996): Developing *Caenorhabditis elegans* neurons may contain both cell-death protective and killer activities. Genes Dev 10:578–91.

Shi L, Chen G, MacDonald G, Bergeron L, Li H, Miura M, Rotello RJ, Miller DK, Li P, Seshadri T, Yuan J, Greenberg A (1996): Activation of an interleukin 1 converting enzyme-dependent apoptosis pathway by granzyme B. Proc Natl Acad Sci USA 93:11002–7.

Shi L, Kam CM, Powers JC, Aebersold R, Greenberg AH (1992): Purification of three cytotoxic lymphocyte granule serine proteates that induce apoptosis through distinct substrate and target cell interactions. J Exp Med 176:1521–29.

Slee EA, Zhu H, Chow SC, MacFarlane M, Nicholson DW, Cohen GM (1996): Benzyloxycarbonyl-Val-Ala-Asp (OMe) fluoromethylketone (Z-VAD.FMK) inhibits apoptosis by blocking the processing of CPP32. Biochem J 315:21–24.

Smale G, Nichols NR, Brady DR, Finch CE, Horton WE Jr. (1995): Evidence for apoptotic cell death in Alzheimer's disease. Exp Neurol 133:225–30.

Smith KGC, Strasser A, Vaux DL (1996): CrmA expression in T lymphocytes of transgenic mice inhibits CD95 (Fas/APO-1)-transduced apoptosis, but does not cause lymphadenopathy or autoimmune disease. EMBO J 15:5167–76.

Song Q, Lees-Miller SP, Kumar S, Zhang Z, Chan DW, Smith GC, Jackson SP, Alnemri ES, Litwack G, Khanna KK, Lavin MF (1996): DNA-dependent protein kinase catalytic subunit: A target for an ICE-like protease in apoptosis. EMBO J 15:3238–46.

Srinavasula SM, Fernandes-Alnemri T, Zangrilli J, Robertson N, Armstrong RC, Wang L, Trapani JA, Tomaselli KJ, Litwack G, Elnemri ES (1996): The Ced-3/interleukin 1β converting enzyme-like homolog Mch6 and the lamin-cleaving enzyme Mch2α are substrates for the apoptotic mediator CPP32. J Biol Chem 271:27099–106.

Steller H (1995): Mechanisms and genes of cellular suicide. Science 267:1445–49.

Sugimoto A, Friesen PD, Rothman JH (1994): Baculovirus p35 prevents developmentally programmed cell death and rescues a ced-9 mutant in the nematode *Caenorhabditis elegans*. EMBO J 13:2023–28.

Sulston JE, Horvitz HR (1977): Post-embryonic cell lineages of the nematode *Caenorhabditis elegans*. Dev Biol 82:110–56.

Sulston JE, Schierenberg E, White JG, Thomson N (1983): The embryonic cell lineage of the nematode *Caenorhabditis elegans*. Dev Biol 100:64–119.

Takahashi A, Alnemri ES, Lazebnik YA, Fernandes-Alnemri T, Litwack G, Moir RD, Goldman RD, Poirier GG, Kaufmann SH, Earnshaw WC (1996): Cleavage of lamin A by Mch2α but not CPP32: multiple interleukin 1β-converting enzyme-related proteases with distinct substrate recognition properties are active in apoptosis. Proc Natl Acad Sci USA 93:8395–400.

Tewari M, Dixit VM (1995): Fas- and tumor necrosis factor-induced apoptosis is inhibited by the poxvirus *crmA* gene product. J Biol Chem 270:3255–60.

Tewari M, Quan LT, O'Rourke K, Desnoyers S, Zeng Z, Beidler DR, Poirier GG, Salvesen GS, Dixit VM (1995a): Yama/CPP32B, a mammalian homolog of CED-3, is a crmA-inhibitable protease that cleaves the death substrate poly(ADP-ribose) polymerase. Cell 81:801–9.

Tewari M, Telford WG, Miller RA, Dixit VM (1995b): CrmA, a poxvirus encoded serpin, inhibits cytotoxic T-lymphocyte-mediated apoptosis. J Biol Chem 270: 22705–8.

Thomas LB, Gates DJ, Richfield EK, O'Brien TF, Schweitzer JB, Steindler DA (1995): DNA end labeling (TUNEL) in Huntington's disease and other neuropathological conditions. Exp Neurol 133:265–72.

Thome M, Schneider P, Hofmann K, Fickenscher H, Meinl E, Neipel F, Mattmann C, Burns K, Bodmer J-L, Schröter M, Scaffidi C, Krammer PH, Peter ME, Tschopp J (1997): Viral FLICE-inhibitory proteins (FLIPs) prevent apoptosis induced by death receptors. Nature 386:517–21.

Thompson CB (1995): Apoptosis in the pathogenesis and treatment of disease. Science 267:1456–62.

Thornberry NA, Bull HG, Calaycay JR, Chapman KT, Howard AD, Kostura MJ, Miller DK, Molineaux SM, Weidner JR, Aunins J, Elliston KO, Ayala JM, Casano FJ, Chin J, Ding GJ-F, Egger LA, Gaffney EP, Limjuco G, Palyha OC, Raju SM, Rolando AM, Salley JP, Yamin T-T, Lee TD, Shively JE, MacCross M, Mumford RA, Schmidt JA, Tocci MJ (1992): A novel heterodimeric cysteine protease is required for interleukin-1β processing in monocytes. Nature 356:768–74.

Thornberry NA, Molineaux SM (1995): Interleukin-1β converting enzyme: A novel cysteine protease required for IL-1β production and implicated in programmed cell death. Protein Sci 4:3–12.

Thornberry NA, Peterson EP, Zhao JJ, Howard AD, Griffin PR, Chapman KT (1994): Inactivation of interleukin-1β converting enzyme by peptide (acyloxy) methyl ketones. Biochemistry 33:3934–40.

Tsujimoto Y, Yunis J, Onarato-Showe L, Erikson J, Nowell PC, Croce C (1984): Molecular cloning of the chromosomal breakpoint of B-cell lymphomas and leukemias with the t(11;14) chromosome translocation. Science 224:1403–6.

Vaux DL, Cory S, Adams JM (1988): Bcl-2 oncogene promotes haemopoietic cell survival and cooperates with c-myc to immortalize pre-B cells. Nature 335:440–442.

Vaux DL, Weissman IL, Kim SK (1992): Prevention of programmed cell death in *Caenorhabditis elegans* by human *bcl-2*. Science 258:1955–57.

Verheij M, Bose R, Lin XH, Yao B, Jarvis WD, Grant S, Birrer MJ, Szabo E, Zon LI, Kyriakis JM, Haimovitz-Friedman A, Fuks Z, Kolesnick RN (1996): Requirement for ceramide-initiated SAPK/JNK signalling in stress-induced apoptosis. Nature 380:75–79.

Vincenz C, Dixit VM (1997): Fas-associated death domain protein interleukin-1β-converting enzyme 2 (FLICE2), an ICE/CED-3 homologue, is proximally involved in CD95- and p55-mediated death signaling. J Biol Chem 272:6578–83.

Walker NPC, Talanian RV, Brady KD, Dang LC, Bump NJ, Ferenz CR, Franklin S, Ghayur T, Hackett MC, Hammill LD, Herzog L, Hugunin M, Houy W, Mankovich JA, McGuiness L, Orlewicz E, Paskind M, Pratt CA, Reis P, Summani A, Terranova M, Welch JP, Xiong L, Möller A, Tracey DE, Kamen R, Wong WW (1994): Crystal structure of the cysteine protease interleukin-1β-converting enzyme: a (p20/p10)$_2$ homodimer. Cell 78:343–52.

Wang L, Miura M, Bergeron L, Zhu H, Yuan J (1994): *Ich-1,* an *Ice/ced-3*-related gene, encodes both positive and negative regulators of programmed cell death. Cell 78:739–50.

Wang S, Miura M, Jung Y-k, Zhu H, Gagliardini V, Shi L, Greenberg AH, Yuan J (1996): Identification and characterization of Ich-3, a member of the interleukin-1β converting enzyme (ICE)/Ced-3 family and an upstream regulator of ICE. J Biol Chem 271:20580–87.

Wang X, Pai J-t, Wiedenfeld EA, Medina JC, Slaughter CA, Goldstein JL, Brown MS (1995): Purification of an interleukin-1B converting enzyme-related cysteine protease that cleaves sterol regulatory element binding proteins between the leucine zipper and transmembrane domains. J Biol Chem 270:18044–50.

Wang X, Zelenski NG, Yang J, Sakai J, Brown MS, Goldstein JL (1996): Cleavage of sterol regulatory element binding proteins (SREBPs) by CPP32 during apoptosis. EMBO J 15:1012–20.

Waterhouse N, Kumar S, Song Q, Strike P, Sparrow L, Dreyfuss G, Alnemri ES, Litwack G, Lavin M, Watters D (1996): Heteronuclear ribonucleoproteins C1 and C2, components of the spliceosome, are specific targets of interleukin 1β-converting enzyme-like proteases in apoptosis. J Biol Chem 271:29335–41.

Weaver VM, Carson CE, Walker PR, Chaly N, Lach B, Raymond Y, Brown DL, Sikorska M (1996): Degradation of nuclear matrix and DNA cleavage in apoptotic thymocytes. J Cell Sci 109:45–56.

Williams MS, Henkart PA (1994): Apoptotic cell death induced by intracellular proteolysis. J Immunol 4247–4255.

Wilson KP, Black J-AF, Thomson JA, Kim EE, Griffith JP, Navia MA, Murcko MA, Chambers SP, Aldape RA, Raybuck SA, Livingston DJ (1994): Structure and mechanism of interleukin-1β converting enzyme. Nature 370:270–75.

Xia Z, Dickens M, Raingeaud J, Davis RJ, Greenberg ME (1995): Opposing effects of ERK and JNK-p38 MAP kinases on apoptosis. Science 270:1326–31.

Xue D, Horvitz HR (1995): Inhibition of the *Caenorhabditis elegans* cell-death protease CED-3 by a CED-3 cleavage site in baculovirus p35 protein. Nature 377:248–51.

Xue D, Shaham S, Horvitz HR (1996): The Caenorhabditis elegans cell-death protein CED-3 is a cysteine protease with substrate specificities similar to those of the human CPP32 protease. Genes Dev 10:1073–83.

Yang E, Korsmeyer SJ (1996): Molecular thanatopsis: A discourse on the BCL2 family and cell death. Blood 88:386–401.

Yu JS, Sena-Esteves M, Paulus W, Breakefield XO, Reeves SA (1996): Retroviral delivery and tetracycline-dependent expression of IL-1β-converting enzyme (ICE) in a rat glioma model provides controlled induction of apoptotic death in tumor cells. Cancer Res 56:5423–5427.

Yuan J (1995): Molecular control of life and death. Curr Opin Cell Biol 7:211–14.

Yuan J, Horvitz HR (1992): The *Caenorhabditis elegans* cell death gene *ced-4* encodes a novel protein and is expressed during the period of extensive programmed cell death. Development 116:309–20.

Yuan J, Shaham S, Ledoux S, Ellis HM, Horvitz HR (1993): The C. elegans cell death gene *ced-3* encodes a protein similar to mammalian interleukin-1B-converting enzyme. Cell 75:641–52.

Zhivotovsky B, Gahm A, Ankarcrona M, Nicotera P, Orrenius S (1995): Multiple proteases are involved in thymocyte apoptosis. Exp Cell Res 221:404–12.

Zhu H, Fearnhead HO, Cohen GM (1995): An ICE-like protease is a common mediator of apoptosis induced by diverse stimuli in human monocytic THP.1 cells. FEBS Lett 374:303–8.

Zou H, Henzel WJ, Liu X, Lutschg A, Wang X (1997): Apaf-1, a human protein homologous to C. elegans CED-4, participates in cytochrome C-dependent activation of caspase-3. Cell 90:405–13.

CHAPTER 7

ENDONUCLEASE ACTIVITIES AND APOPTOSIS

MARIANNA SIKORSKA and P. ROY WALKER
Apoptosis Research Group, Institute for Biological Sciences, National Research Council of Canada, Ottawa, ON K1A 0R6, Canada

INTRODUCTION

It has been 40 years since the first observations that implicated endonucleolytic activity in cell death were made (Cole and Ellis, 1957; Skalka and Matyasova, 1963). During the interim period an essential role for endonuclease activation in apoptosis *in vivo* has been firmly established, but, despite considerable effort, it has not been possible to identify a conserved endonuclease playing a critical role in the death of all cells. Although much of the early work pointed to a Ca^{2+}/Mg^{2+}-dependent endonuclease, it is now apparent that other endonucleases catalyze or contribute to DNA fragmentation in various cell types. Indeed, more than 20 different enzyme activities have been implicated. In this article we will review the involvement of these various nonconserved endonucleolytic activities in the highly conserved process of apoptotic DNA fragmentation and discuss where the conserved steps, if any, may exist.

ENDONUCLEASES IMPLICATED IN APOPTOSIS

Association of an Endonuclease with Cell Death (Research from 1970 to 1990)

The first observation of DNA degradation into discrete oligomeric fragments during cell death was made by Williamson (1970), who observed that DNA fragments, which were multiples of a 400 nucleotide monomer (now known to be approximately 200 bp of duplex DNA), were produced in neonatal liver

When Cells Die, Edited by Richard A. Lockshin, Zahra Zakeri, and Jonathan L. Tilly
ISBN 0-471-16569-7 © 1998 Wiley-Liss, Inc.

primary cultures by the hemopoietic cell population that dies out soon after birth. Four years later, Williams et al. (1974) associated a specific endonuclease activity with cell death *in vitro* using cultured Chang liver and Chinese hamster lung cells. In this study, the cultures were treated with high-specific-activity tritiated thymidine, which damaged DNA and induced cell death. To this day, DNA-damaging agents remain the most powerful class of compounds that induce cell death. The authors concluded that ". . . cells damaged beyond their recuperative capacity may undergo a distinct, nonreversible process that ends coherent metabolism." Similar endonuclease-mediated DNA degradation was observed in lymphocytes dying following ionizing irradiation (Skalka et al., 1976). The data presented in the latter study were significant for two reasons. First, it established that endonuclease-mediated DNA fragmentation also occurred during cell death *in vivo*. Second, the oligomeric DNA fragments were shown to be similar to those generated by the degradation of chromatin-associated DNA in isolated nuclei by exogenous micrococcal nuclease. A subunit structure of chromatin-associated DNA had recently been demonstrated by Hewish and Burgoyne (1973), who activated endogenous nuclease(s) to digest chromatin in isolated liver nuclei. The nucleoprotein monomers were called nucleosomes, each containing approximately 200 bp of DNA wrapped around an octameric histone core (Kornberg, 1974). That work was influential in establishing a link between DNA fragment sizes and chromatin structure. Moreover, since the endogenous nuclease was found to require Ca^{2+} and Mg^{2+} ions for activity, it greatly influenced subsequent work on the nature of the apoptotic endonuclease. In 1978, a similar ladder pattern of DNA degradation in the glucocorticoid-induced death of thymocytes was reported by Zhakarian and Pogosian (1978), who also concluded that endonuclease-mediated internucleosomal cleavage of chromatin played a role in cell death. Similarly, Bachvaroff et al. (1977) demonstrated that DNA fragmentation occurs early in the process of cell death in human splenocytes. Since a Ca^{2+}/Mg^{2+}-dependent endogenous endonuclease was found in the nuclei of the splenocytes, this endonuclease was linked to cell death. Significantly, Appleby and Modak (1977) had shown that a similar pattern of degradation occurred in terminally differentiating lens fiber cells, where the nucleus is destroyed, but the cells do not die.

Although this early work pointed to a role for endonuclease-mediated DNA fragmentation during cell death, little attention had been given to the overall mechanism of cell death in which an endonuclease may play a role. Studies on cell death, particularly during development, had already introduced the concept of an active form of cell death in which cells were specifically deleted by a process called "programmed cell death" (Lockshin, 1969). Furthermore, Kerr et al. (1972) had published a paper that introduced the term *apoptosis* to describe a specific form of cell death, distinct from necrosis, in which the most characteristic feature was a dramatic change in nuclear morphology as a result of the collapse of chromatin into dark electron-dense masses. Wyllie (1980) subsequently tied the glucocorticoid-induced activation of an endogenous endonuclease to the morphological changes occurring in the nucleus of cells undergoing apoptosis by suggesting a link between internucleosomal DNA cleavage and chromatin condensation. The link was further strengthened by Umansky et al. (1981) and Zhivotovsky et al. (1981), who demonstrated internucleosomal DNA cleavage in irradiated and glucocorticoid-treated lymphocytes and tentatively implicated a Ca^{2+}/Mg^{2+}-dependent endogenous endonuclease in the cleavage process.

Following these initial observations a number of publications appeared in which the actual mechanism of internucleosomal DNA cleavage during cell death was studied more directly. At that time, the two major endogenous nuclease activities known to exist in nuclei were the Ca^{2+}/Mg^{2+}-dependent endonuclease, described above, and an acidic endonuclease. Based on studies performed on irradiated thymocytes by Nikonova et al. (1982) and, subsequently, in glucocorticoid-treated thymocytes by Cohen and Duke (1984), the Ca^{2+}/Mg^{2+}-dependent enzyme was considered to be the most likely candidate. Wyllie et al. (1984) and Arends et al. (1990) further cemented the connection between endonuclease activation, internucleosomal DNA fragmentation, and the collapse of chromatin structure in studies on cells treated with A23187, a calcium ionophore, and on isolated nuclei incubated in the presence of micrococcal nuclease. In both cases, DNA cleavage to oligonucleosomes caused a collapse of chromatin structure similar to that seen in apoptosis. Changes in Ca^{2+} ion flux, in the presence of ionophore, were considered to directly activate the nuclease, thus implicating this cation in the process. Furthermore, since the appearance of small DNA fragments was prevented by cycloheximide, an inhibitor of protein synthesis, it was concluded that macromolecular synthesis was required to produce an active endonuclease. Indeed, injection of animals with cycloheximide causes a rapid loss of Ca^{2+}/Mg^{2+}-dependent endonuclease activity (Yamamoto et al., 1984) indicating that, even though it is now considered to be a constitutive nuclear protein, it is turning over rapidly in normal cells.

A significant difference between the way that chromatin is cleaved in apoptosis compared to that occurring in isolated nuclei incubated with exogenous nucleases, such as micrococcal nuclease, was subsequently published by Umansky's group (Umansky et al., 1988; Beletsky et al., 1989). During apoptosis, DNA degradation never proceeds to completion (i.e., all the way to mononucleosomes) as it does in micrococcal nuclease-treated nuclei, indicating that not all linker regions are nuclease sensitive or that, at some point, DNA degradation in apoptosis is specifically turned off. Also in this study, the ends of the small DNA fragments generated in irradiated and glucocorticoid-treated apoptotic thymocytes were studied and found to be predominantly, if not exclusively, 3'-hydroxyl and 5'-P. When DNA is cleaved by an endonuclease, the bond that is broken is either at carbon 5 of the sugar, producing a 5'-hydroxyl and leaving the phosphate on carbon 3 of the adjacent sugar (3'-P), or between the phosphate and carbon 3 to give a 3'-hydroxyl and leaving a 5'-P. This observation was important because it provided a way to better distinguish between candidate endonuclease activities. For example the acid endonuclease produces $3'$-P, whereas the Ca^{2+}/Mg^{2+}-dependent endonuclease produces $3'$-OH. Shortly thereafter, the first candidate Ca^{2+}/Mg^{2+}-dependent endonuclease of 28 kDa was isolated from calf-thymus chromatin by Nikonova et al. (1988) (see Zakeri, Chapter 3, and Trump and Berezesky, Chapter 2, this volume).

In 1989 and 1990 a series of papers on the role of Ca^{2+} ions in DNA fragmentation during cell death were published by Orrenius's group (McConkey et al., 1989a,b,c; Jones et al., 1989). An increase in cytosolic Ca^{2+} levels was observed by McConkey et al. (1989b,c) in thymocytes undergoing apoptosis, as a result of ligation of their T-cell receptor or glucocorticoid treatment. However, the elevated Ca^{2+} level was far below that believed to be required to activate the enzyme based upon *in vitro* assays. Further work by Jones et al. (1989) established that in isolated nuclei fragmentation of DNA could occur at near physiological Ca^{2+} concentrations when an ATP-regenerating system was used. Although demonstrating changes in Ca^{2+} levels during apoptosis, these studies also showed that elevation of intracellular Ca^{2+} alone was insufficient to trigger endonuclease activation (i.e., direct activation of the endonuclease by cation) and other signals must be involved in the activation process. Thus Ca^{2+} is not the ultimate signal for endonuclease activation, and the issue of whether or not it is truly required for actual endonuclease activity was not resolved.

Aurintricarboxylic acid, a putative inhibitor of endonuclease activity, was shown to prevent ionophore-induced DNA fragmentation and cell death. This was taken as further evidence that a Ca^{2+}-activated nuclease must be involved (McConkey et al., 1989b). However, aurintricarboxylic acid is not a specific inhibitor of endonuclease activity and has been shown to have a number of other inhibitory functions. For example, it has been shown to inhibit topoisomerase II (TopoII) and other DNA binding proteins (Catchpoole and Stewart, 1994), NAD(H)/NADP(H)-requiring enzymes (Thompson and Reed, 1995), and calpains (Posner et al., 1995). Moreover, even when DNA fragmentation is inhibited by aurintricarboxylic acid, the cells are not rescued from death (Mizumoto et al., 1994).

In contrast to the work supporting a role of the Ca^{2+}/Mg^{2+}-dependent endonuclease in cell death, Alnemri and Litwack (1990) provided evidence for the activation of a Ca^{2+}-independent neutral endonuclease in glucocorticoid- and novobiocin-treated CEM-C7 lymphocytes. This was the first hint that an endonuclease other than the Ca^{2+}/Mg^{2+}-dependent neutral endonuclease might play a role in cell death, at least in some cells.

In further studies on thymocytes, by far the most extensively studied cell type, Beletsky et al. (1989) provided additional evidence for involvement of the Ca^{2+}/Mg^{2+}-dependent endonuclease as opposed to the acidic endonuclease. This study used internucleosomal DNA fragmentation as the assay of functional activity coupled with an analysis of the ends of the fragments, as well as the effects of zinc ions and cycloheximide, which affect the Ca^{2+}/Mg^{2+}-

dependent endonuclease and not the acidic endonuclease. Moreover, the Ca^{2+}/ Mg^{2+}-dependent endonuclease identified in this study was distinguished from DNaseI, since the latter enzyme produces intranucleosomal DNA fragmentation that is not seen in apoptosis.

In other studies on the overall mechanism of chromatin degradation, the role of proteolysis was examined by Arends et al. (1990), who concluded that there was no large-scale proteolysis of chromatin-associated proteins preceding cleavage of the DNA. Moreover, the pattern and protein composition of the nucleosomes released during apoptosis were indistinguishable from those released from isolated nuclei by digestion with exogenous micrococcal nuclease. This was a significant result, since it showed that internucleosomal DNA cleavage during apoptosis must result from a change in chromatin, rendering this region of DNA more accessible to the apoptotic endonuclease. Such a change was likely to be one of conformation or subtle proteolysis. Micrococcal nuclease has access to the same sites in nuclei isolated from normal cells because of its small size (15 kDa) and because exogenous digestions are usually carried out at low ionic strength, which leads to some relaxation of chromatin structure and may allow simultaneous proteolysis (Walker et al., 1987; Walker and Sikorska, 1987).

From all this early work it became clear that there was an intimate relationship between endonuclease activation and cell death and between the morphological changes in chromatin structure and DNA fragmentation in apoptotic cell death. Furthermore, it was generally assumed that a Ca^{2+}/Mg^{2+}-dependent endonuclease, activated by increases in intracellular Ca^{2+}, catalyzed the major internucleosomal degradative events of apoptosis. However, a number of these conclusions were not firmly established and, indeed, may have been quite misleading. For example, although DNA fragmentation during either apoptosis or endogenous digestion of nuclei at neutral pH requires Ca^{2+} ions, it does not necessarily mean that the actual endonucleolytic cleavage reaction requires Ca^{2+} but that only the *process* as a whole is Ca^{2+}- and Mg^{2+}-dependent. Ca^{2+}-dependent steps could include upstream regulatory events or activation of Ca^{2+}-dependent proteases (Walker and Sikorska, 1994; Walker et al., 1995). Moreover, although increases in intracellular Ca^{2+} have been observed in some cells undergoing apoptosis, the elevated Ca^{2+} does not directly activate the Ca^{2+}/Mg^{2+}-dependent endonuclease. Furthermore, although not discussed at the time, the fact that the same ladder of DNA fragments may be generated by diverse nucleases implies that the ladder *pattern* could be a function of the substrate and not a specific property of a unique apoptotic endonuclease.

Role of Calcium Ions in DNA Fragmentation

Since the early work concluded that elevated Ca^{2+} levels triggered internucleosomal DNA fragmentation during apoptosis, there has been a considerable amount of work carried out on the actual role of Ca^{2+} ions. The work has proceeded along two major lines; first, attempts have been made to identify Ca^{2+}/Mg^{2+}-dependent endonuclease(s), and second, changes in intracellular Ca^{2+} have been measured in cells undergoing apoptosis.

Ca^{2+}/Mg^{2+}-Dependent Endonucleases. Most of the initial studies on the effects of cations on the activation and mechanism of action of endonucleases were carried out on DNaseI, one of the first of this class of enzymes to be purified to homogeneity and crystallized (reviewed by Wiberg, 1958). DNaseI is a Ca^{2+}/Mg^{2+}-dependent endonuclease of approximately 35 kDa found primarily in parotid gland and pancreas. The enzyme is considered to be a secretory protein involved in general nucleic acid degradation. There is also a substantial amount of DNaseI activity in blood, presumably to scavenge DNA from the circulation. Other early work, also independent of the cell death field, identified a number of additional Ca^{2+}/Mg^{2+}-dependent endonuclease activities in nuclei from various tissues (Hewish and Burgoyne, 1973; Ishida et al., 1974; Nakamura et al., 1981; Hashida et al., 1982; Stratling et al., 1984; Nikonova et al., 1988). The enzyme isolated from rat liver by Ishida et al. (1974) had a molecular weight of approximately 27 kDa, the enzyme from porcine liver, isolated by Stratling et al. (1984), was approximately 29 kDa, whereas the enzyme from calf thymus was 25–30 kDa (Nakamura et al., 1981; Nikonova et al., 1988). A similar enzyme of 36 kDa was found in bull seminal plasma (Hashida et al., 1982). All these enzymes were active at neutral pH, inhibited by zinc ions, and shown to be distinct from DNaseI. The enzymes were found to cleave DNA in isolated nuclei to oligonucleosomal DNA fragments, similar to those appearing during apoptosis. High levels of Ca^{2+}/Mg^{2+}-dependent endonuclease activity have subsequently been found in many, but not all, tissues *in vivo* (Giannakis et al., 1991). Liver, thymus, kidney, and spleen have high levels, whereas lung, brain, heart, pancreas, and testes have much less. In addition, Lebedeva et al. (1995a,b) identified Ca^{2+}/Mg^{2+}-dependent endonucleases of 30 and 55 kDa, a Mn^{2+}-dependent enzyme of 30 kDa, and an acidic endonuclease in rat liver extracts, showing the multiplicity of nucleases in the nuclei of some cell and tissue types. Significantly, these authors provided evidence that the Ca^{2+}/Mg^{2+}-dependent endonucleases may be derived from high-molecular-weight precursors of 145 kDa and possibly one of 400 kDa. Multiple nucleases are also found in other tissues. For example, further analysis of the nucleases in thymocytes (Nikonova et al., 1993) also revealed the presence of Mn^{2+}-dependent and acidic nucleases in addition to the Ca^{2+}/Mg^{2+}-dependent endonuclease activity. A number of laboratories have reported the existence of multiple bands of endonuclease activity, ranging in size from 10 to 120–140 kDa, on activity gels in cell and tissue extracts (Lebedeva et al., 1995b; Fraser et al., 1996; Fraser, 1996). These polypeptides may be related by virtue of being active proteolytic fragments of a precursor polypeptide (see Gartner and Hengartner, Chapter 4, this volume).

Thus as a class the Ca^{2+}/Mg^{2+}-dependent endonucleases are typically small (30–50 kDa) nuclear enzymes that are rather ubiquitously distributed. The enzymes are inhibited by zinc ions and by aurintricarboxylic acid. Since as described earlier, the latter compound is a nonspecific inhibitor of all DNA binding proteins, it is however of little use in attempts to distinguish which of these enzymes may be the Ca^{2+}/Mg^{2+}-dependent endonuclease involved in apoptosis. Similarly, zinc ions, which indirectly inhibit the acidic endonucleases, cannot discriminate the apoptotic endonuclease either. Unfortunately, with the exception of DNaseI, work on all these enzymes has not progressed beyond the stage of partial purification and biochemical characterization. Since there are few or no antibodies available for these proteins, and since none of the genes has been cloned, it is impossible to know which enzymes may be the same and which may be related members of a conserved family. This has been a considerable stumbling block in determining the function of these enzymes in general and particularly their role, if any, in apoptosis.

Some recent work has concentrated on those enzymes more likely to be involved in apoptosis by virtue of their presence in the nuclei of apoptotic cells. Thus Peitsch et al. (1993a) have argued that the Ca^{2+}/Mg^{2+}-dependent endonuclease responsible for apoptosis is, in fact, DNaseI since the Ca^{2+}/Mg^{2+}-dependent activity in thymocyte extracts could be immunoprecipitated with DNaseI antibodies. In addition, the apoptotic enzyme was inhibited by actin, a known inhibitor of the DNaseI enzyme. Furthermore, overexpression of the enzyme in COS cells led to DNA fragmentation (Polzar et al., 1993). Previously considered to be only a secretory enzyme of pancreas and salivary gland, a wider tissue distribution of DNaseI has been demonstrated by the same laboratory (Polzar et al., 1994). However, it is not found in all tissues and cells, even though all cells are believed to be capable of undergoing apoptosis.

Several laboratories have identified Ca^{2+}/Mg^{2+}-dependent endonuclease activities other than DNaseI in nuclei of apoptotic cells. For example, Ribeiro and Carson (1993) isolated a Ca^{2+}/Mg^{2+}-dependent endonuclease from apoptotic human splenocytes. The enzyme has a molecular mass of 27 kDa and is inhibited by zinc ions. In addition, a neutral Ca^{2+}/Mg^{2+}-dependent endonuclease with a molecular mass of 31 kDa, producing DNA fragments with 3'-OHs, was isolated from thymocytes (Tanuma and Shiokawa, 1994; Shiokawa et al., 1994) and a similar enzyme (doublet of 32 and 34 kDa) was identified in granulosa and luteal cells (Zeleznik et al., 1989; Boone et al., 1995). Although the latter enzyme was inhibited by zinc ions, it was not inhibited by actin, which is an inhibitor of DNaseI. Two other Ca^{2+}/Mg^{2+}-dependent endonucleases of 40 and 58 kDa, also distinct from DNaseI, have been found in cytotoxic T lymphocytes (Deng and Podack, 1995). Additional Ca^{2+}/Mg^{2+}-dependent endonucleases have been identified in various cells or tissues, including PC12 cells (Villalba et al., 1995) and an 18 kDa enzyme has been found in both immature B cells (Aagaard-Tillery and Jelinek, 1995) and in mouse myoblasts (Fimia et al., 1996). However, none of the genes for these enzymes has been cloned, and few, if any, antibodies are available. This lack of reagents and information has limited interactions between groups, and it is difficult to establish whether or not the various activities reported are actually the same enzyme or if each group is looking at an unique enzyme.

Pandey et al. (1994, 1997) have recently identified a widely distributed nuclear Ca^{2+}/Mg^{2+}-dependent endonuclease with all the properties ascribed to the apoptotic endonuclease. The enzyme has a molecular weight of 97 kDa and is readily distinguishable from either DNaseI or DNaseII. This enzyme cleaves DNA to produce fragments with 3′-OH ends, is also activated by Mn^{2+} ions, and is inhibited by zinc ions. It is active at neutral pH, but may retain sufficient activity in the pH 6.5–7.0 range to effect internucleosomal DNA fragmentation in those cells that become acidic (see the following). The enzyme is easily extractable from nuclei, which then become incapable of internucleosomal DNA fragmentation. When the enzyme is added back, internucleosomal DNA fragmentation resumes in the presence of Ca^{2+} and Mg^{2+} ions. It appears, therefore, that enzymes distinct from the secretory DNaseI, but with similar Ca^{2+}/Mg^{2+}-dependency exist in many cells. Furthermore these are the more likely candidates, particularly the latter enzyme, to be involved in DNA degradation during apoptosis.

In most of the cells mentioned, the Ca^{2+}/Mg^{2+}-dependent enzyme is present constitutively. However, some cells have little or no endogenous Ca^{2+}/Mg^{2+}-dependent endonuclease enzyme, but activity is rapidly induced during apoptosis, suggesting an even more likely involvement in cell death. For example, an inducible Ca^{2+}/Mg^{2+}-dependent endonuclease, consisting of three bands of activity (45, 47, and 49 kDa), has been found in a T-cell hydridoma cell line (Khodarev and Ashwell, 1996). Similarly, normal MCF-7 cells have little or no basal endonuclease activity, but this activity is rapidly induced following treatment with apoptosis-inducing chemotherapeutic drugs (Sokolova et al., 1995). The DNA fragments produced during apoptosis in the latter cells and those produced by this endonuclease *in vitro* contained 5′-P and 3′-OH ends. In addition, a Mg^{2+}-dependent, Ca^{2+}-activated endonuclease was induced in L1210 cells by 1-β-D-arabinofuranosylcytosine (Takauji et al., 1995).

Linking any of these enzymes to apoptosis has relied on the earlier work that demonstrated Ca^{2+}/Mg^{2+}-dependent endonucleases, in general, can degrade chromatin at the linker region to produce DNA ladders and fragments with the same 3′-OH ends seen in apoptotic cell DNA. Unfortunately more discriminatory assays that would distinguish among the various candidate enzymes are not available. Moreover, the possibility of endonuclease activities arising from artifactual sources such as mycoplasma must be taken into account (Paddenberg et al., 1996).

Changes in Intracellular Ca^{2+} Levels during Apoptosis. A number of studies have demonstrated changes in Ca^{2+} levels in cells undergoing apoptosis, or have shown that Ca^{2+} ionophores and other modulators of cellular Ca^{2+} homeostasis can induce apoptosis or apoptotic-like changes.

The Ca^{2+} ionophore, A23187, can induce DNA fragmentation in a number of cells, including thymocytes and PC12 cells (Umansky et al., 1988; Arends et al., 1990; Joseph et al., 1993). Moreover thapsigargin, an inhibitor of the endoplasmic reticulum Ca^{2+}-ATPase, also induces DNA fragmentation in thymocyte primary cultures (Zhivotovsky et al., 1994a). Thapsigargin causes intracellular Ca^{2+} levels to rise by releasing nonmitochondrial stores and triggering capacitative uptake of additional extracellular Ca^{2+} (Jiang et al., 1994). Evidence was provided (Zhu and Loh, 1995) that mobilization of endoplasmic reticulum stores of Ca^{2+}, and not necessarily the increase in cytoplasmic Ca^{2+}, actually acts as the trigger for apoptotic DNA fragmentation. These results were obtained in HL-60 cells and again emphasize that even in cells that contain a Ca^{2+}/Mg^{2+}-dependent endonuclease, an increase in intracellular Ca^{2+} is insufficient to directly activate the endonuclease. Thymocytes are susceptible to adenosine derivatives (Szondy, 1995) and, in response to treatment with these compounds, DNA fragmentation is preceded by a sustained increase in intracellular Ca^{2+} levels. However, even in these experiments, thymocyte subpopulations that can undergo apoptosis in the absence of changes in intracellular Ca^{2+} do exist (Jiang et al., 1995). Moveover, whereas Ca^{2+} levels increase in irradiated thymocytes, there is no change in intracellular Ca^{2+} in irradiated splenocytes, a closely related cell type (Zhivotovsky et al., 1993), although a basal Ca^{2+} level is required for cell death. Similar results have been obtained in etoposide treated HL-60 cells (Yoshida et al., 1993) where the basal level of Ca^{2+} is sufficient for apoptosis to occur. Quite clearly in the latter cell types, if a Ca^{2+}/Mg^{2+}-dependent endonuclease is involved, it must be capable of working at nM levels of free Ca^{2+}.

In a study of Fas-induced apoptosis in B cells, a requirement for elevated intracellular Ca^{2+} was established (Oshimi and Miyazaki, 1995). In the treated cells, the Ca_1^{2+} response was biphasic and the intracellular concentration had to exceed 140–150 nM for fragmentation to commence. However, increases in Ca^{2+} alone were again insufficient to cause fragmentation (i.e., the Ca^{2+} was not directly activating a Ca^{2+}/Mg^{2+}-dependent endonuclease), indicating that other factors must be involved and the calcium flux is an upstream regulatory event.

Evidence against an essential role for Ca^{2+} was presented by Adebodun and Post (1995), who showed that the glucocorticoid, dexamethasone, increased the levels of Ca^{2+} in both apoptosis-sensitive and resistant clones of CEM cells. In addition, they were able to increase intracellular Ca^{2+} levels with ionophore without DNA fragmentation. Therefore, there was no correlation between Ca^{2+} levels and DNA fragmentation in these cells. Morever, elevated Ca^{2+} is not required for DNA fragmentation in PC 12 cells following withdrawal of trophic support (Batistatou and Greene, 1993) or in murine macrophages (Kong et al., 1996). Most notably, Ca^{2+} has an opposite effect in neutrophils (Whyte et al., 1993), where elevation of Ca^{2+} levels by A23187 treatment, actually represses cell death and where Ca^{2+} chelators promote cell death (see Trump and Berezesky, Chapter 2, this volume).

Finally, it should be noted that the bile salt, glycodeoxycholate, induces apoptosis in hepatocytes by increasing intracellular Mg^{2+} levels with no change in Ca^{2+}. The Mg^{2+} ions appeared to be transported into the cell and to activate a Ca^{2+}/Mg^{2+}-dependent endonuclease activity (Patel et al., 1994).

The experiments described, although implicating Ca^{2+} in apoptosis in some cells, established that there was no direct relationship between increased Ca^{2+} and direct endonuclease activation. These findings, therefore, suggest that there are a number of intermediate steps between the two events. This was confirmed by the discovery that Bcl-2 did not prevent changes in Ca^{2+} homeostasis in either spontaneously apoptotic or thapsigargin-treated WEHI17.2 mouse lymphoma cells, but still protected the cells from cell death (Distelhorst and McCormick, 1996). The authors also showed that the elevated Ca^{2+} seen in glucocorticoid-treated cells is unrelated to apoptosis. Similar results were observed in CHO cells (Reynolds and Eastman, 1996). Significantly, Bcl-2 prevented cytoplasmic acidification in CHO cells, suggesting that the latter is more directly linked to DNA fragmentation than Ca^{2+} fluxes (Reynolds et al., 1996). Similarly, Perotti et al. (1990) have shown that, whereas cold shock induced a sustained increase in intracellular Ca^{2+} in both cycling and noncycling McCoy's cells, apoptosis occurred only in the noncycling cells. In the cycling cells, a cycle-dependent increase in PKC activity appeared to repress cell death, further indicating that multiple factors are involved in controlling the ultimate activation of the endonuclease.

In summary, the best evidence of a role for Ca^{2+} in endonuclease activation comes from studies on thymocytes, and is supported by data from a number of other cell types, mainly lymphocytes. However, Ca^{2+} ions are clearly not universal activators or regulators of DNA degradation in all cell types or even in the same cell type depending on its state (cycling vs. noncycling, for example) at the time a death-inducing signal is received.

Acidification and the Role of Acidic Endonucleases

The lack of a universal involvement of Ca^{2+} ions has prompted a search for alternative mechanisms of endonuclease activation. An absolute requirement for Ca^{2+} ions in the activation of the endonuclease responsible for DNA fragmentation in apoptosis was initially questioned by Alnemri and Litwack (1990) and then in a series of papers from Eastman's group (Eastman and Barry, 1992; Barry and Eastman, 1992, 1993; Barry et al., 1993). The latter group showed that there was no correlation between changes in Ca^{2+} and DNA fragmentation in ionomycin (a Ca^{2+} ionophore)-treated HL-60 cells. Instead, they noticed a correlation between intracellular acidification and DNA fragmentation in this cell model (Barry and Eastman, 1992). The pH of the total cell population (a mixture of apoptotic cells and still normal cells) dropped by 0.2–0.3 pH units from an initial value of approximately 7.25. Subsequent studies (Barry et al., 1993) showed that the intracellular pH dropped as low as 6.4 in some individual cells and that these were the ones in the total cell population with fragmented DNA. Similar results were obtained in CHO cells, which do not appear to contain an endogenous Ca^{2+}/Mg^{2+}-dependent endonuclease, but contain large amounts of extractable DNaseII (Barry and Eastman, 1993). DNaseII is cation independent and active at low pH, but it may retain sufficient activity at pH values around 6.5 to catalyze internucleosomal DNA fragmentation in apoptotic cells. The enzyme is inhibited by aurintricarboxylic acid, and although zinc ions do not inhibit DNaseII directly, they do inhibit the fall in intracellular pH that occurs in apoptosis. This event would thereby prevent enzyme activation and DNA fragmentation (Morana et al., 1994). The latter observations were made in ML-1 myeloid leukemia cells, adding one more cell line to the list of those that undergo intracellular acidification. DNaseII may also be involved in the degradation of DNA in lens cells (Torriglia et al., 1995).

The two major arguments still outstanding against the involvement of DNaseII in apoptosis are that the enzyme is primarily lysosomal and that the ends of the DNA fragments produced by DNaseII are 3'-P whereas those occurring during apoptosis, are generally 3'-OH. The subsequent hydrolysis of the 3'-phosphate group by cellular phosphatases, yielding a 3'-OH, however, cannot be ruled out. Recently, a cation-independent acidic nuclease of 40–45 kDa, inhibitable by zinc ions and producing DNA fragments with 3'-OH ends, was isolated from FM3A mouse mammary tumor cells (Hwang et al., 1995). This enzyme is capable of internucleosomal DNA fragmentation when added to target nuclei, and thus this enzyme may make a more likely candidate than DNaseII. Two other acidic nucleases of 32 kDa were identified in thymocyte nuclei, along with the Ca^{2+}/Mg^{2+}-dependent endonuclease activity (Tanuma and Shiokawa, 1994; Shiokawa et al., 1994). Neither of the acidic forms of endonuclease activity was inhibited by zinc ions, and they produced 3'-P at the ends of the DNA. Thus, even in thymocytes where the ends of the DNA fragments produced during apoptosis are 3'-OH, the acidic endonuclease found in thymocyte extracts produces 3'-P (Nikonova et al., 1993). These findings indicate that it is the Ca^{2+}/Mg^{2+}-dependent endonuclease, and not the acidic enzyme, that is involved in apoptosis in these cells.

Most recently, further support for a role of an acidic endonuclease in apoptosis has come from studies by Gottlieb et al. (1995, 1996), who showed that human neutrophils contain only this enzyme and have no detectable Ca^{2+}/Mg^{2+}-dependent endonuclease activity. The human bladder cancer cell line, T24, also possesses only an acidic endonuclease and no Ca^{2+}/Mg^{2+}-dependent endonuclease (Shemtov et al., 1995). Several other cells and cell lines, including hepatocytes, cardiomyocytes, C127, Jurkat, CRF-CEM, HL-60, and HeLa cells, have a similar acidic endonuclease. Unfortunately, the nature of the ends of the DNA fragments produced by either the acidic endonucleases in nuclei *in vitro,* or in these cells as they undergo apoptosis, has not been determined. In further studies on Jurkat cells, Gottleib et al. (1996) showed that intracellular acidification in response to a variety of apoptosis-inducing agents coincided with, or preceded, the appearance of a subdiploid population of cells, suggesting that acidification was responsible for DNA fragmentation. Moreover, buffering of the intracellular pH, to prevent the drop, abolished DNA fragmentation. Similar studies with PC12 cells, carried out by Villalba et al. (1995), identified a cation-independent acid endonuclease that was inhibited by zinc ions. Furthermore, by buffering intracellular Ca^{2+} levels, they were able to show that DNA fragmentation was Ca^{2+} independent. Interestingly, Collins et al. (1996) have isolated a nuclease activity from IL3-dependent BAF3 cells that is activated either by lowering the pH, where it becomes cation independent, or by increasing Ca^{2+} levels at neutral pH. The enzyme cleaves plasmid DNA *in vitro* to produce fragments with 3'-P (5'-OH) ends, and the same ends are seen in the apoptotic DNA fragments produced *in vivo* following IL3 withdrawal. To our knowledge this is the only example of cells producing exclusively fragments with 3'-P ends when undergoing apoptosis.

A number of other laboratories have provided evidence for the activation of an acidic endonuclease in a variety of cell lines during apoptosis. Ostad et al. (1996) examined several cell lines and found that some cells, such as NRK (normal rat kidney), possess an acidic nuclease and no Ca^{2+}/Mg^{2+}-dependent endonuclease. By contrast, other cells, such as mouse C127 cells, possessed only a Ca^{2+}/Mg^{2+}-dependent endonuclease. Since both cell types could be induced to undergo apoptosis by tumor necrosis factor and both cell types produced DNA ladders, it appeared that either of the endonucleases could effect internucleosomal DNA fragmentation. Significantly, overexpression of the Ha-*ras* oncogene protected the NRK cells (containing the acidic nuclease), but not the C127 cells (containing the Ca^{2+}/Mg^{2+}-dependent endonuclease), from DNA degradation. Overexpression of Ha-*ras* prevents the drop in pH that follows exposure of the cells to TNF, thereby preventing the activation of the acidic endonuclease. This is one of the clearest demonstrations of the involvement of different nucleases in apoptosis.

Similar observations were reported by Segal-Bendirdjian et al. (1995), who showed that a drug-resistant L1210 cell line failed to die in the presence of cisplatin or 5-azacytidine because it had lost the Ca^{2+}/Mg^{2+}-dependent endonuclease activity that was present in the apoptosis-sensitive parental cells. Therefore, a nuclear Ca^{2+}/Mg^{2+}-dependent endonuclease appeared to be required for the L1210 cells to die in response to toxic drugs. However, this same drug-resistant cell line, as well as the parental cells, underwent apoptosis in response to staurosporine treatment. In the presence of staurosporine, an acidic endonuclease from the cytoplasm was translocated to the nucleus and effected DNA degradation in lieu of the Ca^{2+}/Mg^{2+}-dependent endonuclease. For unknown reasons the chemotherapeutic drugs were unable to activate the process that effects this translocation. These results also clearly indicate that more than one endonuclease can effect DNA fragmentation in the *same* cell type, albeit a cloned subpopulation, and that different inducers of cell death may activate different nucleases.

In a study on the mechanism of intracellular acidification, Perez-Sala et al. (1995) showed that acidification accompanies DNA fragmentation in HL-60 cells induced to undergo apoptosis in the presence of Lovastatin. Indeed, direct inhibition of the Na^+/H^+ antiporter leading to a decrease in intracellular pH produced DNA fragmentation, whereas activation of the antiporter by phorbol esters protected the cells from death. Significantly, a drop in intracellular pH to 7 can cause significant DNA fragmentation in HL-60 cells. Moreover, peak fragmentation occurs at pH 6.8–6.9, indicating that only small deviations from normal can be sufficient to trigger nuclear fragmentation (Park et al., 1996).

Of note, bcl-2–transfected CHO cells are resistant to staurosporine-induced cell death (Reynolds et al., 1996) and do not undergo intracellular acidification. However, an increase in intracellular Ca^{2+} still occurs, but this cannot directly trigger internucleosomal DNA fragmentation.

Other Endonucleases Implicated in Apoptosis

Nuc18. A Ca^{2+}-dependent endonuclease of 18 kDa, called NUC18, has been isolated from thymus tissue in Cidlowski's laboratory (Gaido and Cidlowski, 1991; Montague and Cidlowski, 1996). The enzyme is inhibited by zinc ions and appears to be part of a higher macromolecular complex of >100 kDa in normal cells. It is released from the complex upon induction of apoptosis by glucocortioids (Gaido and Cidlowski, 1991). NUC18 appears to have some homology to cyclophilins (Montague et al., 1994), which were shown to have an associated nuclease activity, albeit at a very low level (Fraser, 1996). However, recent work has shown that the nuclease activity ostensibly associated with cyclophilin is an artifact of improper protein folding during the renaturation step of the activity gel assay (Schmidt et al., 1996). Since NUC18 cannot reproduce the pattern of DNA fragmentation (high molecular weight, HMW, and ladder formation) typical of apoptosis when added to isolated nuclei (Montague et al., 1997), it is unclear what role, if any, this enzyme plays. Moreover, the enzyme has only been found in thymocytes and S49.1 cells.

Endo–Exonuclease. Endo–exonucleases are a group of ubiquitously distributed enzymes that have a single-strand endonuclease activity, producing fragments with 5'-P and 3'-OH ends, and that also possess a 5' to 3' exonuclease activity (Fraser et al., 1996; Fraser, 1996). They are activated by a variety of cations including Ca^{2+} and Mg^{2+}. In *Neurospora* and yeast they have been shown to be involved in DNA recombination repair (Fraser, 1996). The enzyme appears to reside as an inactive precursor in a variety of cellular compartments and becomes activated by proteolysis. A number of active proteolytic fragments ranging in size from 18 to 145 kDa have been observed, and many of these fragment sizes are the same as those described previously as individual Ca^{2+}/Mg^{2+}-dependent endonucleases. However, the lack of suitable assays to discriminate between the various types of endonucleases has made further correlations difficult. Based upon their properties, a role for endo–exonucleases in apoptosis has been proposed since the appearance of proteo-lytically activated small nuclease proteins coincides with DNA fragmentation in CEM cells (Fraser et al., 1996; Fraser, 1996).

Changes in Endonuclease Activity during Development and Differentiation

It is clear from the discussions to this point that there is accumulating evidence for the involvement of different endonucleases in DNA degradation during apoptosis in various cell types. In addition, a number of studies have shown that the endogenous endonuclease activities found in the same cell type can vary during development and differentiation. For example, Anzai et al. (1995) showed that CD34$^+$ progenitor cells contain only a Mg^{2+}-dependent endonu-clease activity, but when induced to differentiate into polymorphonuclear leukocytes, they lose the Mg^{2+}-dependent endonuclease activity and acquire both an acidic endonuclease and a Ca^{2+}/Mg^{2+}-dependent endonuclease. Inter-estingly, the nonadherent marrow mononuclear cells from normal patients have a Ca^{2+}/Mg^{2+}-dependent endonuclease activity, whereas the cells from patients with acute myelogenous leukemia have both Ca^{2+}-independent Mg^{2+}-dependent and Ca^{2+}/Mg^{2+}-dependent endonuclease activities (Kawabata et al., 1996). Similarly, undifferentiated myoblasts do not produce DNA ladders when they undergo apoptosis, but their differentiated counterparts produce DNA ladders (Fimia et al., 1996). In both cases the morphological condensa-tion still occurs, confirming once again that only HMW DNA fragmentation is required for chromatin condensation (see the following). Interestingly, both undifferentiated and differentiated cells appeared to contain the same nuclear endonuclease activity, indicating that a change in chromatin structure may underlie the different patterns observed.

DNA Sequence Requirements for Fragmentation

If a unique apoptotic endonuclease was being activated to cleave a specific subset of sites on chromatin, some DNA sequence homology at these sites might be expected. On the other hand, if the sites that are being cleaved are generated by changes in chromatin that render them nuclease sensitive, no sequence homology would exist. To examine this Moore et al. (1993) cloned a number of mononucleosomal (200 bp) fragments generated in the IL3-depdendent murine myeloid cell line H7, undergoing apoptosis following IL3 withdrawal, and found no sequence homology between fragments. In addition, there was no preference for transcribed versus nontranscribed regions of DNA. Similar results were obtained by Herrmann et al. (1996), who showed no sequence specificity at sites of cleavage, but instead found a frequent occurrence of tracts of DNA that may cause DNA to bend. Taken together, these results suggest that changes in chromatin that render the substrate more accessible to endonucleolytic attack, as apposed to the activation of a sequence-specific endonuclease, may underlie the mechanism of DNA degradation in apoptosis.

NEW ASPECTS OF THE MECHANISM OF DNA FRAGMENTATION

The work described to this point has implicated a number of enzymes in DNA degradation in apoptosis, suggesting that the degradation of DNA, a hallmark of apoptosis, is not catalyzed by a single conserved endonuclease. However, because of the limitations of conventional agarose gel electrophoresis, all this work relates solely to internucleosomal DNA fragmentation since this is the level of degradation that can be detected on such gels. Recent technological developments in the electrophoresis of DNA molecules has permitted new aspects of DNA fragmentation in apoptosis to be examined. This, in turn, has added new levels of complexity to the DNA degradative process. It is possible that a conserved step exists at the early stages of DNA fragmentation, producing large fragments that are subsequently cleaved into the ladder pattern by nonconserved enzymes (see Zakeri, Chapter 3, Trump and Berezesky, Chapter 2, and Gartner and Hengartner, Chapter 4, this volume).

Generation of High-Molecular-Weight DNA Fragments

During the 1970s and 1980s DNA fragments were separated electrophoretically using agarose as the support medium and a constant voltage in the forward direction. Fragments ranging from <100 bp to approximately 30,000–50,000 bp undergo an ordered logarithmic separation on such gels. Larger fragments cannot be sieved and tend to migrate end on in such a way that migration becomes independent of size. The small DNA fragments produced in apoptosis were well resolved by these gels into the characteristic "ladder" of oligonucleosomal fragments. This pattern became the major biochemical marker for apoptosis and focused undue attention on cleavage of DNA at the linker region of the 10 nm fiber. As such, internucleosomal DNA cleavage became the basis of the assay for detection of candidate apoptotic nucleases, even though micrococcal nuclease, an enzyme completely unrelated to apoptosis, could produce the same pattern. Such an assay, therefore, is of very limited use for trying to understand the mechanism of DNA fragmentation and to identify apoptotic endonucleases.

Following the advent of pulsed field gel electrophoresis (PFGE), a specialized form of electrophoresis that permits the separation of fragments of DNA as large as 2–5 Mbp (Schwartz and Cantor, 1984), DNA fragmentation could be examined at whole new levels (Filipski et al., 1990). This technique, first applied to apoptotic cells in 1991 (Dusenbury et al., 1991; Walker et al., 1991), revealed hitherto unknown levels of complexity in the fragmentation process.

In our study (Walker et al., 1991), the chromatin of thymocytes induced to undergo apoptosis in the presence of chemotherapeutic drugs or glucocorticoids was found to be degraded first into large fragments of 50–300 kbp and subsequently to smaller fragments, eventually producing the oligonucleosomal ladder. A similar study (Dusenbury et al., 1991) showed that large fragments also accumulated in apoptotic colo-rectal cancer cells following treatment with 5-fluorodeoxyuridine. This technique afforded the opportunity to study the earliest stages of DNA fragmentation in apoptosis. Furthermore, these findings raised a series of new questions about the mechanism of degradation, the relationship between high-molecular-weight (HMW) fragmentation and internucleosomal cleavage, and the nature of the endonuclease(s) involved in the various stages. These studies have helped to address a number of problematic aspects of the role of DNA degradation in apoptosis.

Although zinc ions had been shown to inhibit DNA ladder formation in apoptotic cells, there has been controversy over whether or not zinc actually prevents cell death. Cohen and Duke (1984) initially reported the zinc effect on DNA cleavage and indicated that it prevented cell death, but convincing evidence that the cells remained viable was not presented. In a more complete study the effects of zinc ions have been shown to be quite complex and concentration dependent (Provinciali et al., 1995). Indeed, zinc ions can even stimulate cell death at low concentrations. Subsequent studies (Cohen et al., 1992; Brown et al., 1993; Walker et al., 1994) have shown that the cells still died even though internucleosomal DNA fragmentation was blocked, raising some concerns about the role of DNA fragmentation in cell death. However, since HMW DNA fragmentation still occurs in the presence of zinc ions in apoptotic thymocytes (Brown et al., 1993), these data suggest that internucleosomal DNA fragmentation is a dispensible step but that HMW fragmentation is essential for apoptosis.

The discovery of HMW DNA fragmentation helped to solve another problem that had arisen in which cells were apparently dying with the typical nuclear morphological changes (i.e., collapse of chromatin structure) that are associated with apoptosis, but DNA fragmentation could not be observed on conventional gels (Oberhammer et al., 1993a). Further analysis showed that all these cells had degraded their DNA to large fragments (Oberhammer et al., 1993b). Based on these observations, internucleosomal DNA fragmentation is clearly not essential for apoptosis, and the cleavage of DNA at the sites that generate the HMW fragments is sufficient to ensure chromatin collapse and death of the cell. Indeed, it is now apparent that there is a tremendous variation in the extent to which different cell types fragment their DNA (Walker et al., 1995). Many, but not all, lymphocytes degrade their DNA extensively, whereas some cells, such as the DU145 human prostate cell line, produce only very large fragments. However, all cells tested thus far undergo HMW DNA fragmentation.

Studies on a number of systems have suggested that HMW and internucleosomal DNA fragmentation occur in two distinct phases. For example, internucleosomal DNA fragmentation was inhibited by a number of protease inhibitors (Weaver et al., 1996; Hara et al., 1996), including the serine-specific protease inhibitor dichloroisocoumarin, but HMW fragmentation still occurred in the presence of these inhibitors at less than toxic levels (i.e., at high concentrations many of these inhibitors kill the cell immediately, giving the appearance of inhibition of DNA fragmentation). Moreover, the endonuclease activity responsible for generation of the large fragments appears different from that catalyzing internucleosomal DNA fragmentation (Walker et al., 1993; Walker and Sikorska, 1994; Cain et al., 1994a,b; Sun and Cohen, 1994; Walker et al., 1994, 1995). Thus, although the initial stages of DNA fragmentation can be reproduced in hepatocyte or thymocyte nuclei by incubation in the presence of Mg^{2-} ions alone (Walker et al., 1994; Pandey et al., 1994; Zhivotovsky et al., 1994b), the induction of internucleosomal DNA fragmentation at neutral pH requires the presence of Ca^{2+} (Cain et al., 1994b; Walker et al., 1994b).

Chromatin Structure and HMW DNA Fragmentation

The fragment sizes produced during the initial stages of DNA degradation are similar to the size of topological domains or loops into which DNA is organized in the nucleus (Filipski et al., 1990). This has led us (Walker et al., 1995; Walker and Sikorska, 1994) to predict that DNA fragmentation starts at the sites of attachment of the domains to the nuclear matrix (see Cryns and Yuan, Chapter 6, this volume). Fragmentation at these sites has subsequently been confirmed (Lagarkova et al., 1995). Since DNA is initially cleaved at the nuclear matrix attachment points, the possible involvement of topoisomerase II (TopoII) has been raised. Since TopoII binds to these regions and can cut both strands of DNA simultaneously, it could conceivably, in an abortive reaction cycle, generate double-strand breaks. However, a number of studies have shown this not to be the case (Beere et al., 1995, 1996; Rusnak et al., 1996). For example, Rusnak et al. (1996), using monolayers of DU-145 human prostatic cancer cells, which undergo apoptosis without producing DNA ladders, provided evidence that TopoII was not involved in the formation of the initial breaks and that these were almost certainly the product of endonucleolytic activity.

Beere et al. (1995) have also studied the possible role of TopoII in the initial stages of DNA fragmentation in MOLT-4 cells undergoing apoptosis in response to both genotoxic and nongenotoxic agents. MOLT-4 cells show the typical morphological changes of apoptosis, but do not completely degrade their DNA to ladders. TopoII could play a role in DNA fragmentation in two ways. Firstly, as described, it could introduce double-strand breaks at loop attachment regions. Alternatively, it could catalyze alterations in DNA topology, thereby altering chromatin structure and rendering some sites more sensitive to endonucleolytic attack. Beere et al. (1995) were able to show that the direct DNA "cleavage" produced by the TopoII inhibitor, VP16, was reversed upon drug removal, and that this was a distinct step from the initial stages of endonucleolytic DNA cleavage associated with the onset of apoptosis. Thus, as shown earlier (Walker et al., 1991), transient exposure to TopoII inhibitors, such as VM26 and VP16, which stabilize the cleavable complex, is sufficient to trigger apoptosis even though these "breaks" are reversed before endonucleolytic attack on the chromatin. Moreover, since the production of 50 kb fragments in response to TopoII inhibitor-induced apoptosis was the same as that for apoptosis induced by nongenotoxic agents, it is apparent that the inhibition of TopoII has no effect on the endonucleolytic activity that cleaves at these sites. Significantly, these authors (Beere et al., 1995) showed that TopoII protein was specifically degraded in parallel with the production of 50 kb fragments as the cells underwent apoptosis.

To address the second issue that is, whether TopoII activity was required to catalyze an essential change in chromatin structure that initiates DNA fragmentation, Beere et al. (1996) used ICRF-193, a TopoII inhibitor that does not form a cleavable complex, to show that enzyme inhibition had no effect on the induction of apoptosis. This work also showed that TopoII activity is not required to initiate or propagate DNA fragmentation during apoptosis. This raises the question of why TopoII is specifically degraded concomitant with formation of 50 kbp HMW fragments. It is still possible, therefore, that proteolysis of TopoII exposes sites at matrix attachment regions that become accessible to endonucleolytic attack. However, since TopoII appears to be located at only a subset of sites that are attacked by endonuclease, it is unlikely to play a critical role. Significantly, ICRF-193 itself induces apoptosis in thymocytes (Tanimoto et al., 1995) indicating that it is the inhibition of topoII function and not the resultant DNA breaks (that are produced by VM26, VP16, etc.) nor its proteolysis, which is the trigger for cell death.

Formation of Single-Strand Breaks in Apoptotic DNA

Tomei et al. (1993) have provided evidence for a modification to the linker region of chromatin prior to double-strand cleavage. This was observed in C3H/10T1/2 cells that also do not degrade their DNA completely to ladders. The simplest explanation of these data is that there is an accumulation of single-strand breaks (ssb) in the linker regions prior to the appearance of double-strand breaks (dsb). Subsequently, direct evidence for the presence of ssb in the linker region of oligonucleosomal fragments was presented (Peitsch et al., 1993b). Moreover, it had been shown earlier that target cells killed by CTL (cytotoxic cells) accumulate many ssb in their DNA (Gromkowski et al., 1986). The process by which CTL kill cells is now known to be very similar to apoptosis.

There is further evidence that ssb precede or accompany the dsb that result in DNA degradation. For example, Chaudun et al. (1994) and Arruti et al. (1995) have shown that ssb are present in the DNA of lens fiber cells. Lens cells do not undergo apoptosis, but instead they undergo terminal differentiation in which the nucleus is destroyed in a process that resembles, or is even identical to that occurring during apoptosis (see Papermaster, Chapter 12, this volume). Interestingly, the ends of some of these DNA fragments appear to be blocked and only liberate free 3'-OH upon incubation *in vitro*. Furthermore, we have recently shown that many ssb accumulate in HMW DNA during the initial stages of fragmentation (Walker et al., 1997). In this study, a new 2D PFGE-conventional agarose gel electrophoresis technique was used to identify the accumulation of ssb in HMW DNA. In this technique the DNA from apoptotic cells is separated in the first dimension by PFGE and then in the second dimension by conventional agarose gel electrophoresis, run under either normal or denaturing conditions. By comparing the gels run under normal versus denaturing conditions, an accumulation of ssb in duplex DNA fragments >50 kbp was demonstrated. By probing these large fragments with ssb-sensitive nucleases, such as S1 nuclease, it was shown that the ssb were accumulating in the linker regions of chromatin (i.e., at 200 bp intervals). In addition, some evidence was provided for a conformational change in linker DNA that renders it nuclease sensitive. Based upon these observations, it is possible that all chromatin degradation in apoptosis is catalyzed by an endonuclease that generates ssb, initially in the unpaired regions of DNA attached to the nuclear matrix and eventually in the linker regions (see Cryns and Yuan, Chapter 6, this volume).

Matrix Enzymes

The data presented above predict that the initial stages of DNA fragmentation are likely to be catalyzed by an endonuclease, bound to the nuclear matrix, that introduces ssb into DNA. Rzepecki (1994) has identified a nuclear matrix-bound endonuclease capable of recognizing and cleaving nuclear matrix (scaffold)-associated DNA from plant cells. The enzyme is a complex of a 32 kDa endonuclease bound to a 65 kDa protein and is capable of introducing ssb into supercoiled DNA (Szopa et al., 1993; Rzepecki, 1994). We are currently searching for similar nuclear matrix-associated endonuclease(s) in normal and apoptotic mammalian cells.

Initial Morphological Changes to the Nucleus

Extensive internucleosomal DNA fragmentation correlates with the complete collapse of chromatin structure and possibly the formation of apoptotic bodies. Currently, work is being carried out to relate the initial HMW DNA breaks to changes in nuclear morphology. For example, Rusnak et al. (1996), using monolayers of DU-145 human prostatic cancer cells, which undergo apoptosis without producing DNA ladders, were able to establish a temporal correlation between the initial stages of DNA fragmentation and nuclear morphological changes. Fragments of DNA began to accumulate in cells still attached to the dish prior to any observable nuclear morphological change. These fragments were 450–600 kbp and >1 Mbp in size. Some smaller fragments of 30–50 kbp also began to accumulate prior to detachment of the cells. Upon detachment, the nuclear morphology of the cells changed considerably in parallel with the complete degradation of the largest fragments and an increase in the 30–50 kbp fragments.

Although TopoII does not appear to mediate any structural changes in chromatin, there is some evidence that an early change in structure does occur. This became apparent during a study of the relative sensitivity of nuclei to micrococcal nuclease. Lymphocyte cell lines that undergo internucleosomal DNA fragmentation in apoptosis (HL-60, U-937, Ml-1, THP-1) were compared to those that do not (K562, MOLT-4) by Kuribayashi et al. (1996). Nuclei from cells that produce ladders were intrinsically more sensitive to degradation than those that did not, suggesting that an alteration in chromatin structure, as opposed to availability or activity of endonuclease, determined the extent to which cells degraded their DNA. There was no obvious difference in histone H1, the histone associated with the linker region, between the two sets of nuclei. Interestingly, exogenous DNaseI could not detect this difference in structure, and both types of nuclei were degraded equally. These data once again demonstrate that DNaseI has no preference for the internucleosomal linker region of chromatin.

Using confocal microscopy, we have shown that DNA fragmentation, even at the earliest stages, occurred throughout the nucleus, including regions of already condensed chromatin (Weaver et al., 1996; Falcieri et al., 1994b). These findings suggest that the nuclease may already be located at these sites and does not have to penetrate highly condensed chromatin. Moreover, DNA fragments began to accumulate before there was any significant rearrangement or collapse of chromatin structure (Sikorska et al., unpublished observations; Falcieri et al., 1994b). Moreover, in a recent study of the earliest changes in both DNA fragmentation and nuclear morphology, Ghibelli et al. (1995) found a correlation in U937 cells between the appearance of DNA fragments >2 Mbp and a subtle change in overall nuclear morphology. Coincident with the appearance of this very high-molecular-weight band, the nuclei became more rounded, and this occurred in the absence of any detectable chromatin condensation. Ongoing degradation of the DNA to the 50 kbp fragmentation stage was accompanied by chromatin condensation, and, subsequently, complete nuclear fragmentation coincided with complete degradation to DNA ladders.

Site-Specific Proteolysis

In the interphase nucleus each chromosome is anchored to a protein scaffold composed of peripheral nuclear lamina and an intranuclear proteinaceous network or matrix. Within this protein scaffold each chromosome occupies its own three-dimensional space and establishes contacts with the proteins of the scaffold to generate a complex structural organization that reflects the function of that chromosome in each particular differentiated cell type (Cremer et al., 1993). Any alterations to this protein scaffold would be expected to have a direct impact on the structure of chromatin. In apoptosis the collapse of interphase chromosomes into dark, uniformly electron-dense, masses could be brought about not only by the actions of endonucleases but also by nuclear protease(s) acting at the nuclear scaffold. Nuclear proteolysis has been shown to occur at the early stages of apoptosis, which, in turn, may be responsible for nuclear destabilization prior to endonuclease activation. The nuclear lamins, the nuclear mitotic apparatus protein (NuMA), TopoII, and poly(ADP-ribose) polymerase are nuclear proteins specifically targeted by proteolytic enzymes during apoptosis (Kaufmann, 1989; Kaufmann et al., 1993; Oberhammer et al., 1994; Neamati et al., 1995; Weaver et al., 1996) (see Cryns and Yuan, Chapter 6, this volume).

NuMA is an abundant nuclear protein (2×10^5 molecules/cell), which plays an essential role in mitotic spindle formation (Lydersen and Pettijohn, 1980; Cleveland, 1995). Although its function in the interphase nucleus is less clear, the coiled-coil nature of the protein implies that it could form core filaments of the nucleoskeleton or intranuclear matrix (Cleveland, 1995). Therefore, the integrity of this protein, which is compromised in apoptotic cells (Weaver et al., 1996), is absolutely essential for nuclear stability.

The function of the nuclear lamina is also well defined. It provides an attachment site for higher-order chromatin domains, it supports the nuclear envelope structure, and it also maintains a direct link with the cytoplasm through its interaction with intermediate filaments (Gerace et al., 1984; Georgatos and Blobel, 1987; Luderus et al., 1994). Thus this nuclear structure controls not only the organization of interphase chromosome architecture but also the nucleocytoplasmic movement of macromolecules. Therefore, alterations in the nuclear lamina would be expected to contribute to the nuclear morphological changes seen in apoptosis. Indeed, the degradation of lamina components has been reported in HL60 cells in response to anticancer drugs (Kaufmann, 1989), in Simian virus 40 (SV40)-transformed NIH 3T3 fibroblasts in response to CTL-mediated killing (Ucker et al., 1992), in serum-deprived human papiloma virus (HPV)-transformed rat embryo cells (Oberhammer et al. 1994), and in dexamethasme-treated thymocytes (Neamati et al., 1995; Weaver et al., 1996). These studies established that lamin degradation is also an early apoptotic event, possibly preceding chromatin cleavage and collapse.

Therefore, two parallel degradative processes, proteolysis and endonucleolysis, are responsible for nuclear destruction in apoptosis. In fact, nuclear proteolysis might be the rate-limiting factor determining both the onset and the extent of DNA fragmentation. Furthermore, the rapidity of the process of protein degradation, subsequent destabilization of nuclear structure, and, ultimately, its collapse might halt the DNA cleavage process at any given stage of fragmentation. Thus this protease-mediated collapse of proteinaceous structures may create a nuclear microenvironment that first promotes and then inhibits further endonucleolysis. This hypothesis may provide an explanation for the fact that DNA degradation never proceeds to completion in apoptotic cells.

SUMMARY AND FUTURE DIRECTIONS

There is now sufficient evidence to dispel the notion that the internucleosomal phase of DNA fragmentation is a conserved or essential step in apoptosis. Indeed, some cells do not exhibit internucleosomal DNA fragmentation during death. In various cell and tissue types that do produce DNA ladders a number of different enzymes, all capable of cleaving DNA in the linker region of chromatin, have been identified and implicated in apoptosis. However, all cells undergo HMW fragmentation, and it is possible that the endonuclease responsible for this initial stage of fragmentation is the essential and conserved enzyme. Such an enzyme is likely to be located at, or have access to, the sites of initial endonucleolytic attack. Thus it is probably also associated with the nuclear matrix and capable of ssb formation. We are currently trying to identify endonuclease activities associated with the nuclear matrix in mammalian cells. Once DNA cleavage commences, it may lead to conformational changes in the large domain fragments that render more sites on chromatin vulnerable to further endonucleolytic attack. It is evident that a change in chromatin, either a conformational change or one of proteolysis, that renders these sites more sensitive to digestion precedes internucleosomal DNA fragmentation. These secondary stages of DNA degradation could be catalyzed by any one of a number of endonucleases and even by several endonucleases if they exist in the same nucleus at the time of apoptosis. Since the only assay for this process is the formation of a DNA ladder, any number of activities could contribute to the final pattern. Indeed, if there are several nucleases within the nucleus, it is difficult to imagine how only one of them would be active under such catastrophic conditions.

In such a scenario, it is possible that none of the endonucleases is conserved and that a change in chromatin structure brought about, for example, by ion fluxes or site-specific proteolysis of critical nuclear proteins initiates the cascade of degradative events mediated by nonspecific endonucleases. Thus the conserved degradative event may well be one of proteolysis. To understand this process more fully, the nature of early changes in chromatin structure needs to be investigated and correlated with the activation of the nuclear proteolytic process. This requires the development of knowledge about the vulnerable sites in the nuclear lamina and matrix that could lead to changes in chromatin structure. It also requires the development of methodologies to study the earliest stages of fragmentation (>2 Mbp fragment formation) and to detect subtle changes in nuclear morphology.

REFERENCES

Aagaard-Tillery KM, Jelinek DF (1995): Differential activation of a calcium-dependent endonuclease in human B lymphocytes—Role in ionomycin–induced apoptosis. J Immunol 155:3297–307.

Adebodun F, Post JFM (1995): Role of intracellular free (Ca(II) and Zn(II) in dexamethasone-induced apoptosis and dexamethasone resistance in human leukemic CEM cell lines. J Cell Physiol 163:80–86.

Alnemri ES, Litwack G (1990): Activation of internucleosomal DNA cleavage in human CEM lymphocytes by glucocorticoid and novobiocin evidence for a non-calcium ion-requiring mechanisms. J Biol Chem 265:17323–33.

Anzai N, Kawabata H, Hirama T, Masutani H, Ueda Y, Yoshida Y, Okuma M (1995): Types of nuclear endonuclease activity capable of inducing internucleosomal DNA fragmentation are completely different between human CD34(+) cells and their granulocytic descendants. Blood 86:917–23.

Appleby DW, Modak SP (1977): DNA degradation in terminally differentiating lens fiber cells from chick embryos. Proc Natl Acad Sci USA 74:5579–83.

Arends MJ, Morris RG, Wyllie AH (1990): Apoptosis: The Role of the Endonuclease. Am J Pathol 136:593–608.

Arruti C, Chaudun E, De Maria A, Courtois Y, Counis MF (1995): Characterisation of eye-lens DNases: Long term persistence of activity in post apoptotic lens fibre cells. Cell Death Differentiation 2:47–56.

Bachvaroff RT, Ayvazian JH, Skupp S, Rapaport FT (1977): Specific restriction endonuclease degradation of DNA as a consequence of immunologically mediated cell damage. Transplant Proc 9:807 12.

Barry MA, Reynolds JE, Eastman A (1993): Etoposide-induced apoptosis in human HL-60 cells is associated with intracellular acidification. Cancer Res 53:2349–57.

Barry MA, Eastman A (1992): Endonuclease activation during apoptosis: The role of cytosolic calcium and pH. Biochem Biophys Res Commun 186:782–89.

Barry MA, Eastman A (1993): Identification of deoxyribonuclease-II as an endonuclease involved in apoptosis. Arch Biochem Biophys 300:440–50.

Batistatou A, Greene LA (1993): Internucleosomal DNA cleavage and neuronal cell survival-death. J Cell Biol 122:523–32.

Beere HM, Chresta CM, Alejoherberg A, Skladanowski A, Dive C, Larsen AK, Hickman JA (1995): Investigation of the mechanism of higher order chromatin fragmentation observed in drug-induced apoptosis. Mol Pharmacol 47:986–96.

Beere HM, Chresta CM, Hickman JA (1996): Selective inhibition of topoisomerase II by ICRF-193 does not support a role for topoisomerase II activity in the fragmentation of chromatin during apoptosis of human leukemia cells. Mol Pharmacol 49:842–51.

Beletsky IP, Matyasova J, Nikonova LV, Skalka M, Umansky SR (1989): On the role of Ca,Mg-dependent nuclease in the postirradiation degradation of chromatin in lymphoid tissues. Gen Physiol Biophys 8:381–98.

Boone DL, Yan W, Tsang BK (1995): Identification of a deoxyribonuclease I–like endonuclease in rat granulosa and luteal cell nuclei. Biol Reprod 53:1057–65.

Brown DG, Sun X-M, Cohen GM (1993): Dexamethasone-induced apoptosis involves cleavage of DNA to large fragments prior to internucleosomal fragmentation. J Biol Chem 268:3037–39.

Cain K, Inayathussain SH, Kokileva L, Cohen GM (1994a): DNA cleavage in rat liver nuclei activated by Mg^{2+} or Ca^{2+}/Mg^{2+} is inhibited by a variety of structurally unrelated inhibitors. Biochem Cell Biol 72:631–38.

Cain K, Inayathussain SH, Wolfe JT, Cohen GM (1994b): DNA fragmentation into 200–250 and/or 30–50 kilobase pair fragments in rat liver nuclei is stimulated by Mg^{2+} alone and Ca^{2+}/Mg^{2+} but not by Ca^{2+} alone. FEBS Lett 349:385–91.

Catchpoole DR, Stewart BW (1994): Inhibition of topoisomerase II by aurintricarboxylic acid: Implications for mechanisms of apoptosis. Anticancer Res 14:853–56.

Chaudun E, Arruti C, Courtois Y, Ferrag F, Jeanny JC, Patel BN Skidmore C, Torriglia A, Counis MF (1994): DNA strand breakage during physiological apoptosis of the embryonic chick lens-free 3' OH end single strand breaks do not accumulate even in the presence of a cation–independent deoxyribonuclease. J Cell Physiol 158:354–64.

Cleveland DW (1995): NuMA: A Protein involved in nuclear structure, spindle assembly and nuclear re-formation. Trends Cell Biol 5:60–64.

Cohen GM, Sun XM, Snowden RT, Dinsdale D, Skilleter DN (1992): Key morphological features of apoptosis may occur in the absence of internucleosomal DNA fragmentation. Biochem J 286:331–34.

Cohen JJ, Duke RC (1984): Glucocorticoid activation of a calcium-dependent endonuclease in thymocyte nuclei leads to cell death. J Immunol 132:38–42.

Cole LJ, Ellis ME (1957): Radiation-induced changes in tissue nucleic acids: Release of soluble deoxyribonucleotides in spleen. Radiation Res 7:508–17.

Collins MKL, Furlong IJ, Malde P, Ascaso R, Oliver J, Lopez Rivas A (1996): An apoptotic endonuclease activated either by decreasing pH or by increasing calcium. J Cell Sci 109:2393–99.

Cremer T, Kurz A, Zirbel R, Dietzel S, Rinke B, Schrock E, Speicher MR, Mathieu U, Jauch A, Emmerich P, Scherthan H, Ried T, Cremer C, Lichter P (1993): Role of chromosome territories in the functional compartmentalization of the cell nucleus. Cold Spring Harb Symp Quant Biol 58:777–92.

Deng GE, Podack ER (1995): Deoxyribonuclease induction in apoptotic cytotoxic T lymphocytes. FASEB J 9:665–69.

Distelhorst CW, McCormick TS (1996): Bcl-2 acts subsequent to and independent of Ca^{2+} fluxes to inhibit apoptosis in thapsigargin- and glucocorticoid-treated mouse lymphoma cells. Cell Calcium 19:473–83.

Dusenbury CE, Davis MA, Lawrence TS, Maybaum J (1991): Induction of megabase DNA fragments by 5-fluorodeoxyuridine in human colorectal tumor (HT29) cells. Mol Pharmacol 39:285–89.

Eastman A, Barry MA (1992): The origins of DNA breaks: a consequence of DNA damage, or DNA repair, or apoptosis? Cancer Invest 10:229–40.

Falcieri E, Gobbi P, Zamai L, Vitale M (1994a): Ultrastructural features of apoptosis. Scanning Microsc 8:653–66.

Falcieri E, Zamai L, Santi S, Cinti C, Gobbi P, Bosco D, Cataldi A, Betts C, Vitale M (1994b): The behaviour of nuclear domains in the course of apoptosis. Histochem 102:221–31.

Filipski J, Leblanc J, Youdale T, Sikorska M, Walker PR (1990): Periodicity of DNA folding in higher order chromatin. EMBO J 9:1319–27.

Fimia GM Gottifredi V, Passananti C. Maione R (1996): Double-stranded internucleosomal cleavage of apoptotic DNA is dependent on the degree of differentiation in muscle cells. J Biol Chem 271:15575–79.

Fraser MJ (1996): Endo–exonucleases: Actions in the life and death of cells. Bioscience Reports. R. G. Landes Co., Georgetown. Texas. pp 1–73.

Fraser MJ, Tynan SJ, Papaioannou A, Ireland CM, Pittman SM (1996): Endo–exonuclease of human leukaemic cells: Evidence for a role in apoptosis. J Cell Sci 109:2343–60.

Gaido ML, Cidlowski JA (1991): Identification, purification, and characterization of a calcium-dependent endonuclease (NUC18) from apoptotic rat thymocytes. J Biol Chem 266:18580–85

Georgatos G, Blobel G (1987): Lamin B constitutes an intermediate filament attachment site at the nuclear envelope. J Cell Biol 105:117–25.

Gerace L, Comeau C, Benson M (1984): Organization and modulation of nuclear lamina structure. J Cell Sci Supp 1:137–60.

Ghibelli L, Maresca V, Coppola S, Gualandi G (1995): Protease inhibitors block apoptosis at intermediate stages: A compared analysis of DNA fragmentation and apoptotic nuclear morphology. FEBS Lett 377:9–14.

Giannakis C, Forbes IJ, Zalewski PD (1991): Calcium magnesium-dependent nuclease: Tissue distribution, relationship to internucleosomal DNA fragmentation and inhibition by zinc. Biochem Biophys Res Commun 181:915–20.

Gottlieb RA, Giesing HA, Engler RL, Babior BM (1995): The acid deoxyribonuclease of neutrophils: A possible participant in apoptosis-associated genome destruction. Blood 86:2414–18.

Gottlieb RA, Nordberg J, Skowronski E, Babior BM (1996): Apoptosis induced in Jurkat cells by several agents is preceded by intracellular acidification. Proc Natl Acad Sci USA 93:654–58.

Gromkowski SH, Brown TC, Cerutti PA, Cerottini J-C (1986): DNA of human Raji target cells is damaged upon lymphocyte-mediated lysis. J Immunol 136:752–56.

Hara SS, Halicka HD, Bruno S, Gong JP, Traganos F, Darzynkiewicz Z (1996): Effect of protease inhibitors on early events of apoptosis. Exp Cell Res 223:372–84.

Hashida T, Tanaka Y, Matsunami N, Yoshihara K, Kamiya T, Tanigawa Y, Koide SS (1982): Purification and properties of bull seminal plasma Ca^{2-}, Mg^{2+}-dependent endonuclease. J Biol Chem 257:13114–19.

Herrmann M, Voll R, Woith W, Hagenhofer M, Lorenz HM, Manger B, Kalden JR (1996): Small DNA fragments isolated from human T-cell clones are enriched in sequences involved in DNA bending. Cell Death Differentiation 3:391–95.

Hewish DR, Burgoyne LA (1973): Chromatin sub-structure: The digestion of chromatin DNA at regularly spaced sites by a nuclear deoxyribonuclease. Biochem Biophys Res Commun 52:504–10.

Hwang H, Ohtani M, Nakazawa T, Igaki T, Masui O, Kankawa S, Nakayama C, Yoshida S, Yoshiokahiramoto A, Wataya Y (1995): Molecular mechanism of 5-fluoro-2′-deoxyuridine-induced dNTP imbalance cell death: Purification of an endonuclease involved in DNA double strand breaks during dNTP imbalance death. Nucleosides and Nucleotides 14:2089–97.

Ishida R, Akiyoshi H, Takahashi T (1974): Isolation and purification of calcium and magnesium dependent endonuclease from rat liver nuclei. Biochem Biophys Res Commun 56:703–10.

Jiang S, Chow SC, Niotera P, Orrenius S (1994): Intracellular Ca^{2+} signals activate apoptosis in thymocytes: Studies using Ca^{2+}-ATPase inhibitor thapsigargin. Exp Cell Res 212:84–92.

Jiang S, Chow SC, Mccabe MJ, Orrenius S (1995): Lack of Ca^{2+} involvement in thymocyte apoptosis induced by chelation of intracellular $ZnCa^{2+}$ Lab Invest 73:111–17.

Jones DP, McConkey DJ, Nicotera P, Orrenius S (1989): Calcium-activated DNA fragmentation in rat liver nuclei. J. Biol Chem. 264:6398–6403.

Joseph R, Li w, Han E (1993): Neuronal death, cytoplasmic calcium and internucleosomal DNA-fragmentation—Evidence for DNA-fragments being released from cells. Mol Brain Res 17:70–76.

Kaufmann S (1989): Induction of endonucleolytic DNA cleavage in human acute myelogenous leukemia cells by etoposide, camptothecin and other cytotoxic anticancer drugs: A cautionary note. Cancer Res 49:5870–78.

Kaufmann S, Desnoyers S, Ottaviano YL, Davidson NE, Poirier GG (1993): Specific proteolytic cleavage of poly(ADP-ribose) polymerase: An early marker of chemotherapy-induced apoptosis. Cancer Res 53:3976–85.

Kawabata H, Anzai N, Ueda Y, Masutani H, Hirama T, Yoshida Y, Okuma M (1996): High levels of Ca^{2+} -independent endonuclease activity capable of producing nucleosomal-size DNA fragmentation in non-adherent marrow mononuclear cells from patients with myelodysplastic syndromes and acute myelogenous leukemia. Leukemia 10:67–73.

Kerr JFR, Wyllie AH, Currie AR (1972): Apoptosis: A basis biological phenomenon with wide-ranging implications in tissue kinetics. Br J Cancer 26:239–57.

Khodarev NN, Ashwell JD (1996): An inducible lymphocyte nuclear Ca^{2+} /Mg^{2+}-dependent endonuclease associated with apoptosis. J Immunol 156:922–31.

Kong SK, Suen YK, Chan YM, Chan CW, Choy YM, Fung KP, Lee CY (1996): Concanavalin A-induced apoptosis in murine macrophages through a Ca^{2+}-independent pathway. Cell Death Differentiation 3:307–14.

Kornberg R (1974): Chromatin structive: a repeating unit of histones and DNA. Science 184:868–71.

Kuribayashi N, Sakagami H, Iida M, Takeda M (1996): Chromatin structure and endonuclease sensitivity in human leukemic cell lines. Anticancer Res 16:1225–30.

Lagarkova MA, Iarovaia OV, Razin SV (1995): Large-scale fragmentation of mammalian DNA in the course of apoptosis proceeds via excision of chromosomal DNA loops and their oligomers. J Biol Chem 270:20339–41.

Lebedeva LG, Aleksandrova SS, Basnakyan AG, Votrin II (1995a) Characteristics of nuclear endo-DNases from rat liver according to their molecular masses and cation dependence. Biochemistry-English translation of Biokhimiya. 60:597–601.

Lebedeva LG, Alexandrova SS, Votrin II, Basnakian AG (1995b): In vitro proteolysis of endonucleases in rat liver nuclei extracts. Biochem Mol Biol Int 35:433–40.

Lockshin RA (1969): Programmed cell death. Activation of a lysis mechanism involving the synthesis of protein. J Insect Physiol 15:1505–16.

Luderus MEE, Denblaauwen JL, Desmit OJB, Compton DA, Vandriel R (1994): Binding of matrix attachment regions to lamin polymers involves single-stranded regions and the minor groove. Mol Cell Biol 14:6297–305.

Lydersen BK, Pettijohn DE (1980): Human-specific nuclear protein that associates with the polar region of the mitotic apparatus: distribution in a human/hamster hybrid cell. Cell 22:489–99.

McConkey DJ, Hartzell P, Amador-Perez JF, Orrenius S, Jondal M (1989a): Calcium-dependent killing of immature thymocytes by stimulation via the CD3-T cell receptor complex. J Immunol 143:1801–6.

McConkey DJ, Hartzell P, Nicotera P, Orrenius S (1989b): Calcium-activated DNA fragmentation kills immature thymocytes. FASEB J 3:1843–49.

McConkey DJ, Nicotera P, Hartzell P, Bellomo G, Wyllie AH, Orrenius S, (1989c): Glucocorticoids activate a suicide process in thymocytes through an elevation of cytosolic calcium concentration. Arch Biochem Biophys 269:365–70.

Mizumoto K, Rothman RJ, Farber JL (1994): Programmed cell death (apoptosis) of mouse fibroblasts is induced by the topoisomerase II inhibitor etoposide. Mol Pharmacol 46:890–95.

Montague JW, Gaido ML, Frye C, Cidlowski JA (1994): A calcium-dependent nuclease from apoptotic rat thymocytes is homologous with cyclophilin. J Biol Chem 269:18877–80.

Montague JW, Cidlowski JA (1996): Cellular catabolism in apoptosis: DNA degradation and endonuclease activation. Experientia 52:957–62.

Montague JW, Hughes FM Jr, Cidlowski JA (1997): Native recombinant cyclophilins A, B, and C degrade DNA independently of peptidylprolyl cis-trans-isomerase activity. J Biol Chem 272:6677–684.

Moore J, Boswell S, Hoffman R, Burgess G, Hormas R (1993): Mutant H-ras overexpression inhibits a random apoptotic nuclease in myeloid leukemia cells. Leuk Res 17:703–9.

Morana S, Li JF, Springer EW, Eastman A (1994): The inhibition of etoposide-induced apoptosis by zinc is associated with modulation of intracellular pH. Int J Oncol 5:153–58.

Nakamura M, Sakaki Y, Watanabe N, Takagi Y (1981): Purification and characterization of the Ca^{2+} plus Mg^{2+}-dependent endodeoxyribonuclease from calf thymus chromatin. J Biochem Tokyo 89:143–52.

Neamati N, Fernandez A, Wright S, Kiefer J, McConkey DJ (1995): Degradation of lamin B-1 precedes oligonucleosomal DNA fragmentation in apoptotic thymocytes and isolated thymocyte nuclei. J Immunol 154:3788–95.

Nikonova LV, Nelipovich PA, Umansky SR (1982): The involvement of nuclear nucleases in rat thymocyte DNA fragmentation after gamma-irradiation. Biochim Biophys Acta 699:281–89.

Nikonova LV, Zotova RN, Umanskii SR (1988): Isolation of Ca^{2+},Mg^{2+}-dependent nuclease from calf thymus chromatin. Biochimc 54:1397–405.

Nikonova LV, Beletsky IP, Umansky SR (1993): Properties of some nuclear nucleases of rat thymocytes and their changes in radiation-induced apoptosis. Eur J Biochem 215:893–901.

Oberhammer F, Fritsch G, Schmeid M, Pavelka M, Printz D, Purchio T, Lassmann H, Schulte-Hermann R (1993a): Condensation of the chromatin at the membrane of an apoptotic nucleus is not associated with activation of an endonuclease. J Cell Sci 104:317–26.

Oberhammer F, Wilson JW, Dive C, Morris ID, Hickman JA, Wakeling AE, Walker PR, Sikorska M (1993b): Apoptotic death in epithelial cells: Cleavage of DNA to 300 and/or 50 kb fragments prior to or in the absence of internucleosomal fragmentation. EMBO J 12:3679–84.

Oberhammer FA, Hochegger K, Froeschl G, Tiefenbacher R, Pavelka M (1994): Chromatin condensation during apoptosis is accompanied by degradation of lamin A + B, without enhanced activtion of cdc2 kinase. J Cell Biol 126:827–37.

Oshimi Y, Miyazaki S (1995): Fas antigen-mediated DNA fragmentation and apoptotic morphologic changes are regulated by elevated cytosolic Ca^{2+} level. J Immunol 154:599–609.

Ostad M, Shu WP, Kong L, Liu BCS (1996): Ha-ras oncogene expression abrogates a pH-dependent endonuclease activity of apoptosis in normal rat kidney cells. Cancer Lett 98:175–82.

Paddenberg R, Wulf S, Weber A, Heimann P, Beck L, Mannherz HG (1996): Internucleosomal DNA fragmentation in cultured cells under conditions reported to induce apoptosis may be caused by mycoplasma endonucleases. Eur J Cell Biol 71:105–19.

Pandey S, Walker PR, Sikorska M (1994): Separate pools of endonuclease activity are responsible for internucleosomal and high molecular mass DNA fragmentation during apoptosis. Biochem Cell Biol 72:625–29.

Pandey S, Walker PR, Sikorska M (1997): Identification of a novel 97 kDa endonuclease capable of internucleosomal DNA cleavage and its possible role in apoptosis. Biochemistry 36:711–720.

Park HJ, Makepeace CM, Lyons JC, Song CW (1996): Effect of intracellular acidity and ionomycin on apoptosis in HL-60 cells. Eur J Cancer 32A:540–46.

Patel T, Bronk SF, Gores GJ (1994): Increases of intracellular magnesium promote glycodeoxycholate-induced apoptosis in rat hepatocytes. J Clin Invest 94:2183–92.

Peitsch MC, Polzar B, Stephan H, Crompton T, Macdonald HR, Mannherz, HG, Tschopp J (1993a): Characterization of the endogenous deoxyribonuclease involved in nuclear DNA degradation during apoptosis (programmed cell death) EMBO J 12:371–77.

Peitsch MC, Muller C, Tschopp J (1993b): DNA fragmentation during apoptosis is caused by frequent single-strand cuts. Nucleic Acid Res 21:4206–9.

Perez-Sala D, Colladoescobar D, Mollinedo F (1995): Intracellular alkalinization suppresses lovastatin-induced apoptosis in HL-60 cells through the inactivation of a pH-dependent endonuclease. J Biol Chem 270:6235–42.

Perotti M, Toddei F, Mirabelli F, Vairetti M, Bellomo G, McConkey DJ, Orrenius S (1990): Calcium-dependent DNA fragmentation in human synovial cells exposed to cold shock. FEBS Lett 259:331–34.

Polzar B, Peitsch MC, Loos R, Tschopp J, Mannherz HG (1993): Overexpression of deoxyribonuclease I (DNase I) transfected into COS-cells—Its distribution during apoptotic cell death. Eur J Cell Biol 62:397–405.

Polzar B, Zanotti S, Stephan S, Stephan H, Rauchi F, Peitsch MC, Irmler M, Tschopp J, Mannherz HG (1994): Distribution of deoxyribonuclease I in rat tissues and its correlation to cellular turnover and apoptosis (programmed cell death). Eur J Cell Biol 64:200–10.

Posner A, Raser KJ, Hajimohammadreza I, Yuen PW, Wang KKK (1995): Aurintricarboxylic acid is an inhibitor of mu- and m-calpain. Biochem Mol Biol Int 36:291–99.

Provinciali M, Distefano G, Fabris N (1995): Dose-dependent opposite effect of zinc on apoptosis in mouse thymocytes. Int J Immunopharmacol 17:735–44.

Reynolds JE, Li JF, Craig RW, Eastman A (1996): BCL-2 and MCL-1 expression in Chinese hamster ovary cells inhibits intracellular acidification and apoptosis induced by staurosporine. Exp Cell Res 225:430–36.

Reynolds JE, Eastman A (1996): Intracellular calcium stores are not required for Bcl-2-mediated protection from apoptosis. J Biol Chem 271:27739–43.

Ribeiro JM, Carson DA (1993): Calcium-magnesium-dependent endonuclease from human spleen: purification, properties and role in apoptosis. Biochemistry 32:9129–36.

Rusnak JM, Calmels TPG, Hoyt DG, Kondo Y, Yalowich JC, Lazo JS (1996): Genesis of discrete higher order DNA fragments in apoptotic human prostatic carcinoma cells. Mol Pharmacol 49:244–52.

Rzepecki R (1994): The complex of the 32 kD endonuclease and 65 kD protein from plant nuclear matrix preferentially recognizes the plasmid containing SAR DNA element. J Plant Physiol 144:479–84.

Schmidt B, Tradler T, Rahfeld JU, Ludwig B, Jain B, Mann K, Rucknagel KP, Janowski B, Schierhorn A, Kullertz G, Hacker J, Fischer G (1996): A cyclophilin-like peptidyl-prolyl cis/trans isomerase from *Legionella pneumophila*—Characterization, molecular cloning and overexpression. Mol Microbiol 21:1147–60.

Schwartz DC, Cantor CR (1984): Separation of yeast chromosome-sized DNAs by pulsed field gradient gel electrophoresis. Cell 37:67–74.

Segal-Bendirdjian E, Jacqueminsablon A (1995): Cisplatin resistance in a murine leukemia cell line is associated with a defective apoptotic process. Exp Cell Res 218:201–12.

Shemtov MM, Cheng DLW, Kong L, Shu WP, Sassaroli M, Droller MJ, Liu BCS (1995): LAK cell mediated apoptosis of human bladder cancer cells involves a pH-dependent endonuclease system in the cancer cell: Possible mechanism of BCG therapy. J Urol 154:269–74.

Shiokawa D, Ohyama H, Yamada T, Takahashi K, Tanuma S (1994): Identification of an endonuclease responsible for apoptosis in rat thymocytes. Eur J Biochem 226:23–30.

Skalka M, Matyasova J, Cejkova M (1976): DNA in chromatin of irradiated lymphoid tissues degrades *in vivo* into regular fragments. Biochem Biophys Res Commun 72:271–74.

Skalka M, Matyasova J (1963): The effect of low radiation doses on the release of deoxyribonucleotides in hematopoietic and lymphatic tissues. Int J Radiat Biol 7:41–44.

Sokolova IA, Cowan KH, Schneider E (1995): Ca^{2+}/Mg^{2+}-dependent endonuclease activation is an early event in VP-16-induced apoptosis of human breast cancer MCF7 cells *in vitro* Biochim Biophys Acta 1266:135–42.

Stratling WH, Grade C, Horz W (1984): Ca/Mg-dependent endonuclease from porcine liver: Purification, properties and sequence specificity. J Biol Chem 259:5893–98.

Sun XM, Cohen GM (1994): Mg^{2+}-dependent cleavage of DNA into kilobase pair fragments is responsible for the initial degradation of DNA in apoptosis. J Biol Chem 269:14857–60.

Szondy Z (1995): The 2-chlorodeoxyadenosine-induced cell death signalling pathway in human thymocytes is different from that induced by 2-chloroadenosine. Biochem J 311:585–88.

Szopa J, Bode J, Kay V, Kozubek A, Sikorski AF (1993): Does a prototype scaffold matrix-attached region (SAR sequence) affect intrinsic nuclear matrix endonuclease specificity? J Plant Physiol 141:668–72.

Takauji R, Yoshida A, Iwasaki H, Toyama K, Ueda T, Nakamura T, (1995): Enhancement of Ca^{2+}-dependent endonuclease activity in L1210 cells during apoptosis induced by 1-beta-D-arabinofuranosylcytosine: Possible involvement of activating factor(s). Jpn J Cancer Res 86:677–84.

Tanimoto C, Hirakawa S, Kawasaki H, Hayakawa N, Ota Z (1995): ICRF-193 modifies etoposide-induced apoptosis in thymocytes. Acta Med Okayama 49:281–86.

Tanuma S, Shiokawa D (1994): Multiple forms of nuclear deoxyribonuclease in rat thymocytes. Biochem Biophys Res Commun 203:789–97.

Thompson DC, Reed M (1995): Inhibition of NAD(H)/NADP(H)–requiring enzymes by aurintricarboxylic acid. Toxicol Lett 81:141–49.

Tomei LD, Shapiro JP, Cope FO (1993): Apoptosis in C3H-10T1/2 mouse embryonic cell: Evidence for internucleosomal DNA modification in the absence of double-strand cleavage. Proc Natl Acad Sci USA 90:853–57.

Torriglia A, Chaudun E, Chanyfournier F, Jeanny JC, Courtois Y, Counis MF (1995): Involvement of DNase II in nuclear degeneration during lens cell differentiation. J Biol Chem 270:28579–85.

Ucker DS, Obermiller PS, Eckhart W, Apgar JR, Berger NA, Meyers J (1992): Genome digestion is a dispensable consequence of physiological cell death mediated by cytotoxic T lymphocytes. Mol Cell Biol 12:3060–69.

Umansky SR, Korol BA, Nelipovich PA (1981): In vivo DNA degradation in thymocytes of gamma-irradiated or hydrocortisone-treated rats. Biochim Biophys Acta 655:9–17.

Umansky SR, Beletsky IP, Korol BA, Lichtenstein AV, Nelipovich PA (1988): Molecular mechanisms of DNA degradation in dying rodent thymocytes. Mol Cell Biol 7:221–28.

Villalba M, Ferrari D, Bozza A, Delsenno L, Divirgilio F (1995): Ionic regulation of endonuclease activity in PC12 cells. Biochem J 311:1033–38.

Walker PR, Smith C, Youdale T, Leblanc J, Whitfield JF, Sikorska M (1991): Topoisomerase II-reactive chemotherapeutic drugs induce apoptosis in thymocytes. Cancer Res 51:1078–85.

Walker PR, Kokileva L, Leblanc J, Sikorska M (1993): Detection of the initial stages of DNA fragmentation in apoptosis. Biotechniques 15:1032–41.

Walker PR, Weaver VM, Lach B, Leblanc J, Sikorska M (1994): Endonuclease activities associated with high molecular weight and internucleosomal DNA fragmentation in apoptosis. Exp Cell Res 213:100–106.

Walker PR, Pandey S, Sikorska M (1995): Degradation of chromatin in apoptosis. Cell Death Differentiation 2:97–104.

Walker PR, Leblanc J, Sikorska M (1997): Evidence that DNA fragmentation in apoptosis is initiated and propagated by single-strand breaks. Cell Death Differentiation 4:506–515.

Walker PR, Sikorska M (1987): Evidence that the 30 nm fiber is a helical coil with 12 nucleosomes/turn. J Biol Chem 262:12223–27.

Walker PR, Sikorska M (1994): Endonuclease activities, chromatin structure, and DNA degradation in apoptosis. Biochem Cell Biol 72:615–23.

Weaver VM, Carson CE, Walker PR, Chaly N, Lach B, Raymond Y, Sikorska M. (1996): Degradation of nuclear matrix and DNA cleavage in apoptotic thymocytes. J Cell Sci 109:45–56.

Whyte MKB, Hardwick SJ, Meagher LC, Savill JS, Haslett C (1993): Transient elevations of cytosolic free calcium retard subsequent apoptosis in neutrophils in vitro. J Clin Invest 92:446–55.

Wiberg JS (1958): On the mechanism of metal activation of deoxyribonuclease I. Arch Biochem Biophys 73:337–85.

Williams JR, Little JB, Shipley WU (1974): Association of mammalian cell death with a specific endonucleolytic degradation of DNA. Nature 252:754–56.

Williamson R (1970): Properties of rapidly labelled deoxyribonucleic acid fragments isolated from the cytoplasm of primary cultures of embryonic mouse liver cells. J Mol Biol 51:157–68.

Wyllie AH (1980): Glucocorticoid-induced thymocyte apoptosis is associated with endogenous endonuclease activation. Nature 284:555–56.

Wyllie AH, Morris RG, Smith AL, Dunlop D (1984): Chromatin cleavage in apoptosis: Association with condensed chromatin morphology and dependence on macromolecular synthesis. J Pathol 142:67–77.

Yamamoto M, Muratal H, Sumiyoshi H, Endol H (1984): Rapid inactivation of Ca^{2+},Mg^{2-}-dependent endonuclease of rat liver nuclei after cycloheximide treatment. Biochem Biophys Res Commun 119:618–23.

Yoshida A, Ueda T, Takauji R, Liu Y-P, Fukushima T, Inuzuka M, Nakamura T (1993): Role of calcium ion in induction of apoptosis by etoposide in human leukemia HL-60 cells. Biochem Biophys Res Commun 196:927–34.

Zakharian RA, Pogosian RC (1978): Glucocorticoid induction of the degradation of lymphocyte chromatin DNA into regularly repeating fragments. Proc Armenian Acad Sci 67:110–14.

Zhivotovsky B, Nicotera P, Bellomo G, Hanson K, Orrenius S (1993): Ca^{2+} and endonuclease activation in radiation-induced lymphoid cell death. Exp Cell Res 207:163–70.

Zhivotovsky B, Cedervall B, Jiang S, Nicotera P, Orrenius S (1994a): Involvement of Ca^{2+} in the formation of high molecular weight DNA fragments in thymocyte apoptosis. Biochem Biophys Res Commun 202:120–27.

Zhivotovsky B, Wade D, Gahm A, Orrenius S, Nicotera P (1994b): Formation of 50 kbp chromatin fragments in isolated liver nuclei is mediated by protease and endonuclease activation. FEBS Lett 351:150–54.

Zhivotovsky BD, Zvonareva NB, Hanson KP (1981): Characteristics of rat thymus chromatin degradation products after whole body X-irradiation. Int J Rad Biol 39:437–40.

Zhu WH, Loh TT (1995): Roles of calcium in the regulation of apoptosis in HL-60 promyelocytic leukemia cells. Life Sci 57:2091–99.

PART III

CELL DEATH WHEN MITOSIS IS HIGH AND EVANESCENCE IS DESIRABLE

CHAPTER 8

APOPTOTIC SIGNALING PATHWAYS IN LYMPHOCYTES

BARBARA A. OSBORNE
University of Massachusetts, Department of Veterinary Sciences & Program in
Molecular & Cellular Biology, 304 Paige Laboratory, Amherst, MA 01003

INTRODUCTION

Cell death is a critical feature of organisms as varied as plants and animals and has been documented in single-cell organisms (Ameisen, Chapter 1, this volume) as well as complex multicellular species such as mammals. The recognition that cell death is an important aspect of cellular metabolism came from the description of Kerr, Wyllie, and Currie of apoptotic death (1972). In their seminal observations over 20 years ago, these investigators described an active cellular process that leads to the death and removal of unwanted cells. More recently, the work of Horvitz and colleagues has demonstrated a genetic basis for the induction of cell death (Ellis et al, 1991). Working in the small free-living soil nematode, *C. elegans,* these investigators have shown that specific genes control the fate of a cell to live or die.

With the advent of our understanding that some cell deaths are genetically programmed events, a number of different laboratories using a wide variety of systems have demonstrated genetic regulation of apoptosis. One of the more powerful systems of investigation is the immune system. The immune system offers a unique opportunity to isolate large numbers of dying cells in a relatively pure form. The availability of transgenic mice with defined T-cell receptor specificity provides an additional valuable tool for cell death studies. For example, in the thymus, it is possible to induce negative selection, the process by which self-reactive T cells are eliminated by the introduction of antigen to the thymus. Since, in a TCR transgenic mouse, all the thymocytes carry the same TCR, essentially all the cells may be induced to die at the

When Cells Die, Edited by Richard A. Lockshin, Zahra Zakeri,
and Jonathan L. Tilly
ISBN 0-471-16569-7 © 1998 Wiley-Liss, Inc.

same time. Similarly, again through the use of TCR transgenic mice, relatively pure populations of peripheral T cells also can be induced to undergo apoptosis.

During maturation, both T and B cells are exposed to signals generated through receptor crosslinking, and some of these signals lead to apoptosis. In particular, it is well established that immature $CD4^+$ $CD8^+$ thymocytes are subjected to both positive and negative selection. Positive selection is necessary to ensure the presence of mature T cells with the ability to recognize self-major histocompatibility complex molecules (reviewed in von Boehmer, 1994). This is thought to be followed by negative selection, a mechanism ensuring that those T cells with potentially damaging self-reactivity are deleted (reviewed in Nossal, 1994), and recent data demonstrate that negative selection is accomplished by the induction of apoptosis (Smith et al., 1989; MacDonald and Lees, 1990; Murphy et al., 1990). Although the precise molecular mechanism that drives negative selection is not completely understood, current thought is that negative selection involves a high-affinity interaction between the T-cell receptor and antigen (Ashton-Rickart et al., 1994; Hogquist et al., 1994; Sebzda et al., 1994) and costimulatory molecules assist in this process.

In addition to the deletion of autoreactive cells in the thymus, there is a critical need to regulate the proliferation of lymphocytes. One of the unique capabilities of lymphocytes is the ability to proliferate in response to antigen. This feature allows the rapid expansion of cells in an antigen-specific fashion, thus providing the organism with a mechanism for the rapid elimination of pathogens. However, it is clear that a continuously dividing population of cells poses unique hazards. These cells secrete large quantities of inflammatory cytokines useful in combating pathogens but are potentially deleterious to the host if produced in sufficient quantity.

Additionally, any rapidly dividing population of cells are subject to the accumulation of mutation and lymphocytic malignancies are well documented. Thus this advantageous feature of lymphocytes, the ability to respond rapidly to antigen by clonal expansion, must rely on the existence of appropriate check points to control cell division. In lymphocytes, this control appears to rely almost entirely on the activation of a cell death program; Newell and Vincent, Chapter 9, this volume (Thompson, 1995). The signals that induce apoptosis and mechanisms that mediate this cell death in immature and mature population of cells are discussed herein. This chapter will review the signaling pathways that induce apoptosis in cells of the immune system and describe the genes and intracellular events that mediate this process in lymphoid cells, particularly T lymphocytes.

CELL DEATH IN THE THYMUS

A pivotal event, critical to lymphoid development, is the education of immature lymphoid populations to discriminate between self and non-self. In the case of the T-cell lineage, potentially autoreactive immature thymocytes are deleted in the thymus by negative selection (reviewed in Nossal, 1994). Several years ago it became clear that negative selection or clonal deletion occurs through the induction of apoptosis, and this process requires new gene and protein synthesis (Smith et al., 1989; MacDonald and Lees, 1990; Murphy et al., 1990).

Large numbers of precursor cells migrate into the thymus daily and are subjected to selection. However, the majority (90–95%) of these cells die by neglect; that is, they are neither positively nor negatively selected (Surh and Sprent, 1994). Those cells bearing T-cell receptors (TCR) that recognize self major histocompatibility proteins (MHC) are positively selected. A subset of these cells recognize MHC with high affinity and are signaled to undergo negative selection by apoptosis (reviewed in Allen, 1994). These events are illustrated in Fig. 8.1.

What happens to those cells that are neither positively or negatively selected? Death by neglect is apoptotic (Surh and Sprent, 1994), and recent evidence suggests that this form of death occurs via exposure to endogenous glucocorticoids (King et al., 1995). The first clue that glucocorticoids and TCR signaling might intersect comes from experiments in T cells lines. Several years ago data indicated that while TCR signaling and exposure to glucocorticoid both induce apoptosis in T cell lines, simultaneous delivery of both signals

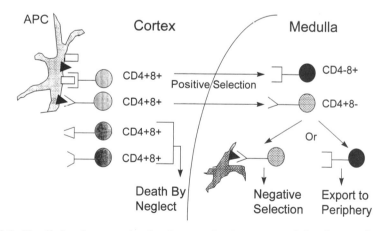

FIG. 8.1. T-cell development in the thymus. In the cortex of the thymus, immature thymocytes develop into CD4$^+$CD8$^+$ thymocytes and become T-cell receptor (TCR) positive. These cells, if the TCR recognizes self-MHC, encounter antigen on an antigen-presenting cell (APC) and undergo positive selection. Those cells with TCRs that do not recognize self-MHC die by neglect, a process thought to be controlled by endogenous glucocorticoids (see text). The positively selected cells migrate to the corticomedullary region of the thymus, where development into single positive CD4$^+$ or CD8$^+$ cells develop and negative selection occurs. Negative selection presumably is induced by a high-affinity TCR interaction with self-MHC plus antigen.

is antagonistic and results in cell survival (Zacharchuk et al., 1990; Iwata et al., 1991). Thymocytes, in addition to T cell lines, are exquisitely sensitive to induction of apoptosis by glucocorticoids, and this process involves induction of new gene expression (Wyllie, 1980). The interaction between TCR signals and endogenous glucocorticoids found in the thymus has been explored by Ashwell and colleagues. Their data demonstrate that a subset of thymic epithelial cells are steroidogenic and that inhibition of steroid synthesis in d17 fetal thymic organ culture results in enhanced deletion of thymocytes in response to TCR ligation (Vacchio et al., 1994). These data provide the first *in vivo* evidence that endogenously produced steroids can antagonize TCR mediated deletion in the thymus and suggest previous results with cell lines mimic the normal events occurring during thymic selection. Recent data from this group extend these observations by the creation of transgenic lines of mice carrying a glucocorticoid receptor antisense construct targeted to the thymus through the use of the *lck* promoter (King et al., 1995). These mice, as might be expected, have an enhanced susceptibility to TCR-meditated apoptosis. In addition, there is an unexpected loss of CD4$^-$ CD8$^-$ cells, suggesting that glucocorticoids may play an important role in the transition from CD4$^-$ CD8$^-$ to CD4$^+$ CD8$^+$ as well as in the survival of cells signaled via the TCR. These studies indicate that thymocytes that do not engage TCR are deleted by endogenous steroids, those cells that engage TCR with moderate avidity are protected via the steroid/TCR antagonism, and the cells that engage TCR at high avidity are deleted, presumably due to overriding the steroid–TCR antagonism.

Do thymocytes require more than a signal through the T-cell receptor to undergo apoptosis? Costimulatory signals are required for activation of peripheral T cells, but recent evidence suggests that costimulation is important in negative selection in the thymus. Two groups have shown that isolated thymocytes do not die in response to TCR crosslinking; however, apoptosis is readily induced in one case by antigen presenting cells (Page et al., 1993) or, in the other instance, by CD28 crosslinking in conjunction with anti-TCR (Punt et al., 1993). The role of CD28 in negative selection has been confirmed in organ culture (Amsen and Kruisbeek, 1996). Also Noelle and colleagues have shown that *in vivo* negative selection of thymocytes requires gp39/CD40 signaling (Foy et al., 1995). Taken together these data indicate that TCR signals in addition to costimulatory signals play an important role in signaling thymocytes. The consequence of costimulatory signals play an important role in signaling thymocytes. The consequence of costimulatory signals is appropriate negative selection, but what signaling pathways are engaged by costimulatory signals have not yet been determined.

GENETIC REGULATION OF APOPTOSIS IN THE THYMUS

Since it is known that new gene synthesis is a requirement for negative selection, it is reasonable to ask what genes mediate apoptosis in thymocytes. While we do not known the details of the complex signaling pathways that induce apoptosis in thymocytes, recent data from several laboratories provide some insight. The first gene shown to play a required role in thymocyte apoptosis was the p53 tumor suppressor gene. Through the use of mice deficient in p53, it was shown that p53 is required for cell death induced by ionizing radiation and many chemotherapeutic agents that cause DNA damage (Lowe et al., 1993; Clarke et al., 1993). However, this gene does not play a role in glucocorticoid or TCR-mediated apoptosis and hence plays no role in negative selection.

To search for genes induced during negative selection, we constructed a cDNA library from thymocytes induced to undergo negative selection. Through the differential screening of this library, we isolated *nur77,* a member of the nuclear hormone receptor superfamily (Liu et al., 1994). *In vitro* studies with antisense (Liu et al., 1994) or dominant negative *nur77* (Woronicz et al., 1994) suggested this gene was required for TCR-mediated cell death. Although the antisense and dominant negative Nur77 data indicated a required role for Nur77 in apoptosis, it was shown that mice carrying a targeted deletion of *nur77* exhibit no defects in TCR-mediated death (Lee et al., 1995). These data suggested either that *nur77* is not required *in vivo* or that a closely related gene is able to compensate in the knockout mice. The latter is supported by the recent observation that transgenic mice carrying a dominant negative *nur77* transgene targeted to the thymus with a thymic-specific promoter have severe defects in negative selection, as measured by deletion with antigen (Calnan et al., 1995; Weih et al., 1996; Zhou et al., 1996). These mice also have increased numbers of thymocytes, supporting the idea that negative selection does not proceed in a normal fashion.

The data described demonstrate that thymocytes are able to die by a variety of different induction strategies. Additionally it is quite clear that each signal initiates a unique signal transduction cascade and that each cascade involves unique sets of genes. For example, as illustrated in Fig. 8.2, glucocorticoids induce apoptosis through a pathway that requires the glucocorticoid receptor (Dieken and Miesfeld, 1992). The induction of *nur77* via TCR ligation is unique to TCR signals, and p53 is required only following induction of apoptosis via ionizing radiation. Whether these unique or private pathways converge on a common path accessed by disparate signals remains to be determined, but it is likely that most, if not all, of these pathways involve the activation of the ICE family of the proteases, the caspases (Sikorska and Walker, Chapter 7, this volume).

CELL DEATH IN PERIPHERAL T CELLS

Peripheral T cells must be able to proliferate rapidly in response to antigenic challenge. However, the ability to proliferate and expand rapidly is a double-edged sword. While this expansion of cells produces populations of clonally related cells with the same antigen specificity, it also results in the production

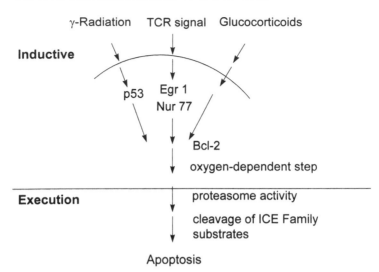

FIG. 8.2. Multiple pathways leading to apoptosis in thymocytes.

of large quantities of several cytokines, some of which are inflammatory. Therefore, it is advantageous to be able to control this proliferative response. At least two strategies are used by mature T cells to control proliferation, and both of these involve the induction of apoptosis. Thus apoptosis appears to be the major mechanism by which mature lymphocytes maintain homeostasis.

The first indications that cell death plays a critical role in the maintenance of lymphoid homeostasis comes from two mutations identified in mice several years ago. The *lpr/lpr* and the *gld/gld* strains of mice were identified as two strains that suffer from lymphoproliferative disorders. In particular, both mice exhibit greatly enlarged lymph nodes and spleen, and the mice die as young adults. The defect in these mice are mutations in Fas and Fas ligand, two very important cell surface molecules that mediate apoptosis in many different cell types (Budd, Chapter 10, this volume).

Fas AND Fas LIGAND AND THE REGULATION OF APOPTOSIS IN T LYMPHOCYTES

Fas belongs to a conserved family of membrane receptors known as the tumor necrosis factor receptor or TNFR family (reviewed in Nagata and Golstein, 1995). In mammals there are at least 10 members of this important family, including the tumor necrosis receptor 1 (TNFR1) and receptor 2 (TNFR2), NGFR, CD40, LtβR, CD30, CD27, 4-1BB, OX40, and Fas (also known as CD95). These receptors interact with eight different ligands, all bearing some similarity to TNF. For example, TNFR1 and TNFR2 interact with the ligand TNF and Fas interacts with Fas ligand. This conserved family of receptor/

ligands has been implicated in a number of cellular responses, including cell proliferation, cell death, and cellular differentiation. While it is not fully understood how these receptor–ligand interactions regulate multiple responses, the role of this family in apoptosis has been extensively investigated (Nagata and Golstein, 1995). Both TNFR1 and Fas contain cytoplasmic domains known as the "death domain" and the ability of these receptors to mediate apoptosis maps to this region of the receptor. Recent data indicate that the "death domains" of Fas and TNFR1 transmit death signals through the interaction with two proteins that themselves contain a death domain and a required death effector domain, known as FADD or TRADD, respectively. The Fas–FADD and TNFR1–TRADD interaction requires trimerization of Fas or TNFR1 by their respective ligands. This receptor trimerization induces the association of Fas with FADD and TNFR1 with TRADD (reviewed by Krammer, 1996). Recently these complexes have been shown to activate the caspases directly through the association of Fas–FADD or TNFR1–TRADD with a molecular hybrid known as FLICE or MACH. FLICE/MACH has a death effector domain that appears to mediate the association with either FADD or TRADD (Boldin et al., 1996; Muzio et al., 1996). But in addition to the death effector domain, this multifunctional protein has a protease activity that directly activates caspases, in particular a CPP32-like caspase. Thus the link from the cell surface to the critical activation of caspases may involve as few as three proteins: Fas, FADD, and FLICE (Fig. 8.3) (Cryns and Yuan, Chapter 6, this volume).

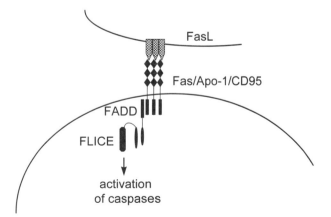

FIG. 8.3. Activation-induced apoptosis in peripheral T cells occurs through Fas–FasL interactions. FasL on a neighboring cell can induce the oligomerization of Fas. This results in the recruitment of FADD to the intracellular death domain of Fas, which, in turn, results in association of FLICE with FADD through the death effector domain. FLICE, through the action of an ICE-like domain, is able to activate caspases directly, resulting in apoptosis (see text for details).

Where is the Fas–FasL pathway used in lymphoid cell death? As indicated, the understanding that defects in Fas give rise to the *lpr/lpr* mouse and the defect in *gld/gld* similarly map to FasL tells us that these molecules are critical in regulating cellular homeostasis. Recent evidence has demonstrated that activation-induced death of mature T cells occurs through Fas–FasL interactions. This has been demonstrated *in vitro* using cell lines or T-cell hybrids that die in response to TCR ligation (Dhein et al., 1995; Brunner et al., 1995; Ju et al., 1995) as well as *in vivo* (Russell et al., 1993; Russell & Wang, 1993) using *lpr/lpr* or *gld/gld* mice that are defective, respectively, in either Fas or FasL. However, due to a low level of expression of Fas in *lpr/lpr* mice, it has been difficult to say without reservation that Fas–FasL interactions are responsible for the deletion of activated mature T cells. To circumvent these problems, Nagata and colleagues have created Fas −/− mice by targeted deletion of the Fas gene (Adachi et al., 1995). These mice display enhanced and accelerated lymphoproliferation when compared with *lpr/lpr* mice (Adachi et al., 1996). By the injection of SEB into Fas −/− mice, group showed that peripheral deletion is dramatically impaired in the absence of Fas. In addition to defects in the clearing of mature T cells, Fas −/− mice show substantial hyperplasia of the liver, suggesting a role for Fas in homoeostatic mechanisms in liver (Adachi et al., 1995).

Fas–FasL interactions also appear to be critical in B-cell homeostasis in the periphery, as suggested by the observation that *lpr* or *gld* mice accumulate high numbers of B cells as well as produce autoantibodies (Geise and Davidson, 1994). Fas −/− mice have higher than normal levels of IgG1, IgG2a, IgG2b, and IgG3 than wild-type mice, and B-cell numbers are increased over 200-fold in these mice. While it is known that B cells can express Fas, FasL is not found either on resting or activated B cells (Wanatabe et al., 1995). The role of Fas in the deletion of autoreactive B cells was shown by Goodnow and colleagues in experiments that demonstrate quite clearly the deletion of such B cells through FasL on CD4+ T cells (Rathmell et al., 1995). Data from Rothstein and colleagues suggest that crosslinking of CD40 on B cells sensitize the cells to Fas-mediated death by FasL+ CD4+ T cells (Rothstein et al., 1995), but crosslinking of both CD40 and Ig on B cells provides protection from Fas-mediated death. These data open the possibility that Fas-mediated death acts to eliminate bystander B cells that are in close proximity to activated T cells and hence susceptible to help (Newell and Vincent, Chapter 9, this volume).

Another clue into the death of peripheral B cells comes from two independent observations (Pulendran et al., 1995; Shokat and Goodnow, 1995) concerning death in germinal centers. Both groups present evidence that antigen-specific, high-affinity germinal center B cells undergo apoptosis following exposure to soluble antigen, and this appears to be a direct effect on the B cell. The physiological significance of these findings is not entirely known, but it seems reasonable to suggest that this mechanism may protect against autoreactive B cells that develop via somatic hypermutation in germinal centers. These results were extended by Han et al., (1995), who demonstrated, through the use of *lpr* mice, that antigen-induced apoptosis in germinal centers is independent of Fas expression. Taken together, the data suggest that autoreactive B cells are removed via Fas–FasL interaction, but B cells activated by antigen in germinal centers do not up-regulate Fas and most probably die by mechanisms yet to be discovered.

Fas–FasL interactions are not the only apoptotic signal in mature T cells. Results from several labs demonstrate a role for TNF-meditated death. In *lpr* mice, peripheral T-cell death can still occur, indicating that molecules other than Fas mediate T cell death. Two different groups provide convincing evidence (Sytwu et al., 1996; Zheng et al., 1995) that in the absence of Fas, TNF mediates apoptosis in mature T-cell populations. Both groups also provide data that indicate TNF is a normal *in vivo* modulator of apoptosis in mature T cells. Whether TNF and Fas-mediated apoptosis account for all mature T- and B-cell death is still a question, but since new TNF and TNFR family members continue to be discovered, it may be that other family members will be found to participate in lymphoid cell death. In particular, TRAMP, a TNFR1-related surface protein, has recently been cloned and found to be expressed at high levels in lymphoid tissue, and TRAMP can deliver a potent death signal after *in vitro* transfection into cell cultures (Bodmer et al., 1997).

CTL-INDUCED TARGET CELL DEATH

In addition to the examples of apoptosis in mature lymphoid cells, cytotoxic T lymphocytes (CTLs) are well-known inducers of apoptosis in target cells. CTLs are specialized T cells that form an important arm of the cellular immune response to intracellular pathogens and transformed cells. CTLs recognize viral or bacterially infected cells as well as transformed cells and efficiently kill such cells. Several years ago it was recognized that CTLs kill by inducing apoptosis in the target cell (reviewed in Berke, 1995). CTL-induced apoptosis can occur by two distinct pathways. One pathway utilized by CTLs is a secretory pathway that involves the exocytosis of performed granules that are found in the cytosol of the CTL (Henkart and Sitkovsky, 1994). These granules contain two critical proteins, granzymes and perforin. In the majority of instances, CTLs kill through the action of granzymes and perforin. In particular, granzyme B and perforin, when added as purified proteins kill cells (Berke, 1995). The mechanism by which granzyme and perforin kills is not completely understood, but it is known that granzyme B can activate the G_2 cell cycle kinase, cdc2, and activated cdc2 is sufficient to kill cells (Shi et al., 1994). Greenberg and colleagues recently demonstrated that cdc2 inactivation by wee1 kinase cells from granzyme B–induced death, directly linking the level of activation of cdc2 with apoptosis (Chen et al., 1995). There is evidence that granzyme B activates at least on ICE-like protease, CPP32, perhaps linking the activation of a cell cycle kinase with the activation of ICE proteases (Darmon et al., 1995).

Not all targets of CTLs are killed by the granzyme-mediated pathway. In an elegant series of experiments utilizing mice lacking perforin (Kagi et al., 1994) or granzyme B (Heusel et al., 1994), a low level of CTL killing was observed, indicating a mechanism of killing distinct from the perforin/granzyme-mediated death must exist (Lowin et al., 1994). The existence of a CTL hybridoma that lyses Fas positive but not Fas negative targets (Rouvier et al., 1993) suggested a role of Fas–FasL in CTL induced death. Consistent with this notion is the observation that activated T cells from *gld* mice do not lyse Fas⁺ target cells (Ramsdell et al., 1994). Whether or not all CTL killing is due to the perforin–granzyme pathway in addition to Fas–FasL killing remains to be seen. It seems likely, however, that these two unique pathways may account for the majority of CTL-induced apoptosis (Sarin et al., 1997).

SURVIVAL SIGNALS IN LYMPHOCYTES

Growth factors such as IL-3 and IL-4 provide signals for survival in lymphoid populations, but the best characterized survival factor in lymphocytes is Bcl-2. The *bcl-2* gene was first described as a 14:18 translocation in neoplastic B cells (Tsujimoto et al., 1984) and later found to be a negative regulator of cell death (Vaux et al., 1988). The role of Bcl-2 in survival of lymphoid cells has been reviewed quite recently (Strasser, 1995; Cory, 1995). Briefly, both immature B and T cells express bcl-2 during maturation processes in the bone marrow and thymus, respectively. In particular, Linette and colleagues have shown that Bcl-2 is up-regulated at the CD4⁺CD8⁺ stage of thymic development, probably to protect these cells as they undergo positive selection, and expression of Bcl-2 in the thymus can promote thymocyte differentiation (Linette et al., 1994). However, bcl-2 does not protect thymic T cells from negative selection (Sentman et al., 1991; Strasser et al., 1991) and is not up-regulated in cells undergoing negative selection (Linette et al., 1994). Bcl-2 expression is high in single positive CD4 or CD8 cells. During B-cell development, Bcl-2 is expressed at high levels in the most immature pro-B cells, is down-regulated in pre B cells undergoing Ig gene rearrangements, and is up-regulated in mature B cells (Cory, 1995). In summary, the observed expression patterns of bcl-2 suggest that bcl-2 acts to extend longevity in lymphoid cell populations, and this conclusion is supported by the observation that mature lymphocytes in mice lacking bcl-2 are short-lived and highly susceptible to apoptosis (Veis et al., 1993).

Bcl-2 is a member of a family of genes characterized by homology within the BH1 and BH2 domains, and various members of this family affect the life and death of lymphocytes. For example, BAX, a protein that heterodimerizes with Bcl-2, may counteract the effects of Bcl-2 and promote apoptosis (Oltavi et al., 1993). Mice that are deficient for BAX display significant levels of lymphoid hyperplasia (Knudson et al., 1995). Another important family member, Bcl-x$_L$ acts in a similar fashion as Bcl-2 to promote cell survival (Boise et al., 1993), and mice deficient in Bcl-x die at day 13 of gestation and display massive cell death in hematopoietic tissue as well as in neuronal tissue of the spinal cord, brain, and dorsal root ganglia (Motoyama et al., 1995). The mechanism by which Bcl-2 or any of its related family members act to perturb cell death is unknown, but it is thought that these proteins exert their effects through complex interactions among family members (Yang et al., 1995; Sedlak et al., 1995).

Quite recently, the crystal structure of Bcl-x has been determined, and striking conformational similarity to pore-forming proteins such as bacterial toxins has been found (Muchmore et al., 1996). This conformational homology led the authors to propose that Bcl-x, Bcl-2, and other death-protecting family members might function by the ability to regulate pores in intracellular membranes, particularly the mitochondrial membrane. Such a notion is supported by the data from Kroemer and colleagues (to be discussed) that indicate a membrane permeability transition occurs during apoptosis in many different cell types. It may be that Bcl-2 acts to control the passage of molecules from this critical organelle to the cytosol. Experimental evidence from the Thompson laboratory supports this by demonstrating that in artificial lipid bilayers Bcl-2 can form an ion channel in synthetic lipid membranes (Minn et al., 1997: Birge et al., Chapter 13, this volume).

Yet another clue to Bcl-2 function comes from the recent and very exciting observation that the *C. elegans,* Bcl-2 homologue CED-9 interacts with a required worm cell death gene CED-4 and this CED-9/CED-4 complex interacts with and inactivates CED-3, the worm caspase protein. Thus in worms it appears that if CED-9 is present in a cell, it binds to CED-4, which can in turn bind CED-3, leading to the prevention of apoptosis through the prevention of caspase activity. In the absence of CED-9, CED-3 is free to mediate apoptosis (Chinnaiyan et al., 1997; Wu et al., 1997). This fits with the genetic evidence from worms that indicate death is a result of inactivation or lack of expression of CED-9 (Hengartner and Horvitz, 1992; Gartner and Hengartner, Chapter 4, this volume).

In addition to bcl-2 and its related family members, other proteins that can inhibit apoptosis have been described. For example, in the insect virus, baculovirus, genes that encode antiapoptotic proteins known as IAPs (for inhibitor of apoptosis protein) have been described. Recent data report the isolation of IAPs from mammalian cells, and it is reasonable to assume that these mammalian IAP-like proteins may also act to prevent or delay apoptosis in cells of the immune system (Uren et al., 1996).

OTHER MOLECULAR MEDIATORS OF LYMPHOID CELL DEATH

In addition to the genes described that affect primary and peripheral lymphoid cell death, other mediators of cell death are important to mention. For example, Kroemer and colleagues have shown that an early and uniform event that follows induction of apoptosis in primary and secondary T and B cells is the loss of mitochondrial membrane potential (reviewed in Kroemer et al., 1997). These authors also describe a convenient fluorescence assay for mitochondrial membrane potential that is quite useful as a measure of cell death. This group has gone on to show that if the loss of mitochondrial membrane potential is abrogated, cells can be rescued from apoptosis. Interestingly and perhaps relatedly, we found in our search for genes that are regulated during apoptosis in thymocytes that transcription of the mitochochondrial genome is profoundly and rapidly down-regulated following induction of death (Osborne et al., 1994; Trump and Berezesky, Chapter 2, this volume). Whether the effects of mitochondrial membrane transcription are causative or are side effects of the induction of death is unknown, but recent evidence from Kroemer and colleagues strongly suggest that loss of mitochondrial membrane potential is a required event during many forms of lymphoid cell death.

Another mediator of cell death in many systems including the immune system is molecular oxygen. The suggestion that bcl-2 may act as an antioxidant (Hockenberry, 1993) as well as data demonstrating that antioxidants protect many cells from apoptosis suggests that oxygen may be a required component of many cell-death pathways. We have shown that thymocytes require oxygen for the induction of cell death by glucocorticoids, PMA + Ca^{2+} ionophore, or radiation, and inhibition of oxygen-dependent deaths by the antioxidant N-acetylcysteine protects thymocytes from deletion by the superantigen Staphylococcus enterotoxin B (SEB) both *in vitro* and *in vivo* (McLaughlin et al., 1996).

PROTEASES AND CELL DEATH IN THE IMMUNE SYSTEM

The involvement of the CED-3/ICE family of cysteine proteases in apoptosis in both invertebrates and vertebrates suggest these proteases are common effectors of many forms of cell death (reviewed in Yuan, 1995). Henkart and colleagues have shown that inhibition of both cysteine and serine proteases will block cell death in mature T cells as well as T-cell hybridomas (Sarin et al., 1993; Sarin et al., 1995). It also is known that thymocytes from mice deficient in ICE are not susceptible to death induced by anti-Fas (Kuida et al., 1995); however, the lack of ICE in these mice does appear to affect physiological cell death processes in the thymus. The precise identity of the ICE family member active in T-cell death is controversial. Granzyme B is known to activate one family member, CPP32 (Darmon et al., 1995), while other data indicate granzyme B also can activate ICE (Shi, 1996). In extracts of Fas-triggered Jurkat cells, ICE has been identified by one group (Enari et al., 1995), but others have found CPP32 activity (Schlegel et al., 1995). Yet another group has data indicating a new ICE family member, ICE-LAP3, is induced, activated, and required for Fas-mediated apoptosis (Duan et al.,

1996). The role of ICE-like proteases in apoptosis of freshly isolated T- and B-cell populations is not known at present, but data indicate that CPP32-like activity is expressed in the thymus. Thus, while it is not known precisely what ICE family members are important in cell death in lymphoid cells, it is highly likely, indeed certain, that one or more of these important proteases contribute to the cell death of immunocytes.

The ICE family of proteases is a required component of most, if not all, death pathways, but the mechanism by which these proteins, always present in cells, is activated so that they can function in apoptosis is elusive. In an attempt to explore the activation of ICE proteases, the caspases, we asked whether the multicatalytic protease known as the proteasome might play an important role in this critical step during apoptosis. The 26S proteasome is a conserved multicatalytic proteolytic complex present in all eukaryotic cells (Goldberg et al., 1995; Coux et al., 1996) that is responsible for the degradation of most cellular proteins (Rock et al., 1994). The 26S proteasome is composed of a 20S catalytic core and associated regulatory proteins. This complex plays a critical role in the ubiquitin-proteasome-dependent proteolytic pathway, where it catalyzes the rapid degradation of proteins covalently linked to chains of ubiquitin. This pathway is highly regulated and selective and regulates many important cellular processes such as cell cycle progression (Glotzer et al., 1991, Pagano et al., 1995).

In the proteasome, proteins are generally hydrolyzed to small peptides, most of which are degraded further to amino acids by cellular exopeptidases, while some are utilized in antigen presentation. However, it has also recently been demonstrated that the 26S proteasome can also catalyze limited proteolytic processing of inactive precursors to active forms. For example, the generation of the active form of NFκB involves ubiquitination and the proteolytic processing of inactive 105K precursors to the active 50K form by the 26S proteasome (Palombella et al., 1995). Thus this structure may also play a role in the signal-induced proteolytic cascades that mediate cell death. Because of the involvement of proteolytic events in cell death and the known function of the proteasome in proteolytic processes, we asked whether the proteasome also plays a central role in cell death.

Our data (Grimm et al., 1996) show that the proteasome is required for most forms of apoptosis in thymocytes. Furthermore our data suggest that the proteasome functions upstream of the activation of the caspases since PARP, a direct target of CPP32, is not cleaved in cells where the proteasome is inhibited. More recent and unpublished data show that in the presence of proteasome inhibitors, CPP32 is not activated, providing further evidence that the proteasome is functioning upstream of the activation of CPP32. Similar complementary data were reported by Sadoul et al. (1996), who showed that the proteasome is required for the induction of apoptosis in neuronal cells following nerve growth factor withdrawal, indicating that proteasome function may be a common feature of many kinds of cell death (Sikorska and Walker, Chapter 7, and Cryns and Yuan, Chapter 6, this volume).

CELL DEATH AND DISEASE IN THE IMMUNE SYSTEM

As might be predicted, defects in normal cell death can lead to disease. Perhaps the best example of this comes from studies of two groups of patients, each with mutations in Fas. The individuals in both groups are, for the most part, quite ill and have lymphoproliferative disorders reminiscent of those observed in *lpr/lpr* mice (Fisher et al., 1995; Rieux et al., 1995). Another role played by Fas in disease is the function performed by FasL in sites of immune privilege such as the eye and the testis. Recent data have shown that FasL expression in these tissues is high and suggest that such expression can prevent damage caused by activated T cells to these tissues (Griffin et al., 1995; Bellgrau et al., 1995). For example, although most tissue can tolerate the nonspecific damage caused by inflammatory responses, delicate tissues such as the eye and the testis can suffer irreparable damage from inflammation. One mechanism to prevent this damage is through the prevention of inflammation. This seems to occur in these two tissues, since the high level of FasL expression can cause the induction of apoptosis in any cell in the vicinity that expresses elevated Fas.

Finally, the death of CD4+ T cells observed in HIV-infected individuals is known to occur by apoptosis (Gougeon et al., 1993; Ameisen et al., 1995). (Finkel and Casella, Chapter 11, this volume). Cell death has been observed in both infected and uninfected cells, and recent data from Finkle and colleagues suggest that, *in vivo,* apoptosis is induced primarily in bystander cells by infected neighboring cells (Finkel and Casella, Chapter 11, this volume). The role of cell death in the pathogenesis of AIDS has not yet been determined.

CONCLUSIONS AND SUMMARY

This chapter has focused on the events that occur when lymphoid cells are induced to undergo apoptosis. What the accumulated data indicate is that lymphocytes can be killed by a variety of induction strategies and each of these inducing signals engages unique pathways that ultimately lead to apoptosis (Fig. 8.2). We suggest that private entries into the apoptotic pathway probably intersect and these multiple private pathways merge into one or a few shared series of events. The ability to manipulate private pathways offers us unique advantages to orchestrate cell death in a very controlled manner. For example, one can envision clinical strategies where one might induce cell death in T lymphocytes through the manipulation of the TCR pathway. Such a strategy could induce death in a subset of activated peripheral T cells and leave intact hematopoietic precursors. Indeed, the employment of more general strategies might induce apoptosis in large population of cells both lymphoid and nonlymphoid and result in inappropriate deaths. The accumulating data regarding cell-death pathways in lymphocytes promise to provide the information necessary to allow appropriate clinical control of cell death in this important population of cells.

ACKNOWLEDGMENTS

I thank Richard Goldsby for his thoughtful and constructive comments and Lisa Grimm for her design of Figs. 8.1 and 8.2.

REFERENCES

Adachi M, Suematsu S, Suda T, Watanabe D, Fukuyama H, Ogasawara J, Tanaka T, Yoshida N, Nagata S (1996): Enhanced and accelerated lymphoproliferation in Fas-null mice. Proc Natl Acad Sci USA 93:2131–36.

Adachi M, Suematsu S, Kondo T, Ogasawara U, Tanaka T, Yoshida N, Nagata S (1995): Targeted mutation in the Fas gene cause hyperplasia in the peripheral lymphoid organs and liver. Nature Genetics 11:294–300.

Allen PM (1994): Peptides in positive and negative selection: A delicate balance. Cell 76:593–96.

Ameisen JC, Estaquier J, Idziorek T, De Bels F (1995): The relevance of apoptosis to AIDS pathogenesis. Trends Cell Biol 5:27–32.

Amsen D, Kruisbeek AM (1996): CD28-B7 interactions function to co-stimulate clonal deletion of double-positive thymocytes. Int Immunol 8:1927–36.

Ashton-Rickardt PG, Bandeira A, Delaney JR, Van Kaer L, Pircher HP, Zinkernagel RM, Tonegawa S (1994): Evidence for a differential avidity model of T cell selection in the thymus Cell 76:651–63.

Bellgrau D, Gold D, Delawry H, Moore J, Franzosoff A, Duke RD (1995): A role for CD95 ligand in preventing graft rejection. Nature 337:630–32.

Berke G: The CTL's kiss of death. Cell 1995, 81:9–12.

Bodmer, J-L, Burns K, Schneider P, Hofman K, Steiner V, Thome M, Bornard T, Hahne M, Schroter M, Becker K, Wilson A, French LE, Browning JL, MacDonald HR, Tschopp J (1997): TRAMP, a novel apoptosis-mediating receptor with sequence homology to tumor necrosis factor receptor 1 and Fas(Apo-1)/CD95. Immunity 6:79–88.

Boise LH, Gonzalez-Garcia M, Postema CE, Ding L, Lindsten T, Turka LA, Mao X, Nunez G, Thompson C (1993): Bcl-x, a bcl-2-related gene that functions as a dominant regulator of apoptotic cell death. Cell 74:597–608.

Boldin MP, Goncharov TM, Goltzev YV, Wallach D (1996): Involvement of MACH, a novel MORT1/FADD-interacting protease, in Fas/APO-1- and TNF receptor-induced cell death. Cell 85:803–15.

Brunner T, Mogil RJ, LaFace D, Yoo NJ, Mahoubl A, Echeverri F, Martin SJ, Force WR, Lynch DH, Ware CF, Green DR (1995): Cell-autonomous Fas (CD95)/Fas-ligand interaction mediates activation-induced apoptosis in T-cell hybridomas. Nature 373:441–44.

Calnan BJ, Szychowski S, Chan FK-M, Cado D, Winoto A (1995): A role for the orphan steroid receptor Nur77 in apoptosis accompanying antigen-induced negative selection. Immunity 3:273–82.

Chen G, Shi L, Litchfield DW, Greenberg AH (1995): Rescue from granzyme B-induced apoptosis by Wee1 kinase. J Exp Med 181:2295–300.

Chinnaiyan AM, O'Rouke K, Lane BR, Dixit VM (1997): Interaction of CED-4 with CED3 and CED-9: A molecular framework for cell death. Science 275:1122–26.

Clarke AR, Purdie CA, Harrison DJ, Morris RG, Bird CC, Hooper ML, Wyllie AH (1993): Thymocyte apoptosis induced by p53-dependent and independent pathways. Nature 362:849–52.

Cory S: BCL-2 family and lymphocyte survival (1995): Ann Rev Immunol 13:513–43.

Coux O, Tanaka K, Goldberg AL (1996): Structure and functions of the 20S and 26S proteasomes. Ann Rev Biochem 65:801–47.

Darmon AJ, Nicholson DW, Bleackley RC (1995): Activation of the apoptotic protease CPP32 by cytotoxic T-cell-derived granzyme B. Nature 377:446–48.

Dhein J, Walczak H, Baumler C, Debatin K-M, Krammer PH (1995): Autocrine T-cell suicide mediated by APO-1/(Fas/CD95). Nature 373:438–41.

Dieken ES, Miesfeld RL (1992): Transcriptional transactivation functions localized to the glucocorticoid receptor N terminus are necessary for steroid induction of lymphocyte apoptosis. Mol Cell Biol 12:589–97.

Duan H, Chinnaiyan AM, Hudson PL, Wing JP, He WW, Dixit VM (1996): ICE-LAP3, a novel mammalian homolog of the Caenorhabditis elegans cell death protein CED-3 is activated during Fas- and tumor necrosis factor-induced apoptosis. J Biol Chem 271:1621–25.

Ellis R, Yuan J, Horvitz HR (1991): Mechanisms and functions of cell death. Ann Revs Cell Biol 7:663–96.

Enari M, Hase A, Nagata S (1995): Apoptosis by a cytosolic extract from Fas-activated cells. EMBO J 14:5201–8.

Finkel TH, Tudor-Williams G, Banda NK, Cotton MF, Curiel T, Monks C, Baba TW, Ruprecht RM, Kupfer A (1995): Apoptosis occurs predominately in bystander cells and not in productively infected cells of HIV- and SIV-infected lymph nodes. Nature Medicine 1:129–34.

Fisher G, Rosenberg FJ, Straus SE, Dale JK, Middleton LA, Lin AY, Strober W, Lenardo MJ, Puck JM (1995): Dominant interferring Fas gene mutations impair apoptosis in human autoimmune lymphoproliferative syndrome. Cell 81:935–46.

Foy TM, Page DM, Waldschmidt TJ, Schoneveld A, Laman JD, Masters SR, Tygrett L, Ledbetter JA, Aruffo A, Claassen E, Xu JC, Flavell RA, Oehen S, Hedrick SM, Noelle RJ (1995): An essential role for gp39, the ligand for CD40, in thymic selection. J Exp Med 182:1377–88.

Giese T, Davidson WF (1994): Chronic treatment of C3H-lpr/lpr and C3H-gld/gld mice with anti-CD8 monoclonal antibody prevents the accumulation of double negative T cells but not autoantibody production. J Immunol 152:2000–10.

Glotzer M, Murray A, Kirschner, M (1991): Cyclin is degraded by the ubiquitin pathway. Nature 349:132–38.

Goldberg AL, Ross S, Adams, J (1995): New insights into proteasome function: From archeaebacteria to drug development. Chem Biol 2:503–8.

Gougeon ML, Laurent CA, Hovanessian AG, Montagnier L (1993): Direct and indirect mechanisms mediating apoptosis during HIV infection: Contribution to in vivo CD4 T cell depletion. Semin Immunol 5:187–94.

Griffin TS, Brunner T, Fletcher SM, Green DR, Ferguson TA (1995): Fas ligand-induced apoptosis as a mechanism of immune privilege. Science 270:1189–92.

Grimm LA, Goldberg AL, Poirier GG, Schwartz LM, Osborne BA (1996): Proteasomes play an essential role in thymocyte apoptosis. EMBO J 15:3835–44.

Han S, Zheng B, Dal Porto J, Kelsoe G (1995): In situ studies of the primary immune response to (4-hydroxy-3-nitrophenyl)acetyl IV. Affinity dependent, antigen driven B cell apoptosis in germinal centers as a mechanism for maintaining self-tolerance. J Exp Med 182:1635–44.

Hengartner MO, Horvitz HR (1992): Caenorhabditis elegans gene ced-9 protects cells from programmed cell death. Nature 356:494–99.

Henkart PA, Sitkovsky MV (1994): Cytotoxic lymphocytes. Two ways to kill target cells. Curr Biol 4:923–25.

Heusel JW, Wesselschmidt RL, Shresta S, Russell JH, Ley TJ (1994): Cytotoxic lymphocytes require granzyme B for the rapid induction of DNA fragmentation and apoptosis in allogeneic target cells. Cell 76:977–87.

Hockenberry DM, Oltvai ZN, Yin XM, Milliman CL, Korsmeyer SJ (1993): Bcl-2 functions in an antioxidant pathway to prevent apoptosis. Cell 75:241–51.

Hogquist KA, Jameson SC, Heath WR, Howard JL, Bevan MJ, Carbone FR (1994): T cell receptor antagonist peptides induce positive selection. Cell 76:17–27.

Iwata M, Hanaoka S, Sato K (1991): Rescue of thymocytes and T cell hybridomas from glucocorticoid-induced apoptosis by stimulation via the T cell receptor/CD3 complex: A possible in vitro model for positive selection of the T cell repertoire. Eur J Immunol 21:643–48.

Ju S-T, Panka DJ, Cul H, Ettinger R, El-Khatib M, Sherr DH, Stanger BZ, Marshak-Rothstein A (1995): Fas (CD95)/FasL interactions required for programmed cell death after T-cell activation. Nature 373:333–48.

Kagi D, Lederman B, Bürki K, Seiler P, Odermatt B, Olsen KJ, Podack ER, Zingernagel RM, Hengartner H (1994): Cytotoxicity mediated by T cells and natural killer cells is greatly impaired in perforin-deficient mice. Nature 369:31–37.

Kerr JFR, Wyllie AH, Currie A (1972): Apoptosis: A basic biological phenomenon with wide-ranging implications in tissue kinetics. Br J Cancer 26:239–57.

King L, Vacchio MS, Hunziker R, Margulies DH, Ashwell JD (1995): A targeted glucocorticoid receptor antisense transgene increases thymocyte apoptosis and alters thymocyte development. Immunity 5:647–56.

Knudson CM, Tung KSK, Tourtellotte WG, Brown GAJ, Korsmeyer SJ (1995): Bax-deficient mice with lymphoid hyperplasia and male germ cell death. Science 270:96–99.

Krammer PH (1996): The CD95 (APO-1/Fas) receptor/ligand system: death signals and diseases. Cell Death Differ 3:159–60.

Kroemer G, Zamzami N, Susin SA (1997): Mitochondrial control of apoptosis. Immunol Today 18:44–51.

Kuida K, Lippke JA, Ku G, Harding MW, Livingston DJ, Su MS, Flavell RA (1995): Altered cytokine export and apoptosis in mice deficient in interleukin-1 beta converting enzyme. Science 267:2000–3.

Lee SL, Wesselscmidt RL, Linette GP, Kanagawa O, Russell JH, Milbrandt J (1995): Unimpaired thymic and peripheral T cell death in mice lacking the nuclear receptor NGF1-B (Nur77). Science 269:532–35.

Linette GP, Grusby MJ, Hedrick SM, Hansen TH, Glimcher LH, Korsmeyer SJ (1994): Bcl-2 is upregulated at the CD4+ CD8+ stage during positive selection and promotes thymocyte differentiation at several control points. Immunity 1:197–205.

Liu ZG, Smith SW, McLauglin KA, Schwartz LM, Osborne BA (1994): Apoptotic signals delivered through the T-cell receptor require the immediate-early gene nur77. Nature 367:281–84.

Lowe SW, Schmidtt EM, Smith SW, Osborne BA, Jack T (1993): p53 is required for radiation-induced apoptosis in mouse thymocytes. Nature 362:847–49.

Lowin B, Hahne M, Mattman C, Tschopp J (1994): Cytolytic T-cell cytotoxicity is mediated through perforin and Fas lytic pathways. Nature 370:650–52.

MacDonald HR, Lees RK (1990): Programmed death of autoreactive thymocytes. Nature 43:624–46.

McLaughlin K, Osborne BA, Goldsby RA (1996): The role of oxygen in thymocyte apoptosis. Eur J Immunol 26:1170–75.

Minn AJ, Velez P, Schendel SL, Liang H, Muchmore SW, Fesik SW, Fill M, Thompson CB (1997): Bcl-x(L) forms anion channel in synthetic lipid membranes. Nature 385:353–57.

Motoyama N, Wang F, Roth KA, Sawa H, Nakayama K, Nakayama K, Negishi I, Senju S, Zhang Q, Fujii S, Loh DY (1995): Massive cell death of immature hematopoietic cells and neurons in Bcl-X-deficient mice. Science 267:1506–10.

Muchmore SW, Sattler M, Liang H, Meadows RP, Harlan JE, Yoon HS, Nettesheim D, Chang BS, Thompson CB, Wong SL, Ng SL, Fesik SW (1996): X-ray and NMR structure of human Bcl-xL, an inhibitor of programmed cell death. Nature 381:335–41.

Murphy KM, Heimberger AB, Loh DY (1990): Induction by antigen of intrathymic apoptosis of CD4+CD8+TCRlo thymocytes in vivo. Science 250:1720–23.

Muzio M, Chinnaiyan AM, Kischkel FC, O'Rourke K, Shevchenko A, Ni J, Scaffidi C, Bretz JD, Zhang M, Gentz R, Mann M, Krammer PH, Peter ME, Dixit VM (1996): FLICE, a novel FADD-homologous ICE/CED-3-like protease, is recruited to the CD95 (Fas/APO-1) death-inducing signaling complex. Cell 85:817–27.

Nagata S, Golstein P (1995): The Fas death factor. Science 267:1449–56.

Nossal GJV (1994): Negative selection of lymphocytes. Cell 76:229–39.

Oltvai ZN, Milliman CL, Korsmeyer SJ (1993): Bcl-2 heterodimerizes in vivo with a conserved homolog, Bax, that accelerates programmed cell death. Cell 74:609–19.

Osborne BA, Smith SW, McLauglin KA, Grimm L, Kallinch T, Schwartz LM (1994): Identification of genes induced during apoptosis in T lymphocytes. Immunol Rev 142:301–20.

Pagano M, Tam SW, Theodoras AM, Beer-Romero, Del Sal G, Chau V, Yew PR, Draetta GF, Rolfe M (1995): Role of the ubiquitin-proteasome pathway in regulating abundance of the cyclin-dependent kinase inhibitor p27. Science 269:682–85.

Page DM, Kane, LP, Allison JP, Hedrick SM (1993): Two signals are required for negative selection of CD4+ CD8+ thymocytes. J Immunol 151:1868–80.

Palombella VJ, Rando OJ, Goldberg AL, Mantiatis T (1994): The ubiquitin-proteasome pathway is required for processing the NF-κB1 precursor protein and the activation of NF-κB. Cell 78:773–85.

Pulendran B, Kannourakis G, Nouri S, Smith KGC, Nossal GJV (1995): Soluble antigen can cause enhanced apoptosis of germinal-centre B cells. Nature 375:331–34.

Punt JA, Osborne BA, Takahama Y, Sharrow SO, Singer A (1994): Negative selection of CD4$^+$CD8$^+$ thymocytes by T-cell receptor-induced apoptosis requires a costimulatory signal that can be provided by CD28. J Exp Med 179:709–13.

Ramsdell F, Seaman MS, Miller RE, Tough TW, Alderson MR, Lynch DH (1994): gld/gld mice are unable to express a functional ligand for Fas. Eur J Immunol 24:928–33.

Rathmell JC, Cooke MP, Ho WY, Grein J, Townsend SE, Davis MM, Goodnow CC (1995): CD95 (Fas)-dependent elimination of self-reactive B cells upon interaction with CD4$^+$ T cells. Nature 376:181–84.

Rieux-Laucat F, Le Deist F, Hivroz C, Roberts IAG, Debatin KM, Fisher A, de Villartay JP (1995): Mutations in Fas associated with human lymphoproliferative syndrome and autoimmunity. Science 268:1347–50.

Rock KL, Gramm C, Rothstein L, Clark K, Skin R, Dick L, Hwang D, Goldberg AL (1994): Inhibitors of the proteasome block the degradation of most cell proteins and the generation of peptides presented in MHC class I molecules. Cell 78:761–71.

Rothstein TL, Wang JKM, Panka DJ, Foote LC, Wang Z, Stanger B, Cul H, Ju S-T, Marshak-Rothstein A (1995): Protection against Fas-dependent Th1-mediated apoptosis by antigen receptor engagement in B cells. Nature 374:163–65.

Rouvier E, Luciani M-F, Golstein P (1993): Fas involvement in Ca^{2+}-independent T cell-mediated cytotoxicity. J Exp Med 177:195–200.

Russell JH, Wang R (1993): Autoimmune gld mutation uncouples suicide and cytokine/proliferation pathways in activated mature T cells. Eur J Immunol 23:2379–82.

Russell JH, Rush B, Weaver C, Wang R (1993): Mature T cells of autoimmune lpr/lpr mice have a defect in antigen-stimulated suicide. Proc Natl Acad Sci USA 90:4409–13.

Sadoul R, Fernandez PA, Quiquerez AL, Martinov I, Maki M, Schroter M, Becherer JD, Immler M, Tschopp J, Martinou JC (1996): Involvement of the proteasome in the programmed cell death of NGF-deprived sympathetic neurons. EMBO J. 15:3845–52.

Sarin A, Nakajima H, Henkart PA (1995): A protease dependent TCR-induced death pathway in mature lymphocytes. J Immunol 154:5806–12.

Sarin A, Adams DH, Henkart PA (1993): Protease inhibitors selectively block T-cell receptor-triggered programmed cell death in a murine T cell hybridoma and activated peripheral T cells. J Exp Med 178:1693–1700.

Sarin A, Williams MS, Alexander-Miller MA, Berzofsky J, Zacharchuk CM, Henkart PA (1997): Target cell lysis by CTL granule exocytosis is independent of ICE/Ced-3 family proteases. Immunity 6:209–15.

Schlegel J, Peters I, Orrenius S, Miller DK, Thornberry NA, Yamin TT, Nicholson DW (1995): CPP32/Apopain is the ICE-like protease involved in Fas-mediated apoptosis. J Biol Chem 271:1841–44.

Sebzda E, Wallace VA, Mayer J, Young RSM, Mak TW, Ohashi PS (1994): Mature T cell reactivity altered by peptide agonist that induces positive selection. Science 263:1615–18.

Sedlak TW, Oltavi ZN, Yang E, Wang K, Boise LH, Thompson CB, Korsmeyer SJ (1995): Multiple Bcl-2 family members demonstrate selective dimerizations with Bax. Proc Natl Acad Sci USA 92:7834–38.

Sentman CL, Shutter JR, Hockenberry D, Kanagawa O, Korsmeyer SJ (1991): bcl-2 inhibits multiple forms of apoptosis but not negative selection in thymocytes. Cell 67:879–88.

Shi L, Nishioka WK, Th'ng J, Bradbury EM, Litchfield DW, Greenberg AH (1994): Premature p34^{cdc2} activation required for apoptosis. Science 263:1143–46.

Shi L, Chen G, MacDonald G, Bergeron L, Li H, Miura M, Rotello RJ, Miller DK, Li P, Seshadri T, Yuan J, Greenberg AH (1996): Activation of an interleukin 1 converting enzyme-dependent apoptosis pathway by granzyme B. Proc Natl Acad Sci USA 93:11002–7.

Shokat KM, Goodnow CG (1995): Antigen-induced B-cell death and elimination during germinal-centre immune responses. Nature 375:334–38.

Smith CA, Williams G, Kingston R, Jenkinson EJ, Owen JT (1989): Antibodies to CD3/T-cell receptor complex induces death by apoptosis in immature T cells in thymic culture. Nature 338:181–83.

Strasser A (1995): Life and death during lymphocyte development and function: Evidence for two distinct killing mechanisms. Curr Opin Immunol 7:228–34.

Strasser A, Harris AW, Cory S (1991): bcl-2 transgene inhibits T cell death and perturbs thymic self-censorship. Cell 67:889–99.

Surh CD, Sprent J (1994): T-cell apoptosis detected in situ during positive and negative selection in the thymus. Nature 372:100–3.

Sytwu HK, Liblau RS, McDevitt HO (1996): The roles of Fas/APO-1 (CD95) and TNF in antigen-induced programmed cell death in T cell receptor transgenic mice. Immunity 5:17–30.

Thompson CB (1995): Apoptosis in the pathogenesis and treatment of disease. Science 267:1456–62.

Tsujimoto Y, Finger LR, Yunis J, Nowell PC, Croce CM (1984): Cloning of the chromosome breakpoint of neoplastic B cells with the t(14:18) translocation. Science 226:1097–99.

Uren AG, Pakusch M, Hawkins CJ, Puls KL, Vaux DL (1996): Cloning and expression of apoptosis inhibitory protein homologs that function to inhibit apoptosis and/or bind tumor necrosis factor receptor-associated factors. Proc Natl Acad Sci USA 93:4974–78.

Vacchio MS, Papadopoulos V, Ashwell JD (1994): Steroid production in the thymus: Implications for thymocyte selection. J Exp Med 179:1835–46.

Vaux DL, Cory S, Adams JM (1988): Bcl-2 gene promotes haemopoietic cell survival and cooperates with c-myc to immortalize pre-B cells. Nature 335:440–42.

Veis DJ, Sorenson CM, Shutter JR, Korsmeyer SJ (1993): Bcl-2-deficient mice demonstrate fulminant lymphoid apoptosis, polycystic kidneys, and hypopigmented hair. Cell 75:229–40.

von Boehmer H (1994): Positive selection of thymocytes. Cell 76:219–28.

Watanabe D, Suda T, Nagata S (1995): Expression of Fas in B cells of the mouse germinal center and Fas-dependent killing of activated B cells. Int Immunol 7:1949–56.

Weih F, Ryseck RP, Chen L, Bravo R (1996): Apoptosis of nur77/N10-transgenic thymocytes involves the Fas/Fas ligand pathway. Proc Natl Acad Sci 93:5533–37.

Woronicz JD, Calnan B, Ngo V, Winoto A (1994): A requirement for the orphan steroid receptor Nur77 in apoptosis of T cell hybridomas. Nature 367:277–81.

Wu D, Wallen HD, Nunez G (1997): Interaction and regulation of subcellular localization of CED-4 by CED-9. Science 275:1126–29.

Wyllie AH (1980): Glucocorticoid induced thymocyte apoptosis is associated with endogenous endonuclease activation. Nature 284:555–56.

Yang E, Zha J, Jockel J, Boise LH, Thompson CB, Korsmeyer SJ (1995): Bad, a heterodimeric partner for Bcl-x_L and Bcl-2, displaces Bax and promotes cell death. Cell 80:285–91.

Yuan J (1995): Molecular control of life and death. Curr Opin Cell Biol 7:211–14.

Zacharchuk CM, Mercep M, Chakaborti PK, Simons S Jr, Ashwell JD (1990): Variations in thymocytes susceptibility to clonal deletion during: Implications for neonatal tolerance. J Immunol 145:4037–45.

Zamzami N, Marchetti P, Castedo M, Zanin C, Vayssière JL, Petir PX, Kroemer G (1995): Reduction in mitochondrial potential constitutes an early irreversible step of programmed lymphocyte death in vivo. J Exp Med 181:1661–72.

Zheng L, Fisher G, Miller RE, Peschon J, Lynch DH, Lenardo MJ (1995): Induction of apoptosis in mature T cells by tumor necrosis factor. Nature 377:348–51.

Zhou T, Cheng J, Yang P, Wang Z, Liu C, Su X, Bleuthmann H, Mountz JD (1996): Inhibition of Nur77/Nurr1 leads to inefficient clonal deletion of self-reactive T cells. J Exp Med 183:1879–92.

CHAPTER 9

REGULATION OF IMMUNE RESPONSES BY IMMUNE DIRECTED CELL DEATH

M. KAREN NEWELL and MICHAEL S. VINCENT
Division of Immunology, Dept. of Medicine, University of Vermont College of
Medicine, Burlington VT 05405

AN INTRODUCTION TO THE IMMUNE SYSTEM

The primary function of an immune response is self-versus non-self-discrimination. The evolutionary advantage is survival from invading pathogens. However, an equally important selective advantage may be the ability of the immune system to recognize changes in self when damage has occurred. Damage or infection can then be dealt with by removal of that which is recognized, either pathogens or damaged self-tissue. Taken together, the immune system has been compared to a sensory organ, not responding to taste, light, or sound, but rather to changes in the cellular and molecular environment (Cohen, 1996). The focus of this chapter is on how immune-directed cell death may provide a mechanism that is the basis for self-versus non-self-discrimination and potentially the immunorecognition and removal of damaged cells.

It is necessary to begin with an overview of how the immune system functions. Clearly, any vertebrate born without a competent immune system soon dies unless measures are immediately taken to isolate it from its environment. Invertebrates also have strategies to avoid pathogens, including the use of phagocytic cells that engulf either invading agents or dying cells. The same kinds of cells, the phagocytes, also play an important role in protecting vertebrates from infection. In fact, two types of cells divide the work of the vertebrate immune system, phagocytes and lymphocytes.

When Cells Die, Edited by Richard A. Lockshin, Zahra Zakeri,
and Jonathan L. Tilly
ISBN 0-471-16569-7 © 1998 Wiley-Liss, Inc.

Lymphocytes provide specific recognition in an immune response. They develop in the central lymphoid organs, which include the thymus and the bone marrow. From there, the lymphocytes exit into the peripheral compartment, where the lymph nodes, spleen, and gut, as well as tonsils and appendix, provide the peripheral lymphoid tissues. Peripheral lymphocytes express receptors that very specifically recognize antigens. While there may be millions of antigenic "specificities" that can be recognized by lymphocytes and thousands of copies of the receptor on a single cell, all the receptors on a given cell are identical (Cohen, 1996). Therefore, to generate a functional repertoire of receptors for the number of antigens that an animal may encounter necessitates many different receptors and the ability to expand the number of those receptors that can recognize the antigens of interest.

Lymphocytes can be grouped into two categories by the manner in which the cell recognizes antigen: B lymphocytes, which are the precursors of antibody secreting cells; and T lymphocytes, which recognize changes at the cellular level (Claman and Chaperon, 1969). B cells recognize antigens that are not necessarily cell associated. The B-cell antigen receptor is a cell-bound version of the antibody molecule that the B cell will eventually secrete. Antibodies are soluble products that can specifically neutralize toxins, viruses, and coat bacteria in preparation for engulfment by phagocytic cells. The binding of antigen to the B cell causes the B cell to become activated and, in the right milieu, to proliferate, thereby expanding the number of cells that will eventually secrete antibody. The T-cell receptor for antigen recognizes antigens only when the antigen is associated with self molecules on surrounding cells.

CELL DEATH IN THE DEVELOPING IMMUNE SYSTEM

The bone marrow and the thymus were the first organs where immune-directed cell death was proposed to play a role in preserving self- versus non-self-discrimination. B lymphocytes mature in the bone marrow. Nossal proposed, and he and others eventually demonstrated, that binding to self-antigens in the bone marrow resulted in B-cell death, thereby removing self-reactive B cells from the pool of B cells entering the periphery (Nossal, 1994; Hasbold and Klaus, 1990). This process has been termed "clonal abortion." The T-cell antigen receptor repertoire in the periphery is determined during maturation of the developing cells, which occurs primarily in the thymus (von Boehmer, 1994). Experimentally, immature thymocytes can be induced to undergo apoptosis by engagement of the T-cell receptor, irradiation, or treatment with glucocorticoids (Zacharchuk et al., 1991; Sellins and Cohen, 1987; Cohen and Duke, 1984). These observations mark the first demonstration of cell death as pivotal in shaping an immune response. A summary of this process which can serve as a reference for the ensuing discussion, is presented as Figure 9.1. These processes are extensively discussed by Osborne (Chapter

8, this volume), Budd (Chapter 10, this volume), and Finkel and Casella (Chapter 11, this volume).

THE ROLE OF MAJOR HISTOCOMPATIBILITY ENCODED MOLECULES IN CELL DEATH

The discovery, over 50 years ago, of major histocompatibility complex (MHC)-encoded molecules defined these cell surface antigens as the elements that controlled tumor rejection between distinct members of the same species (Snell, 1981). In the intervening years, the same molecules were demonstrated to restrict the activation of T lymphocytes to antigens that were associated with them, thus serving to define self- versus non-self at the level of the T cell (Doherty and Zinkernagel, 1975). In fact, by genetic mapping of responses to a variety of infectious agents, these molecules were described as immune response, or *ir,* genes. More recently, it has become apparent that MHC molecules serve an additional function, that being signal transducing receptors for the cell on which they are expressed. Perhaps not surprisingly, these molecules are also critically involved at many levels in life and death signals for cells during an immune response.

When phagocytic cells engulf antigens, the cells break down and digest the products of their engulfment, which results in the re-expression and association of breakdown products with newly expressed MHC (in this case, MHC class II) molecules on the cell surface. This process, known as antigen presentation (Allen and Unanue, 1987), appears to be required for the activation of CD4⁺, or helper, T cells. The engagement of the T-cell receptor for antigen and MHC molecules on the antigen-presenting cell provides for the first signal to the T cell in a cell-mediated immune response. A second signal is required for T-cell activation. It has been termed the requirement for "costimulation" (Bretscher and Cohn, 1970; Linsley and Ledbetter, 1993). The newly activated T cell thus begins to proliferate and differentiate into a cell that produces lymphokines or soluble mediators of growth and activation of other cells. Some helper T cells provide lymphokines and/or other co-factors necessary for B-cell activation and antibody secretion. Others provide lymphokines and/or other cofactors for the activation of "killer" T cells, CD8⁺ T lymphocytes (see Fig. 9.1).

CD8⁺ T-CELL-MEDIATED KILLING OF INFECTED TARGET CELLS

CD8⁺ T cells express T-cell receptors that recognize antigens processed differently than those antigens phagocytized by professional phagocytic cells. Specifically, CD8⁺ T cells recognize products of viral infections associated with any MHC (in this case MHC class I) bearing cell (Henkart, 1985). All nucleated cells express MHC class I and thus are capable of being recognized by CD8⁺

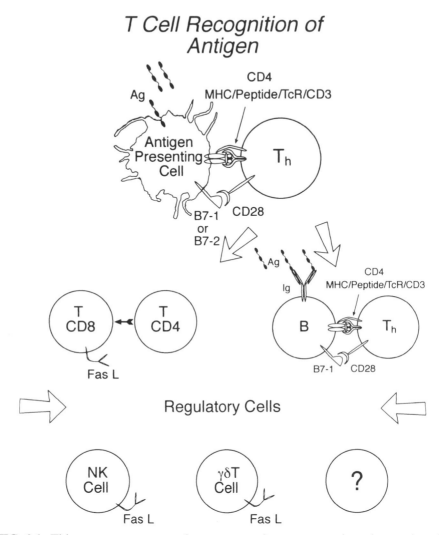

FIG. 9.1. This cartoon represents the sequence of events occurring when antigen is engulfed by phagocytes in the immune system, termed "professional" antigen-presenting cells. These cells have been ascribed the name because they have the capacity to engulf and break down antigens and re-express those changed molecules in association with MHC molecules on the antigen-presenting cell surface. T cells then recognize the antigens. Also required for T-cell activation are encounters between co-stimulatory molecules, here termed B7-1 or B7-2 and CD28. The result of T-cell activation is the expansion in numbers of antigen-specific T cells and a subdivision of labor. The helper, CD4+ T cell, then can go on to "help" B cells make antibody or "help" other T cells, the killer CD8+ T cell, to kill virally infected targets. In the lower panel, we have listed regulatory cells that may play a role in determining the outcome of a response by regulating which type of response occurs.

TABLE 9.1. Differing Responses of Proliferation and Death in Response to Stimulation

Target Cell	Effector	Response	Effector	Result
Chronic antigen stimulation or B-cell bystander	T-cell CD40L	B cell with increased Fas	T-cell FasL	B-cell death
Acute B-cell receptor stimulation		B cell with B7	T-cell cytokines	B-cell proliferation
CD4$^+$ T cell and $\gamma\delta$ T cell + antigen		$\gamma\delta$ T cell ↑ FasL/ CD4$^+$ T cell		CD4$^+$ T-cell death

T cells when infected. This is important to insure the removal of any infected tissue. CD8$^+$ T cell-mediated killing of targets requires this type of recognition. It is not known if the death of target cells results from CD8$^+$ T-cell-mediated induction of apoptosis or by release of perforin or granzyme by the CD8$^+$ T cell (Granger and Kolb, 1968; Kagi et al., 1994). The actual process may include elements of both or be tissue dependent. Conversely, it is also possible that the processes converge. What is clear is that the killer CD8$^+$ T cell can kill more than one target without dying itself. The CD4$^+$ T cell appears to take a different tack (see Table 9.1. and Flow Diagram below).

CELL DEATH IN PERIPHERAL IMMUNE FUNCTION

Evidence for a regulatory role being played by the death of peripheral T cells came in large part from studies of mouse strains that have lymphoproliferative defects. These mice accumulate large numbers of lymphocytes and manifest dysregulated immune responses. The study of these mice, termed *lpr* for

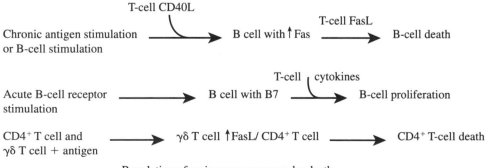

Regulation of an immune response by death

lyphoproliferative, and *gld* for generalized lymphoproliferative disorder, led to the discovery of the molecules responsible for their phenotype, namely Fas (CD95) and Fas ligand (FasL), respectively (Watanabe-Fukunaga et al., 1992; Suda et al., 1993; Budd, Chapter 10, this volume). The study of how these molecules shape an immune response has been explosive. It is now clear that Fas and its ligand are actively involved in apoptotic death of peripheral lymphoid, and potentially other, cells (Suda et al., 1995). In fact, not only can a cell bearing FasL, such as a CD8$^+$ T cell, a natural killer (NK), or a $\gamma\delta$ T cell, kill Fas-bearing targets; a single cell that has both molecules will also die (Brunner et al., 1995; Dhein et al., 1995). The regulation of the function of these molecules is only now being unravelled and will be discussed elsewhere (see Osborne, Chapter 8; Finkel and Casella, Chapter 11; Budd, Chapter 10; and Benitez, Chapter 17, this volume).

Given the numbers of lymphocytes that result from an immune response, it seems only logical that there must be a way of removing antigen-specific T cells once the response is complete. In fact, experimental evidence now suggests that a common fate for the CD4$^+$ T cell is apoptotic death once a response has subsided. However, there clearly exists a form of memory that is centrally important for becoming immune to a variety of pathogens. This would suggest that not all CD4$^+$ T cells die at the end of a response. The complex regulation of life and death signals that determine the fate of antigen specific CD4$^+$ T cells is one of vital interest.

Recent studies suggest that both T-cell receptor-mediated activation (Alderson et al., 1993) and T-cell activation induced death may involve Fas and FasL (Alderson et al., 1995). However, during the course of an immune response antigen-activated T cells express the Fas molecule, but are refractory to Fas-induced death for several days. This argues strongly in support of the notion that the presence of antigen may protect the responding T cell from Fas-mediated death. Likewise, growth factor withdrawal, which could also result from decreased antigen receptor engagement, results in apoptotic cell death of responding T cells (Duke and Cohen, 1986). These issues are currently under investigation in our laboratory and others.

The issue of lymphocyte death also involves many other receptor ligand pairs, including the CD4 molecule on the CD4$^+$ T cell as well as the MHC molecules *per se* on the cells on which these molecules are expressed. The role of these molecules cannot be excluded from a discussion of cell death of immune cells because the pathophysiological implications of these molecular interactions, or, alternatively, of failure of these interactions, including those occurring in HIV infections or in autoimmune diseases, directly implicate these receptor–ligand pairs (Budd, Chapter 10; Finkel and Casella, Chapter 11, this volume). In fact, our early work aimed at addressing the role of CD4 engagement in the activation of T cells led to the discovery that separation of signals generated by CD4 engagement and T-cell receptor engagement results in T-cell receptor-mediated death of the T cell (Newell et al., 1990). This was followed by experiments that indicated that parallel processes may account for the loss of CD4$^+$ T cells in HIV when gp120 from the virus has bound CD4 and subsequent antigen–receptor engagement occurs (Banda et al., 1992). This result will be discussed at great length by Finkel and Casella (Chapter 11, this volume).

We have also established that CD4 cross-linking increases the sensitivity of a subset of CD4$^+$ T cells to Fas-induced death signals, as well as rapidly increasing cell surface Fas expression. Our results demonstrate much earlier CD4$^+$ T-cell-specific effects of CD4 cross-linking. Both the rapid induction of Fas expression and Fas-mediated cell death occur selectively in CD4$^+$ T cells, since CD4$^-$ cells in unfractionated spleen cultures are unaffected. Furthermore, experiments with isolated CD4$^+$ T cells show that CD4-induced changes in Fas can occur independently of other cell types and their products (Desbarats et al., 1996).

MHC class II engagement on B lymphocytes prior to antigen–receptor engagement likewise results in death of B cells (Newell et al., 1993). However, in this case, at least for B cells on which the antigen receptor has not been engaged, no additional receptor engagement is necessary. Thus engagement of MHC class II on a B lymphocyte prior to antigen–receptor engagement would provide a safeguard against polyclonal, non-antigen-specific B-cell activation. This phenomenon is discussed further in the following. The prototype for this phenomenon occurs in autoimmune disease. Our laboratory is currently addressing the possibility that this is a predisposing factor in mouse models of systemic lupus erythematosus.

REGULATION OF IMMUNE-DIRECTED CELL DEATH

An emerging theme in the regulation of cell death during an immune response is the involvement of specific effector cells that may modulate the response by promoting the death of subsets of responding immune cells. The effector cells implicated include conventional T-cell subsets, as well as $\gamma\delta$ T cells and NK cells. In most cases it is unclear what the qualitative effect of this directed cytotoxicity is on the eventual immune response, but obviously the potential to delete subsets of immune cells is a powerful mechanism by which the immune response can be regulated.

Both CD4$^+$ and CD8$^+$ T cells can mediate the destruction of autologous immune cells, thereby modulating an immune response. The CD4$^+$ subset has been shown to play a critical immunomodulatory role in eliminating autoreactive B cells. In a B- and T-cell transgenic system, Goodnow and colleagues have shown that chronically activated B cells specific for hen egg lysozyme (HEL) are eliminated by HEL-specific CD4$^+$ T cells in a Fas-dependent fashion upon activation. In contrast, naive B cells proliferate and secrete Ig when they encounter HEL-specific T cells and antigen (Rathmell et al., 1995). These authors went on to show that the critical determinant of B-cell susceptibility to apoptosis was the timing of B-cell receptor stimulation along with costimulatory information from the T cell. With chronic antigenic stimulation, such as an autoantigen, cognate T-cell help provides a limited costimulus to the B cell, with subsequent B-cell up-regulation of Fas. The CD4[+] T cell provides this costimulus to the B cell by up-regulation of the tumor necrosis factor (TNF) superfamily member, CD40 ligand, which signals through B cell CD40. Ultimately the B cell undergoes Fas-mediated apoptosis as a result of the adjacent cognate T cell expressing FasL. A similar sequence ensues if the B-cell receptor is not engaged, resulting in the death of bystander B cells. When the B-cell receptor is acutely engaged, however, the B cell provides costimulation via up-regulation of B7-1. The T cell is then fully activated and secretes cytokines, and the B cell undergoes clonal expansion rather than deletion, despite the presence of Fas and FasL (Rathmell et al., 1996). In this way, a CD4[+] T cell recognizing its T-cell receptor ligand can effect diametrically opposed fates for the presenting B cell: death if the B cell recognizes nothing or self, and proliferation if it acutely recognizes a cognate ligand.

NATURAL KILLER-MEDIATED CELL DEATH

Early work on NK cell-dependent cytotoxicity was limited by the enigmatic nature of the activating ligands. It is now clear that the recognition process for the NK subset often involves MHC class I recognition by a unique subset of cell surface receptors (Lopez-Botet et al., 1996). These NK receptors function by preventing cytolytic activity towards targets encountered by the NK cell, provided that these inhibitory receptors recognize the appropriate MHC class I ligand. Consequently, cells will be eliminated if they have been mutated or altered in some fashion so that they do not express the same MHC molecules that the remainder of the organism's cells express.

The characterization of NK inhibitory receptors has moved at a rapid pace in the last year (Lopez-Botet et al., 1996). Remarkably, the predominant human NK receptor family is composed of type I transmembrane glycoproteins containing Ig superfamily domains, while the predominant rodent receptor family consists of type II transmembrane glycoproteins containing a C-type lectin domain. Despite their differences, members of both families mediate fine specificity among various MHC class I molecules (Lopez-Botet et al., 1996). Interestingly, many surface molecules traditionally associated with the NK subset are also found on subsets of CD8[+] T cells and γδ T cells, implying the capability of recognizing different class I molecules in these cells as well. It is not yet clear how information emanating from T-cell receptor will be integrated with NK receptor signals in these latter subsets. The activating receptors for NK cells are not as well characterized, but a conserved set of activating receptors is encoded within the NK genetic locus (Brown et al., 1997). Furthermore, a subset of the MHC specific human NK type I transmembrane receptors may be activating (Lopez, Botet, 1996). The mechanism of NK cytolytic activity must be triggered through the activation of the target's apoptotic machinery because DNA fragmentation precedes chromium release in cytotoxicity assays (Berke, 1994). In addition to their well-known release of cytolytic granules, NK cells can also upregulate FasL upon engagement of Fc receptor (Eischen, 1996).

$\gamma\delta$ T CELLS AND CD4$^+$ T-CELL DELETION

In the course of our work exploring the T-cell response among Lyme synovial fluid lymphocytes to *Borrelia burgdorferi,* an early observation was a pronounced expansion of the Vδ1 subset of $\gamma\delta$ T cells in the presence of antigen (see Chapter 10, this volume). Interestingly, the expansion of the $\gamma\delta$ subset seemed to correlate inversely with a simultaneous decline in the CD4$^+$ subset. This led us to speculate that the $\gamma\delta$ T cells were mediating the destruction of the CD4$^+$ T-cell population. Cell surface staining confirmed that the CD4$^+$ T-cell population was selectively enriched for Fas, whereas the $\gamma\delta$ population expressed high levels of FasL. Staining for nicked DNA indicative of apoptosis revealed a distinct CD4^{lo} population positive for apoptosis (CD4^{lo}, expressing low levels of CD4), and the proportion of CD4$^+$ cells comprising this population was modulated by the percentage of $\gamma\delta$ T cells in the co-cultures (Vincent et al., 1996).

The importance of the Fas–FasL system in this bulk culture model was confirmed using $\gamma\delta$ T-cell clones. A number of Vδ1 $\gamma\delta$ clones derived from Lyme synovial fluid were tested for *in vitro* cytotoxicity to a known Fas-sensitive target, the *Jurkat* T-cell line. Not only were the $\gamma\delta$ clones efficient killers of *Jurkat,* mutagenized Fas^{lo} (expressing low levels of Fas) variants of *Jurkat* were highly resistant to $\gamma\delta$ killing. Furthermore, *in vitro* blocking experiments showed that nonlytic, anti-Fas antibodies or a Fas–Fc construct could inhibit a portion of the lytic activity towards *Jurkat,* and that the inhibition was nearly complete if the perforin system was additionally blocked by the chelation of calcium. These data indicate an immunoregulatory loop, by which $\gamma\delta$ T cells are expanded in response to a foreign antigen, and subsequently restrict the nature of the CD4$^+$ T-cell response by Fas–FasL-mediated apoptosis.

A number of models of immune regulation have utilized small numbers of antigen-driven $\gamma\delta$ cells to alter the outcome of an immune response (McMenamin et al., 1994; Huber et al., 1996; Harrison et al., 1996). The usual explanation proffered for these potent $\gamma\delta$ regulatory effects is cytokine-mediated immune deviation. However, equally plausible is the hypothesis that $\gamma\delta$ regulatory cells act via induction of apoptosis in susceptible cellular targets. Crosses between mice deficient in various T-cell subsets and Fas provide additional evidence of a role for $\gamma\delta$ T cells in mediating regulatory functions via cell death. In Fas-mutant (*lpr*) mice deficient in $\gamma\delta$ cells, mortality, renal disease, and gammaglobulinemia are all increased relative to littermate controls, and conventional $\alpha\beta$ T cells expand polyclonally. This is consistent with a non-Fas-mediated limitation of autoaggressive T cells, most likely via a perforin pathway. Interestingly, only the CD4$^+$ subset of $\alpha\beta$ T cells appears to expand (Peng et al., 1996a). These mice also provide evidence for regulation of B-cell death in that *lpr* mice totally deficient in T cells develop B-cell lymphomas, and reconstitution with either $\alpha\beta$ or $\gamma\delta$ T cells alone is sufficient to prevent lymphoma development (Peng et al., 1996b).

In summary, the collective data on cell death in the immune system support the notion that just as cell death shapes tissue modeling or development, cell death likely shapes the course and nature of an immune response. This is evidenced by the proponderance of immune defects in genetic conditions characterized by defects in the apoptotic cascade. Likewise, the importance of cell death in maintaining homeostasis is reflected by the number of pathophysiological conditions arising as a consequence of acquired defects in cell death, such as HIV infection or cancer.

REFERENCES

Alderson MR, Armitage RJ, Maraskovsky E, Tough TW, Roux E, Schooley K, Ramsdell F, Lynch DH (1993): Fas Transduces Activation Signals in Normal Human T Lymphocytes. J Exp Med 178:2231–2235.

Alderson MR, Tough TW, Davis-Smith T, Braddy S, Falk B, Schooley KA, Goodwin RG, Smith CA, Ramsdell F, Lynch DH (1995): Fas Ligand Mediates Activation-induced Cell Death in Human T Lymphocytes. J Exp Med 181:71–77.

Allen PM, Unanue, ER (1987): Antigen processing and presentation at a molecular level. Adv Exp Med Biol 225:147–154.

Banda NK, Bernier J, Kurahara DK, Kurrle R, Haigwood N, Sekaly RP, Finkel TP (1992): Crosslinking CD4 by human immunodeficiency virus gp120 primes T cells for activation-induced apoptosis. J Exp Med 176:1099–1106.

Berke G (1994) The Binding and Lysis of Target Cells by Cytotoxic Lymphocytes. Ann Rev Immunol 12:735–773.

Bretscher PA, Cohn M (1970): A theory of self discrimination. Science 169:1042–1049.

Brown MG, Scalzo AA, Matsumoto K, Yokoyama WM (1997): The natural killer gene complex. Immunol Revs 155:53–65.

Brunner T, Mogil RJ, LaFace D, Yoo NJ, Mahboubl A, Echeverri F, Martin SJ, Force WR, Lynch DH, Ware CF, Green DR (1995): Cell-autonomous Fas (CD95)/Fas-ligand interaction mediates activation-induced apoptosis in T-cell hybridomas. Nature 373:342–345.

Claman HN, Chaperon EA. (1969): Immunological complementation between thymus and marrow cells—a model for the two cell theory of immunocompetence. Transplantation Reviews 1:92–99.

Cohen JJ (1996): Immunology Course Book. University of Colorado Health Science Center Press: pp. 1–6.

Cohen JJ, Duke RC (1984): Glucocorticoid activation of a calcium-dependent endonuclease in thymocyte nuclei leads to cell death. J Immunol 132:38–42.

Desbarats J, Freed JH, Campbell PA, Newell M. K. (1996): Fas (CD95) expression and death-mediating function are induced by CD4 crosslinking on CD4⁺ T cells. Proc Natl Acad Sci 93:11014–11018.

Dhein J, Walczak H, Baumler C, Debatin K-M, Krammer PH (1995): Autocrine T-cell suicide mediated by APO-1/(Fas/CD95). Nature 373:438–440.

Doherty PC, Zinkernagel RM (1975) A Biological Role for the Major Histocompatibility Antigens. Lancet:406–414.

Duke RC, Cohen JJ (1986): IL-2 Addiction: Withdrawal of growth factor activates a suicide program in dependent T cell lines. Lymphokine Res 5:289–299.

Eischen CM, Schilling JD, Lynch DH, Krammer PH, Leibson PJ (1996): Fc receptor-induced expression of fas ligand on activated NK cells facilitates cell-mediated cytotoxicity and subsequent autocrine NK cell apoptosis. J Immunol 156:2693–2699.

Granger GA, Kolb WP (1968): Lymphocyte in vitro cytotoxicity: mechanisms of immune and nonimmune small lymphocyte mediated target cell destruction. J Immunol 101:111–120.

Henkart PA (1985): Mechanism of lymphocyte-mediated cytotoxicity. Ann Rev Immunol 3:31–58.

Hasbold J, Klaus GGB (1990): Anti-immunoglobulin antibodies induce apoptosis in immature B cell lymphomas. Eur J Immunol 20:1685–1690.

Harrison LC, Dempsey-Collier M, Kramer DR, Takahashi K (1996): Aerosol insulin induces regulatory CD8 gd cells that prevent murine insulin-dependent diabetes. J Exp Med 184:2167–2174.

Huber S, Mortenson A, Moulton G (1996): Modulation of cytokine expression by CD4⁺ T cells during Coxsackievirus B3 infection of Balb/c mice initiated by cells expressing the γδ T cell receptor. J Virol 70:3039–3044.

Kagi D, Vignaux F, Lederman B, Burki K, Depraetere V, Nagata S, Hengartner H, GolsteinP (1994): Fas and perforin pathways as major mechanisms of T cell-mediated cytotoxicity. Science 265:528–530.

Linsley P, Ledbetter JA (1993): Role of the CD28 receptor during T cell responses to antigen. Ann Rev Immunol 11:191–212.

Lopez-Botet M, Moretta L, Strominger J (1996): NK-cell receptors and recognition of MHC class I molecules. Immunol Today 17:212–214.

McMenamin C, Pimm C, McKersey M, Holt PG (1994): Regulation of IgE responses to inhaled antigens in mice by antigen-specific gd T cells. Science 265:1869–1871.

Newell MK, Maroun CR, Haughn L, Julius M (1990): Death of mature T cells by separate ligation of CD4 and the T cell receptor for antigen. Nature 347:286–289.

Newell MK, VanderWall JB, Beard KS, Freed JH (1993): Ligation of MHC class II mediates apoptotic cell death in resting B lymphocytes. Proc Nat Acad Sci 90:10459–10463.

Nossal GJV (1994): Negative selection of lymphocytes. Cell 76:229–239.

Peng SL, Madaio MP, Hayday AC, Craft J (1996): Propagation and Regulation of Systemic Autoimmunity by $\gamma\delta$ T Cells. J Immunol 157:5689–5698.

Peng SL, Robert ME, Hayday AC, Craft J (1996): A Tumor-suppressor Function for Fas Revealed in T Cell-deficient Mice. J Exp Med 184:1146–1154.

Rathmell JC, Cooke MP, Ho WY, Grein J, Townsend SE, Davis MM, Goodnow CC (1995): CD95 (Fas)-dependent elimination of self-reactive B cells upon interaction with CD4+ T Cells. Nature (London) 376:181–184.

Rathmell JC, Townsend SE, Xu JC, Flavell RA, Goodnow CC (1996): Expansion or Elimination of B cells in vivo: Dual Roles for CD40- and Fas (CD95)-Ligands Modulated by the B cell antigen receptor. Cell 87:319–329.

Sellins KS, Cohen JJ (1987): Gene induction by g-irradiation leads to DNA fragmentation in lymphocytes. J Immunol 139:3199–3206.

Snell G (1981): Studies in Histocompatibility. Science 213:172–177.

Suda T, Okazaki Y, Naito T, Yokota N, Arai S, Ozaki K, Nakao K, Nagata S (1995): Expression of the Fas ligand in cells of the T lineage. J Immunol 154:3806–3813.

Suda T, Takahashi T, Golstein P, Nagata S (1993): Molecular cloning and expression of the Fas ligand, a novel member of the tumor necrosis factor family. Cell 75:1169–1178.

Vincent M, Roessner K, Lynch D, Cooper SM, Sigal LH, Budd RC (1996): Apoptosis of Fas high CD4+ Synovial T cells by Borrelia reactive Fas Ligand high gamma delta T cells in Lyme Arthritis. J Exp Med 184:1149–1154

von Boehmer H (1994): Positive selection of lymphocytes. Cell 76:219–228.

Watanabe-Fukunaga R, Brannan CI, Copeland NG, Jenkins NA, Nagata S (1992): Lymphoproliferative disorder in mice explained by defects in Fas antigen that mediates apoptosis. Nature (London) 356:314–317.

Zacharchuk, CM, Mercep M, Ashwell JD (1991): Thymocyte activation and death: a mechanism for molding the T cell repertoire. Ann NY Acad Sci 636:52–70.

CHAPTER 10

APOPTOSIS IN AUTOIMMUNITY

RALPH C. BUDD

Immunobiology Program, The University of Vermont College of Medicine,
Burlington, VT 05405-0068

APOPTOSIS IN THE IMMUNE SYSTEM

One of the most important processes during immune development is the removal of self-reactive lymphocytes through apoptosis (Wyllie, 1980; Cohen and Duke, 1984; Jenkinson et al., 1989). Without this process of negative selection, the random generation of the myriad antigen receptors on lymphocytes would far too frequently give rise to an autoimmune diathesis. As a result, developing T and B lymphocytes are subject to a rigorous deletion process. Mutations in the pathways leading to lymphocyte apoptosis can therefore quickly lead to a state of autoimmunity.

Resting T lymphocytes are not very susceptible to apoptosis. However, upon activation via the T-cell antigen receptor (TCR), T cells express receptors for the critical growth factor, interleukin-2 (IL2), and enter a committed step to undergo cell cycling in the presence of IL2 (Smith, 1988). To achieve maximal production of IL2, T cells must obtain a second costimulatory signal via the CD28 molecule, which is engaged by two molecules upregulated on activated B lymphocytes or macrophages known as B7-1 and B7-2 (Lenschow et al., 1994). However, a second receptor exists for B7-1 and B7-2 known as CTLA-4, which is only transiently expressed by activated T cells (June et al., 1994). Compared to CD28, CTLA-4 has a 20-fold greater affinity for B7-1 and B7-2, and it confers a negative regulatory signal to T cells that appears to result in apoptosis (Gribben et al., 1995). Entry into cell cycle induced by IL2 also leads to an upregulation of surface Fas expression and sensitization to Fas ligation (Altman et al., 1981). As a consequence, actively cycling T cells are more vulnerable to Fas-induced apoptosis, creating a situation of a balance of power between the forces of proliferation and those of apoptosis.

When Cells Die, Edited by Richard A. Lockshin, Zahra Zakeri,
and Jonathan L. Tilly
ISBN 0-471-16569-7   © 1998 Wiley-Liss, Inc.

Exposure to Fas ligand (FasL)-bearing cells during this period results in deletion of sensitized T cells. The source of the FasL can be the activated T cells themselves (i.e., autonomous cell suicide) or other FasL positive cells (i.e., homicide), as has been demonstrated for the sertoli cells of the testis and certain components of the eye (Bellgrau et al., 1995; Griffith et al., 1995; Papermaster, Chapter 12; Zakeri, Chapter 3; Tilly, Chapter 16, this volume). Cells of the immune system are therefore among the most prone to clonal expansion followed by rapid deletion through apoptosis (Newell, Chapter 9, this volume). This has value during an infectious process where rapid expansion of immune cells is required, but this marked proliferative phase must quickly resolve itself following clearance of the infection, lest self-reactive lymphocytes linger at the scene. Thus defects at many levels in the process of lymphocyte apoptosis can lead to autoimmune tendencies.

AUTOIMMUNITY RESULTING FROM GENETICS DEFECTS IN THE APOPTOTIC MACHINERY

Four examples of mutations in genes of the apoptotic machinery illustrate how these can impact on deletion of lymphocytes and provoke autoimmune features. The classical examples of mice with defective apoptosis in the immune system are the *lpr* and *gld* mice, which bear, respectively, mutations in the genes for *fas* or *fasL* (Watanabe-Fukunaga et al., 1992; Suda et al., 1993). With age these mice develop an enormous accumulation of lymphocytes, the majority of which have an unusual phenotype of expressing the TCR in the absence of markers of mature T cells, namely CD4 (helper T cells) or CD8 (cytolytic T cells) (Davignon et al., 1985). This actually represents the phenotype of T cells undergoing apoptosis in the normal immune system and is discussed further below. Concomitant with the adenopathy is the development of an autoimmune syndrome resembling human systemic lupus erythematosus, including glomerulonephritis, vasculitis, and dermatitis (Theophilopoulos and Dixon, 1981). Lymphocyte infiltration into the salivary gland also occurs, resembling Sjogren's Syndrome, as does synovitis resembling rheumatoid arthritis. Presumably, self-reactive lymphocytes are not deleted through apoptosis at a sufficient rate, and their retention provokes various types of autoimmunity (Budd et al., 1987). A similar syndrome has been observed in humans who have mutations in the *fas* gene.

Fas-mediated apoptosis is associated with up-regulation of the hematopoieic cell phosphatase (*Hcph*) gene during apoptosis in human cell lines (Su et al., 1994). This gene is mutated in the motheaten (*me*) mouse, which also manifests delayed lymphocyte deletion and develops an autoimmune syndrome (Schultz et al., 1993).

As noted earlier, ligation of CTLA-4 by activated T cells confers a negative signal that may lead to apoptosis (Gribben et al., 1995). Mice that are geneti-

PROLIFERATION

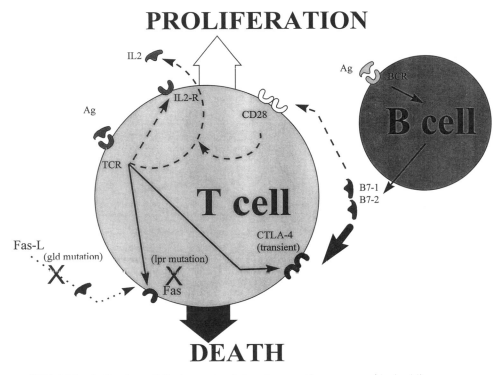

FIG. 10.1. Activation elicits two sets of signals: growth responses (dashed lines, upper half) and responses that lead to death (solid lines, lower half). Transiently, CTLA-4 can trap B7-1 and B7-2, or in the presence of Fas-L the cell will die.

cally deficient in CTLA-4 manifest a profound lymphadenopathy and autoimmune syndrome, much like the *lpr* mouse (Tivol et al., 1995). Although the phenotype of the T lymphocytes that accumulate in CTLA-4-deficient mice are mature in phenotype, rather than the unusual CD4⁻CD8⁻ T cells in *lpr* mice, this may merely reflect the level in the apoptotic signal cascade at which CTLA-4 and Fas are involved.

The T-cell growth cytokine, IL2, is not only central to the promotion of T-cell proliferation, since the resulting induction of cell cycling also primes T cells to undergo apoptosis if the TCR is engaged during this period (Lenardo, 1991). Thus blocking of IL2 function *in vivo* with the use of anti-IL2 or anti-IL2R (IL2 receptor) antibodies prevents the deletion of T cells (Lenardo, 1991). Not unexpectedly, mice in which the IL2 gene has been disrupted by homologous recombination also have decreased apoptosis of T lymphocytes, and these mice also exhibit a variety of autoimmune features, particularly an inflammatory bowel syndrome (Sadlack et al., 1993). Thus disruption of a variety of genes that govern apoptosis of T lymphocytes can precipitate a tendency toward autoimmune features.

APOPTOSIS AS A SOURCE OF AUTOANTIGENS

One of the great enigmas in autoimmunity is why autoantibodies are frequently found to react to biosynthetic nuclear and cytoplasmic components rather than structural and membrane components. Thus, in diseases like lupus, scleroderma, or dermatomyositis, antibodies are often seen that react to molecules involved with activities such as splicing of messenger RNA (Sm, snRNPs), relaxation of supercoiled DNA from topoisomerase I activity (scl-70), or RNA transcription (SS-B/La), to mention a few (Tan, 1994). The question to be addressed is how these seemingly sequestered self-proteins become immunogenic. A clue to this puzzle may be in the recent observation that clustering of these autoantigens to membrane blebs appears to occur preferentially during apoptosis (Casciola-Rosen et al., 1994; Trump and Berezesky, Chapter 2, this volume). For example, UV irradiation of keratinocytes induces apoptosis, and clusterings of nucleosomal DNA, Ro, La, and the small nuclear ribonucleoproteins (snRNPs) could be identified within the blebs at the surface of these cells. This anatomic packaging may make these autoantigens considerably more available to the immune system, particularly from apoptotic cells. In situations where apoptosis may be more prominent, as has been described in human lupus (Huang et al., 1994), this may provide fuel to the fire.

AUTOIMMUNITY DUE TO APOPTOSIS OF THE INNOCENT BYSTANDER

The concept of "autoimmune-mediated tissue injury" is often a broadly applied clinical term. It likely encompasses a wider array of tissue damage than might pass close scrutiny under the strict immunologic rigor of requiring an immune response to an identifiable autoantigen. Thus, while some autoimmune disorders, such as Graves Disease, which affects the thyroid gland, may have a clear autoantibody response to a tissue component that directly causes the syndrome, most autoimmune diseases do not have an identified causative antigen. As a result, many of the recognized autoantibodies are merely correlative epiphenomena and have little if anything to do with pathogenesis. Another feature is that immune injury of a specific organ may result from activation of the immune system in an area anatomically distant from the site of damage. Histologically this might resemble an intense infiltration of lymphocytes with corresponding apoptosis of the surrounding parenchymal cells. Collectively, this could be viewed as an immune or even autoimmune-mediated injury. This might occur in the pancreas in diabetes, in the synovium in rheumatoid arthritis, or in the liver in autoimmune hepatitis (Huang et al., 1994). This section tenders the view that many autoimmune syndromes may result from the infiltration of activated lymphocytes bearing FasL into tissues that express cell-surface Fas.

A particularly striking example of this form of "autoimmunity" might occur in the liver. The liver contains a population of lymphocytes resembling the apoptotic phenotype, TCR-$\alpha\beta^+$/CD4$^-$/CD8$^-$/B220$^+$, that was mentioned earlier (Huang et al., 1994). Indeed these resident hepatic lymphocytes contain subdiploid DNA, consistent with apoptosis. An enormous influx into the liver of similar T cells can be induced in mice that are transgenic for a single TCR following the administration of the specific antigen for that TCR (Huang et al., 1994). These infiltrating T cells are highly activated and bear surface expression of FasL. Concomitant with this we have observed the rapid onset of apoptosis of hepatocytes, which have high surface levels of Fas (R. Budd, unpublished observations). In fact, with a sufficiently high dose of antigen, it is possible to induce fulminant hepatic failure and death. This scenario closely parallels clinical conditions of autoimmune hepatitis or fulminant hepatic failure during certain infectious states where there is a strong immune stimulus. It is of interest that in *lpr* mice, lacking Fas expression, the livers are considerably larger than normal mice. Perhaps the constant low-level, daily influx of FasL positive lymphocytes maintains a homeostasis of hepatic cell number. This would then represent an example of the "innocent bystander" hypothesis, in which lymphocytes, which were activated in the spleen or lymph nodes and express FasL, migrate to an organ containing Fas positive and parenchymal cells (e.g., hepatocytes) with resulting apoptosis (see Finkel, Chapter 11, this volume).

SUPERANTIGENS

The above example in the liver might seem somewhat artificial given that the mice involved are essentially clonal for a single TCR transgene, whereas the frequency of responding T cells to conventional antigens is much lower. However, T-cell response frequencies approaching those seen in the TCR transgenic mice do occur in certain infectious diseases characterized by the production of so-called "superantigens." These compounds, produced mostly by bacteria but also by some viruses, can directly bind both the major histocompatibility complex (MHC) molecule and the TCR outside the traditional antigen binding groove of the MHC (Herman et al., 1991; Newell, Chapter 9, this volume). This serves to crosslink MHC and TCR and results in responding T-cell frequencies that are orders of magnitude greater than with traditional antigens. The binding of superantigens to the TCR is almost exclusively governed by the TCR β-chain (Herman et al., 1991). As a result, superantigens provoke a profound initial expansion of T cells bearing a particular TCR β-chain. For example, in mice that receive a superantigen known as *Staphylococcus* enterotoxin B (SEB), there is a dramatic increase in TCR-V$\beta8^+$ cells within two days followed by their rapid apoptosis and deletion in a partially Fas-dependent manner. Superantigen exposure results in at least two processes that can contribute to autoimmune features. First, there is a massive release of cytokines, which can promote the expansion of self-reactive T cells that might be present at a low and clinically irrelevant frequency. Second, as a result of the pronounced T-cell activation, FasL expression is up-regulated (Ettinger et al., 1995). This would lead to a condition much like that of the TCR transgenic mice (Osborne, Chapter 8, this volume).

There are a growing number of superantigens described from various microorganisms that are pathogenic in humans. Perhaps the best two described to date are toxic shock syndrome and Kawasaki syndrome. While not necessarily autoimmune by a strict definition, they have features that closely resemble autoimmune syndromes and have often been viewed as such. Toxic shock syndrome results from a *Staphylococcus aureus* toxin known as toxic shock syndrome toxin-1 (TSST-1) (Choi et al., 1990). It is most frequently associated with menstruation and tampon use, although similar manifestations have been increasingly recognized in other situations involving focal *S. aureus* infections. Fever, rash, and hypotension are the hallmarks of this syndrome, but multiorgan involvement can occur, including liver dysfunction and arthritis, among others. There is also a pronounced increase in TCR-Vβ2+ cells in affected individuals (Choi et al., 1990). Kawasaki syndrome is also a multisystem disease that affects primarily infants and young children. Kawasaki syndrome patients manifest an increase in Vβ2+ cells in their peripheral blood (Abe et al., 1992). This feature combined with the overlapping clinical picture of Kawasaki syndrome and toxic shock syndrome, as well as the known seasonal variation in the incidence of this disease, provides strong circumstantial evidence for an infectious, perhaps superantigen, etiology of Kawasaki syndrome.

A process similar to the liver infiltration by activated T cells bearing FasL might occur in other organs, such as in joints in rheumatoid arthritis. Synovium lacks a basement membrane, and this results in easy access of activated FasL positive lymphocytes to joint tissue. Many cells types in synovium express Fas and are susceptible to Fas-mediated apoptosis (Asahara et al., 1996). A parallel situation to this may occur via $\gamma\delta$ T cells in Lyme arthritis. Lyme disease is caused by the spirochete, *Borrelia burgdorferi,* which is transmitted by *Ixodes* ticks that live on white-tailed deer (Burgdorfer et al., 1982). A portion of affected individuals can develop a chronic arthritis that is resistant to antibiotics and can resemble rheumatoid arthritis, both clinically and histologically, in which a CD4+ helper T-cell infiltrate is observed (Steere et al., 1988). An unusual population of T cells known as $\gamma\delta$ T cells also reside in Lyme arthritis synovial fluid (Vincent et al., 1996). Although little is known about the function of $\gamma\delta$ T cells, as they comprise only about 2–3% of peripheral blood T cells (Groh et al., 1989), they accumulate at many sites of chronic inflammation. In response to a sonicate of *B. burgdorferi,* the $\gamma\delta$ T cells in Lyme arthritis synovial fluid expand rapidly at the expense of the resident CD4+ T cells (Vincent et al., 1996). Further inspection reveals that the $\gamma\delta$ T cells express high levels of FasL nearly constitutively, whereas the CD4+ T cells have high surface Fas. The $\gamma\delta$ T cells appear to function by inducing apoptosis of the proinflammatory CD4+ cells. Upon first analysis, this situation sounds like an immunoregulatory suppressive mechanism for CD4$^+$ cells, and indeed, it has been observed in a few models of autoimmune arthritis or orchitis where removal of $\gamma\delta$ T cells results in more severe inflammation (Pelegri et al., 1996; Mukasa et al., 1995; Newell and Vincent, Chapter 9, this volume).

The $\gamma\delta$ T-cell story may have a dark side, however, that is being actively studied. As these cells may bear the highest and most prolonged levels of FasL, one can easily imagine that they might prove damaging to Fas-expressing parenchymal cells of various organs, similar to the hepatic lymphocytes. Second, it would appear that the Fas-induced apoptosis of CD4$^+$ T cells by $\gamma\delta$ T cells is selective. A brief immunologic diversion is warranted. CD4$^+$ T cells segregate themselves into two types, depending upon their pattern of cytokine expression (Mosmann, 1995). Type I helper T cells (Th1) produce primarily IL2 and Interferonδ that are required for activation of cytolytic T cells and macrophages. Type 2 helper T cells (Th2) produce IL4, IL5, and IL6, which are vital to B-cell growth and production of antibodies. Most inflammatory sites of infection or autoimmunity contain a predominence of Th1 cells. In fact, mice that are genetically predisposed to a Th2-biased repertoire are susceptible to certain infections, as occurs with *Leishmania major* (Chakkalath and Titus, 1994). These Th2-biased mice may also be predisposed to certain autoimmune conditions characterized by autoantibody production. Of note, $\gamma\delta$ T cells are highly lytic toward Th2 cells over Th1 cells, as are FasL-transfected fibroblasts (R. Budd, unpublished observation). This may be one of the mechanisms by which selective apoptosis maintains a Th1 cytokine profile at many sites of inflammation. It may also influence the course of certain autoimmune diseases. An example is found in Coxsackie virus B3 (CVB3)-induced autoimmune myocarditis. CVB3 induces a Th1-type CD4$^+$ T-cell infiltrate in myocardium, resulting in autoimmune myocarditis. A non-pathogenic variant of CVB3 induces an equally strong immune response, but it is Th2-biased and there is no resulting myocarditis. However, with the transfer of as few as 5,000 $\gamma\delta$ T cells, the myocardial infiltrate is now switched to Th1 and myocardial damage ensues (Huber et al., 1996). Thus the sustained expression of FasL at sites of inflammation is a two-edged sword, causing diminished autoimmune reactions in some situations such as arthritis, but potentially exacerbating inflammation by biasing toward a Th1 infiltrate and causing apoptosis of "innocent" Fas-expressing parenchymal cells.

REFERENCES

Abe J, Kotzin BL, Melish ME, Glode ME, Kohsaka T, Leung DY (1992): Selective expansion of T cells expressing T cell receptor variable regions Vβ2 and Vβ8 in Kawasaki disease. Proc Natl Acad Sci 89:4066–69.

Altman A, Theophilopoulos AN, Weiner R, Katz DH, and Dixon FJ (1981): Analysis of T cell functions in autoimmune murine strains. Detects in production of and responsiveness to IL-2. J Exp Med 154:791–801.

Asahara H, Hasumuna T, Kobata T, Yagita H, Okumura K, Inoue H, Gay S, Sumida T, Nishioka K (1996): Expression of Fas antigen and Fas ligand in the rheumatoid synovial tissue. Clin Immunol Immunopathol 81:27–34.

Bellgrau D, Gold D, Selawry H, Moore J, Franzusoff A, and Duke RC (1995). A role for CD95 ligand in preventing graft rejection. Nature 377:630–32.

Budd RC, Schreyer M, Miescher GC, MacDonald HR (1987): T cell lineages in the thymus of lpr/lpr mice: Evidence for parallel pathways of normal and abnormal T cell development. J Immunol 139:2210–18.

Burgdorfer W, Barbour AG, Hayes FF, Benach JL, Grunwaldt BE, David JP (1982): Lyme disease—a tick-borne spirochetosis? Science 216:1317–19.

Casciola-Rosen LA, Anhalt G, Rosen A (1994): Autoantigens targeted in systemic lupus erythematosus are clustered in two populations of surface structures on apoptotic keratinocytes. J Exp Med 179:1317–30.

Chakkalath HR, Titus RG (1994): Leishmaniamajor-parasitized macrophages augment Th2-type T cell activation. J Immunol 153:4378–87.

Choi Y, Lafferty JA, Todd J, Gelfand EW, Kappler J, Marrack P, Kotzin BL (1990): Selective expansion of T cells expressing Vβ2 in toxic shock syndrome. J Exp Med 172:981–91.

Cohen JJ, Duke RC (1984): Glucocorticoid activation of a calcium-dependent endonuclease in thymocyte nuclei leads to cell death. J Immunol 132:38–42.

Davignon JL, Budd RC, Ceredig R, Piguet PF, MacDonald HR, Cerottini JC, Vasalli P, Izui S (1985): Functional analysis of T cell subsets from mice bearing the *lpr* gene. J Immunol 135:3704–9.

Ettinger R, Panka DJ, Wang JK, Stanger BZ, Ju ST, Marshak-Rothstein A (1995): Fas ligand-mediated cytotoxicity is directly responsible for apoptosis of normal CD4$^+$ T cells responding to a bacterial superantigen. J Immunol 154:4302–8.

Gribben JC, Freeman GJ, Boussiotis VA, Renner P, Jellis CL, Greenfield E, Barber M, Restivo VA, Ke X, Gray GS, Nadler LM (1995):CTLA4 mediates antigen-specific apoptosis of human T cells. Proc Natl Acad Sci USA 92:811–15.

Griffith TS, Brunner T, Fletcher DG, Green DG, Ferguson TA (1995). Fas ligand-induced apoptosis as a mechanism of immune privilege. Science 270:1189–92.

Groh V, Porcelli S, Fabbi M, Lanier LL, Picker LJ, Anderson T, Warnke RA, Bhan AT, Strominger JL, Brenner MB (1989): Human lymphocytes bearing T cell receptor *gd* are phenotypically diverse and evenly distributed throughout the lymphoid system. J Exp Med 169:1277–94.

Herman A, Kappler J, Marrack P, Pullen AM (1991): Superantigens: Mechanisms of T-cell stimulation and role in immune responses. Ann Rev Immunol 9:745–72.

Huang L, Soldeville G, Leeker M, Flavell R, Crispe N (1994): The liver eliminates T cells undergoing antigen-triggered apoptosis in vivo. Immunity 1:741–49.

Huber S, Mortenson A, Moulton G (1996): Modulation of cytokine expression by CD4+ T cells during Coxsackie virus B3 infection of Balb/c mice initiated by cells expressing the $\gamma\delta$ T cell receptor. J Virol 70:3039–44.

Jenkinson EJ, Kingston R, Smith CA, Williams GT, Owen JJT (1989): Antigen-induced apoptosis in developing T cells: A mechanism for negative selection of the T cell receptor. Eur J Immunol 19:2175–77.

June CH, Bluestone JA, Nadler LM, Thompson CB (1994): The B7 and CD28 receptor families. Immunol Today 15:321–30.

Lenardo MJ (1991): Interleukin 2 programs mouse $\alpha\beta$ T lymphocytes for apoptosis. 353:858–61.

Lenschow DJ, Sperling AJ, Cooke MP, Freeman G, Rhee L, Decker DC, Gray G, Nadler LM, Goodnow CC, Bluestone JA (1994): Differential up-regulation of the B7-1 and B7-2 costimulatory molecules after Ig receptor engagement by antigen. J Immunol 153:1990–97.

Mosmann TR (1995): Differentiation of subsets of CD4+ and CD8+ T cells. CIBA Found Symp 195:42–50.

Mukasa A, Hiromatsu G, Matzusaki R, O'Brien R, Borne W, Nomoto K (1995): Bacterial infection of the testis leading to autoagressive immunity triggers apparently opposed responses of $\alpha\beta$ and $\gamma\delta$ T cells. J Immunol 155:2047–56.

Pelegri C, Kuhlein P, Buchner CB, Schmidt A, Franch H, Castell M, Hunig T, Emmrich F, Kinne RW (1996): Depletion of g/d T cells does not prevent or ameliorate, but rather aggravates, rat adjuvant arthritis. J Immunol 38:204–15.

Sadlack B, Merz H, Schorle H, Schimpl A, Feller AC, Horak I (1993): Ulcerative colitis-like disease in mice with a disrupted interleukin-2 gene. Cell 75:253–61.

Schultz LD, Schweitzer PA, Rajan TV, Yi T, Ihle JN, Matthews J, Thomas ML, Beier DR (1993): Mutations at murine motheaten locus are within the hematopoietic cell protein-tyrosine phosphatase (Hcph) gene. Cell 73:1445–54.

Smith KA (1988): Interleukin-2: Inception, impact, and implications. Science 240:1169–76.

Steere AC, Duray PH, Butcher EC (1988): Spirochetal antigens and lymphoid cell surface markers in Lyme synovitis: Comparison with rheumatoid synovium and tonsillar lymphoid tissue. Arthritis Rheum 31:487–97.

Su X, Zhou T, Wu J, Jope R, Mountz JD (1994): Dephosphorylation of a 65 kD protein delivers a signal for Fas-mediated apoptosis. FASEB J 8:218–22.

Suda T, Takahashi T, Golstein P, Nagata S (1993): Molecular cloning and expression of the Fas ligand, a novel member of the tumor necrosis factor family. Cell 75:1169–78.

Tan EM (1994): Autoimmunity in apoptosis. J Exp Med 179:1083–86.

Theophilopoulos AN, Dixon FJ (1981): Etiopathogenesis of murine SLE. Immunol Rev 55:179–216.

Tivol EA, Borriello F, Schweitzer AN, Lynch WP, Bluestone JA, Sharpe AH (1995): Loss of CTLA-4 leads to massive lymphoproliferation and fatal multiorgan tissue destruction, revealing a critical negative regulatory role of CTLA-4. Immunity 3:541–47.

Vincent M, Roessner K, Lynch D, Cooper SM, Sigal LH, Budd RC (1996): Apoptosis of Fas high CD4+ synovial T cells by Borrelia reactive Fas ligand high gamma delta T cells in Lyme arthritis. J Exp Med 184:2109–2117.

Watanabe-Fukunaga R, Brannan CI, Copeland NG, Jenkins NA, Nagata S (1992). Lymphoproliferative disorder in mice explained by defects in *Fas* antigen that mediates apoptosis. Nature 356:314–17.

Wyllie AH (1980): Glucocorticoid-induced thymocyte apoptosis is associated with endogenous endonuclease activation. Nature 284:555–56.

CHAPTER 11

AIDS AND CELL DEATH

TERRI H. FINKEL AND CAROLYN R. CASELLA

Basic Sciences, National Jewish Center, Denver, CO 80206

INTRODUCTION

Over the past several years, remarkable progress has been made in our understanding of the pathogenesis of human immunodeficiency virus (HIV) infection. Years of previous work had shown that HIV was associated with depletion of CD4[+] helper T cells and with immune system destruction. However, the mechanisms of this depletion were puzzling, since levels of virus were often low to unmeasurable in peripheral blood. Breakthroughs in our understanding came with studies showing high viral turnover throughout the course of disease, measurable by more sensitive polymerase chain reaction (PCR) techniques. In addition, large amounts of virus were found to disseminate to lymphoid organs, particularly early in disease. Over the past year, several reports have linked viral load to disease progression, strengthening the conclusion that the virus is responsible for CD4[+] T-cell depletion.

Despite these breakthroughs, a remaining mystery of HIV infection is how the virus causes the destruction of the CD4[+] (and ultimately the CD8[+]) T cells. We discuss several possible mechanisms of T-cell death, including the killing of uninfected CD4[+] and CD8[+] T cells. We also discuss the possibility that the virus protects the cell it infects at least until viral replication can be completed (Cryns and Yuan, Chapter 6; Amiesen, Chapter 1, this volume).

THE IMMUNE SYSTEM

The human immune system is designed to recognize and eliminate foreign invaders such as bacteria and viruses (Gelfand and Finkel, 1996; Braun and Stiehm, 1996; Janeway and Travers, 1996). Key players in this drama are B

When Cells Die, Edited by Richard A. Lockshin, Zahra Zakeri,
and Jonathan L. Tilly
ISBN 0-471-16569-7   © 1998 Wiley-Liss, Inc.

lymphocytes (which mature in the bone marrow) and T lymphocytes (which mature in the thymus). Each B cell expresses a unique antibody on its cell surface that is specific for a particular antigen. This antibody and its associated molecules constitute the B-cell receptor for antigen (BCR). When engaged by antigen, in the presence of helper T cells (see the following), the BCR signals the cell to divide and differentiate into an antibody-secreting cell. Soluble antibody can then recognize and bind to foreign antigen in the tissues or circulation and target this antigen for destruction. The involvement of antibodies in immune surveillance is called "humoral" immunity (Osborne, Chapter 8; Newell and Vincent, Chapter 9; Budd, Chapter 10, this volume).

T lymphocytes, on the other hand, are involved in "cellular" immunity because antigen recognition and binding occur during cell–cell interactions. Critical to these cell–cell interactions are membrane glycoproteins of the major histocompatibility complex (MHC), which bind peptide fragments derived from foreign proteins (Fig. 11.1). There are two kinds of MHC molecules, MHC class I molecules and MHC class II molecules. B cells and other antigen-presenting cells (APCs) degrade foreign antigen into peptide fragments. This foreign antigen may be internalized from the environment in intracellular vesicles, as in the case of a bacterium that infects a macrophage or a bacterial protein bound by antibody on the surface of B cells. Peptides derived from such antigens are bound by MHC class II molecules and transported to the cell surface (Fig. 11.2). Alternatively, this foreign antigen may be synthesized in the cytosol, as in the case of an infecting virus. Peptide fragments of viral proteins are bound by MHC class I molecules and transported to the cell

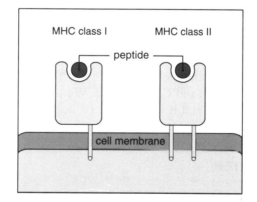

FIG. 11.1. MHC molecules display peptide fragments of antigens on the surface of cells. MHC molecules are membrane proteins whose outer extracellular domains form a cleft in which a peptide fragment can bind. These fragments, which are derived from proteins degraded inside the cell, including foreign protein antigens, are bound by the newly synthesized MHC molecule before it reaches the surface. There are two kinds of MHC molecules, MHC class I molecules and MHC class II molecules, which differ in structure and function. Adapted from Janeway and Travers, 1996, with permission.

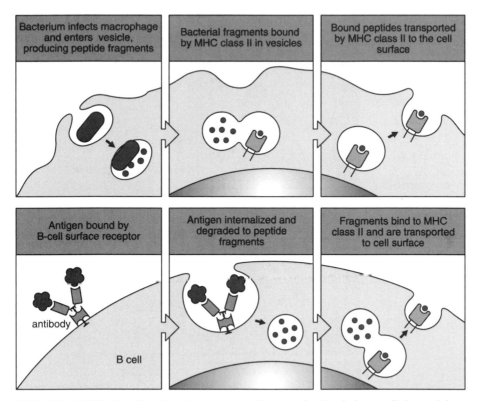

| Bacterium infects macrophage and enters vesicle, producing peptide fragments | Bacterial fragments bound by MHC class II in vesicles | Bound peptides transported by MHC class II to the cell surface |
| Antigen bound by B-cell surface receptor | Antigen internalized and degraded to peptide fragments | Fragments bind to MHC class II and are transported to cell surface |

FIG. 11.2. MHC class II molecules present antigen originating in intracellular vesicles. Some bacteria infect cells and grow in intracellular vesicles. Peptides derived from such bacteria are bound by MHC class II molecules and transported to the cell surface (top row). MHC class II molecules also bind and transport peptides derived from antigen that has been bound and internalized by endocytosis into intracellular vesicles (bottom row). Adapted from Janeway and Travers, 1996, with permission.

surface (Fig. 11.3). Not surprisingly, these two different classes of MHC molecules on APCs (class I and class II) present antigen to two different classes of T cells, CD4$^+$ or CD8$^+$ T cells (Fig. 11.4).

Both subsets of T cells are characterized by a T-cell receptor for antigen (TCR), analogous to the BCR. Each T cell expresses a unique TCR on its cell surface that is specific for a particular peptide–MHC complex. The CD4 or CD8 coreceptors on the T cells complete the antigen recognition complex by interacting with the TCR and binding to class II or class I, respectively, on the APC.

Two classes of CD4$^+$ T cells, inflammatory (T_H1) and helper (T_H2) T cells, recognize and bind to antigen from the environment that has been processed by a macrophage or B cell, and then complexed with class II MHC (Fig. 11.4). As discussed, B cells capture antigen by their surface antibody, process this internalized antigen to peptides, and then express peptide antigen bound to

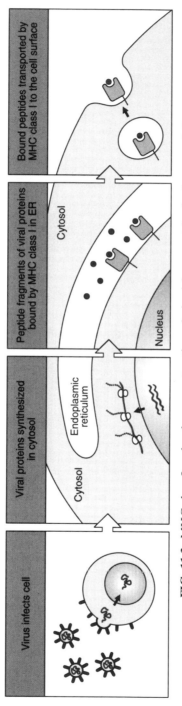

FIG. 11.3. MHC class I molecules present antigen derived from proteins in the cytosol. In cells infected with viruses, viral proteins are synthesized in the cytosol. Peptide fragments of viral proteins are transported into the endoplasmic reticulum, where they are bound by MHC class I molecules, which then deliver the peptides to the cell surface. Adapted from Janeway and Travers, 1996, with permission.

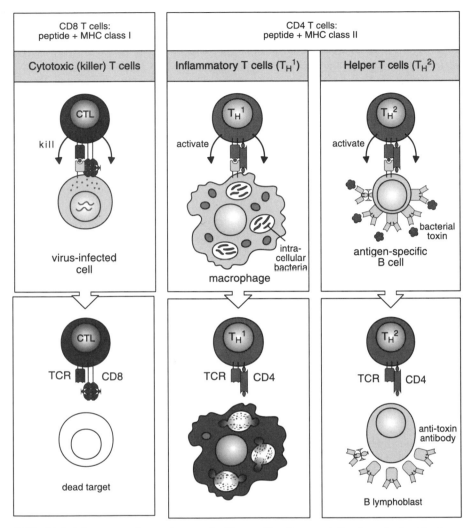

FIG. 11.4. There are three classes of effector T cells, specialized to deal with three classes of pathogens. CD8$^+$ cytotoxic cells or cytotoxic T lymphocytes (CTL) (left panels) kill target cells that display antigenic fragments of cytosolic pathogens, most notably viruses, bound to MHC class I molecules at the cell surface. Inflammatory CD4$^+$ T cells (T$_H$1) (middle panels) and helper CD4$^+$ T cells (T$_H$2) (right panels) both express the CD4 coreceptor and recognize fragments of antigens degraded within intracellular vesicles, displayed at the cell surface by MHC class II molecules. The inflammatory CD4$^+$ T cells, upon activation, activate macrophages, allowing them to destroy intracellular microorganisms more efficiently. Helper CD4$^+$ T cells, on the other hand, activate B cells to differentiate and secrete immunologlobulins, the effector molecules of the humoral immune response. Adapted from Janeway and Travers, 1996, with permission.

class II. When CD4 and TCR on a T_H2 cell are engaged by peptide antigen and class II on the B cell, the T cell is activated and signals the B cell to differentiate and secrete immunoglobulin (Fig. 11.4). When CD4 and TCR on a T_H1 cell are engaged by peptide antigen and class II on the macrophage, the T cell is activated and signals the macrophage to destroy the intracellular microorganisms (Fig. 11.4). A critical component of this signaling is the production of growth factors or cytokines required for immune cell proliferation and differentiation. CD4$^+$ T cells therefore function as "helper" T cells, to help B cells and other APCs (e.g., dendritic cells, monocytes, and macrophages) divide and differentiate.

CD8$^+$ T cells recognize and bind to processed viral antigens complexed with class I on the surface of an APC. When CD8 and TCR on the T cell are engaged by peptide antigen and class I on the APC, the TCR signals the T cell to kill the infected APC (Fig. 11.4). CD8$^+$ T cells therefore function as "cytotoxic" T lymphocytes (CTLs), to kill cells harboring virus. Cytokines produced by "helper" CD4$^+$ T cells are also critical to the elimination of virally infected cells by CD8$^+$ CTLs.

Thus B lymphocytes and T lymphocytes (CD4$^+$ and CD8$^+$ T cells) function in concert to survey and clear the body of foreign invaders, be they in the circulation, in tissues, or living inside cells. How does this elegant system of immune surveillance become disrupted in HIV infection?

HIV INFECTION

The HIV Genome

HIV is one of the most complex viruses known to man. It is one of a family of retroviruses and, like other retroviruses, contains two strands of RNA, as well as the proteins essential for packaging and replication, encoded by the *gag, pol,* and *env* genes (Coffin, 1990). Unlike other retroviruses, HIV also encodes several other so-called "accessory" proteins, such as Tat, Nef, Vpr, and Vpu, that function at various points in its life cycle (see the following). The HIV genome is so complex that the function of many of its genes is still incompletely understood, although more work has gone into the study of HIV than that of any other human pathogen.

HIV Replication

HIV, like other retroviruses, is a positive-sense RNA virus that replicates through a DNA intermediate (Coffin, 1990). Briefly, the viral envelope protein, gp120, binds to the CD4 coreceptor on the surface of CD4$^+$ cells or macrophages. Recent work suggests that this binding induces a conformational change in gp120, and subsequent binding to one of a family of chemokine receptors (Wu et al., 1996; Trkola et al., 1996). The viral core then enters the cell by fusion of the cellular and viral membranes. Uncoating occurs, and the viral RNA is reverse transcribed to a double-stranded DNA that is integrated into the host cell chromosome. New viral RNA transcripts and proteins are synthesized using host cellular machinery. The viral proteins form new particles, encapsidate genomic length viral RNA, and bud from the cells. Finally, the particle matures and is then competent to begin a new round of replication.

The Clinical History of HIV Infection

HIV infection follows a similar clinical course, whether acquired *in utero,* perinatally, by sexual intercourse, blood transfusion, or IV drug abuse. Initial infection is often accompanied by a flulike illness and is associated with an acute high-titer viremia. It is during this period of acute viremia that virus is disseminated widely to the lymphoid tissue. This viremia is quickly controlled by mechanisms that are not yet understood, although cytotoxic $CD8^+$ T lymphocytes (i.e., "cellular" immunity) are thought to be more critical to this control than anti-HIV antibodies (i.e., "humoral" immunity). However, although circulating viral loads decrease significantly (from $\sim10^7$ to $<10^4$ copies/ml), the virus is not eliminated, and in fact remains at much higher levels (10^6 to $>10^8$ copies/g) in the lymphoid tissue (Haase et al., 1996).

   In most individuals, the circulating viral load remains at low levels for years and the individual remains asymptomatic, without any of the clinical features of AIDS. However, this "clinical latency" is not also a "viral latency," since viral production and destruction are ongoing at a rapid pace. In addition, this "clinical latency" is not also an "immunologic latency," since $CD4^+$ T-cell production and destruction are also ongoing at a rapid pace. For reasons that are not yet clear, these cycles of production and destruction continue for years until the numbers of $CD4^+$ T cells (and/or $CD8^+$ T cells) decline below a critical level. It is at this point that the circulating viral load again skyrockets, $CD4^+$ T cell numbers fall precipitously, and the clinical symptoms of AIDS (acquired immune deficiency syndrome) appear, progressing ultimately to death.

   How does HIV cause destruction of $CD4^+$ T cells?

HIV PATHOGENESIS

As discussed, HIV causes immunodeficiency in its host via an ongoing, virally induced process. Several studies in recent years have shown that HIV-infected individuals have actively replicating virus throughout all stages of disease (Pantaleo et al., 1993; Embertson et al., 1993; Piatak et al., 1993; Ho et al., 1995; Wei et al., 1995; Perelson et al., 1996). In addition, the level of viral RNA in the peripheral blood is a predictor of time until disease onset (Furtado et al., 1995; Ioannidis et al., 1996; O'Brian et al., 1996; Mellors et al., 1996): The higher the viral load, the faster the disease progression. These observations argue that the virus is responsible for the depletion of $CD4^+$ T cells and the destruction of the immune system. Less well understood is the mechanism by which HIV depletes its host of CD4[+] T cells.

HIV may kill the CD4[+] T cell that it infects, and there are data to suggest that the HIV-infected cell is deleted *in vivo* (Ho et al., 1995; Wei et al., 1995; Perelson et al., 1996). On the other hand, it has been argued that HIV infection of CD4[+] T cells is noncytopathic (Zinkernagel and Hengartner, 1994). We propose that HIV acts to protect the infected cell from death, at least until the cell has produced high levels of virus. After this has occurred, the cell may die as a result of immune clearance or as a result of cytopathic effects of the virus (Fig. 11.5).

Results from several laboratories suggest that HIV is also capable of killing uninfected CD4[+] cells (Fig. 11.5; Su et al., 1995; Maldarelli et al., 1995; Nardelli et al., 1995; Heinkelein et al., 1995; Finkel et al., 1995). We hypothesize that death of uninfected cells occurs *in vivo* and is an important feature of the pathogenesis of the virus. Another important feature of the pathogenesis of the virus is the decline in HIV-specific CD8[+] cytotoxic T lymphocytes (CTLs), which occurs just before the rapid increase in viremia and the accelerated decline of CD4[+] T cells that signals progression to AIDS (reviewed in Feinberg, 1996). We hypothesize that the decline in CD8[+] T-cell function, which results in high virus production, is instrumental in the decline of CD4[+] T cells.

APOPTOSIS IN HIV INFECTION

Apoptosis may play an important role in the pathogenesis of AIDS.

Definition of Apoptosis

Apoptosis is a regulated form of cell suicide that is characterized by progressive cell shrinkage, nuclear condensation, cleavage of DNA into nucleosome-sized fragments, loss of membrane integrity, membrane blebbing, and, finally, engulfment of intact cell fragments ("apoptotic bodies") by phagocytic cells. Apoptosis is necessary for normal immune function and development (reviewed in Touchette, 1995). Immature T cells die by apoptosis in the thymus if their TCRs are specific for self-antigens and therefore autoreactive. Mature T cells die by apoptosis following antigenic stimulation, during selection for high-affinity T cells in the peripheral lymphoid tissues (Zheng et al., 1996), and as a means of down-regulating the immune response (Lenardo, 1996) (Osborne, Chapter 8; Newell and Vincent, Chapter 9; Budd, Chapter 10, this volume).

Apoptosis Assays

Apoptosis of lymph node (LN) cells and peripheral blood lymphocytes (PBLs) has been measured in HIV-infected individuals using techniques that assay

Fate of Infected and Uninfected T Cells in HIV Disease

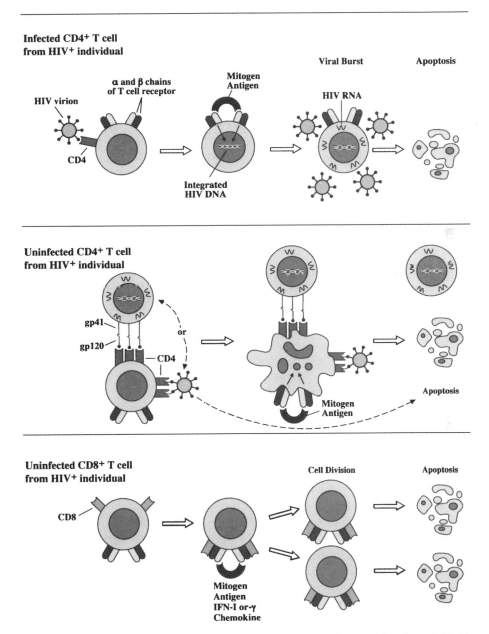

FIG. 11.5. Fate of infected and uninfected T cells in HIV disease. (a) Infected CD4$^+$ T cells are protected from death until virus is released. (b) Uninfected CD4$^+$ T cells are primed for apoptosis by binding virus or gp120 from an infected cell. These primed CD4$^+$ T cells then undergo apoptosis in response to antigenic stimulation. Alternatively, uninfected CD4$^+$ T cells may be induced to undergo apoptosis directly by binding gp120 from an infected cell. (c) In HIV infection, CD8$^+$ T cells expand in response to several possible stimuli and then undergo apoptosis.

one or another of the identifying characteristics of apoptotic cells (Duke and Cohen, 1992). These techniques include (Zakeri, Chapter 3; Benitez-Bribiesca, Chapter 17, this volume):

1. *DNA binding dyes.* These dyes are used to stain live cells for subsequent microscopic analysis or for subsequent flow cytometric analysis. When used for microscopic analysis of live cells, this technique assays nuclear morphology and membrane permeability. It provides one of the best ways of assessing apoptotic morphology in a quantitative manner and of discriminating live from dead cells and apoptotic from necrotic cells. The disadvantages of this technique are that it cannot be done on fixed samples (and therefore HIV-infected samples must be handled in a containment facility), and it is time consuming.

When used for flow cytometric analysis, this technique assays membrane permeability. While easy and quantitative, this application does not discriminate apoptotic from necrotic cells, and, as above, containment is an issue with unfixed samples.

These dyes also bind DNA and are widely used to flow cytometrically assay stages of the cell cycle. Resting cells contain a 2N amount of DNA and form a "G_1" peak, while cells in mitosis contain a 4N amount of DNA and form a "G_2" peak. DNA binding dyes have also been used at lower concentrations in fixed cells to identify an apoptotic "G_0" peak below the resting G_1 peak. This lower peak may be due to loss of DNA fragments with apoptosis or to altered conformation of apoptotic DNA. While this technique has shown good discrimination between the G_0 and G_1 peaks in murine cells, the results have been less than satisfactory in the human system and may depend upon the cell type used.

2. *DNA laddering.* This technique assays endonucleolytic cleavage of DNA into 200 base pair multimers. Although this assay is reasonably definitive if a DNA ladder is seen, it requires large amounts of DNA and therefore large numbers of cells, which can be problematic when working with patient samples. In addition, this assay is not quantitative, despite densitometric tricks that have been used to make comparisons between lanes on a single gel.

3. *Scatter changes.* This technique identifies apoptotic cells by their decreased forward scatter (due to cell shrinkage) and their increased side scatter (due to DNA condensation) (Swat et al., 1991; Darzynkiewicz et al., 1992; Dive et al., 1992; Cotton et al., 1997). Although concerns have been raised that this technique may not discriminate between apoptotic and necrotic cells, the latter typically have both decreased forward *and* side scatter. The great virtues of this technique are (1) it is easy; (2) it is quantitative; (3) it can be done retrospectively on any cell sample that has been analyzed flow cytometrically, since all samples are routinely analyzed for forward and side scatter; (4) it can be done on fresh or fixed cells, and, best of all, (5) it can be done in conjunction with cell surface phenotyping or with immunofluorescent labeling of an intracellular marker (or two or three!).

4. *TUNEL or in situ nick translation.* This assay again takes advantage of that classic feature of apoptosis, DNA fragmentation. The enzyme terminal deoxynucleotidyl transferase (TdT) adds labeled nucleotides to the ends of nicked DNA. The label is then revealed in different ways depending on the application; a radioactive label allows development of grains over cells in fixed tissue, while a biotin–avidin conjugate allows immunohistochemical labeling of cells in tissue or flow cytometric analysis of cells in suspension. This technique is quantitative and allows for concurrent phenotyping of cells and (at least in fixed tissue) for analysis of expressed RNAs by *in situ* hybridization. The technique is, however, technically tricky, the reagents are expensive, and the TdT is often contaminated with RNase, making *in situ* hybridization unreliable.

Apoptosis in HIV-Infected Individuals and in Models of HIV Disease

Apoptosis is increased in the lymph nodes (LNs) and PBLs of HIV-infected individuals (Groux et al., 1992; Gougeon ct al., 1993; Estaquier et al., 1994; Meyaard et al., 1994; Sarin et al., 1994; Finkel et al., 1995; Carbonari et al., 1995; Muro-Cacho et al., 1995; Cotton et al., 1996; Gougeon et al., 1996; Cotton, et al., 1997) in both the CD8$^+$ and CD4$^+$ subsets of T cells (Gougeon et al., 1993; Meyaard et al., 1994; Muro-Cacho et al., 1995; Bofill et al., 1995; Carbonari et al., 1995; Gougeon et al., 1996; Cotton et al., 1997. T-cell apoptosis has also been studied in simian models of HIV disease. HIV-infected chimpanzees, and African green monkeys infected with simian immunodeficiency virus-$_{agm}$(SIV$_{agm}$), do not develop disease and do not show abnormal levels of CD4$^+$ T-cell apoptosis. However, rhesus macaques infected with a pathogenic strain of SIV develop simian AIDS and show an increase in CD4$^+$ T cell apoptosis (Estaquier et al., 1994). Thus numbers of apoptotic cells are increased in HIV infection, and this increase is correlated with *in vivo* pathogenicity.

In addition, some studies (Finkel et al., 1995; Cotton et al., 1996; Gougeon et al., 1996; Cotton et al., 1997) have shown a correlation between the extent of apoptosis and disease progression. While these data argue for a pathogenic role for apoptosis in the progression to AIDS, other studies have not found a correlation between apoptosis and disease (Carbonari et al., 1995; Muro-Cacho, 1996), and it has been argued that apoptosis is correlated instead with the general state of activation of the LN (Muro-Cacho et al., 1996). Thus, while it is generally accepted that increased apoptosis occurs in HIV-infected individuals, many questions remain about whether this apoptosis occurs as a result of a normal or aberrant immune response to a chronic viral infection, or whether the virus directly induces apoptosis of immune cells.

Apoptosis of Uninfected CD4$^+$ T Cells

As discussed, HIV$^+$ individuals carry a high viral burden throughout the course of infection. Despite this, it is only a small fraction ($<0.1\%$) of CD4$^+$ T cells that are productively infected at any point in time (Embretson et al., 1993). Thus the question arises in HIV infection as to who is dying—the infected cells or the uninfected cells (or both)? Notably, the number of apoptotic CD4$^+$ T cells from the peripheral blood of HIV-infected individuals is greater than the number of infected cells, suggesting that uninfected cells are dying by apoptosis (Carbonari et al., 1995). This is supported by recent data of Haase and co-workers showing that calculated CD4$^+$ T cell turnover rates well exceed observed numbers of productively infected T cells (Haase et al., 1996), again suggesting that uninfected cells die in HIV-infected individuals (Budd, Chapter 10, this volume).

Direct evidence that uninfected cells are dying comes from LN samples from HIV-infected children and SIV-infected macaques that were examined for both apoptosis and infection (Finkel et al., 1995). Apoptosis was measured by *in situ* nick translation, and infection was measured by *in situ* hybridization for full-length HIV RNA. While a correlation was found between apoptosis and infection, the majority of apoptotic cells were found to be uninfected.

How does HIV induce apoptosis?

MECHANISMS OF APOPTOSIS IN HIV INFECTION

Fas–Fas-L Interactions

One major pathway of apoptosis is mediated through the tumor necrosis factor (TNF) family of receptors. Fas, in particular, has been extensively studied in recent years. Ligation of Fas by Fas ligand (Fas-L) present on the same or on a neighboring cell can induce either proliferation or apoptosis (Anderson et al., 1993; Brunner et al., 1995; Dhien et al., 1995; Ju et al. 1995). HIV$^+$ people have a higher percentage of PBL positive for Fas expression (Katsikis et al., 1995; Silvestris et al., 1996) and more Fas expressed per cell (Silvestris et al., 1996; Gougeon et al., 1996). More important, the CD4$^+$ and CD8$^+$ T cells from HIV-infected people are more susceptible to death induced by Fas ligation (Katsikis et al., 1995; Estaquier et al., 1995; Silvestris et al., 1996). Fas-induced death of CD4$^+$ T cells has been found by some to correlate with disease stage, with more death seen in symptomatic than in asymptomatic people (Katsikis et al., 1995). An intriguing new report shows that macrophages express Fas-L and that infection of macrophages by HIV-1 increases their level of Fas-L expression (Bradley et al., 1996). Hence, macrophages may be a source of Fas-L that is killing susceptible CD4$^+$ cells (Budd, Chapter 10, this volume).

Regulation of Fas/Fas-L Expression by HIV-Encoded Proteins

What causes the increase in Fas expression during HIV-1 infection? The increased Fas expression could be a result of chronic immune activation (reviewed in Lynch et al., 1995). In addition, crosslinking of CD4 has been shown to increase the expression of Fas (Oyaizu et al., 1994). Thus it is possible that gp120 binding to CD4 upregulates Fas expression on uninfected cells rendering them more susceptible to Fas-induced apoptosis. In support of this view, soluble gp120 has also been shown to augment Fas-induced death of CD4$^+$ T cells (Westendorp et al., 1995).

What causes the increase in Fas-L expression during HIV-1 infection? The Tat protein of HIV-1 has been shown to increase the level of Fas-L expression on T cells when added exogenously (Westendorp et al., 1995). Tat is secreted from infected cells (Ensoli et al., 1990) and may upregulate Fas-L on uninfected cells. These cells could then either kill themselves by binding Fas expressed on the same cell or kill another cell that has upregulated Fas.

These data are intriguing, and suggest the involvement of Fas–Fas-L interactions in HIV-induced apoptosis. In addition, recent data show that inhibition of the interleukin 1β-converting enzyme (ICE) family of proapoptotic proteases, which acts downstream in the Fas pathway, inhibits apoptosis of cells in HIV-infected cultures (Glynn et al., 1996). However, surprisingly, blocking anti-Fas mAbs had no effect on HIV-induced apoptosis in these cultures. Clearly, the roles of other members of the TNF receptor superfamily (such as TNF receptor itself and the recently described DR3, Chinnaiyan et al., 1996), which also kill via activation of ICE-like proteases, require investigation.

What are other candidate mechanisms for apoptosis in HIV infection?

The CD4–gp120 Interaction

Apoptosis occurs when infected and uninfected cells are cultured together (Maldarelli et al., 1995; Nardelli et al., 1995; Heinkelein et al., 1995). Viral replication in the cultures is not required because reverse transcriptase and protease inhibitors fail to prevent apoptosis (Maldarelli et al., 1995; Corbeil et al., 1995). However, soluble CD4 or antibodies that block CD4–gp120 interactions can block apoptosis (Cohen et al., 1992; Heinkelein et al., 1995; Maldarelli et al., 1995; Corbeil et al., 1995). Furthermore, cells that express only Env (gp160/gp120) can cause the death of uninfected CD4$^+$ cells (Cohen et al., 1992; Kolesnitchenko et al., 1995; Heinkelein et al., 1995). This evidence indicates that interaction of gp120 on an infected cell with CD4 on an uninfected cell can cause the apoptosis of the uninfected cell (Fig. 11.5). Multinucleated syncytium formation is not a prerequisite for this death, as killing in single cells is seen (Maldarelli et al., 1995; Nardelli et al., 1995; Heinkelein et al., 1995; Kolesnitchenko et al., 1995). Soluble gp120 can also cause the apoptosis of uninfected cells by binding to CD4 on preactivated cells (Forster et al., 1995), or by binding to CD4, followed by subsequent activation of the cell through the T-cell receptor (Banda et al., 1992; Radrizzani et al., 1995). Interestingly, CD4$^+$ T cells from chimpanzees do not undergo activation-induced apoptosis following CD4 ligation by gp120 (Banda et al., 1996). This lack of bystander cell death is thus correlated with a general lack of pathogenicity of HIV infection in this primate species.

Signaling for Death

What is known about the signal transduction mechanisms triggered by HIV gp120 or Tat that lead to cell death? The induction of apoptosis by gp120 binding to CD4 may be caused by an increase in Fas expression or by activation of the transcription factor AP-1 (Chirmule et al., 1995), or the recruitment of the tyrosine kinase, Lck (Corbeil et al., 1996), causing improper activation of the cell (reviewed in Kabelitz et al., 1993; Green and Scott, 1994). The cytoplasmic tail of CD4 is required for induction of apoptosis by HIV, and the ability of CD4 to bind Lck is correlated with enhanced HIV-induced apoptosis (Corbeil et al., 1996; Birge et al., Chapter 13, this volume).

Recent data suggest that interaction between HIV-1 or gp120 and CD4 induces activation of both Lck and the serine kinase, Raf-1, with resultant activation of binding of several nuclear transcription factors (Popik and Pitha, 1996). Interestingly, this activation of Raf-1 is unusual in that Ras activation is not also involved. Thus the interaction between CD4 and the HIV envelope may trigger aberrant signaling pathways.

As mentioned, Tat is secreted from infected cells and augments Fas-L expression (Ensoli et al., 1990; Westendorp et al., 1995). The ability of Tat to increase Fas-L expression may increase apoptotic death of uninfected cells expressing Fas. Soluble Tat has also been shown to cause the apoptosis of PBLs by increasing the activity of cyclin A–dependent kinases (Li et al., 1995). Thus gp120 and Tat may both augment the destruction of uninfected CD4$^+$ cells.

Apoptosis of Thymocytes in HIV Disease

Apoptosis is seen in the thymic tissue of HIV-infected individuals, as well as in the peripheral lymphoid organs. Infection within an implanted human thymus in the SCID (severe combined immunodeficiency)-hu mouse model causes apoptosis of all thymocyte populations (Su et al., 1995). The number of cells infected did not account for the amount of apoptosis seen, and it was concluded that uninfected cells were dying (Su et al., 1995). Although controversial, bone marrow (BM) progenitor CD34$^+$ cells from HIV$^+$ individuals have been shown to be uninfected (Re et al., 1993; Neal et al., 1995) and undergo apoptosis upon culture (Re et al., 1993). The effect of HIV on the thymus and BM may explain the inability of the immune system to replace indefinitely the lost CD4$^+$ and CD8$^+$ T cells.

PROTECTION OF HIV-INFECTED CELLS FROM APOPTOSIS

Recent reports of viral kinetics during HIV infection suggest that the half-life of an infected cell is between 1.0 and 2.6 days (Perelson et al., 1996).

These analyses suggest that the ultimate fate of the infected cell is death. However, the cause of this death is unknown. The infected cell may be cleared by the immune system or may die from cytopathic effects of the virus. Apoptosis has been estimated to take only 4 hours (Howie et al., 1994). If apoptosis is initiated during the early phase of infection, the virus might not have time to complete its replication cycle. Thus it would be beneficial for the virus to prevent apoptosis until high levels of virus are produced (Fig. 11.5; Table 11.1; Cryns and Yuan, Chapter 6; Ameisen, Chapter 1; and Mittler, Chapter 5, this volume).

Down-Regulation of CD4

The most obvious line of defense for the infected cell is to reduce the level of CD4 on the cell surface, because of the role that CD4 plays in initiating apoptosis (see the preceding). Reduction of surface CD4 would prevent infection by new viruses (a process called "superinfection interference"), reduce the chance that soluble gp120 would bind to the cell, and reduce the chance that improper signaling through CD4 would occur (Salghetti et al., 1995; reviewed in Kabelitz et al., 1993; Green and Scott, 1994; Ameisen et al., 1994). This function is, apparently, so important that three viral proteins contribute to CD4 down-regulation. Nef, a protein predominantly expressed early in infection, reduces CD4 on the surface of the cell by displacing Lck and thereby allowing CD4 to be endocytosed (Salghetti et al., 1995). Envelope glycoprotein (Env) binds CD4 in the endoplasmic reticulum (ER) and thereby sequesters it (Crise et al., 1990). Vpu, which is expressed predominantly late in the life cycle, disrupts the CD4-Env complex in the ER, resulting in the release of Env and the degradation of CD4 (Willey et al., 1992).

Inhibition of Apoptosis by HIV-Encoded Proteins

The capsid protein of HIV (CA) binds to cyclophilin A (Luban et al., 1993). Cyclophilin A has been found to function as an endonuclease that induces DNA fragmentation (Montague et al., 1994) and binds several immunosuppressive drugs, one of which is cyclosporin A (CsA). Jurkat T cells incubated with CsA do not increase Fas-L expression upon TCR cross-linking (Dhien et al., 1995). The cyclophilin A–binding activity of HIV-1 CA may inhibit Fas-L expression in a similar manner. If so, CA expression could protect cells from apoptosis by decreasing the availability of Fas-L or by preventing cyclophilin A from functioning as an endonuclease, thereby protecting the infected cell from DNA degradation.

HIV-Tat and -Nef have properties that may prevent or delay death of infected cells. Tat has been shown to prevent apoptosis caused by serum withdrawal (Zauli et al., 1993; Caputo et al., 1990) and by TNF-α or anti-Fas antibodies (Gibellini et al., 1995). These functions of Tat might help the cell survive *in vivo* in the absence of cytokines needed for the survival of CD4$^+$ T cells, or in the presence of apoptotic stimuli. Tat has also been reported to

TABLE 11.1. HIV Gene Products Implicated in the Control of Cell Death

I. Means by which HIV may Prevent or Delay Apoptosis

HIV Gene Product	Alteration of Cellular Function	Reference
Tat	Inhibition of apoptosis	Caputo et al., 1990; Zauli et al., 1993; Gibellini et al., 1995
	Inhibition of antigen-induced proliferation	Viscidi et al., 1989
	Decreased MHC class I expression	Howcroft et al., 1993
	Increased bcl-2 expression	Zauli et al., 1995
	Decreased p53 expression	Li et al., 1995
Nef	Increased endocytosis of CD4	Salghetti et al., 1995
	Decreased Lck activation	Greenway et al., 1995; Collette et al., 1996
	Binding of p53, Hck, pp44$^{mapk/erk1}$	Greenway et al., 1995; Saksela et al., 1995
	Decreased expression of IL-2Rα chain	Greenway et al., 1995; Collette et al., 1996
	Inhibition of proliferation in response to IL-2	Greenway et al., 1995
	Decreased MHC class I surface expression	Howcroft et al., 1993
Vpu	Degradation of CD4	Willey et al., 1992
Vpr	Inhibition of activation of pp34^{cdc2}–cyclin B	He et al., 1995; Jowett et al., 1995; Re et al., 1995; Rogel et al., 1995
Env (gp120/gp160)	Inhibition of surface expression of CD4 by binding to CD4 in the ER	Crise et al., 1990
Capsid	Binding to cyclophilins A and B	Luban et al., 1993

304

II. Means by which HIV may Cause Apoptosis or Necrosis in Highly Infected Cells

HIV Gene Product	Alteration of Cellular Function	Reference
Vpr	Cell cycle arrest in G2	Kolesnitchenko et al., 1995
Tat	Increased activity of cyclin A–dependent kinases	Li et al., 1995
	Increased Fas-L expression	Ensoli et al., 1990; Westendorp et al., 1995
Env (gp120–gp160)	Increased Fas expression	Oyaizu et al., 1994
	Induction of CD4-Lck interaction	Corbeil et al., 1992
	Increased membrane permeability	Fermin and Garry, 1992
	Syncytia formation	Cohen et al., 1992
	Formation of intracellular CD4–gp160 complexes	Koga et al., 1990
	Activation of AP-1	Chirmule et al., 1995
	Activation of Lck and Raf-1	Popik and Pitha, 1996
Unintegrated viral DNA	Induction of cytolysis	reviewed in Levy, 1993
Many	Targeting for cytotoxic T lymphocyte-mediated killing	Cheynier et al., 1994
Many	Targeting for lymphokine-activated killers	Brenner et al., 1990
Many	Targeting for antibody-dependent cellular cytotoxicity	Brenner et al., 1990
Many	Targeting for complement-mediated lysis	Spear et al., 1990
Unknown	Induction of necrosis	Cao et al., 1996
Unknown	Activation of ICE-like proteases	Glynn et al., 1996
Unknown	Increased susceptibility to Fas-induced cell death	Katsikis et al., 1995; Estaquier et al., 1995; Silvestris et al., 1996
Unknown	Increased expression of Fas-L on macrophages	Bradley et al., 1996

TABLE 11.1. (Continued)

III. Means by which HIV may Induce Death in Uninfected Bystander Cells

HIV Gene Product	Alteration of Cellular Function	Reference
Tat	Increased Fas-L expression	Ensoli et al., 1990; Westendorp et al., 1995
Env (gp120/gp160)	Induction of "mitotic catastrophe"	Cohen et al., 1992; Kolesnitchenko et al., 1995
	Increased Fas expression	Oyaizu et al., 1994
	Induction of cytolysis	Weinhold et al., 1989; Heinkelein et al., 1995; Nardelli et al., 1995
	Induction of apoptosis of preactivated cells	Forster et al., 1995
	Priming for activation-induced apoptosis	Banda et al., 1992; Radrizzani et al., 1995; Banda et al., 1996
	Increased membrane permeability	Fermin and Garry, 1992
	Activation of AP-1	Chirmule et al., 1995
	Activation of Lck and Raf-1	Popik and Pitha, 1996
Many	Cytokine-induced cell death or cytokine withdrawal	reviewed in Levy, 1993
Many	Exhaustive activation	Zinkernagel and Hengartner, 1994
Unknown	Telomere shortening in CD8[+] T cells	Effros et al., 1996; Wolthers et al., 1996
Unknown	Activation of ICE-like proteases	Glynn et al., 1996
Unknown	Increased susceptibility to Fas-induced cell death	Katsikis et al., 1995; Estaquier et al., 1995; Silvestris et al., 1996
Unknown	Increased expression of Fas-L on macrophages	Bradley et al., 1996

decrease transcription of *p53* (Li et al., 1995) and to increase the transcription of *bcl-2* (Zauli et al., 1995). In addition, Nef has been reported to bind p53 (Greenway et al., 1995). High levels of the bcl-2 protein can block apoptosis in response to several apoptotic stimuli (reviewed in Penninger and Mak, 1994). High levels of p53 induce apoptosis if DNA damage is incurred, and p53 has been shown to repress *bcl-2* expression (reviewed in Solary et al., 1996). The reduction of p53 and the increase in *bcl-2* induced by these HIV gene products would both have potent antiapoptotic effects. In addition, Tat and Nef have been shown to down-regulate surface expression of MHC class I (Howcroft et al., 1993; Schwartz et al., 1996). This may prevent targeting of infected cells by $CD8^+$ CTLs.

These mechanisms, designed to prevent death of the infected cells, are also seen in other viral infections. Epstein–Barr virus is known to increase the transcription of *bcl-2,* while adenovirus, Epstein–Barr, and African swine fever virus encode genes that mimic *bcl-2* (reviewed in Thompson, 1995). Human papilloma and adenovirus inactivate p53, and Herpes simplex viruses 1 and 2 down-regulate MHC class I surface expression (Hill et al., 1994), while cytomegalovirus uses multiple tactics to prevent MHC class I molecules from exposing its antigens on the cell surface (Ahn et al., 1996; Jones et al., 1996; Wiertz et al., 1996; Hengel et al., 1996). Thus inhibition of cellular apoptosis may be a general mechanism used by viruses to slow the death of the host cells upon which they depend for reproduction.

Inhibition of Cellular Activation by HIV-Encoded Proteins

Several viral gene products affect the proliferative or activation status of cells in ways that may alter the cells' susceptibility to apoptosis. Vpr causes the infected cell to arrest in the *G2* phase of the cell cycle by inhibiting activation of $p34^{cdc2}$-cyclin B (He et al., 1995; Jowett et al., 1995; Re et al., 1995; Rogel et al., 1995). It is possible that this arrest in the cell cycle increases the level of virus production by preventing the cell from reentering a resting/G0 state. It has been postulated that this cell cycle arrest could protect infected cells from apoptosis (Shi et al., 1994; He et al., 1995; Jowett et al., 1995).

Proliferation of PBLs in response to IL-2 is greatly reduced in the presence of Nef (Greenway et al., 1995). This may be due to the ability of Nef to down-regulate the alpha chain of the IL-2R and to prevent the phosphorylation and activation of the tyrosine kinase Lck (Greenway et al., 1995; Collette et al., 1996). In addition, Nef binds to $pp44^{mapk/erk1}$ and Hck (Greenway et al., 1995; Saksela et al., 1995), proteins that can mediate proliferative signals (Ruderman, 1993; Kafalas et al., 1995). Furthermore, Tat has been shown to inhibit proliferation in response to antigen (Viscidi et al., 1989). These functions of Nef and Tat may help infected cells avoid apoptosis by preventing improper activation of the cell (reviewed in Kabelitz et al., 1993; Green and Scott, 1994).

How Do Infected Cells Die?

In the presence of these postulated antiapoptotic mechanisms of the virus, why does the infected cell have on average a half-life of only 1.5 days? The immune system has several mechanisms to combat viral infection. HIV-1 specific CTL have been shown to infiltrate the infected white pulp of the spleen (Cheynier et al., 1994). Even though Tat and Nef have been shown to down-regulate MHC class I surface expression, these proteins are expressed at highest levels early in infection and might not be able to suppress MHC class I expression throughout infection. Reduction of MHC class I surface expression might be most beneficial to the virus early in infection, so that replication can be completed. Thus it is possible that, in the short term, the virus inhibits apoptosis. In the long term, however, the virus might cause apoptosis in the infected cell by arresting the cell cycle in *G2* (Kolesnitchenko et al., 1995), by increasing the activity of cyclin A–dependent kinases (Li et al., 1995), or by subverting the cell machinery. It is also possible that the virus could result in cell killing by means other than apoptosis, for example, by necrosis (Cao et al., 1996). A direct effect of viral proteins on membrane permeability has been described (Fermin and Garry, 1992). In addition, antibody binding to surface expressed Env could cause cell lysis by antibody-dependent cell-mediated cytotoxicity (ADCC; Brenner et al., 1990) or by the complement cascade (Spear et al., 1990).

CD8⁺ T CELLS IN HIV INFECTION

The CTL response to infection by HIV is an important tool used by the immune system to keep the level of virus low, and, hence, the progression to disease slow. The CTL response declines before the onset of disease, in which a higher virus load and a faster decline in CD4⁺ T cells is seen (Feinberg, 1996). In addition to killing infected cells, CD8⁺ T cells have also been shown to secrete suppressive factors that reduce the ability of HIV to replicate (Laurence et al., 1983; Walker et al., 1988). Several suppressive factors have been identified in the past year. These include IL-16, RANTES, MIP-1α, and MIP-2α (Baier ct al., 1995; Cocchi et al., 1995). It appears that the chemokines RANTES, MIP-1α, and MIP-2α act by blocking the second receptor for HIV, the chemokine receptor CC CKR5, on T cells and monocytes/macrophages (Alkhatib et al., 1996; Deng et al., 1996; Dragic et al., 1996). Thus CD8⁺ T cells have an important function in controlling disease. By killing infected cells and reducing viral replication they keep the virus in check and reduce viral load.

Why do HIV-specific CD8⁺ CTLs decline? CD8⁺ PBLs in infected individuals undergo spontaneous apoptosis upon overnight culture (Meyaard et al., 1994; Muro-Cacho et al., 1995; Carbonari et al., 1995; Cotton et al., 1996; Cotton et al., in press) and apoptotic CD8⁺ T cells can be found in the LN from HIV$^+$ people (Muro-Cacho et al., 1995). CD8$^+$ T-cell apoptosis occurs at all stages of disease. Apoptosis of CD8$^+$ T cells is also seen in other viral infections (Bofill et al., 1995) and may be a general feature of the immune response to many kinds of viruses. If so, the extent of loss of CD8 cells during progression to AIDS may be magnified due to the chronic nature of HIV infection. Immune activation that subsequently leads to apoptosis could be caused directly by antigenic stimulation of the CD8$^+$ T cells (Fig. 11.5); high levels of virus are present in the lymph nodes throughout infection (Pantaleo et al., 1993; Embertson et al., 1993; Finkel et al., 1995; Muro-Cacho et al., 1995). Alternatively, the CD8$^+$ T cells may be activated through nonspecific mechanisms such as exposure to type I interferon (Tough et al., 1996) or the RANTES chemokine (Bacon et al., 1995). Finally, recent data show infection of CD8$^+$CD4$^-$ T cells with HIV-2, via the chemokine receptor, CXCR4 (also known as LESTR or fusin) (Endres et al., 1996). It has been suggested that HIV evolves *in vivo* to bypass CD4 and to use CXCR4 as an alternative receptor. This raises the intriguing possibility that CD8$^+$ T cells could be destroyed late in disease by direct infection.

Several possibilities have been postulated to explain the specific loss of CTLs in HIV disease. One possibility is that a switch from a cell-mediated T$_H$1 response to a humoral T$_H$2 response during disease progression results in less help for the CD8$^+$ CTLs and a diminution of the response (Clerici and Shearer, 1994). It is also possible that the destruction of the CD4$^+$ T cells leads to less help for the CTLs and a reduction in their response. Another possibility is clonal exhaustion of the CTLs due to the high virus burden in HIV infection (reviewed in Zinkernagel and Hengartner, 1994). Two recent studies report that the telomere length of CD8$^+$ T cells (though, surprisingly, not CD4$^+$ T cells) from HIV-infected individuals is shorter than in age-matched controls (Effros et al., 1996; Wolthers et al., 1996). Shortened telomere length is a marker for increased cell turnover and may indicate replicative exhaustion of the CD8$^+$ cells. This finding is consistent with chronic immune activation in HIV infection and is, likely, not only due to increased turnover of HIV-specific cells, but may be caused by nonspecific activation, as discussed.

Although the mechanism of CD8$^+$ CTL decline is disputed, the effect of this decline on disease progression is likely to be an important factor in the pathogenesis of HIV disease. Left unchecked by the CTL and suppressor CD8$^+$ T-cell activity, the viral load increases, causing faster destruction of the CD4$^+$ T cells and ultimately immune destruction.

SUMMARY AND FUTURE DIRECTIONS

In this chapter, we have discussed the role of apoptosis in HIV disease. Evidence indicates that apoptosis in HIV infection is, at least in part, mediated by the Fas–FasL pathway. We also discuss mechanisms by which HIV could cause the apoptosis of uninfected cells, in particular apoptosis mediated by

CD4-gp120 interactions. Finally, we hypothesize that the virus protects the infected cell from apoptosis until a high level of progeny virus can be produced, and examine mechanisms that the virus may use to prevent apoptosis.

What predictions can we make based on the hypothesis that HIV encodes genes that protect the infected cell from HIV-induced apoptosis? One prediction is that host cells infected with a virus encoding antiapoptotic genes would produce more virus (compared to infection with a virus lacking these genes), since these cells would live longer and thus support more rounds of replication. In fact, this prediction is supported by experiments in which a cell line transfected with the antiapoptotic gene, E1B, produced significantly more virus than the mock transfected control (Antoni et al., 1995).

A second prediction is that a cell infected with a virus encoding antiapoptotic genes would live longer than an uninfected bystander cell. This prediction is supported by analyses of HIV- and SIV-infected lymph nodes. As discussed earlier in this chapter, while increased numbers of apoptotic cells are found in lymph nodes from HIV seropositive, compared to seronegative, individuals, these apoptotic cells are only rarely productively infected (Finkel et al., 1996). Conversely, HIV- or SIV-infected cells are only rarely apoptotic. While this is obviously not a kinetic study (and therefore does not address how *fast* infected versus uninfected cells die), the data suggest that HIV-infected cells are relatively protected from apoptosis.

In contrast, a virus that does *not* encode antiapoptotic genes would kill predominantly infected, rather than uninfected cells. *In vivo,* this virus would produce only low titer and transient viremia, and would not induce disease. By analogy with the mouse virus, lymphocytic choriomeningitis virus, the phenotype of a virus that rapidly kills the cell it infects might be that of an *attenuated* virus. Thus HIV mutants deleted in candidate antiapoptotic genes could be tested in culture and in animal models and analyzed for viral production, apoptosis of infected and uninfected cells, and chronicity of infection. For example, chronic infection *in vitro* has been described to require intact nef, as well as deletions of vpr or vif (Nishino et al., 1994). Intriguingly, the nef-defected virus is attenuated *in vivo,* although it is not known whether this is due to effects on cell death.

If our hypothesis proves to be true, antiretroviral drugs targeted to viral proteins that prevent apoptosis might hasten the death of the cell, reduce viremia, and prolong the life of the infected individual. Alternatively, this type of therapy might provide a useful adjunct to the triple combination therapy that at the time of this writing holds the most promise for eliminating virus in infected individuals. Although this therapy has been shown to reduce viral load to undetectable levels (even by sensitive PCR techniques), it is not known whether latent virus remaining in "Trojan Horse" reservoirs might make cessation of these toxic drugs impossible. Drugs that would specifically induce apoptosis of these infected cells might be the key to destroying this hidden reservoir.

REFERENCES

Ahn K, Angulo A, Ghazal P, Peterson PA, Yang Y, Fruh K (1996): Human cytomegalovirus inhibits antigen presentation by a sequential multistep process. Proc Nat Acad Sci 93:10990–95.

Alkhatib G, Combadiere C, Broder CC, Feng V, Kennedy PE, Murphy PM, Berger EA (1996): CC CKR5: A RANTES, MIP-1α, MIP-1β receptor as a fusin cofactor for macrophage-tropic HIV-1. Science 272:1955–58.

Ameisen JC, Estaquier J, Idziorek T (1994): From AIDS to parasite infection: Pathogen-mediated subversion of programmed cell death as a mechanism for immune dysregulation. Immunol Rev 142:13–50.

Anderson MR, Armitage RJ, Naraskovsky E, Tough TW, Roux E, Schooley K, Ramsdell F, Lynch DH (1993): Fas transduces activating signals in normal human T lymphocytes. J Exp Med 178:2231–35.

Antoni BA, Sabbatini P, Rabson AB, White E (1995): Inhibition of apoptosis in human immunodeficiency virus-infected cells enhances virus production and facilitates persistent infection. J Virol 69:2384–92.

Bacon KB, Premack BA, Gardner P, Schall TJ (1995): Activation of dual T cell signaling pathways by the chemokine RANTES. Science 269:1727–30.

Baier M, Werner A, Bannert N, Metzner K, Kurth R (1995): HIV suppression by interleukin-16. Nature 378:563.

Banda NK, Bernier J, Kurahara DK, Kurrle R, Haigwood N, Sekaly RP, Finkel TH (1992): Crosslinking CD4 by human immunodeficiency virus gp120 primes T cells for activation-induced apoptosis. J Exp Med 176:1099–106.

Banda NK, Satterfield WC, Steimer K, Kurrle R, Finkel TH (1996): Resistance to AIDS in chimpanzees is correlated with lack of gp120-induced anergy and priming for apoptosis. Apoptosis 1:49–62.

Bofill M, Gombert W, Borthwick NJ, Akbar AA, McLaughlin JE, Lee CA, Johnson MA, Pinching AJ, Janossy G (1995): Presence of CD3$^+$CD8$^+$Bcl-2^{low} lymphocytes undergoing apoptosis and activated macrophages in lymph nodes at HIV-1$^+$ patients. Am J Pathol 146:1542–55.

Bradley AD, McElhinny JA, Leibson PJ, Lynch DH, Alderson MR, Paya CV (1996): Upregulation of Fas ligand expression by human immunodeficiency virus in human macrophages mediates apoptosis of uninfected T lymphocytes. J Virol 70:199–206.

Braun J, Stiehm ER (1996): The B-lymphocyte system. In Stiehm ER (ed): Immunologic Disorders in Infants and Children. WB Saunders Co., Philadelphia pp 35–74.

Brenner BG, Gryllis C, Wainberg MA (1990): Role of antibody-dependent cellular cytotoxicity and lymphokine-activated killer cells in AIDS and related diseases. J Leuk Biol 50:628–40.

Brunner T, Mogil RJ, LaFace D, Yoo NJ, Mahboubi A, Echeverri F, Martin SJ, Force WR, Lynch DH, Ware CF, Green DR (1995): Cell autonomous Fas(CD95)/Fas-ligand interaction mediates activation-induced apoptosis in T cell hybridomas. Nature 373:441–44.

Cao J, Park I-W, Cooper A, Sodroski J (1996): Molecular determinants of acute single-cell lysis by human immunodeficiency virus type 1. J Virol 70:1340–54.

Caputo A, Sodroski JG, Haseltine WA (1990): Constitutive expression of HIV-1 tat protein in human Jurkat T cells using a BK virus vector. J AIDS 3:372–79.

Carbonari M, Cibati M, Pesce AM, Sbarigia D, Grossi PDO, Luzi GG, Fiorilli M (1995): Frequency of provirus-bearing CD4$^+$ cell in HIV type 1 infection correlates with extent of in vitro apoptosis of CD8$^+$ but not of CD4$^+$ cells. AIDS Res Hum Retroviruses 11:789–94.

Cheynier R, Henrichwark S, Hadida F, Pelletier E, Oksenhendler E, Autran B, Wain-Hobson S (1994): HIV and T cell expansion in splenic white pulps is accompanied by infiltration of HIV-specific cytotoxic T lymphocytes. Cell 78:373–87.

Chinnaiyan AM, O'Rourke K, Yu G-L, Lyons RH, Garg M, Duan DR, Xing L, Gentz R, Ni J, Dixit VM (1996): Signal transduction by DR3, a death domain-containing receptor related to TNFR-1 and CD95. Science 274:990–92.

Chirmule N, Goonewardena H, Pahwa S, Pasieka R, Kalyanaraman VS, Pahwa S (1995): HIV-1 envelope glycoproteins induce activation of activated protein-1 in CD4$^+$ T cells. J Biol Chem 270:19364–69.

Clerici M, Shearer GM (1994): The Th1-Th-2 hypothesis of HIV infection: New insights. Immunol Today 15:575–81.

Cocchi F, DeVico AL, Garzino-Demo A, Arya SK, Gallo RC, P. Lusso P (1995): Identification of RANTES, MIP-1α, and MIP-1β as the major HIV-suppressive factors produced by CD8$^+$ T cells. Science 270:1811–15.

Coffin J (1990): Retroviridae and their replication. In Fields BN, Knipe DM (eds): Virology. New York: Raven Press pp 1437–500

Cohen DI, Tani Y, Tian H, Boone E, Samelson LE, Lane HC (1992): Participation of tyrosine phosphorylation in the cytopathic effect of human immunodeficiency virus-1. Science 256:542–45.

Collette V, Dutartre H, Benziane A, Roman-Morales R, Benarous R, Harris M, Olive D (1996): Physical and functional interaction of Nef with Lck. HIV-1 Nef-induced T-cell signaling defects. J Biol Chem 271:6333–41.

Corbeil J, Richman DD (1995): Productive infection and subsequent interaction of CD4-gp120 at the cellular membrane is required for HIV-induced apoptosis of CD4$^+$ T cells. J Gen Virol 76:681–90.

Corbeil J, Tremblay M, Richman DD (1996): HIV-induced apoptosis requires the CD4 receptor cytoplasmic tail and is accelerated by interaction of CD4 with p56^{lck}. J Exp Med 183:39–48.

Cotton MF, Ikle, DN, Rapaport, EL, Marschner, S Kurrle, R, Finkel TH (1997): CD4$^+$ and CD8$^+$ T cell apoptosis correlates with disease severity in HIV-1 infection. Ped Res, 42:656–64.

Cotton MF, Casella C, Rapaport ER, Marschner S, Finkel TH (1996): Apoptosis in HIV-1 infection. In Krammer PH, Nagata S (eds): Mechanisms of Cell Death and Apoptosis, No. 97. Marburg, Germany: Behring Institute Mitteilungen, pp 1–12.

Crise B, Buonocore L, Rose JK (1990): CD4 is retained in the endoplasmic reticulum by the human immunodeficiency virus type 1 glycoprotein precursor. J Virol 64:5585–93.

Darzynkiewicz Z, Bruno S, Del Bino G, Gorczyca W, Hotz MA, Lassota P, Traganos F (1992): Features of apoptotic cells measured by flow cytometry. Cytometry 13:795–808.

Deng HK, Liu R, Ellmeier W, Choe S, Unutmaz D, Burkhart M, Marzio PD, Marmon S, Sutton RE, Hill CM, Davis CB, Peiper SC, Schall TJ, Littman DR, Landau

NR (1996): Identification of a major co-receptor for primary isolates of HIV-1. Nature 381:661–66.

Dhien J, Walczak H, Baumler C, Debatin K-M, Krammer PH (1995): Autocrine T cell suicide mediated by APO-1 (Fas/CD95). Nature 373:438–41.

Dive C, Gregory CD, Phipps DJ, Evans DL, Miller AI, Wyllie AH (1992): Analysis and discrimination of necrosis and apoptosis (programmed cell death) by multiparameter flow cytometry. Biochim Biophys Acta 1133:275–85.

Dragic T, Litwin V, Allaway GP, Martin SR, Huang V, Nagashima KA, Cayanan C, Maddon PJ, Koup RA, Moore JP, Paxton WA (1996): HIV-1 entry into CD4+ cells is mediated by the chemokine receptor CC-CKR-5. Nature 381:667–73.

Duke RC, Cohen JJ (1992): Morphological and biochemical assays of apoptosis. In Coligan JE, Kruisbeek AM, Margulies DH, Shevach EM, Strober W (eds): Current Protocols in Immunology. New York, NY: John Wiley and Sons, pp 3.17.1–3.17.16.

Effros RB, Allsopp R, Chiu C-P, Hausner MA, Hirji K, Wang L, Harley CB, Villeponteau B, West MD, Giorgi JV (1996): Shortened telomers in the expanded CD28-CD8+ cell subset in HIV disease implicate replicative senescence in HIV pathogenesis. AIDS 10:F17–F22.

Embertson J, Zupancic M, Ribas JL, Burke A, Racz P, Tenner-Racz K, Hasse AT (1993): Massive covert infection of helper T lymphocytes and macrophages by HIV during the incubation period of AIDS. Nature 362:359–62.

Endres MJ, Clapham PR, Marsh M, Ahuja M, Turner JD, McKnight A, Thomas JF, Stoebenau-Haggarty B, Choe S, Vance PJ, Wells TNC, Power CA, Sutterwala SS, Doms RW, Landau NR, Hoxie JA (1996): CD4-independent infection by HIV-2 is mediated by fusin/CXCR4. Cell 87:745–56.

Ensoli B, Barillari G, Salahuddin SZ, Gallo RC, Wong-Staal F (1990): Tat protein of HIV-1 stimulates growth of cells derived from Kaposi's sarcoma lesions of AIDS patients. Nature 345:84–86.

Estaquier J, Idziorek T, Zou W, Emilie D, Farber C-M, Bourez J-M, Ameisen JC (1995): T helper type 1/T helper type 2 cytokines and T cell death: Preventive effect of interleukin 12 on activation-induced and CD95 (Fas/Apo-1)-mediated apoptosis of CD4+ T cells from human immunodeficiency virus-infected persons. J Exp Med 182:1759–67.

Estaquier JB, Idziorek FD, Barre-Sinoussi F, Hurtrel B, Aubertin-A-M, Venet A, Mehtali M, Muchmore E, Michel P, Mouton V, Girard M, Ameisen JC (1994): Programmed cell death and AIDS: Significance of T-cell apoptosis in pathogenic and nonpathogenic primate lentiviral infection. Proc Natl Acad Sci USA 91:9431–35.

Feinberg MB (1996): Changing the natural history of HIV disease. Lancet 348:239–46.

Fermin CD, Garry RF (1992): Membrane alterations linked to early interactions of HIV with the cell surface. Virology 191:941–46.

Finkel TH, Tudor-Williams G, Banda NK, Cotton MF, Curiel T, Monks C, Baba TW, Ruprecht RM, Kupfer A (1995): Apoptosis occurs predominantly in bystander cells and not in productively infected cells of HIV- and SIV-infected lymph nodes. Nature Med 1:129–34.

Forster S, Berverley P, Aspinall R (1995): gp120-induced programmed cell death in recently activated T cells without subsequent ligation of the T cell receptor. Eur J Immunol 25:1778–82.

Furtado MR, Kingsley LA, Wolinsky SM (1995): Changes in the viral mRNA expression pattern correlate with a rapid rate of CD4$^+$ T-cell number decline in human immunodeficiency virus type 1–infected individuals. J Virol 69:2092–100.

Gelfand EW, Finkel TH (1996): The T-lymphocyte system. In Stiehm ER (ed): Immunologic Disorders in Infants and Children. WB Saunders Co., Philadelphia, pp. 14–34.

Gibellini D, Caputo A, Celeghini C, Bassini A, La Placa M (1995): Tat-expressing Jurkat cells show an increased resistance to different apoptotic stimuli, including acute human immunodeficiency virus-type 1 (HIV-1) infection. Brit J Heam 89:24–44.

Glynn JM, McElligott DL, Mosier DE (1996): Apoptosis induced by HIV infection in H9 T cells is blocked by ICE-family protease inhibition but not by a Fas(CD95) antagonist. J Immunol 157:2754–58.

Gougeon M-L, Garcia S, Heeney J, Tschopp R, Lecoeur H, Guetard D, Rame V, Daughet C, Montagnier L (1993): Programmed death of T cells in AIDS-related HIV and SIV infection. AIDS Res Hum Retroviruses 9:553–63.

Gougeon ML, Lecoeur H, Dulioust A, Enouf M-G, Crouvoisier M, Goujard C, Debord T, Montagnier L (1996): Programmed cell death in peripheral lymphocytes from HIV-infected persons. J Immunol 156:3509–20.

Groux HG, Torpier D, Monte D, Mouton Y, Capron A, Ameisen JC (1992): Activation-induced death by apoptosis in CD4$^+$ T cells from human immunodeficiency virus-infected asymptomatic individuals. J Exp Med 175:331–40.

Green DR, Scott DW (1994): Activation-induced apoptosis in lymphocytes. Curr Opin Immunol 6:476–87.

Greenway A, Azad A, McPhee D (1995): Human immunodeficiency virus type 1 Nef protein inhibits activation pathways in peripheral blood mononuclear cells and T-cell lines. J Virol 69:1842–50.

Haase AT, Henry K, Zupancic M, Sedgewick G, Faust RA, Melroe H, Cavert W, Gebhard K, Staskus K, Zhang Z-Q, Dailey PJ, Balfour HH Jr, Erice A, Perelson AS (1996): Quantitative image analysis of HIV-1 infection in lymphoid tissue. Science 274:985–89.

He J, Choe S, Walker R, Marzio P, Morgan DO, Landau NR (1995): Human immunodeficiency virus type 1 protein R (Vpr) arrests cells in the G2 phase of the cell cycle by inhibiting p34^{cdc2} activity. J Virol 69:6703–11.

Heinkelein M, Sopper S, Jassoy C (1995): Contact of human immunodeficiency virus type 1–infected and uninfected CD4$^+$ T lymphocytes is highly cytolytic for both cells. J Virol 69:6925–31.

Hengel H, Flohr T, Hammerling GJ, Koszinowski UH, Momburg F (1996): Human cytomegalovirus inhibits peptide translocation into the endoplasmic reticulum for MHC class I assembly. J Gen Vir 77:2287–96.

Hill AB, Barnett BC, McMichael AJ, McGeoch DJ (1994): HLA class I molecules are not transported to the cell surface in cells infected with herpes simplex virus types 1 and 2. J Immunol 152:2736–41.

Ho DD, Neumann AU, Perelson AS, Chen W, Leonard JM, Markowitz M (1995): Rapid turnover of plasma virions and CD4 lymphocytes in HIV-1 infection. Nature 373:123–26.

Howcroft TK, Strebel K, Martin MA, Singer DS (1993): Repression of MHC class I gene promoter activity by two-exon Tat of HIV. Science 260:1320–22.

Howie SE, Sommerfield AJ, Gray E, Harrison DJ (1994): Peripheral T lymphocyte depletion by apoptosis after CD4 ligation in vivo: Selective loss of CD44⁻ and "activating" memory T cells. Clin Exp Immunol 95:195–200.

Ioannidis JPA, Cappelleri JC, Lau J, Sacks HS, Skolnik PR (1996): Predictive value of viral load measurements in asymptomatic untreated HIV-1 infection: A mathematical model. AIDS 10:255–62.

Janeway CA Jr, Travers P (1996): Immunobiology. London, San Francisco and Philadelphia: Current Biology Ltd, Second Edition.

Jones TR, Wiertz EJ, Sun L, Fish KN, Nelson JA, Ploegh HL (1996): Human cytomegalovirus US3 impairs transport and maturation of major histocompatibility complex class I heavy chains. Proc Nat Acad Sci 93:11327–33.

Jowett JBJ, Planelles V, Poon B, Shah NP, Chen M-L, Chen ISV (1995): The human immunodeficiency virus type 1 vpr gene arrests infected T cells in the G2 + M phase of the cell cycle. J Virol 69:6304–13.

Ju S-T, Panka DJ, Cui H, Ettinger R, El-Khatib M, Sherr DH, Stanger BZ, Marshak-Rothstein A (1995): Fas(CD95)/FasL interactions required for programmed cell death after T cell activation. Nature 373:444–48.

Kabelitz D, Pohl T, Pechhold K (1993): Activation-induced cell death (apoptosis) of mature peripheral T lymphocytes. Immunol Today 14:338–39.

Kafalas P, Brown TR, Brickell PM (1995): Signalling by the p60c-src family of protein-tyrosine kinases. Inter J Biochem Cell Biol 27:551–63.

Katsikis PD, Wunderlich ES, Smith CA, Herzenberg LA (1995): Fas antigen stimulation induces marked apoptosis of T lymphocytes in human immunodeficiency virus-infected individuals. J Exp Med 181:2029–36.

Koga Y, Sasaki M, Yoshida H, Wigzell H, Kimura G, Nomoto K (1990): Cytopathic effect determined by the amount of CD4 molecules in human cell lines expressing envelope glycoprotein of HIV. J Immunol 144:94–102.

Kolesnitchenko V, Wahl LM, Tian H, Sunila I, Tani V, Hartmann C-P, Cossman J, Raffeld M, Orenstein J, Samelson LE, Cohen DJ (1995): Human immunodeficiency virus 1 envelope-initiated G2-phase programmed cell death. Proc Natl Acad Sci USA 92:11889–93.

Laurence JA, Gottlieb AB, Kunkel HG (1983): Soluble suppressor factors in patients with acquired immunodeficiency syndrome and its prodrome. J Clin Invest 72:2072–81.

Lenardo MJ (1996): Fas and the art of lymphocyte maintenance. J Exp Med 183:721–24.

Levy JA (1983): Pathogenesis of human immunodeficiency virus infection. Microbiol Rev 57:183–289.

Li CJ, Wang C, Friedman DJ, Pardee AB (1995): Reciprocal modulations between p53 and Tat of human immunodeficiency virus type 1. Proc Natl Acad Sci USA 92:5461–64.

Li CJ, Friedman DJ, Wang C, Metelev V, Pardee AB (1995): Induction of apoptosis in uninfected lymphocytes by HIV-1 Tat protein. Science 268:429–31.

Luban J, Bossolt KL, Franke EK, Kalpana GV, Goff SP (1993): Human immunodeficiency virus type 1 Gag protein binds to cyclophilins A and B. Cell 73:1067–78.

Lynch DH, Ramsdell F, Alderson MR (1995): Fas and FasL in the homeostatic regulation of immune responses. Immunol Today 16:569–74.

Maldarelli F, Sato H, Berthold E, Orenstein J, Martin MA (1995): Rapid induction of apoptosis by cell-to-cell transmission of human immunodeficiency virus type 1. J Virol 69:6457–65.

Mellors JW, Rinaldo CR Jr, Gupta P, White RM, Todd JA, Kingsley LA (1996): Prognosis in HIV-1 infection predicted by the quantity of virus in plasma. Science 272:1167–70.

Meyaard L, Otto SA, Keet PM, Roos MTL, Miedema F (1994): Programmed death of T cells in human immunodeficiency virus infection. No correlation with disease progression to disease. J Clin Invest 93:982–88.

Montague JW, Gaido ML, Frye C, Cidlowski JA (1994): A calcium-dependent nuclease from apoptotic rat thymocytes is homologous with cyclophilin. Recombinant cyclophilins A, B, and C have nuclease activity. J Biol Chem 269:18877–80.

Muro-Cacho CA, Pantaleo G, Fauci AS (1995): Analysis of apoptosis in lymph nodes of HIV-infected persons. Intensity of apoptosis correlates with the general state of activation of the lymphoid tissue and not with stage of disease or viral burden. J Immunol 154:5555–66.

Nardelli B, Gonzalez CJ, Schechter M, Valentine FR (1995): CD4$^+$ blood lymphocytes are rapidly killed in vitro by contact with autologous human immunodeficiency virus-infected cells. Proc Natl Acad Sci USA 92:7312–16.

Neal TF, Holland HK, Baum CM, FVillinger F, Ansari AA, Saral R, Wingard JR, Fleming WH (1995): CD34$^+$ progenitor cells from asymptomatic patients are not a major reservoir for human immunodeficiency virus-1. Blood 86:1749–56.

Nishino Y, Nakaya T, Fujinaga K, Kishi M, Azuma I, Ikuta K (1994): Persistent infection of MT-4 cells by human immunodeficiency virus type 1 becomes increasingly likely with *in vitro* serial passage of wild-type but not *nef* mutant virus. J Gen Virology 75:2241–51.

O'Brian WA, Hartigan PM, Martin D, Esinhart J, Hill A, Benoit S, Rubin M, Simberkoff MS, Hamilton JD (1996): Changes in plasma HIV-1 RNA and CD4$^+$ lymphocyte counts and the risk of progression to AIDS. N Engl J Med 334:426–31.

Oyaizu N, McCloskey TW, Than S, Hu R, Kalyanaraman VS, Pahwa S (1994): Crosslinking of CD4 molecules upregulates Fas antigen expression in lymphocytes by inducing interferon-gamma and tumor necrosis factor-alpha secretion. Blood 84:2622–31.

Pantaleo G, Graziozi C, Dermarest JF, Butini L, Montroni M, Fox C, Orenstein JM, Kotler DP, Fauci AS (1993): HIV infection is active and progressive in lymphoid tissue during the clinically latent stages of disease. Nature 362:355–63.

Penninger JM, Mak TW (1994): Signal transduction, mitotic catastrophes, and death in T-cell development. Immunol Rev 142:231–72.

Perelson AS, Neumann AU, Markowitz M, Leonard JJM, Ho DD (1996): HIV-1 dynamics *in vivo:* Virion clearance rate, infected cell life-span and viral generation time. Science 271:1582–86.

Piatak M, Saag MS, Yang LC, Clark SJ, Kappes JC, Luk K-C, Hahn BH, Shaw GM, Lifson JD (1993): High levels of HIV-1 in plasma during all stages of infection determined by competitive PCR. Science 259:1749–54.

Popik W, Pitha PM (1996): HIV-1 binding to CD4 activates Raf-1 by Ras-independent pathway. J AIDS & Hum Retrovir, Abstracts of the 1996 Annual Meeting of the Institute of Human Virology:6.

Radrizzani M, Accornero P, Amidei A, Aiello A, Delia D, Kurrle R, Colombo MP (1995): IL-12 inhibits apoptosis induced in a human Th1 clone by gp120/CD4 cross-linking and CD3/TCR activation or by IL-2 deprivation. Cell Immunol 161:14–21.

Re F, Braaten D, Franke EK, Luban J (1995): Human immunodeficiency virus type 1 arrests the cell cycle in G2 by inhibiting the activation of p34^{cdc2}–cyclin B. J Virol 69:6859–64.

Re MC, Zauli G, Gibellini D, Furlini G, Ramazzotti R, Monari P, Ranieri S, Capitani S, La Placa M (1993): Uninfected haematopoietic progenitor (CD34$^+$) cells purified from the bone marrow of AIDS patients are committed to apoptotic cell death in culture. AIDS 7:1049–55.

Rogel ME, Wu LI, Emmerman M (1995): The human immunodeficiency virus type 1 vpr gene prevents cell proliferation during chronic infection. J Virol 96:882–88.

Ruderman JV (1993): MAP kinases and the activation of quiescent cells. Curr Opin Coll Biol 5:207 13.

Saksela K, Cheng G, Baltimore D (1995): Proline-rich (PxxP) motifs in HIV-1 Nef bind to SH3 domains of subset of Src kinases and are required for the enhanced growth of Nef$^+$ viruses but not for down-regulation of CD4. EMBO J 14:484–91.

Salghetti S, Mariani R, Skowronski J (1995): Human immunodeficiency virus type 1 Nef and p56^{lck} protein-tyrosine kinase interact with a common element in CD4 cytoplasmic tail. Proc Natl Acad Sci USA 92:349–53.

Sarin A, Clerici M, Blatt SP, Hendrix CW, Shearer GM, Henkart PA (1994): Inhibition of activation-induced programmed cell death and restoration of defective immune responses of HIV$^+$ donors by cysteine protease inhibitors. J Immunol 153:862–72.

Schwartz O, Marechal V, Le Gall S, Lemonnier F, Heard JM (1996): Endocytosis of major histocompability complex class I molecules is induced by the HIV-1 Nef protein. Nature Med 2:338–42.

Shi L, Nishioka J, Th'ng JB, Litchfield DW, Greenberg AH (1994): Premature activation required for apoptosis. Science 263:1143–45.

Silvestris F, Carrorio P, Frassanito MA, Tucci M, Romito A, Nagata S, and Dammacco F (1996): Overexpression of Fas antigen of T cells in advanced HIV-1 infection: Differential ligation constantly induces apoptosis. AIDS 10:131–41.

Solary E, Dubrez L, Eymin B (1996): The role of apoptosis in the pathogenesis and treatment of diseases. Eur Respir J 9:1293–305.

Spear GT, Landay AL, Sillivan BL, Dittel B, Lint TF (1990): Activation of complement on the surface of cells infected by human immunodeficiency virus. J Immunol 144:1490–96.

Su L, Kaneshima H, Bonyhadi M, Salimi S, Kraft D, Rabin L, McCune JM (1995): HIV-1 induced thymocyte depletion is associated with indirect cytopathicity and infection of progenitor cells in vivo. Immunity 2:25–36.

Swat W, Ignatowica L, Kisielow P (1991): Detection of apoptosis of immature CD4$^+$8$^+$ thymocytes by flow cytometry. J Immunol Meth 137:79–87.

Thompson CB (1995): Apoptosis in the pathogenesis and treatment of disease. Science 267:1456–62.

Touchette N (1995): Evolutions: Apoptosis. J NIH Res 7:9193.

Tough DF, Borrow P, Sprent J (1996): Induction of bystander T cell proliferation by viruses and type I interferon in vivo. Science 272:1947–49.

Trkola A, Dragic T, Arthos J, Binley JM, Olson WC, Allaway GP, Cheng-Mayer C, Robinson J, Maddon PJ, Moore JP (1996): CD4-dependent, antibody-sensitive interactions between HIV-1 and its co-receptor CCR-5. Nature 384:184–87.

Viscidi RP, Mayur K, Lederman HM, Frankel AD (1989): Inhibition of antigen-induced lymphocyte proliferation by Tat protein from HIV-1. Science 246:1606–8.

Walker CM, Moody DJ, Stitles DP, Levy JA (1988): CD8$^+$ lymphocytes can control HIV infection in vitro by suppressing virus replication. Science 234:1563–66.

Wei X, Ghosh SK, Taylor ME, Johnson VA, Emini EAD, Lifson JD, Bonhoeffer S, Nowak MA, Hahn BH, Saag MS, Shaw GM (1995): Viral dynamics in human immunodeficiency virus type 1 infection. Nature 373:117–22.

Weinhold KJ, Lyerly HK, Stanley SD, Austin AA, Matthews TJ, Bolognesi DP (1989): HIV-1 gp120-mediated immune suppression and lymphocyte destruction in the absence of viral infection. J Immunol 142:3091–97.

Westendorp MO, Frank R, Ochsenbauer C, Stricker K, Dien J, Walcazk H, Debatin K-M, Krammer PH (1995): Sensitization of T cells to CD95-mediated apoptosis by HIV-1 Tat and gp120. Nature 375:497–500.

Wiertz EJ, Jones TR, Sun L, Bogyo M, Geuze HJ, Ploegh HL (1996): The human cytomegalovirus US11 gene product dislocates MHC class I heavy chains from the endoplasmic reticulum to the cytosol. Cell 84:769–79.

Willey RL, Maldarelli F, Martin MA, Strebel K (1992): Human immunodeficiency virus type 1 Vpu protein induces rapid degradation of CD4. J Virol 66:7193–200.

Wolthers KC, Wisman GBA, Otto SA, de Roda Husman A-M, Schaft N, de Wolf F, Goudsmit J, Coutinho RA, van der Zee AGJ, Meyaard L, Miedema F (1996): T cell telomere length in HIV-1 infection: No evidence for increased CD4$^+$ T cell turnover. Science 274:1543–47.

Wu L, Gerard NP, Wyatt R, Choe H, Parolin C, Ruffing N, Borsetti A, Cardoso AA, Desjardin E, Newman W, Gerard C, Sodroski J (1996): CD4-induced interaction of primary HIV-1 gp120 glycoproteins with the chemokine receptor CCR-5. Nature 384:179–83.

Zauli G, Gibelini D, Milani D, Mazzoni M, Borgatti P, La Placa M, Capitani S (1993): Human immunodeficiency virus type 1 Tat protein protects lymphoid, epithelial, and neuronal cell lines from death by apoptosis. Cancer Res 53:4481–85.

Zauli G, Gibellini D, Caputo A, Bassini A, Negrini M, Monne M, Mazzoni M, Capitani S (1995): The human immunodeficiency virus type-1 Tat protein upregulates Bcl-2 gene expression in Jurkat T-cell lines and primary peripheral blood mononuclear cells. Blood 86:3823–34.

Zheng B, Han S, Zhu Q, Goldsby R, Kelsoe G (1996): Alternative pathways for the selection of antigen-specific peripheral T cells. Nature 384:263–66.

Zinkernagel RM, Hengartner H (1994): T-cell-mediated immunopathology versus direct cytolysis by virus: Implications for HIV and AIDS. Immunol Today 15:262–68.

PART IV

CELL DEATH IN LONG-LIVED CELLS

CHAPTER 12

APOPTOSIS OF TISSUES OF THE EYE DURING DEVELOPMENT AND DISEASE

DAVID S. PAPERMASTER

Neuroscience Program, Department of Pharmacology, University of Connecticut Health Center, Farmington, CT 06030-3205

INTRODUCTION

Even a casual look at the eye reveals its elegant structure. The anterior compartments achieve transparency by an extraordinary variety of cellular and molecular strategies that, when altered in disease, immediately report themselves by their loss of their crucial feature—they become cloudy. In physical terms, they begin to scatter light. Consequently, a clear image of the environment is lost. In the lens, an opacity is called a cataract, and its formation usually arises as the proteins in the cells become less soluble and spatially ordered. While cataracts may be readily replaced surgically, their treatment still consumes enormous amounts of the cost of health care because its development is so common, especially in the elderly.

In the back of the eye, the light-sensitive retina records the color and intensity of our world with amazing resolution and responsiveness—the dark-adapted rod responds to a single photon! There is simply no other sensory system capable of this level of amplification. While for most of us, this tissue faithfully goes about its lifelong work, asking only that it be provided with 1% of our cardiac output (although it weighs only about 1 gram) and that we not injure it by looking directly at the sun, for some, the retinal neurons die prematurely, that is, before the death of their host. Again, this problem is becoming increasingly common as our populations age. Retinal neurons have no capacity for renewal in adulthood; the result is blindness. Many disorders of

When Cells Die, Edited by Richard A. Lockshin, Zahra Zakeri, and Jonathan L. Tilly
ISBN 0-471-16569-7 © 1998 Wiley-Liss, Inc.

the retina, both acquired and inherited, have started to reveal their underlying causes, and most of them involve the activation of programmed cell death cascades with all the classical features of apoptosis.

The anterior compartments, the cornea, the lens, and the fluids and supporting structures of the anterior and posterior chambers differ from each other and from the neural retina in the back of the eye by as great a structural and developmental gap as exists in any tissue of our bodies. Remarkably, however, their entire coordinated development is under the control of only one or a few genes, and failures in the formation of any part leads to major consequences for the other parts. Two examples will illustrate this point. In the fruit fly, Drosophila melanogaster, a mutated gene, eyeless, is a master gene controlling the formation of all the structures of the eye. So, transgenic flies, with an extra copy of the normal gene under the control of genes controlling leg structure, place extra eyes on their legs (Halder et al., 1995). Eyeless is conserved, and its human homologue, Pax-6, when mutated also leads to malformation of the eye. Human Pax-6 mutation leads to aniridia, the absence of the iris, and to more serious failures of eye development (Cvekl and Piatigorsky, 1996). This conservation of structure and function of these genes was not expected, for the structures of the compound eye of the fly and the human eye were thought to have emerged by two independent evolutionary steps (Quiring et al., 1994).

A second example is a rare inherited malignancy, retinoblastoma, which primarily afflicts very young children from birth to two years of age. It was the source of the concept of tumor suppressor genes based on its pattern of inheritance (Knudsen, 1971). The analysis of the gene's structure and function revealed its central role in control of the mitotic cycle of all cells, not just in the eye (see reviews by Jacks and Weinberg, 1996, Lee et al., 1995, and Weinberg, 1996). Several oncogenic viruses gain their capacity by binding to and inactivating the retinoblastoma gene product, pRb. Loss of pRb activity allows the infected cells to enter the proliferative cycle (Nevins, 1992). The transgenic expression of these viral oncoproteins in the eye has generated some provocative findings about cell death of photoreceptors and lens fiber cells and has revealed new complexities about the molecular pathogenesis of cell death. Thus the study of the structures of the eye, its diseases, and its specialized gene products have provided insights not only of value to vision research, but also to general aspects of the role of apoptosis in developmental biology and disease.

THE DEVELOPMENT OF THE EYE AND THE DIFFERENTIATION OF THE LENS

As the neural tube begins to elongate and thicken, two bulging masses of neuroectoderm arise along the anterior-lateral walls that progressively enlarge to form a distinct pair of protrusions that collapse upon themselves to form cup-shaped structures after approaching the lateral ectoderm. At that point, the ectoderm invaginates to form a vesicle that detaches to become the primitive lens while the overlying ectoderm prepares to become the cornea and eyelids. The overlying ectoderm remains continuous and shrouds the entire structure until apoptosis of the epithelium separates the two lids from each other. Lateral neuroectodermal cells migrate into the primitive cornea to form the stromal keratocytes that lay down unique highly ordered arrays of collagen between the overlying corneal squamous epithelium and the endothelium, which faces the aqueous (Figure 12.1).

Meanwhile, the lens vesicle enlarges and undergoes a dramatic transformation into a two-component tissue. The anterior epithelium remains cuboidal, relatively undifferentiated and can proliferate throughout life at the equator of the lens. As the cells divide, they penetrate into the interior of the lens and differentiate into long fiber cells that undergo a unique transformation. They synthesize cytoplasmic proteins at high concentration, including special proteins that appear to keep their neighbors from denaturing, the αA- and αB-crystallins (Rao et al., 1995). They also select molecules from the repertoire of different enzymes in each species (e.g. enolase), to be synthesized at high concentration to serve refraction (Piatigorsky and Wistow, 1989). As a consequence of the high protein concentration, these cells become transparent. Interference in the orderly packing or folding of these proteins leads to light scattering—a cataract.

After completion of synthesis of the crystallins and the establishment of tight junctions with their neighbors, the lens fiber cells lose their organelles, including nuclei, endoplasmic reticulum, and mitochondria by a process that in part resembles a form-fruste of apoptotis. But the fiber cells don't die and never are removed from the lens. Thus we look through lens cells that were born during our embryogenesis, and we continue to add mature fiber cells at the lens equator like the layers of an onion for our entire lives. Manipulation of the function of the Rb protein induces abnormalities in the proliferation or differentiation of the fiber cells and leads to fiber cell apoptosis and cataract formation. These studies offer provocative insights into the role of the cell cycle and apoptosis and will be summarized in the following.

APOPTOSIS OF THE RETINA DURING DEVELOPMENT

Shortly after the original descriptions of apoptosis by Kerr, Wyllie, and Currie (1972), and Wyllie (1980), Young (1984) and Penfold and Provis (1986) studied the histology of the developing retina and quantitated dying cells by counting pyknotic nuclei and demonstrated the phagocytosis of the dying cells by neighboring photoreceptors. This approach revealed that apoptosis was an important component of the formation of the mature population of retinal cells.

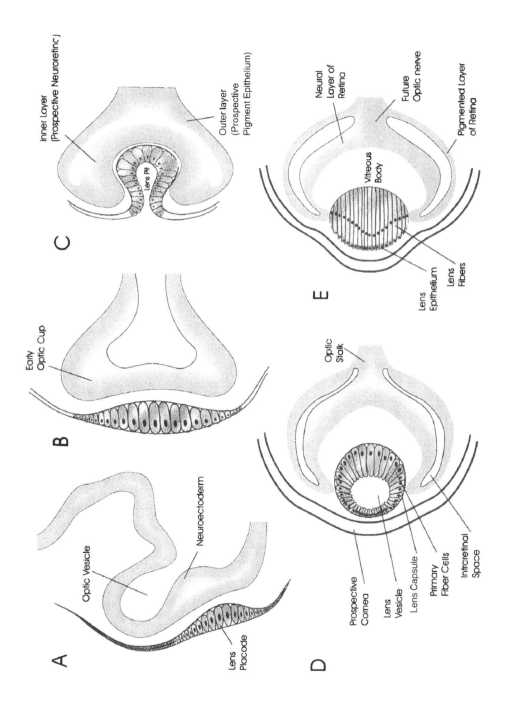

A

Optic Vesicle

Neuroectoderm

Lens Placode

B

Early Optic Cup

C

Inner Layer (Prospective Neuroretina)

Lens Pit

Outer layer (Prospective Pigment Epithelium)

D

Prospective Cornea

Lens Vesicle

Lens Capsule

Primary Fiber Cells

Intraretinal Space

Optic Stalk

E

Neural Layer of Retina

Future Optic nerve

Vitreous Body

Pigmented Layer of Retina

Lens Epithelium

Lens Fibers

Since the lifetime of the dying nuclei could not be determined, the method, *ipso facto* limits the capacity to estimate death rates accurately. By marking dividing cells with tritiated thymidine and locating them by autoradiography, Young (1984) traced their birth dates and differentiated fates in detail. A precise order in the movement and differentiation of these cells from an undifferentiated ventricular neuroectoderm leads successively to the birth of ganglion cells, cones, amacrine, Mueller and horizontal cells, and finally to the rods and bipolar neurons. Not all the cells survive to become mature components of the adult retina. Over half the retinal ganglion cells die by apoptosis (Wong et al., 1987). Their fate is apparently tied to their success or failure in wiring to contacts in the lateral geniculate body, the next station in the pathway of visual transmission to the central nervous system. If the ganglion cells begin to die again in adulthood, the ensuing blindness is called glaucoma. It is usually but not always associated with elevated intraocular pressure.

Upon completion of the separation of the ganglion cells from the rest of the developing retina, the remaining neuroblasts, now termed the ventricular layer, continue to divide rapidly to generate the remaining cell types and layers of the retina. During this stage, apoptosis of the ventricular neuroepithelial cells continues, and conventional histologic stains reveal the nuclear pyknosis. DNA cleavage in the nuclei during apoptosis is especially easily revealed by the TUNEL assay (Fig. 12.2a). Finally, as photoreceptor cells and bipolar cells complete differentiation, the entire retina becomes post-mitotic with no detectable apoptosis (Fig. 12.2b). With good health, most of these cells survive the entire lifetime of their host so that the healthy elderly human, for example, need not experience any significant loss of visual acuity. Why then are we so familiar with middle-aged and elderly humans whose vision is impaired? What has happened to their retinas so that they cannot read or recognize faces or lose lateral vision?

FIG. 12.1. Highly schematic diagram of the early events in mouse eye development. (A) At embryonic day 9 to 9.5 (E 9–9.5), the optic vesicle is attached to the ventral wall of the prosencephalon via the optic stalk. The lens placode (prospective lens) becomes apparent as a thickened area of the surface ectoderm. (B) At E 9.5–10, the area of the lens placode has enlarged. (C) At E 10.5, the central part of the lens placode indents to form the lens pit and the optic vesicle invaginates to form the optic cup. (D) At about E 11.5, the lens pit is converted into the lens vesicle which is surrounded by a capsule. (E) At E 13.5, the lens comprises the anterior cuboidal epithelial cells and the posterior elongating fiber cells. The neural retinal layer behind the lens begins to differentiate and the primitive cornea develops in front of the lens. (Modified from Cveki and Piatigorsky, 1996, with permission.)

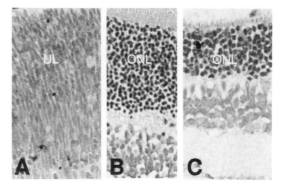

FIG. 12.2. Apoptosis of the retinal neurons in the developing and mature eye. Apoptotic nuclear DNA that has been cleaved during apoptosis is 3′ end-labeled with biotinyl-dUTP by terminal deoxyribonucleotidyl transferase. Biotins are detected with streptavidin-peroxidase and diaminobenzidine in this TUNEL assay (Gavrieli et al., 1992). (A) Mouse retina at post-natal age 4 days (P4). The ganglion cell layer has partially separated from the differentiating ventricular layer. A few apoptotic cells (dark nuclei) are scattered throughout the deeper ventricular layer. (B) Mature mouse retina. No TUNEL-labeled cells are seen. (C) P14 *rds/rds* mouse retina. Despite the complete maturation of the photoreceptors, no outer segments form. The photoreceptors die at a slow rate revealed both by the TUNEL assay and the thinning of the outer nuclear layer of photoreceptor nuclei.

APOPTOSIS OF THE HUMAN RETINA IN DISEASE

Like most recent studies of apoptosis in other tissues, research into the role of apoptosis in eye disease awaited the presentation of convincing evidence of the presence of apoptosis-promoting and antiapoptotic genes. Since 1993, a flurry of activity led quickly to the documentation of apoptosis of photoreceptors affected by a variety of inherited degenerations or environmental stimuli (for recent references, see reviews by Adler, 1996, Papermaster, 1997, and Papermaster and Windle, 1995).

Photoreceptor Cell Structure and Visual Cell Excitation

To understand the photoreceptor degenerations, we must first make a small diversion into their special microanatomy and the organelles they possess that give them their capacity to capture light. Rods and cones are elongated neurons that completely compartmentalize their specialized functions. In the center of the cell, they resemble other neuronal cell bodies and contain the usual machinery for biosynthesis, including, nuclei, rough endoplasmic reticulum, Golgi apparatus, and mitochondria. They elongate an axon and synapse with neurons in the adjacent layer. At the other end, however, they project a nonmotile (9 + 0) cilium and then progressively enlarge the plasma membrane of the tip of the cilium into an exquisitely arrayed stack of membranous disks, termed the outer segment, that are laden with the light-sensitive protein, rhodopsin, containing its bound vitamin A, retinal, in its interior (Fig. 12.3).

Rhodopsin, a heptahelical transmembrane protein, is coupled to a trimeric G protein, transducin. Upon light capture, retinal undergoes a cis–trans isomerization, exciting rhodopsin, which activates 500 molecules of transducin before phosphorylation of its C-terminal domain and binding of arrestin shuts rhodopsin off. Transducin dissociates and its α-subunits bind to inhibitory γ-subunits of cyclic-GMP phosphodiesterase (PDE), releasing activated α- and β-PDE subunits which degrade cGMP to low levels. This removes cGMP from the active site of a plasma membrane cGMP-gated Na^+–Ca^{2+} channel, closing the channel, hyperpolarizing the membrane, generating a graded potential that propagates to the axon tip and blocks synaptic transmitter release. Thus light shuts off the photoreceptor (Hargrave, 1991; Hargrave et al., 1993; Palczewski, 1994). The horizontal and bipolar cells of the adjacent layer in the retina register this change in the rate of release of neurotransmitter and alter their signals to the ganglion cell appropriately. These cells initiate image processing to deeper layers of the retina and the CNS.

Mutations Causing Retinal Degenerations in Humans Have Parallels in Mice

Mutations of some specialized proteins involved in the photoactivation cascade, including rhodopsin, both α- and β-PDE chains, and the cGMP channel, induce the death of photoreceptors. Curiously, most of the rhodopsin mutations induce a slow, dominantly inherited degeneration so that vision is lost in mid-life. Other mutations localized to nearly every human chromosome induce a variety of retinal degenerations (Bird, 1995).

In addition, mutations arise in the structural proteins involved in generating the special disk-shaped structure of the outer segment membranes. One of these proteins, peripherin/rds, lies at the rim of the disk, and its name reflects both its site of function and its association with a slow retinal degeneration. Detailed studies by Sanyal and colleagues (Sanyal t al., 1980; Jansen et al., 1987) showed that homozygous mice could not form outer segments. Immunocytochemical studies and determinations of the rate of synthesis of photoreceptor proteins showed that these rods synthesize rhodopsin at normal rates and transport it normally to the tip of the cilium (Agarwal et al., 1990). There, unable to become organized into a disk, the membranes fall off the tip as small rhodopsin-laden vesicles that are avidly phagocytosed by the adjacent retinal pigment epithelium cells (Nir and Papermaster, 1986). These imperfect photoreceptor cells die slowly by apoptosis (Fig. 12.2c) (Papermaster and Nir, 1994). Further molecular studies of the protein showed that it forms homodimers that also link to homodimers of a similar molecule, named rom-1. Mutations interfere with homodimerization and disk formation (Goldberg and Molday, 1996). In an extraordinary twist, a mutation of peripherin/*RDS*

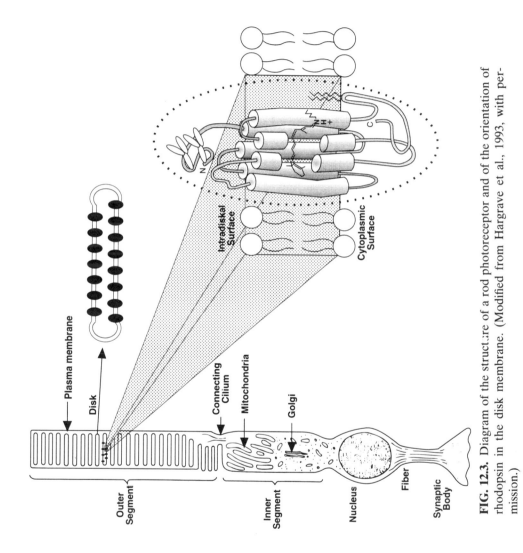

FIG. 12.3. Diagram of the structure of a rod photoreceptor and of the orientation of rhodopsin in the disk membrane. (Modified from Hargrave et al., 1993, with permission.)

Plasma membrane

Disk

Intradiskal Surface

Cytoplasmic Surface

Connecting Cilium

Mitochondria

Golgi

Outer Segment

Inner Segment

Nucleus

Fiber

Synaptic Body

that by itself caused no disease (i.e., was recessive), when coupled to a mutated *ROM-1* that also, by itself, caused no disease (also recessive), generated defective disk formation in a double heterozygote—a digenic retinal degeneration (Kajiwara et al., 1994).

Why should mutations of a light-sensitive protein, rhodopsin, or a disk-forming protein, peripherin/rds or rom-1, cause a cell to die? Clearly, all other neurons need no outer segment to survive. Do these mutations cause a fundamental alteration in the metabolism of photoreceptor cells that compromises their viability in some unique way? If so, several important questions remain. First, the cells do not die uniformly. Rods are generally more sensitive than cones if the mutations affect both cells (only rods express rhodopsin; cones use homologous color-sensitive proteins). The death of the cells can either be widely distributed in a uniform scattered pattern over the entire retina, or, as is usually true in humans, cell death seems to begin in small islands and spreads to adjacent domains as well as dropping out throughout the retina. Finally, if these mutations are intrinsically damaging, why is the death of these cells delayed for years, even decades?

Two interesting families illustrate these dilemmas dramatically. In one family, a middle-aged sister was completely blind, and loss of both rod and cone function was documented electrophysiologically. She had a Pro-23-His mutation in the N-terminal domain of her rhodopsin. Her older sister, who drove a truck at night to earn a living, thought she was spared and was irritated when genetic studies revealed she had the same mutation. Her vision exam revealed a loss of rod function. That meant her children might not be spared the disease (Berson, 1993). Why could she see? Preservation of cone function in her retina and use of headlights and street lights provided her with sufficient acuity in bright light. She was therefore unaware that she had lost the use of 90% of her photoreceptors, her rods, which serve vision in dim light. Why did her cones survive and her sister's cones die? If we could answer that question, this family's disability could be ameliorated at a practical level without ever preserving their rods.

In a second family, a 68-year-old man with a Thr-17-Met mutation in his rhodopsin's N-terminal domain, lost all rods and cones in the bottom half of each eye. His rods survived perfectly in the top half. Accordingly, since the visual image is inverted, he could not see above the horizon but had no difficulty avoiding objects while walking. Unfortunately, the cones in his fovea were not spared so that he lost high-acuity vision and was legally blind. His 35-year-old son is developing a similar "altitudinal" degeneration (Li et al., 1994). Since the father's dorsal rod photoreceptors were dying so slowly, it took a clever application of the TUNEL assay in an *en face* view of a large area of retina to reveal the rare apoptotic photoreceptor in the boundary between the healthy and injured parts of his retina (Li and Milam, 1995). If all rods bear the same mutation, why do half live for so long, and others die in mid-life? Since cones do not express the rhodopsin gene, why should they die at all in such families?

These clinical dilemmas led me to propose recently that the rhodopsin mutations are a necessary but insufficient cause of these degenerations (Paper-master, 1995). These families show that cones do not have to die just because the surrounding rods have died and that it is possible for some rods to survive throughout an entire lifetime with a mutation in the rhodopsin gene. Other factors, operating in ways that are not yet understood, are either adequate or inadequate to prevent the stimulus initiated by a rhodopsin mutation to start an apoptotic cascade. The first trials of a systemic therapy, taking large doses of vitamin A, seem to retard the rate of retinal degeneration in some families (Berson et al., 1993) and are an example of the potential for this sort of reasoning.

Not only is there heterogeneity in the level of expression of the diseases of the retina from one family member to another (so-called incomplete pene-trance) and variation of expression within the retina leading to a region of visual loss (a scotoma), but also mutations of the same gene can cause diseases that were thought to be separate entities clinically (Dryja and Berson, 1995). For example, a human peripherin/RDS missense mutation can be associated with retinitis pigmentosa, while a null mutation results in a different pattern of retinal degeneration called retinitis punctata albescens (Kajiwara et al., 1993).

One obvious possibility for sorting out the genetic basis of incomplete penetrance and disease heterogeneity is to understand that mutations of differ-ent parts of the gene can markedly affect the phenotype. This appears to be the case in the peripherin/RDS mutation just described. In addition, the outbred genetics of humans contributes other genes that may affect photoreceptor survival and modify the phenotype of these dominant disorders. Most experi-mental studies of retinal degenerations in rats and mice involve inbred strains that eliminate genetic heterogeneity. Studies of disease-causing mutations in mice with different genetic backgrounds are just beginning and may reveal an important contribution in the retinal disease phenotype. Precedence for such effects in cancers induced in inbred mice offer insight. Strains that are resistant or permissive for the tumor's growth are well known.

As more of our population ages, a major disability, age-related macular degeneration (ARMD), becomes a national crisis, both socially and in the cost of care of the blind. These individuals lose the function of their cones in the central retina, termed the macula (Latin for "spot"). Here, predominantly red- and green-sensitive cones gather at very high density to serve high-acuity central vision, such as reading. As one of my patients with ARMD put it, "I don't bump into things, but I can't see the face of my grand-daughter." You can somewhat reproduce the impact of this disease by placing your thumb a few inches in front of the center of your eye and imagine life without this central field of view. By contrast, patients with retinitis pigmentosa, a heterogeneous collection of disorders usually affecting the mid-peripheral rods, have tunnel vision. You can imagine this loss by trying to look through your partially closed hand, forming a circle of the thumb and forefinger in front of your eye. In later stages, when cones die, they are blind.

Macular degeneration already affects 12 million Americans. In the next two decades, this number could double. Only Alzheimer's disease affects a part of the nervous system of so many individuals. Every other neuronal degeneration pales by comparison to this huge number, yet, because it primarily impacts on the elderly, who are taught that it is just another example of the loss of function associated with aging, little public arousal has surrounded this devastating disease. Its victims lose the opportunity to care for themselves, to read, to watch television and many other aspects of socialization just when visually oriented activities are among the most accessible ways for them to enjoy their senior years fruitfully. Because so few of the affected had elderly parents, we have little information concerning the inheritability of this late-onset photoreceptor degeneration. The limited histologic studies of donated eyes do not reveal a clear pathology, and there are active debates about the significance of several of the observed changes. A few kindreds with an inherited form of macular degeneration in North Carolina and in France are being intensively studied to determine the molecular defect (Small et al., 1997). Three recent studies, starting from different directions converged to reveal a surprising result about macular degeneration. A rare form of severe macular degeneration of children, Stargardt's disease, is a consequence of a mutation of a gene coding for a very large protein (210-240 kDa) which is a member of the ABC-transporter family and was termed *ABCR* (Allikmets et al, 1997a, Azarian and Travis, 1997, Illing et al, 1997). This gene family includes such genes as the cystic fibrosis Cl⁻ transporter and the multiple drug resistance gene. The protein had actually been discovered over 20 years previously in the author's lab as a consequence of studies of membrane biosynthesis in the retina and was the first protein localized immunocytochemically to the rim of the disk (Papermaster et al, 1976, 1979, 1982), in the same domain as the rds/peripherin and rom-1 proteins which were found a decade later. In our studies, using polyclonal antibodies, the large "rim protein" was localized not only to rod disks but also to the edges of cone lamellae (Papermaster et al, 1979, 1982). Yet a monoclonal antibody to the ABCR rim protein localizes it only to the rims of rod disks (Illing et al, 1997). When the grandparents of children with Stargardt's macular degeneration were studied, they had age related macular degeneration (Allikmets et al, 1997b). Mutations of this gene account for 16% of all cases of age related macular degeneration. Thus the same mutation, in the child with two defective copies causes photoreceptor cell death in a few years, while the heterozygous grandparents retain their vision for six or more decades. Again, as we have already discussed above in studies of rhodopsin mutations, a gene expressed in rods causes cone death. Is there a second, as yet undiscovered, homologous gene expressed in cones which can account for the remaining cases of age related macular degeneration? This dramatic outcome raises enormous new challenges for analysis of this late-onset form of neuronal cell death but also forshadows a fascinating new era in the studies of these special photoreceptor proteins and their role in the visual cells. Whether this will provide insight into the age-related form of the disease awaits completion of those studies.

EXPERIMENTAL STUDIES OF RETINAL DEGENERATIONS

Inherited retinal degenerations of *rds* mutant mice, mice with PDE (*rd*) mutations, and transgenic mice bearing mutant rhodopsin genes all show features that lead to the proposal that apoptosis is the "final common pathway" of retinal degeneration. Besides studies of mutant RCS (*rdy*) rats, albino rats exposed to bright light for 24 hours, rats with retinal detachment, and rat pups born to mothers exposed to lead also show all the criteria of apoptosis in ultrastructural images and by the usual biochemical approaches. Thus both inherited and environmentally induced injuries of the retina initiate the programs of cell death. These experiments have been the subjects of recent reviews (Adler, 1996, Papermaster, 1997; Papermaster and Windle, 1995).

Once this early phase of descriptive analysis established the broad aspects of retinal apoptosis, investigations began to clarify which mechanisms and pathways either induce or execute cell death. In contrast to the increasing precision of description arising from studies of various cells in tissue culture, the studies of retinal apoptosis have not provided simple answers and raise important questions about the degree to which the studies *in vitro* can be extrapolated to tissues residing in their natural site. Since the eye is an "accessible piece of brain," we hope that insights gained from studying its special neurons may provide insights into the neuronal degenerations that plague other parts of the nervous system.

Perturbation of the Cell Cycle by Viral Oncoproteins Causes Retinal Apoptosis and the Development of Retinal Tumors

DNA tumor viruses contain genes that promote cell growth by generating products that bind to cellular proteins involved in the cell cycle. The SV40 T antigen (Tag) binds pRb and p53. Human papilloma virus (HPV) oncoproteins E6 and E7, and adenovirus proteins E1A, E1B19K, and E1B55K, also bind to or otherwise alter the functions of p53 and pRb and related proteins, respectively. The combined actions of these oncoproteins constitute "two hits" in oncogenesis. SV40 Tag or HPV16 E7 inactivation of pRb enhances entry into the S phase of the cell cycle; the inactivation of p53 by SV40 Tag or HPV16 E6 blocks its function as a checkpoint that would normally induce apoptosis if the cell were proliferating inappropriately (Levine, 1997). Support for this paradigm developed using temperature-sensitive mutants of p53 and oncoproteins of adenovirus with similar properties (Debbas and White, 1993).

When SV40 Tag was expressed in the retina under the control of the rhodopsin promoter, it was fully expected that a tumor would form (a retinoblastoma). Instead, the photoreceptors died! If the retinas were removed shortly after birth, just as the photoreceptors began to dfiferentiate, the cells proliferated in tissue culture, something post-mitotic photoreceptors never do (Al-Ubaidi et al., 1992b). The native environment was somehow antiproliferative and conducive to apoptosis of these perturbed cells *in vivo*.

The rhodopsin promoter drives expression starting a few days before birth and for the remainder of the rod photoreceptor cell's life. An alternative model, which uses the interstitial retinoid binding protein (IRBP) upstream sequence, was evaluated. IRBP is a protein secreted by photoreceptors that function to transport the water-insoluble cis- and trans-vitamin A forms of retinal and retinol from the retinal pigmented epithelium (RPE) to the photoreceptor. It provides the 11-cis retinal needed by rhodopsin to regenerate after bleaching by light. IRBP-SV40 Tag transgenic mice develop retinoblastoma in the entire retina and in the pineal, where IRBP is also normally expressed (Al-Ubaidi et al., 1992a; Howes et al., 1993). IRBP expression begins a few days earlier than rhodopsin, just as ventricular neuroblasts are undergoing their last division before becoming post-mitotic. Thus proliferation or death of photoreceptors was determined by the developmental onset of expression of the oncoprotein, and only a few days made all the difference.

Because SV40 Tag is a large, complex molecule, the smaller E6 and E7 proteins of HPV16, an oncogenic strain of the virus, were chosen as perturbants. The same IRBP promoter driving expression of HPV16 E7 generated apoptosis of photoreceptors instead of a retinoblastoma. Since HPV16 E7 binds to pRb, and to its relatives, p107 and p130, it should have functionally recapitulated the defect in the eyes of small children with inherited or acquired defects in their Rb genes. If that tumor required only one genetic hit on each copy of the Rb gene, why was the IRBP-E7 construct not a sufficient inducer of retinoblastoma? The p53 gene is normal in most human retinoblastomas. Could the presence of an intact p53 gene be the source of the apoptosis in the IRBP-E7 mice? To test that question, we bred IRBP-E7 transgenic mice to p53 knockout mice to generate IRBP-E7/p53-/- offspring. After one month, they developed tumors (Howes et al., 1994b). Multifocal retinoblastomas, undifferentiated pigmented epithelial tumors, adenocarcinomas of the RPE, and papillomas of the ciliary body formed simultaneously. An important finding was that the photoreceptor layered died by apoptosis prior to tumor development (Fig. 12.4). Thus the apoptosis appeared to be independent of p53 function, while tumor formation, stimulated by E7 expression, was largely controlled by p53, a role not assigned to it in human retinoblastomas. The precursors of the tumors did not label with short exposure to BrdU, indicating that the premalignant cells were slow growing. Yet they survived the p53-independent apoptotic collapse of the photoreceptor layer. In addition, the IRBP promoter clearly could direct extraretinal expression of the oncogene in the eye (Papermaster et al., 1995). The mice also develop cataracts secondary to apoptosis of their lens fiber cells, a topic addressed in more detail in the following.

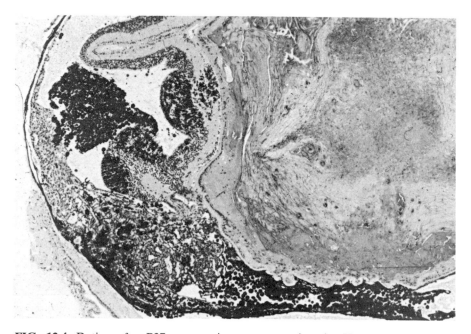

FIG. 12.4. Retina of a P37 transgenic mouse carrying the IRBP transgene on a p53-/- genetic background. The retina is distorted by the presence of multifocal retinoblastomas and numerous other tumors of other parts of the retina and ciliary body. Note that the detached retina adjacent to the tumor has only two layers—the ganglion cells and the inner nuclear layer. All photoreceptors have died by apoptosis prior to the formation of the tumor despite the absence of the p53 gene.

The multicentric retinoblastomas of IRBP-E7p53-/-transgenic mice were similar to, but not the same as, the uniform tumors formed by IRBP-SV40 Tag transgenic mice. The tumors of the other ocular tissues did not form in the IRBP-SV40 Tag transgenic mice. Clearly, while these oncoproteins have similar properties, they also perturb other cellular components. For example, E7 binds pRb, which releases bound E2F1, E2F2, and E2F3, each of which participates downstream in transcription of cell cycle genes. E7 also binds p107, which releases E2F4 and binds p130, which releases E2F5. Thus its effects are neither an exact complement of loss of function of the Rb gene alone nor of the binding of pRB by a truncated SV40 Tag. The IRBP-SV40 Tag transgenic mice never develop a multilayered retina, yet the IRBP-E7p53-/-mice form well-demarcated ganglion cell and inner nuclear layers despite the earlier onset of expression of the oncoprotein and the loss of their photoreceptors. These differences indicate that more than pRb and p53 are involved in controlling the cell cycle and the apoptosis associated with inappropriate proliferation in the retina. The stage of differentiation of the retina and the perturbation of other components are also decisive in determining the outcome *in vivo.*

As was discussed, normal retinas employ apoptosis as part of their developmental program. The retinas of the p53 nullizygous mice are normal; thus the apoptotic component of this developmental pathway is clearly p53-independent. The HPV-16 E6 protein inactivates p53, in part by directing p53 protein into the ubiquitin pathway of catabolism. Since the effects of the absence of p53 are so benign in the retina, it was interesting to find that IRBP-E6 transgenic mice also develop apoptotis of their photoreceptors (Howes et al., unpublished results). The action of this oncoprotein, therefore, also is more complex than the simple interference with the functions of p53. The elegant dissection of its role in differentiating lens cells has further clarified this issue (see the following).

The Role of the bcl-2 Family in Retinal Apoptosis

The genetics of cell death emerged from studies of Caenorhabditis elegans and human tumors. Mammalian homologues of the worm genes were soon discovered. The antiapoptotic function of bcl-2, initially discovered as a protein that contributed to development of human B-cell lymphoid tumors, was found to be functionally and structurally related to ced-9 (Hengartner and Horvitz, 1994). Ced-3 mutations revealed the central role of the proteolytic caspase genes of the IL-1β converting enzyme (ICE) family (Yuan et al., 1993). Exactly how the antiapoptotic members of the bcl-2 gene family interact with the caspase pathway is still unclear. Recent evidence favors an activation of caspases by release of cytochrome c from mitochondria. Bcl-2 and bcl-xL appear to block the release of cytochrome c (Kluck et al., 1997; Yang et al., 1997). The antiapoptotic effects of bcl-2 and bcl-xL are widespread in many tissues; so they were obvious candidates to be studied for their capacity to inhibit retinal apoptosis.

Several laboratories sought to exploit the antiapoptotic activity of these genes in transgenic mice using the opsin or IRBP promoters. The first results reported by Chen et al. (1996) were promising. Using the mouse opsin promoter to drive high expression of human Bcl-2, there was delay of rod degeneration in mice with the *rd* mutation of the β-PDE chain, in transgenic mice carrying opsin mutations, and in albino mice exposed to bright light for 24 hours. Curiously, high levels of Bcl-2 expression alone were themselves harmful to the photoreceptors, and these transgenic mice lost photoreceptor cells within a few months. These results were also exceptional in that the protein localized immunocytochemically near the synapse of the photoreceptors rather than in the inner segment, where most of the cell's mitochondria reside.

Using the human opsin promoter and Bcl-2 gene, in a different genetic background, Joseph and Li (1996) found no protection from its overexpression in rods and no retinal degeneration induced by the transgene. Our own results were similar; we used the IRBP promoter to generate high levels of immunocytochemically detectable human Bcl-2 and bcl-xL in photoreceptor inner segments, yet there was no rescue of *rd* retinas or of IRBP-E7 induced photoreceptor apoptosis. High levels of expression of the transgenes also caused no injury to the retina (Phipps et al., unpublished results). Since all three groups document the production of the transgenic proteins in the retinas, we are not currently able to explain the different results. Different mouse strains were used to generate the transgenic mice. Phosphorylation of Bcl-2 and bcl-xL inactivates the proteins. The analysis of the level of phosphorylation of these proteins is still underway. Since this is still an early stage in the study of these molecules, it is too soon to conclude any generalizations. Their inclusion in this chapter is designed solely to illustrate the complexity to be expected from *in vivo* studies. There is ultimately no substitute for *in vivo* analysis, long term, since any therapy would be tested in this context. Consequently, we are at least learning some boundaries to success that, hopefully, will collapse with further investigation.

Some Soluble Factors Ameliorate Retinal Degenerations in Rats and Mice

Numerous growth factors prevent neuronal apoptosis in tissue culture. Since it is relatively straightforward to administer these factors by intravitreal injection or by injection into the interphotoreceptor space, many of them were studied for their capacity to arrest or retard retinal degeneration. The RCS rat retina has a defect in phagocytosis of shed rod outer segment disks that leads to a buildup of undigested debris between the RPE and the rods. Beginning about postnatal day 24, the retina begins to die rapidly by apoptosis so that the entire layer disappears in about one week. Injection of acidic or basic fibroblast growth factors (FGF) and ciliary neurotrophic factor and a host of others seem to retard the degeneration in both RCS rats and in albino rats exposed to bright constant light for 24 h. (LaVail et al., 1992; Steinberg, 1994). Attempts to attain similar success in mouse models failed, indicating there are some species differences of response (Yasamura et al., 1995). Recent preliminary reports of partial success in retarding the *rd* retinal degeneration by injection of adenovirus vectors carrying the CNTF gene show there may be a potential therapeutic approach based on these ideas.

Studies in Drosophila Melanogaster Offer Hope of a Cure

Apoptosis in fruit flies is controlled by a small set of genes given colorful names, as usual, by their discoverers. The first, named reaper (*rpr*), is activated in all cells that undergo apoptosis. Loss-of-function mutations block apoptosis, leading to larvae with enlarged nervous systems that die in the pupal stage (White et al., 1994, 1996). A second mutation, head involution defective (*hid*) (Grether et al., 1995) and a third named grim (*grm*) act similarly. Expression of *hid* controlled by the fly eye-specific promoter *GMR* leads to failure of compound eye formation. The antiapoptotic Baculovirus gene, *p35,* appears to act by the inhibition of the caspase cascade—a fragment of the cleaved p35 binds to the caspase as an inhibitor (Xue and Horvitz, 1995). Expression of *p35* in fly eyes has no deleterious effects, but in double transgenic flies expressing both *hid* and *p35* genes or *rpr* and *p35* in the retina, a nearly normal compound eye forms (White et al., 1996). In a recent experiment designed to model the human disease, Steller's group generated flies with a retinal degeneration initiated by a mutation of the fly opsin gene at a site homologous to a deleterious human rhodopsin mutation. When these flies were crossed to flies expressing p35 in the retina, the offspring developed normal compound eyes that functioned physiologically (Davidson and Steller, submitted for publication). This is the first genetic cure of a retinal degeneration by antiapoptotic therapy. It provides great encouragement that the theory of these approaches is appropriate despite the complexity of the outcomes of the first round of experiments in mice.

APOPTOSIS OF THE LENS IS INDUCED BY EXPERIMENTAL DYSREGULATION OF THE CELL CYCLE

The development of the Normal Lens and the Effect of Loss of Function of Rb and p53 Genes

The lens has become the object of elegant studies of cell differentiation and the control of cell death because it is not required for viability. Its unique cell types and the availability of many differentiation specific markers readily allow alterations in its development to be quantitatively studied. During normal differentiation, lens fiber cells lose their nuclei and organelles once synthesis of the crystallins and formation of gap junctions is completed. Some aspects of this process resemble apoptosis since agarose electrophoresis reveals defined DNA fragments and the fiber cell nuclei usually label by the TUNEL assay. But this is a form fruste of apoptosis since the cells do not die and are not phagocytosed by their neighbors. Localization of p53 mRNA by *in situ* hybridization revealed restriction of its expression to the anterior and lateral epithelium and not in the central lens, where cells were denucleating (Pan and Griep, 1995). Finally, this process of nuclear destruction is p53 independent since it proceeds normally in p53-/-mice. Although lens fiber cell differentiation and denucleation were normal in p53-/-mice, the small vessels lining the back of the developing lens did not die normally by apoptosis as they matured, suggesting that the development of vascular network was perturbed by the loss of this gene (Pan and Griep, 1995).

The same genes investigated in retinal apoptosis were also evaluated in lens cell differentiation and apoptosis. *Rb* knockout mice do not survive beyond the thirteenth embryonic day (E13). The histology of their lenses revealed an accumulation of abnormally proliferating fiber cells throughout the anterior of the lens. Cell division was not restricted to the lateral equatorial domain as seen in the normal lens. Pyknotic nucleated fiber cells were scattered among the proliferating cells in the lens interior. Was the apoptosis a consequence of activation of p53? To test this, double knockout mice were bred that lacked both *Rb* and *p53* genes. While the *Rb-/-*embryos still died at E13, lens fiber cell apoptosis in the absence of *p53* was nearly entirely abrogated. The cells still did not differentiate properly and still proliferated abnormally (Morgenbesser et al., 1994). This result fit with the cell culture studies that indicated there was a close link between loss of pRb and cell proliferation and the function of p53 as a gatekeeper of the cell cycle. p53 promoted apoptosis in the face of abnormal proliferation.

Perturbation of Cell Cycle Genes by Using Viral Oncoproteins in Transgenic Mice and Lens-Specific Promoters

HPV16 E7 and E6 genes, when controlled by the α-crystallin promoter, were expressed in the lens fiber cells and affected their proliferation, differentiation, and apoptosis in a pattern that supported the conclusions drawn from the *Rb–p53* double knockout mice. The α-crystallin gene is expressed early in embryonic life in mouse lens fiber cells, and its promoter induces high levels of expression of the transgenes in the developing lens. α-crystallin-E7 mice had one advantage over the Rb knockout mice: They did not die during embryogenesis because the transgene's expression was restricted to the lens, a nonvital organ. Consequently, both embryonic and perinatal mice could be studied before the lens became so abnormal that it could not form a differentiated tissue. The transgenic lenses also accumulated proliferating lens fiber cells in the center that rapidly died by apoptosis (Pan and Griep, 1994). When these mice were crossed to a p53-/-background, however, the early apoptosis was reduced, but now apoptosis could be evaluated after E13. As the embryos aged, lens fiber cells continued to die by apoptosis after E15, showing that the same perturbation by E7 in the same tissue had different effects as the tissue matured. At early stages, apoptosis was largely p53 dependent, while later, it became p53 independent (Pan and Griep, 1995).

Interesting changes were found in α-crystallin-E6 transgenic mice. They had a nearly normal lens except that their fiber cells did not lose their nuclei and that generated a cataract. Since p53 knockout mice have a normal lens, this showed that the loss of nuclei involved a pathway independent of p53 and that E6 had effects beyond its impact on breakdown of p53 (Pan and Griep, 1994). Crossing α-crystallin-E6 to α-crystallin-E7 transgenic mice largely recapitulated the results in Rb-/-X p53-/-hybrids; apoptosis declined dramatically, but the cells still proliferated abnormally and then progressed to form a lens fiber cell tumor in 29% of the mice by 6–7 months of age, a disease never described in humans (Pan and Griep, 1994). E7-p53-/- mice similarly developed lens tumors in a third of the mice by 3 months of age, showing that the effect of E6 in tumor formation was largely a consequence of its effects on p53 function. The tumors expressed α-crystallin, proving their lenticular origin (Pan and Griep, 1995). Using the same promoter in transgenic mice to drive expression of a truncated form of SV40-Tag that retained the N-terminal pRb binding domain but lacked the C-terminal p53 binding domain generated results comparable to α-crystallin-E7 experiments. Crossing these mice to a p53 nullizygous background also greatly reduced lens fiber cell apoptosis at early stages, but tumor formation was less frequent than in the double transgenic E6 × E7 mice (Fromm and Overbeek, 1994).

Extraretinal Expression of IRBP Transgenes in the Lens

Using the IRBP promoter coupled to the E7 gene, we found an unexpected generation of cataracts. Prior studies of this promoter using various constructs of reporter genes, such as chloramphenicol acetyl-transferase, a bacterial gene with an easily traced activity, revealed the predicted expression in embryonic and adult photoreceptors as befits a protein whose function is to transport retinal to the rods and cones within the interphotoreceptor space. Not only did IRBP-E7 mice develop tumors of several ocular tissues when crossed to p53-/- mice as described; they continued to develop cataracts because of abnormal proliferation and apoptosis of posterior fiber cells in a p53 null

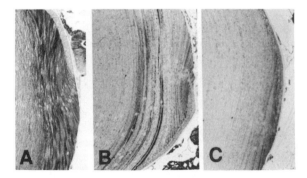

FIG. 12.5. Transient expression of the IRBP-Bcl-2 transgene during lens maturation. This experiment demonstrates the capacity of the IRBP promoter to direct expression in an extraretinal site and helps to explain the apoptosis of lens fiber cells expressing IRBP-E7. (A) At birth, lateral fiber cells are heavily labeled, while central fiber cells do not contain detectable Bcl-2. (B) Three days later, the labeled fiber cells have moved from the lateral lens as new unlabeled cells are generated there. Labeling density is decreased. (C) By P8, the formerly labeled fiber cells have lost their Bcl-2 label, and newly formed fiber cells at the periphery of the lens are unlabeled. Antigen was revealed in paraffin sections of the mouse lens by bound anti-Bcl-2 antibody, which was detected using a second-stage antibody conjugated to horse radish peroxidase and diaminobenzidine to generate a dark brown precipitate at the site of the antigen.

background. Thus their apoptosis was largely p53 independent (Papermaster et al., 1995).

This extraretinal expression of IRBP transgenes was clarified by two experiments. Since we were also using the IRBP promoter to induce synthesis of Bcl-2 and bcl-xL, we looked for the gene products by immunocytochemistry in young mice. We found abundant expression of these trangenes in the newborn lens—as well as the retina as expected, and in the ciliary body and RPE, which partially explained the variety of tumors in these tissues described. Clearly, the 1.5-Kb promoter contained information for expression in other ocular tissues. The time course of expression of the IRBP transgenes in the lens was informative. Bcl-2 protein was detectable by immunocytochemistry in lateral fiber cells at birth, but by 3 days after birth, the labeled cells had shifted toward the center of the lens and the lateral fiber cells differentiating at these later ages were free of the transgenic protein. By eight days after birth, the protein was no longer detectable in the lens (Figure 12.5). Thus the lens controls its expression in a transient wave, while in the retina once it turns on it stays on (Papermaster et al., 1996).

Evaluating these results in the context of the studies by Pan and Griep (1995) described above suggests that IRBP drives expression of E7 in the lens later than the α-crystallin promoter during a more mature developmental stage when fiber cell apoptosis is p53 independent. When IRBP-E7 begins expression in perinatal fiber cells, their abnormal structure and continued proliferation apparently so disrupts the closely regulated junctions of fiber cells that they begin to die and may not need continuous expression of the transgene. This hypothesis awaits detailed analysis of the time course of E7 expression in the lenses of these transgenic mice. Unfortunately, good antibodies to E7 that are suitable for immunocytochemistry are not yet available, and in situ hybridization studies, while adequate to demonstrate the E7 mRNA, would not answer the questions as readily as direct demonstration of the protein product in the lens cells.

Studies of Lens Development in Chimeric Mice

A new model for evaluation of cell cycle genes in differentiation of fiber cells was developed by DePinho and colleagues. They exploited targeted gene disruption in embryonic stem cells (ES cells) and the aphakia (ak/ak) mouse in a blastocyst lens complementation system (LCS). The ak/ak mouse fails to develop a lens because of a mutation on chromosome 19 that is not yet characterized. Injecting ES cells into an aphakia blastocyst generates chimeric mice with ES cells contributing to many tissues. The presence of a lens in the LCS chimeras is unequivocal evidence of lens formation induced by the donated ES cells. Since aphakic mice have small eyes, the successful LCs mice could be easily selected by simple inspection of those with normal appearing eyes. Complex cellular studies were not needed for identification of the successful chimeras. About a third of the chimeric mice developed one or two normal appearing eyes. Using strain-specific markers, they showed that the ES cells were the sole donors to lens cells in these chimeras. Thus it became possible to generate a lens *in vivo* from cultured cells with any mutation one desired to study.

Using ES cells with targeted disruption of the Rb gene, they generated mice where the effect of the loss of this gene could be studied beyond E13, since the mice survived. At E16.5, the LCS Rb-/- lens fiber cells were proliferating abnormally and arrayed in a disordered pattern. They incorporated BrdU in the interior of the lens, showing a disorderly progression through the S phase of the cell cycle. TUNEL assays revealed persisting apoptosis among the proliferating fiber cells (Liegeois et al., 1996). Studies in that laboratory are now in progress to evaluate different elements of the cell cycle to examine their role in regulation of proliferation and differentiation.

THE NEED TO EVALUATE THE EYE AS AN ENTIRE ORGAN

One consequence of these studies of apoptosis in the retina and lens is just starting to unfold—a unification of studies of the anterior and posterior of the eye because of the evaluation of factors that regulate differentiation and cell death in both lens and retina. The developmental unity revealed by the studies of the Pax-6 gene and its relatives in flies and man illustrate an onto-genic unity paralleling the cross-fertilization of the studies of cell death in the eye. The advantages of examining such questions in the eye are many. Viability and reproduction proceed well in blind mice; thus perturbation of the eye *in vivo* can be followed for months or years to evaluate models that proceed slowly. The presence of cataracts or small eyes reliably reports a disturbance of lens function. Blind mice do not respond normally to hand motions by jumping away. Reduction or loss of retinal function in the living mouse can be established conclusively by determining the magnitude of an electrical photoresponse that is measured physiologically with a cotton electrode on the cornea and a second electrode on the ear after a flash of light (the electroretino-gram). Thus selection of living mice for further study does not require detailed cellular analysis initially so that appropriate matings can be established. Finally, both the lens and the retina have many cell-specific promoters that provide the investigator with probes that are both spatially and temporally controlled.

These early studies of retinal degeneration and of cataract formation are important models of human diseases, especially since many of them are age related. We do not yet know to what extent the middle-aged photoreceptor can be compared developmentally with a youthful cell. Because photoreceptors become post-mitotic during the third trimester of pregnancy in the human except in the periphery, we are tempted to think that these specialized neurons simply go throughout post-mitotic life renewing their molecules rather than themselves. Similarly in the lens, the adult lens fiber cells need to protect their integrity for the entire life of the host to remain transparent. Yet the human diseases clearly show that there are both temporal and spatial controls of cell death. There must be more subtle differences in the fiber cell as it ages, perhaps under genetic control. As more mouse strains enter eye research, we may uncover some of these heterogeneous elements that modulate the effects of damaging mutations.

CONCLUSIONS AND ANTICIPATION OF NEW AREAS OF RESEARCH

Even at this early stage in the study of apoptosis in the eye, new insights into the pathogenesis of retinal degenerations and the development of the lens have appeared. They show the importance of the control of the cell cycle in regulating differentiation and death in both the retina and the lens, especially the role of the *Rb* and *p53* genes and their downstream targets. The role of apoptosis in retinitis pigmentosa is clear and excellent models exist in experimental animals. Recent work suggests that another major vision disorder, glaucoma, also may be a consequence of apoptosis of ganglion cells.

Many pathways for inducing the apoptotic cascade in the eye are already apparent, as they are in other tissues. This chapter has emphasized the retinal degenerations induced by rhodopsin, PDE, and rds/peripherin mutations as models because some aspects of their molecular function are most clear. Other mutations induce retinal degeneration by acting on proteins involved in the phototransduction cascade, on structural elements of the photoreceptor membrane, and on widely expressed genes with special and unique functions in the retina. Thus the likely success of antiapoptotic therapy is most likely to be demonstrated here soon, given the encouragement of the success in flies. Evidence for a narrow final pathway for cell death, while likely, has not yet been established in the eye. The means for delivering vectors carrying antiapoptotic genes is an area of active research in both academe and in pharmaceutical laboratories. The fundamental research outlined here will need to progress so that the genes to be inserted are clearly understood and well chosen. Because the eye is post-mitotic, it is likely that these approaches will not crash on the reefs of tumor formation as they might in other tissues.

REFERENCES

Adler R (1996): Mechanisms of photoreceptor death in retinal degenerations. From the cell biology of the 1990s to the ophthalmology of the 21st century? Arch Ophthalmol 114:79–83.

Agarwal N, Nir I, Papermaster DS (1990): Opsin synthesis and mRNA levels in dystrophic retinas devoid of outer segments in retinal degeneration slow (rds) mice. J Neuroscience 10:3275–85.

Allikmets R, Singh N, Sun H, Shroyer NF, Hutchinson A, Chidambaram A, Gerrard B, Baird L, Stauffer D, Peiffer A, Rattner A, Smallwood P, Li Y, Anderson KL,

Lewis RA, Nathans J, Leppert M, Dean M, and Lupski JR (1997a). A photoreceptor cell-specific ATP-binding transporter gene (ABCR) is mutated in recessive Stargardt macular dystrophy. Nature Genet 15:236–46.

Allikmets R, Shroyer NF, Singh N, Seddon JM, Lewis RA, Bernstein PS, Peiffer A, Zabriskie NA, Li Y, Hutchinson A, Dean M, Lupski JR and Leppert M. 1997b, Mutation of the Stargardt Disease Gene (ABCR) in Age-related Macular Degeneration. Science 277:1805–7.

Al-Ubaidi MR, Font RL, Quiambao AB, Keener MJ, Liou GI, Overbeek PA, Baehr W (1992a): Bilateral retinal and brain tumors in transgenic mice expressing simian virus 40 large T antigen under control of the human interphotoreceptor retinoid-binding protein promoter. J Cell Biol 119:1681–87.

Al-Ubaidi MR, Hollyfield JG, Overbeek PA, Baehr W (1992b): Photoreceptor degeneration induced by the expression of simian virus 40 large tumor antigen in the retina of transgenic mice. Proc Natl Acad Sci USA. 89(4):1194–98.

Azarian SM, Travis GH (1997): The photoreceptor rim protein is an ABC transporter encoded by the gene for recessive Stargardt's disease (ABCR). FEBS Lett 409:247 52.

Berson EL (1993): Retinitis pigmentosa. The Friedenwald Lecture. Invest Ophthalmol Vis Sci 34:1659–76.

Berson EL, Rosner B, Sandberg MA, Hayes KC, Nicholson BW, Weigel-DiFranco C, Willet W (1993): A randomized trial of vitamin A and vitamin E supplementation for retinitis pigmentosa. Arch Ophthalmol 111:761–72.

Bird AC (1995): Retinal photoreceptor dystrophies: LI Edward Jackson Memorial Lecture. Am J Ophthalmol 119:543–62.

Cvekl A, Piatigorsky J (1996): Lens development and crystallin gene expression: Many roles for Pax-6. Bioessays 18:621–30.

Chen J, Flannery JG, LaVail MM, Steinberg RH, Xu J, Simon MI. (1996): bcl-2 overexpression reduces apoptotic photoreceptor cell death in three different retinal degenerations. Proc Natl Acad Sci USA 93:7042–47.

Debbas M, White E (1993): Wild-type p53 mediates apoptosis by E1A, which is inhibited by E1B. Genes Dev 7:546–54.

Dryja TP, Berson EL (1995): Retinitis pigmentosa and allied diseases. Invest Ophthalmol Vis Sci 36:1197–200.

Fromm L, Shawlot W, Gunning K, Butel JS, Overbeek PA (1994): The retinoblastoma protein-binding region of simian virus 40 large T antigen alters cell cycle regulation in lenses of transgenic mice. Mol Cell Biol 14:6743–54.

Gavrieli Y, Sherman Y, Ben-Sasson S (1992): Identification of programmed cell death in situ via specific labeling of nuclear DNA fragmentation. J Cell Biol 119:493–501.

Goldberg AF, Molday RS (1996): Subunit composition of the peripherin/rds-rom-1 disk rim complex from rod photoreceptors: Hydrodynamic evidence for a tetrameric quaternary structure. Biochem 35:6144–49.

Grether ME, Abrams JM, Agapite J, White K, Steller H (1995): The head involution defective gene of Drosophila melanogaster functions in programmed cell death. Genes Dev 9:1694–708.

Halder G, Callaerts P, Gehring WJ (1995): Induction of ectopic eyes by targeted expression of the eyeless genes in Drosophila. Science 267:1788–92.

Hargrave PA (1991): Seven-helix receptors. Curr Opin Struct Biol 1:575–81.

Hargrave PA, Hamm PE, Hofmann KP (1993): Interaction of rhodopsin with the G-protein, transducin. Bioessays 15:43–50.

Hay BA, Wassarman DA, Rubin GM (1995): Drosophila homologs of baculovirus inhibitor of apoptosis proteins function to block cell death. Cell 83:1253–62.

Hay BA, Wolff T, Rubin GM (1994): Expression of baculovirus P35 prevents cell death in Drosophila. Development 120:2121–29.

Hengartner MO, Horvitz HR. (1994): C. elegans cell survival gene ced-9 encodes a functional homolog of the mammalian proto-oncogene bcl-2. Cell 76:665–76.

Howes KA, Lasudry JGH, Albert DM, Windle JJ (1994a): Photoreceptor cell tumors in transgenic mice. Invest Ophthamol Vis Sci 35:342–51.

Howes KA, Ransom N, Papermaster DS, Lasudry JGH, Albert DM, Windle JJ (1994b): Apoptosis or retinoblastoma: Alternative fates of photoreceptors expressing the HPV-16 E7 gene in the presence or absence of p53. Genes Dev 8:1300–10.

Illing M, Molday LL, and Molday RS, 1997: The 220-kDa Rim Protein of retinal rod outer segments is a member of the ABC transporter superfamily. J Biol Chem 272:10303–10.

Jacks T, Weinberg RA (1996): Cell-cycle control and its watchman. Nature 381:643–44.

Jansen HG, Sanyal S, de Grip WJ, Schalken JJ (1987): Development and degeneration of retina in rds mutant mice: Ultraimmunohistochemical localization of opsin. Exp Eye Res 42:363–73.

Joseph RM, Li T (1996): Overexpression of bcl-2 or bcl-xL transgenes and photoreceptor degeneration. Invest Ophthalmol Vis Sci 37:2434–46.

Kajiwara K, Berson EL, Dryja TP (1994): Digenic retinitis pigmentosa due to mutations at the unlinked peripherin/RDS andROM1 loci. Science 264:1604–8.

Kajiwara K, Sandberg MA, Berson EL, Dryja TP (1993): A null mutation in the human peripherin/RDS gene in a family with autosomal dominant retinitis punctata albescens. Nature Genetics 3:208–12.

Kerr JFR, Wyllie AH, Curie AR (1972): Apoptosis: A basic biological phenomenon with ranging implications in tissue kinetics. Br J Cancer 26:239–57.

Kluck RM, Bossy-Wetzel E, Green DR, Newmeyer DD (1997): The release of cytochrome c from mitochondria: A primary site for Bcl-2 regulation of apoptosis. Science 275:1132–36.

Knudson AG, Jr (1971): Mutation and cancer: Statistical study of retinoblastoma. Proc. Natl Acad Sci USA 68:820–23.

LaVail MM, Yasumura D, Matthes MT, Yancopoulos GD, Steinberg RH (1992): Multiple growth factors, cytokines, and neurotrophins rescue photoreceptors from the damaging effects of constant light. Proc Natl Acad Sci USA 89:11249–53.

Lee WH, Chen PL, Riley DJ (1995): Regulatory networks of the retinoblastoma protein. Ann NY Acad Sci 752:432–45.

Levine AJ (1997): p53, the cellular gatekeeper for growth and division. Cell 88:323–31.

Li Z-Y, Jacobson SG, Milam AH (1994): Autosomal dominant retinitis pigmentosa caused by the threonine-17-methionine rhodopsin mutation: Retinal histopathology and immunocytochemistry. Exp Eye Res 58:397–408.

Li Z-Y, Milam AH (1995): Apoptosis in retinitis pigmentosa. In Anderson RE, Hollyfield JG, LaVail MM (eds): New York: Plenum Press, pp 1–8.

Liegeois NJ, Horner JW, DePinho RA (1996): Lens blastocyst and complementation for the genetic analysis of growth, differentiation and death in vivo. Proc Natl Acad Sci USA 93:1303–7.

Morgenbesser SD, Williams BO, Jacks T, DePinho RA (1994): p53-dependent apoptosis produced by Rb-deficiency in the developing mouse lens. Nature 371:72–74.

Nevins JR (1992): The E2F transcription factor—A link between the Rb tumor suppressor protein and viral oncoproteins. Science 258:424–29.

Nir I, Papermaster DS (1986): Immunocytochemical localization of opsin in the inner segment and ciliary plasma membrane of photoreceptors in retinas of rds mutant mice. Invest Ophthal Vis Sci 27:836–40.

Palczewski K (1994): Is vertebrate phototransduction solved? New insights into the molecular mechanism of phototransduction. Invest Ophthalmol Vis Sci 35:3577–81.

Pan H, Griep AE (1994): Altered cell cycle regulation in the lens of HPV-16 E6 or E7 transgenic mice: Implications for tumor suppressor gene function in development. Genes Dev 8:1285–99.

Pan H, Griep AE (1995): Temporally distinct patterns of p53-dependent and p53-independent apoptosis during mouse lens development. Genes Dev 9.2157–69.

Papermaster DS (1995): Necessary but insufficient Nature. Medicine 1:874–75.

Papermaster DS (1997): Apoptosis of the retina and lens. Cell Death Differentiation 4:21–28.

Papermaster DS, Converse CA, and Zorn M. 1976: Biosynthetic and immunochemical characterization of a large protein in frog and cattle rod outer segment membranes. Exp Eye Res 23:105–15.

Papermaster DS, Howes K, Ransom N, Windle JJ (1995): Apoptosis of photoreceptors and lens fiber cells with cataract and multiple tumor formation in the eyes of transgenic mice lacking the p53 gene and expressing the HPV 16 E7 gene under the control of the IRBP promoter. In Anderson RE, Hollyfield JG, LaVail MM (eds.): Degenerative Diseases of the Retina. New York: Plenum Press, pp 39–49.

Papermaster DS, Nir I (1994): Apoptosis in inherited retinal degenerations. In Mihich E, Schimke RH (eds): Apoptosis. New York: Plenum Press, pp 15–30.

Papermaster DS, Reilly P, and Schneider BG. 1982: Cone lamellae and red and green rod outer segment disks contain a large intrinsic membrane protein in their margins. Vision Res 22:1417–28.

Papermaster DS, Windle JJ (1995): Death at an early age: Apoptosis in inherited retinal degenerations. Invest Ophthalmol Vis Sci 36:977–83.

Papermaster DS, Schneider BG, Zorn M, and Kraehenbuhl JP. 1978. Immunocytochemical localization of a large intrinsic membrane protein to the incisures and margins of frog rod outer segment disks. J Cell Biol 78:415–25.

Penfold PL, Provis JM (1986): Cell death in the development of the human retina: Phagocytosis of pyknotic and apoptotic bodies by retinal cells. Graefe's Arch Clin Exp Ophthalmol 224:549–53.

Piatigorsky J, Wistow G (1989): Enzyme-crystallins: Gene sharing as an evolutionary strategy. Cell 57:197.

Quiring R, Walldorf U, Kloter U, Gehring WJ (1994): Homology of the eyeless gene of Drosophila to the small eye gene in mice and Aniridia in humans. Science 265:785–89.

Rao PV, Huang Q, Horwitz J, Zigler JS Jr. (1995): Evidence that a-crystallin prevents non-specific protein aggregation in the intact eye lens. Biochim Biophys Acta 1245:439–47.

Sanyal S, de Ruiter A, Hawkins RK (1980): Development and degeneration of retina in rds mutant mice: Light microscopy. J Comp Neurol 194:193–207.

Small KW, Puech B, Mullen L, Yelchits S (1997): North Carolina macular dystrophy phenotype in France maps to the MCDR1 locus. Mol Vis 3:1–3.

Steinberg RH (1994): Survival factors in retinal degenerations. Curr Opin Neurobiol 4:515–24.

Weinberg RA (1996): E2F and cell proliferation: A world turned upside down. Cell 85:457–59.

White K, Grether ME, Abrams JM, Young L, Farrell K, Steller H (1994): Genetic control of programmed cell death in Drosophila. Science 264:677–83.

White K, Tahaoglu E, Steller H (1996): Cell killing by the Drosophila gene reaper. Science 271:805–7.

Wong ROL, Hughes A (1987): Role of cell death in the topogenesis of neuronal distributions in the developing cat retinal ganglion cell layer. J Comp Neurol 262:496.

Wyllie AH (1980): Cell death: The significance of apoptosis. Int Rev Cytol 68:251–307.

Xue D, Horvitz HR (1995): Inhibition of the Caenorhabditis elegans cell-death protease CED-3 by a CED-3 cleavage site in baculovirus p35 protein. Nature 377:248–51.

Yang J, Liu X, Bhalla K, Kim CN, Ibrado AN, Cai J, Peng TI, Jones DP, Wang X (1997): Prevention of apoptosis by bcl-2: Release of cytochrome c from mitochondria blocked. Science 275:1129–1132.

Yasumura D, Matthes MT, Lau C, Unoki K, Steinberg RH, LaVail (1995): Attempts to rescue photoreceptors with survival factors in mice with inherited retinal degenerations or constant lightdamage. Invest Ophthalmol Vis Sci 36: Suppl 1, S252.

Young RW (1984): Cell death during differentiation of the retina in the mouse. J Comp Neurol 229:362–73.

Yuan J, Shaham S, Ledoux S, Ellis HM, Horvitz HR (1993): The C. elegans cell death gene ced-3 encodes a protein similar to mammalian interleukin-1 beta-converting enzyme. Cell 75:641–52.

CHAPTER 13

SIGNAL TRANSDUCTION AND NEURONAL CELL DEATH DURING DEVELOPMENT AND DISEASE

RAYMOND B. BIRGE AND J. EDUARDO FAJARDO
Laboratory of Molecular Oncology, The Rockefeller University, 1230 York Avenue, New York, NY 10021

BARBARA L. HEMPSTEAD
Department of Medicine and Division of Neuroscience, Cornell University Medical College, New York, NY 10021

Over the past decade the fields of programmed cell death and apoptosis have emerged from descriptive morphological considerations into the mechanistic study of intracellular signal transduction. Analogous to signaling events that regulate proliferation and differentiation, signals mediating cell death in neurons require extracellular factors to activate transmembrane receptors. Ligand binding to receptors, such as Fas ligand (FasL) and tumor necrosis factor (TNF) binding to Fas and TNF receptors, respectively, orchestrates the assembly of multiprotein complexes and the formation of second messengers that convey cell death signals to a functional, healthy cell. During the past few years, several of the pathways for apoptosis have been identified, and many of the participating proteins characterized. The focus of this chapter is on the mechanisms of neuronal cell death, reviewing the recent literature on neurotrophic factors and the effector pathways that regulate both neuronal survival and apoptosis.

When Cells Die, Edited by Richard A. Lockshin, Zahra Zakeri, and Jonathan L. Tilly
ISBN 0-471-16569-7 © 1998 Wiley-Liss, Inc.

NEUROTROPHIN RECEPTORS

Role of Neurotrophins and Neurotrophin Receptors During Neuronal Differentiation and Survival

The neurotrophins consist of a family of five related polypeptide growth factors that promote neuronal differentiation and survival: nerve growth factor (NGF), brain derived neurotrophic factor (BDNF), and neurotrophins 3, 4/5 and 6 (NT-3, NT4/5, and NT-6) (reviewed in Lindsay et al., 1994; Snider, 1994; Lewin et al., 1996). The neurotrophins are structurally similar proteins with relative molecular masses of 13–14 kD, that share approximately 50% homology among family members. Each neurotrophin mediates its action on responsive neurons by binding to two classes of cell surface receptors. The low-affinity neurotrophin receptor (p75) is a 75 kD glycosylated type I integral membrane protein (Chao et al., 1995) and a member of the Fas/TNF family, based on structural homologies in the extracellular domains. The p75 receptor binds all neurotrophins and acts to influence high-affinity NGF binding and ligand internalization; it may play a role in apoptosis (reviewed in Chao et al., 1995). The second class of neurotrophin receptors are the Trk family of receptor tyrosine kinases, composed of TrkA, TrkB, and TrkC. The full-length Trk receptors share approximately 50% sequence homology in their extracellular domains, and 85% in the cytoplasmic domains. The TrkA, TrkB, and TrkC tyrosine kinases serve as the receptors for NGF, BDNF, and NT-3, respectively, and TrkA and TrkB can also act as receptors for NT-4/5 (Fig. 13.1) (reviewed in Korsching, 1993; Snider, 1994). Other molecules that have been shown to influence the differentiation and survival of neurons in the peripheral (PNS) and central nervous systems (CNS) include members of the neurokine family, including ciliary neurotrophic factor (CNTF) and leukemia inhibitory factor (LIF), and other neurotrophic growth factors such as fibroblasts growth factor (FGF), insulin-like growth factor (IGF), glial cell line–derived neurotrophic factor (GDNF), and bone morphogenic protein (BMP) (see Snider, 1994; Bothwell, 1995).

Although neurotrophin receptors are also present in non-neuronal tissue including heart, intestine, lung, testis, kidney, and vascular smooth muscle cells (TrkC) (Klein et al., 1989; Tessarollo et al., 1993), their function and distribution in neuronal populations is best understood, and will be the focus of this chapter. Elucidation of the critical role of NGF during the development of the sympathetic nervous system led to the formulation of the classic "target-derived neurotrophin hypothesis" (Levi-Montalcini, 1987; Thoenen, 1991). Accordingly, in the rat model, neural crest cells proliferate, migrate, and differentiate to form several types of cells, including ganglia of the sympathetic chain, first identifiable at embryonic day 12 (E12) and reaching a peak in neuronal number by birth (Rubin, 1985). NGF, produced in limiting amounts by target tissues, promotes process formation and maintains innervation by responsive neurons in order to match the number of innervating neurons to

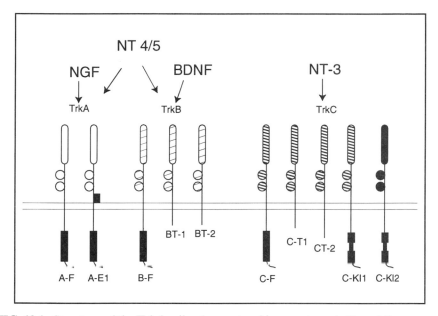

FIG. 13.1. Structure of the Trk family of neurotrophin receptors: A, B, and C represent the TrkA, TrkB, and TrkC receptors. F denotes the full-length, kinase-active isoform; T denotes the truncated isoform; EI denotes the extracellular insert; and KI denotes the kinase insert. The dark blocks within the intracellular domains indicate the tyrosine kinase domains. The specific neurotrophins that interact with each Trk receptor are indicated by arrows.

the needs of the proximal target area (Fig. 13.2). However, in the perinatal period, approximately 50% of these neurons die by programmed cell death when their axons reach target fields with a limiting supply of neurotrophin (Barde, 1989). This model is supported by classic experiments in which administration of exogenous NGF to the peripheral target field increases the density of innervation (Levi-Montalcini, 1987). In more recent experiments, sympathetic neuron death could be largely prevented when a peripheral target field overexpressed NGF, after the introduction of an NGF transgene in keratinocytes (Takami et al., 1995) (Fig. 13.2).

The importance of neurotrophin receptors in promoting survival and regulating neuronal number and plasticity has been further delineated using germline loss-of-function mutations in the *trk* and *p75* receptor genes (see Snider, 1994). These studies formally demonstrate that Trk and p75 receptors mediate trophic signaling, and suggest that each neurotrophin–receptor system has distinctive roles in the developing mammalian nervous system. For example, targeted disruptions of TrkA/NGF receptor cause extensive neuronal death in trigeminal, sympathetic, and dorsal root ganglia (DRG) neurons (three well-known targets for NGF) and decrease the number of cholinergic basal

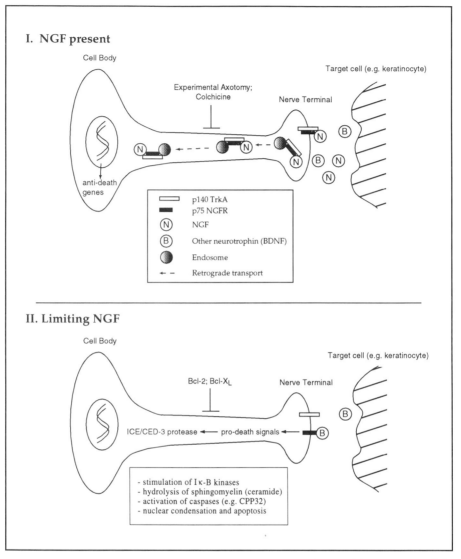

FIG. 13.2. Topology of neuronal survival and death in TrkA-positive neurons of the peripheral nervous system: In the developing nervous system, neurons depend on limiting quantities of target-derived NGF for the survival (panel I). When a neuron encounters a target field with a surplus of NGF, NGF binds with high affinity to the p75/TrkA complex and is retrogradely transported to the cell body (see text for details). Experimental axotomy or microtubule-destabilizing drugs (e.g., colchicine) that inhibit retrograde transport induce apoptosis in the presence or absence of NGF, indicating that neuronal survival requires the axonal transport of NGF. When redundant neurons come in contact with target fields that have depleted NGF (panel II), p75 appears to signal independently from TrkA, either alone or in the presence of a second neurotrophin such as BDNF, to induce apoptosis (see text for details). Overexpression of Bcl-2 or Bcl-x_L, or inhibitors of ICE/CED3 protease activation, are capable of blocking apoptosis during NGF withdrawal, suggesting that p75 is upstream to the ICE/CED3 pathway. Other pathways that are activated by p75 are indicated in the solid box.

forebrain projections to the hippocampus and cortex (Smeyne et al., 1994). Similarly, targeted disruptions of TrkB and TrkC result in specific peripheral nervous system deficiencies, but CNS defects have been difficult to evaluate as these animals die shortly after birth (Klein et al., 1993; 1994). Mice lacking TrkB show decreases in the numbers of trigeminal ganglia and motor neurons of the facial nucleus, while ablation of TrkC in null mutant mice does not impair motor neuron function but causes defects in muscle afferents, resulting in abnormal posture and proprioception (Snider, 1994; Lindsay, 1996). However, the interpretation of null mutant animals is complicated by the demonstration that a single class of neurons may coexpress several Trk family members (Davies, 1994). In addition, at different stages of development, maturation, and aging, specific neurons can acquire or lose dependence on particular neurotrophins. Therefore, the role of neurotrophins during the life and death of a particular neuron are subject to dynamic regulatory changes.

While Trk receptors are best characterized in their ability to promote survival and differentiation during embryonic development, these receptors are also expressed in adults, where they may also modulate neuronal function (McMahon et al., 1994). Recent studies indicate that BDNF can function to enhance synaptic transmission, suggesting that neurotrophic factors might be potential therapeutic agents to treat diseases of the adult nervous system (Patterson et al., 1996). In the adult CNS, TrkA is up-regulated in basal forebrain neurons near sites of neural injury, and NGF increases choline acetyltransferase (ChAT) activity and promotes neurite outgrowth in cultured cholinergic basal forebrain neurons from adult rats (Kordower et al., 1994). The potential role of NGF in maintaining cholinergic neurons is especially important since atrophy and degeneration of these neurons correlates with the memory loss and cognitive impairment associated with Alzheimer's disease (Olson, 1993; Rylett et al., 1994). Since intraventricular injections of NGF can diminish age-related degeneration of cholinergic neurons (Venero et al., 1994) and intracerebral grafts of NGF-secreting Schwann cells can rescue axotomized basal forebrain neurons in rodents (Olson, 1993), it is likely that loss of target-derived support, or an insensitivity of basal forebrain cholinergic neurons to NGF, may contribute to the cholinergic cell death seen in Alzheimer's disease (Cotman et al., Chapter 14, this volume). Clearly, a comprehensive understanding of the signal transduction components during neurotrophin signaling should provide new insights regarding experimental strategies to treat neurodegenerative diseases and spinal cord injuries.

Signal Transduction by the Trk Tyrosine Kinase

The Trk receptors are members of the larger family of receptor tyrosine kinases that includes the EGF, FGF, and PDGF receptors (reviewed in Schlessinger et al., 1992) (Fig. 13.1). Following receptor binding, the TrkA receptor undergoes NGF-dependent oligomerization, followed by tyrosine autophosphorylation (reviewed in Greene et al., 1995). Activation of Trk receptors can result in proliferation, differentiation, and cessation of cell division, or chemotaxis, functions that are in part cell-type specific (Raffoni et al., 1993; Chao et al., 1995). As for other receptors, the precise molecular mechanisms utilized by the activated Trk receptors to promote diverse biological effects in different cell types is unknown. Following NGF binding, tyrosine phosphorylation of the Trk receptor initiates a cascade of signal transduction. Several proteins, including phosphoinositol 3-kinase (PI 3-kinase) (Soltoff et al., 1992), phospholipase C (PLC-γ) (Kaplan et al., 1994), Shc (Geer et al., 1995), Nck (Park et al., 1992), and Crk (Hempstead et al., 1994) contain Src homology 2 (SH2) or protein tyrosine binding (PTB) domains that bind to specific phosphorylated sequences in Trk receptor to propagate intracellular signals (reviewed in Kaplan et al., 1994). Using mutational analysis of specific tyrosine residues in the intracellular domain of Trk, basic insights about the signaling pathways that regulate the diverse actions of NGF have been obtained. For example, mutation of tyrosine 785, which blocks binding of PLC-γ to TrkA, inhibits the induction of the gene for the neural-specific cytoskeletal protein peripherin but does not compromise $p21^{ras}$ activation (Loeb et al., 1994). Moreover, mutagenesis of both the Shc (tyrosine 490) and PLC-γ (tyrosine 785) binding sites on TrkA abrogates NGF-mediated neurite outgrowth. These studies indicate that numerous biological functions regualted by NGF could result from activation of a specific subset of signaling proteins recruited by the phosphorylated receptor (Obermeier et al., 1994; Stevens et al., 1994).

TrkA activates the $p21^{ras}$ pathway by first binding Shc, which rapidly becomes phosphorylated on tyrosine 317, and thus creates a binding site for the SH2 domain of Grb2 (McGlade et al., 1992) (Fig. 13.3). Grb2 then recruits

---

FIG. 13.3. TrkA promotes neuronal survival by the activation of intracellular signaling pathways: Binding of NGF to TrkA stimulates several signaling pathways. TrkA tyrosine phosphorylation results in the recruitment of the cytoplasmic Shc protein via its PTB domain, resulting in Shc phosphorylation. Activated Shc stimulates Ras activity by binding to Grb2, which activates the SOS guanine nucleotide exchange protein. Activation of MEK1 and ERK may be involved in suppressing apoptosis, possibly via the stimulation of gene transcriptional events involving AP-1 in the nucleus (see text for details). Raf-1 can also bind Bcl-2 in the mitochondrial membrane and act as a Bcl-2 effector to suppress apoptosis, possibly by phosphorylating death-antagonizing Bcl-2 members such as BAD. Activation of PI 3-kinase results in production of phosphatidylinositol 4,5 bisphosphate (PIP2), which in turn binds to and activates the Akt (PKB), a serine/theonine kinase that has been postulated to phosphorylate substrates involved in suppressing neuronal apoptosis. Additionally, the SH2/SH3-domain-containing adaptor protein Crk binds the TrkA receptor and subsequently increases the tyrosine phosphorylation of focal adhesion proteins FAK and paxillin. Overexpression of Crk in PC12 cells suppresses apoptosis, which may be involved in signaling to focal adhesions.

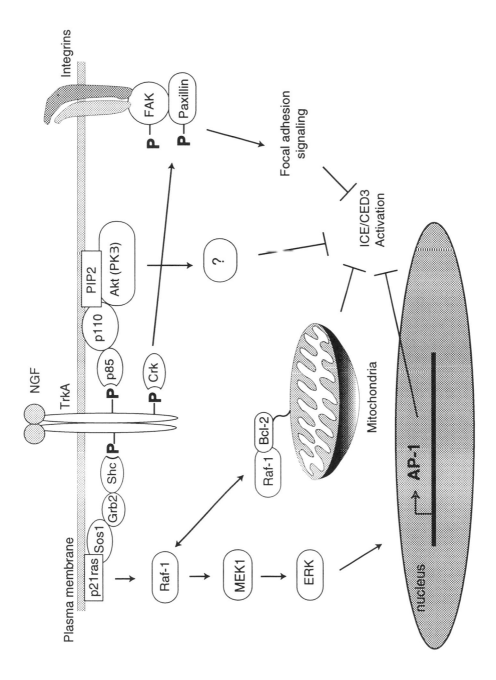

the guanine nucleotide exchange protein for p21^{ras}, mSOS, to the plasma membrane, resulting in the exchange of GTP for GDP on p21^{ras}. Overexpression of Shc and Grb2, or the oncogenic p21^{ras} protein, in PC12 cells can recapitulate many aspects of NGF signaling, including neurite outgrowth (Rozakis-Adcock et al., 1992), while overexpression of dominant negative N17^{ras} mutant protein can inhibit NGF-mediated neurite outgrowth (Hagag et al., 1986). These and other studies have clearly demonstrated that p21^{ras} activation is critical for NGF-induced neurite outgrowth formation and differentiation.

TrkA Effector Pathways Involved in Neuronal Survival

While p21^{ras} activation appears to be essential for NGF-induced differentiation, it is still uncertain what role Ras plays during survival and apoptosis. Some recent experiments have demonstrated that up-regulation of the Ras/Raf/MAP kinase signaling pathway (by expressing of constitutively active MKK1) has an apoptosis-suppressing effect on neuronal PC12 cells (Xia et al., 1995) (Fig. 13.3). Moreover, expression of Grb3-3, a splice variant of Grb2 that has a deleted SH2 domain and acts as a dominant negative mutant to prevent Grb2 from binding Shc and activating the p21^{ras} pathway, rapidly induces apoptosis in PC12 cells (Fath et al., 1994). In contrast to these findings, however, experiments on differentiated postmitotic neurons or PC12 cells have shown that down-regulation of p21^{ras} can similarly suppress apoptosis during NGF withdrawal, suggesting a paradigm in which p21^{ras} causes rather than suppresses apoptosis (Table 13.1). It may be possible to explain these discrepancies based upon the fact that: (1) p21^{ras} can utilize different effector pathways in different cellular situations, such as for cytoskeletal signaling or cellular proliferation and transformation (Khosravi-Far et al., 1996); and (2) p21^{ras} can mediate different responses in neuronal cells depending on the strength and duration of activation (Traverse et al., 1995). In this latter scenario, the EGF and NGF effects on proliferation and differentiation, respectively, have been correlated with differential p21^{ras} activation. Sustained or persistent activation is often correlated with a differentiation response, while transient activation is correlated with proliferation (Dikic et al., 1994; Traverse et al.,

TABLE 13.1. The Central Role of p21^{ras} in Neuronal Signaling

Effect	Reference
Induces differentiation/suppresses growth	Hagag et al., 1986; Rozakis-Adcock et al., 1992
Induces mitogenesis	Barbacid, 1987
Induces apoptosis	Ferrari et al., 1994
Prevents apoptosis	Borasio et al., 1989
Regulates voltage-dependent ion channels	Hemmick et al., 1992

1995). Therefore, p21^{ras} activation may suppress apoptosis under conditions when NGF is present, but induce apoptosis when NGF is absent (Table 13.1) (see Budd, Chapter 10; Tilly, Chapter 16, this volume).

Another Trk effector pathway that appears to play an important role during neuronal survival is mediated by PI 3-kinase. PI 3-kinase exists as a heterodimer of an 85 kD regulatory subunit (p85 α or β) and a 110 kDa catalytic subunit (Otsu et al., 1991). The p85 has a multidomain structure, consisting of an N-terminal SH3 domain, two proline-rich motifs that can bind other SH3 containing proteins, a p110 binding domain, and two SH2 domains that mediate binding to TrkA as well as to other receptors (Hiles et al., 1992). The p110 subunit functions as a lipid kinase that phosphorylates, phosphatidyl inisitol (PI), PI-4 phosphate (PI-4P) and PI-4,5 diphosphate (PI-4,5P2) on the D3 position of the inositol ring (Dyand et al., 1994), giving rise to phosphatidyl-inositol-3,4 bisphosphate (PIP2) and phosphatidylinositol-3,4,5 triphosphate (PIP3) lipid products that have potent effects of mitogenesis, differentiation, and apoptosis (Carpenter et al., 1996). Results by Cooper and colleagues have identified the activation of the PI 3-kinase pathway as a critical event for neuronal survival, as NGF-mediated survival of PC12 cells can be blocked by inhibitors of PI 3-kinase (Yao et al., 1995). In other studies, overexpression of wild-type PDGF receptor in PC12 cells, which binds to and activates PI 3-kinase, results in survival, while overexpression of PDGF receptors carrying a point mutant in the PI 3-kinase binding site, and thus incapable of activating PI 3-kinase, abolish this effect (Yao et al., 1995). Although PI 3-kinase signaling is extremely complex and appears to activate a number of divergent signaling cascades (reviewed in Carpenter et al., 1996), a serine/threonine kinase called Akt or RAC-PK, when activated by PIP2, can directly suppress apoptosis, while mutants of Akt that cannot be activated by PI3-kinase induce apoptosis (Dudek et al., 1997) (Fig. 13.2). Elucidation of the signaling by Akt should therefore provide insights into the role of protein kinase pathways during neuronal apoptosis.

Another serine/threonine kinase, Raf-1, has also been implicated in suppressing apoptosis in a number of cell systems. Though Raf-1 is best understood for the role it plays in growth factor–mediated signaling in the p21^{ras} pathway at the plasma membrane, Raf-1 has also been shown to dimerize with Bcl-2 in the mitochondrial membrane (Wang et al., 1996) (Fig. 13.3). In fact, subcellular targeting of Raf-1 to mitochondria effectively abrogates apoptosis while targeting of Raf-1 to the plasma membrane, which constitutively activates ERK/MAP kinase, has no effect in inhibiting apoptosis (Wang et al., 1996). This suggests that Raf-1 may be a *bona fide* Bcl-2 effector kinase when tethered to the mitochondrial membrane. Although the substrates in the mitochondria phosphorylated by Raf-1 have not been clearly identified, the death inducing Bcl-2 family member, BAD, may be phosphorylated by mitochondrial tagged Raf-1, an event that has been suggested to interfere with the ability of BAD to cause apoptosis (Wang et al., 1996; Zha et al., 1996). Other TrkA effector proteins, including the SH2/SH3-domain-containing adaptor proteins Nck and Crk, have also been implicated in apoptosis. For example, the Crk protein binds TrkA, delays apoptosis in PC12 cells, and enhances tyrosine kinase signaling in focal adhesions (Glassman et al., 1997). Consistent with this idea, activation of focal adhesion proteins, such as focal adhesion kinase (FAK), have been shown to suppress apoptosis in several systems (Frisch et al., 1996; Tidball and Albrecht, Chapter 15, this volume).

p21^{ras} Activation and Defective Cell Cycle

As alluded to, there is some controversy regarding the role of p21^{ras} in neuronal survival (Table 13.1). In non-neuronal cells, such as embryonic fibroblasts, NIH3T3 cells, and myoblasts, aberrant p21^{ras} activation can induce apoptosis if cells are growth arrested by either serum starvation or by cytostatic drugs such as vinblastine (Tanaka et al., 1994; Wang et al., 1995). This has served as the basis for the hypothesis that simultaneous and conflicting growth promoting and growth arresting signals may trigger apoptosis (Budd, Chapter 10; Birge et al., Chapter 13; Tilly, Chapter 16, this volume). A similar apoptotic trigger has been proposed for neuronal cells, including superior cervical ganglia (SCG) neurons and NGF-differentiated PC12 cells, based upon the observation that expression of dominant negative N17^{ras} protein can block the cell death occurring after NGF withdrawal (Ferrari et al., 1994). Moreover, mouse embryos lacking rasGAP exhibit a striking increase in apoptotic death in cells that make up the anterior neural tube and cranial neural crest (Henkemeyer et al., 1995). Cells isolated from embryonic day 9.5 embryos exhibit a greater sensitivity to growth factor–induced Ras-GTP exchange and fail to down-regulate Ras-GTP after growth factor stimulation, suggesting that aberrant p21^{ras} signaling may indeed underlie a mechanism for neural apoptosis.

Although it is unclear how aberrant activation of the p21^{ras} pathway could result in neuronal apoptosis, p21^{ras} induces the synthesis of specific proteins and transcription factors, such as c-Jun/AP-1, that are involved in proliferation and apoptosis (Barbacid, 1987). One interesting possibility is that during NGF deprivation, post-mitotic neuronal cells attempt to re-enter the cell cycle, resulting from the activation of p21^{ras}. This is referred to as a "defective cell cycle," since the factors required for successful cell cycle traversal are not synthesized in post-mitotic cells (Ferrari et al., 1994). Although this model is largely descriptive, it is supported by the fact that: (1) several cell cycle genes known to be induced in the G1 to S transition, including c-*fos,* c-*jun,* *cyclin A,* and *cyclin D,* are up-regulated within hours following NGF withdrawal in cultured sympathetic neurons (Estus et al., 1994; Freeman et al., 1994; Meikrantz et al., 1994); and (2) cell death in SCG neurons or NGF-differentiated PC12 cells is suppressed by inhibition of *de novo* protein or RNA synthesis (Mesner et al., 1992; see Zakeri, Chapter 3, this volume).

Importantly, overexpression of the cyclin D–dependent kinase inhibitor p16^{INK4} protects neuronal cells from death during NGF withdrawal, suggesting that activation of cyclin D1–dependent kinases are essential, rather than correlative, during apoptosis (Kranenburg et al., 1996). However, what remains puzzling with this model is that overexpression of wild-type or oncogenic p21^{ras} in PC12 cells does not induce apoptosis, but rather results in differentiation and survival (Bar-Sagi et al., 1985; Borasio et al., 1989). It should also be pointed out that cell death in rapidly proliferating cells can be similarly induced by rapid exit from the cell cycle (Evan et al., 1992; Lowe et al., 1993). Thus the defective cell cycle–mediated apoptotic pathway for post-mitotic neurons may simply reflect the inability of any cell to survive a drastic and sudden change in environmental cues, and p21^{ras} activation may represent a consequence resulting from such a change. Clearly, credibility of either the p21^{ras} effector hypothesis or the defective cell cycle hypothesis would be strengthened if p21^{ras} substrates proved to be effectors or modulators of ICE/CED-3 proteases.

Trk Isoforms and Modulation of Receptor Function

The preceding arguments suggest that NGF-induced survival pathways are mainly initiated by the binding of specific intracellular signaling proteins to TrkA. However, each *trk A, B,* or *C* transcript can undergo alternative splicing, resulting in the production of several isoforms for each *trk* gene that can modulate Trk signaling when expressed with the full-length receptors (Fig. 13.1). TrkA encodes two proteins, one full-length TrkA, and one with an additional 8 amino acids in the extracellular domain, which displays modest changes in ligand affinity. TrkB and TrkC encode full-length kinase-active proteins, as well as truncated forms lacking a kinase domain that are generated via alternative splicing of exons encoding unique amino acids and termination sequences (reviewed in Barbacid, 1994). The truncated forms of TrkB and TrkC can inhibit signaling when coexpressed with full-length TrkB and TrkC receptors by interfering with the oligomerization of the kinase-active isoforms (Canossa et al., 1996; Ninkina et al., 1996). The *trk* C gene also encodes additional isoforms that contain 14 or 25 amino acid inserts within the tyrosine kinase domain (Tsoulfas et al., 1996). The isoforms with kinase inserts display an attenuated tyrosine phosphorylation response to ligand, and fail to promote cell differentiation when expressed in a neural crest cell line. This system represents the first receptor tyrosine kinase system that expresses variable isoforms that retain ligand binding and ligand-inducible kinase activation, yet differ in biological capabilities (Tsoulfas et al., 1996). It is also noteworthy that NGF signaling pathways involved in regulation of gene expression, differentiation, and survival can be controlled by non-tyrosine-containing receptor domains. Deletion of one such motif, called the KFG sequence, which is conserved in all Trk family members and is located in the juxtamembrane region, impairs NGF-induced neurite outgrowth and the induction of c-*fos,* c-*jun,* and TIS1 immediate early genes (Peng et al., 1995). Thus, during NGF-induced homodimerization and initiation of signal transduction cascades, both phosphotyrosine-dependent and phosphotyrosine-independent events are likely critical to contribute to NGF actions.

Although neurotrophin-induced receptor oligomerization and recruitment of signaling proteins initiate signal transduction cascades, long-term neuronal survival requires TrkA internalization and retrograde transport (Fig. 13.2). Both chemical inhibition of retrograde axonal transport and experimental axotomy result in degeneration effects in neurotrophin-sensitive neurons (Barde, 1989; Oppenheim, 1991). Thus, unlike the case for receptors such as those for EGF, which requires internalization for receptor down-regulation via lysosomal degradation, it appears that Trk receptors are sorted differently, such that trophic signaling continues to occur intracellularly (Eveleth et al., 1992). In fact, for several tyrosine kinase receptor systems, including TrkA and the insulin receptor (DiGuglielmo et al., 1994; Baass et al., 1995), tyrosine kinase activity is retained, and perhaps even enhanced, within internalized endosomal-like vesicles. This suggests that critical TrkA targets involved in neuronal survival are localized in an internalized subcellular compartment.

THE P75 NEUROTROPHIN RECEPTOR

Role of p75 in Neurotrophin Internalization and Trk-Mediated Survival

Although two distinctive classes of cell surface receptors for NGF have been identified and molecularly characterized (Chao et al., 1995), the function of p75 is less well defined than that of the Trk receptors. Site-directed mutagenesis of NGF shows that TrkA and p75 interact with independent domains of NGF, and mutant NGF molecules can be generated to exhibit some selectivity towards either p75 or TrkA. NGF molecules that bind predominantly TrkA induce immediate-early genes and neurite outgrowth in cultured sympathetic neurons, supporting the hypothesis that TrkA is the principal signal-transducing neurotrophin receptor for the aforementioned pathways (Weskamp et al., 1991; Ibanez et al., 1992). However, coexpression of p75 can enhance the sensitivity of sensory neurons to NGF and augments the proportion of high-affinity binding sites for NGF (Hempstead et al., 1991; Barker et al., 1994). Interestingly, the p75 neurotrophin receptor, which has been implicated in neural apoptosis (see the following), appears to play an important role in retrograde axonal transport of BDNF, NT3, and NGF (Curtus et al., 1995). In fact, administration of anti-p75 neutralizing antibodies or dominant negative forms of p75 molecules that can bind NGF but do not interact with Trks, completely block retrograde transport of NT4 and BDNF in sensory neurons (Curtus et al., 1995). Therefore, in the presence of TrkA, p75 may potentiate survival by sensitizing neurons towards neurotrophic factors (Table 13.2).

TABLE 13.2. Multiple Roles of the p75 Low-Affinity NGF Receptor During Neurotrophin Signaling

Biological Functions of p75	Reference
Presents neurotrophins (e.g., NGF, BDNF, NT3) to Trk molecules	Barker et al., 1994
Increases the proportion of high-affinity NGF binding sites	Hempstead et al., 1991; Weskamp et al., 1991
Retrograde transport of neurotrophins	Curtus et al., 1995
Induces apoptosis (ligand dependent)	Rabizadeh et al., 1993

Role of p75 in Ligand-Dependent Apoptosis

In contrast to its ability to potentiate Trk function, p75 may function independently to promote apoptosis when TrkA is absent or is present at low levels, or when TrkA positive cells switch during development to express only TrkB or TrkC (Barrett et al., 1994; Casaccia-Bonnefil et al., 1996; Frade et al., 1996; Zee et al., 1996) (Table 13.2). Thus TrkA expression may exert a dominant effect on neuronal survival, such that even when TrkA levels are as low as 10% of those of p75, as is the case for PC12 cells, NGF acts as a survival factor (Hempstead et al., 1991). This is supported by recent studies by Barde and colleagues utilizing the developing chick retina as a model system (Frade et al., 1996). Retinal ganglia cells express *NGF* and *p75* RNA, but not *trkA* RNA at embryonic day 4 (E4). Application of antibodies that block NGF binding to p75 into the vitreal cavity of E4.5 embryos results in a substantial reduction in the neuronal death, implicating p75 in inducing apoptosis. Similar results have been obtained with oligodendrocytes, which express p75 but not TrkA (Casaccia-Bonnefil et al., 1996). These primary postnatal cells die in culture when treated with NGF and survive when exposed to anti-NGF antibodies.

A signaling role for p75 in cell death during NGF-mediated target cell innervation of peripheral neurons has also been proposed (Bothwell, 1996). Thus, while limiting quantities of target-derived NGF may regulate neuronal number, the excess TrkA/p75 expressing cells (those that do not become stimulated by NGF) may die upon contact with a different neurotrophin such as BDNF or NT3, both of which can bind the p75 but not TrkA (Fig. 13.2). Thus, in cultured SCG neurons that express both p75 and TrkA, BDNF treatment results in cell death (Bothwell, 1996). Furthermore, p75 levels are up-regulated during target innervation, as determined by indirect immunofluorescence, and it has been hypothesized that this facilitates active cell death in redundant neurons. Interestingly, transgenic mice expressing the intracellular domain of p75 under a pan-neural α-tubulin promoter results in a profound and selective loss in both sympathetic and sensory neurons that require NGF for survival. By comparison, those neurons that do not require

NGF for survival, such as facial motor neurons, are not depleted by apoptosis (Parker et al., 1996). Collectively, these observations suggest that p75 can cause cell death when TrkA activity is compromised. However, these data must also be reconciled with the observation that sympathetic neurons derived from p75-null mutant mice die normally in culture during NGF deprivation (Lee et al., 1992). Moreover, p75-null mutant animals can survive into adulthood with a significant loss, rather than gain, of sensory and sympathetic neurons (Bothwell, 1996). These results also suggest that redundant or p75-independent apoptotic mechanisms exist for neuronal cell death *in vivo*.

Signal Transduction by the p75 NGF Receptor: Relation to TNFR1 and Fas

An important clue into the function of p75 was revealed in the early 1990s when the apoptosis-inducing TNFR1 and Fas receptor genes were molecularly cloned and the encoded proteins shown to share structural homology with the extracellular domain of p75 (Schall et al., 1990; Itoh et al., 1991) (Fig. 13.4). Although the advances in TNFR1 and Fas signaling have provided a basic framework for ligand-mediated apoptosis, comparably very little prog-

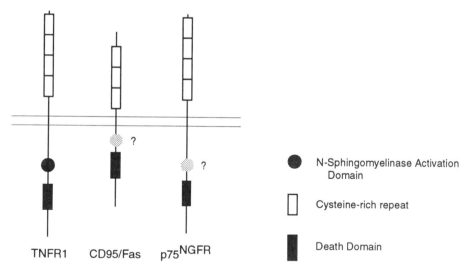

FIG. 13.4. Structure of receptors mediating cell death: The p75 shares sequence homology with the death-inducing CD95/Fas and TNFR1 in both the extracellular and intracellular domains. All three receptors contain extracellular cysteine-rich motifs, as well as death domains in the C-terminal cytoplasmic regions. The TNFR1 also contains an N-sphingomyelonase activation domain that is involved in the activation of neutral sphingomyelenase. Although sphingomyelonase myelin hydrolysis is observed following both CD95/Fas and p75 activation, corresponding activating sequence has not been defined for these latter receptors.

ress has been made on the mechanism of p75 signaling. In part this reflects the fact that, while TNFR1 and Fas signaling require ligand stimulation, initial reports suggested that p75 could mediate apoptosis in either the presence or absence of ligand (Rabizadeh et al., 1993). Second, the ligand for TNFR1 and Fas receptor belong to a superfamily that includes TNF, LT-α, Fas ligand (FasL), OX40L, CD40L, CD27L, CD30L, 4-IBBL, and LTβ, many of which are membrane tethered or are believed to behave as trimers in solution (Baker et al., 1996). NGF does not share structural homologies with these members of the TNF superfamily (Baker et al., 1996). Third, while structural homology exists among extracellular domains of each receptor, as characterized by 40-amino acid cysteine-rich motifs, the more limited homology in the intracellular domains was not recognized (Hofmann et al., 1995). However, recent studies suggest that the extreme C-terminus of p75 is homologous to the death domains of TNFR1 and Fas, and molecular modeling of this region of p75 suggests that it can adopt a structure similar to the death domains of TNFR1 and Fas (Chapman, 1995; Feinstein et al., 1996). Thus, in light of the recent findings that p75 can directly mediate cell death, it is likely that p75 may mediate cell killing in a fashion analogous to TNFR1 and Fas (Fig. 13.5).

Because of rapid progress on TNFR1 and Fas signaling, it may be instructive to review some of their signaling features in the context of p75 signal transduction events. For Fas, which is perhaps the simplest and best understood cytotoxic pathway, a model has emerged, reminiscent of receptor tyrosine kinase signaling, whereby ligand binding results in receptor clustering (probably trimerization) followed by the subsequent recruitment of intracellular proteins, some of which contain death domains (Chinnaiyan et al., 1995; Boldin et al., 1996; Muzio et al., 1996) (Fig. 13.5). Using the yeast two-hybrid system, several novel genes have been cloned and the encoded proteins shown to interact with the Fas and TNFR1 death domains, including TRAF1, TRAF2, TRAF3, TRADD, FADD/MORT/, and RIP (reviewed in Baker et al., 1996). Proteins with death domains can either form homomeric receptor clusters or heterodimerize with proteins that could initiate downstream signaling pathways (see Cryns and Yuan, Chapter 6, this volume). The binding of FasL to Fas recruits FADD (Fas-associated protein with a death domain) (Chinnaiyan et al., 1995) (Fig. 13.5), which, in turn, recruits another death domain–containing protein called MORT1/FLICE. MORT1 contains a domain with extensive homology with ICE/CED-3, including a canonical QACRC motif and an aspartate cleavage site, implying that FLICE may serve as a cysteine protease for the activation of CPP32 or other ICE family members (Muzio et al., 1996). Thus Fas activation may result in a series of protein–protein interactions leading directly to caspase activation and apoptosis (Fig. 15.5). Does an analogous pathway exist for p75-induced killing? Perhaps, although to date no death domain–containing substrates for p75 have been reported using analogous yeast two-hybrid screening approaches that have been successful in cloning TNFR1- and Fas-associated proteins. Direct proof of this hypothesis should be testable by generating receptor chimeras that

A. Fas-mediated apoptosis

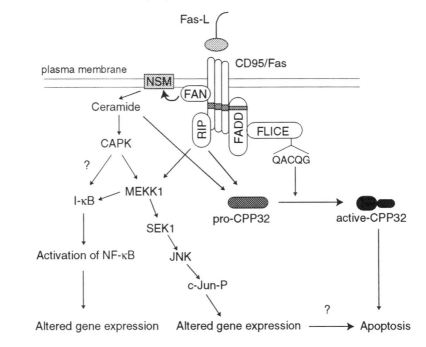

B. p75-mediated apoptosis

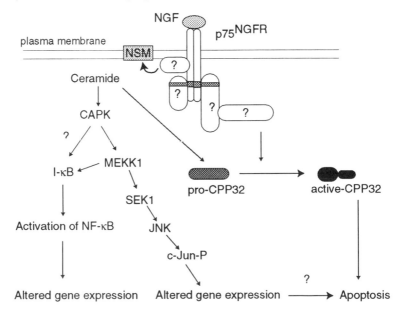

express the intracellular domains of p75 with the extracellular domains of TNFR1 or Fas, and then observe whether these chimeric receptors recapitulate TNF or Fas signaling.

An additional similarity in TNFR1, Fas, and p75 NGFR signaling pathways involves the activation of a kinase pathway capable of phosphorylating I-κB and stimulating NF-κB activity (Hsu et al., 1995; Baker et al., 1996; Carter et al., 1996). All three receptor systems also activate the sphingomyelin (SM) pathway (Dobrowsky et al., 1994; Wiegmann et al., 1994; Cuvillier et al., 1996), resulting in the production of ceramide, a potent intracellular mediator of apoptosis in many cell types, including embryonic neurons (Chao, 1995; Tepper et al., 1995), Wiesner et al., 1996) (Fig. 15.5). Recently, a novel WD-repeat protein called FAN (factor associated with neutral-sphingomyelinase activation) was isolated that binds to a nine amino acid sequence motif in the cytoplasmic domain of the TNFR1 (Adam-Klages et al., 1996). Overexpression of full-length FAN enhances sphingomyelinase activity, suggesting that FAN regulates endogenous ceramide production. When applied exogenously, membrane-permeant analogs of ceramide induce NF-κB and c-Jun N-terminal kinase (JNK) activation (Chao, 1995; Verheij et al., 1996), similar to that seen with TNFR1 and p75 activation, suggesting that ceramide may be a common downstream effector of both apoptotic pathways (Fig. 15.5). Support for this model comes from studies demonstrating that endogenous inhibitors of ceramide production, such as sphingosine I phosphate, activation of Protein Kinase C (PKC) by phorbol esters or certain growth factors, or stereospecific ceramide isomers that do not activate JNK, dramatically suppress apoptosis (Cuvillier et al., 1996).

At present, it is not clear how ceramide activates Stress-Activate Protein Kinase (SAPK)/Jun N-terminal Kinase (JNK) and whether or not this causes apoptosis. The SAPK/JNK pathway involves sequential activation of MEKK1, SEK1, SAPK/JNK and culminates with serine phosphorylation of the transcription factor c-Jun (Kyriakis et al., 1994) (Fig. 15.5). Catalytically inactive

---

FIG. 13.5. Fas and p75-mediated apoptosis: Is there common ground? A hypothetical model for the signaling of p75 (panel B) is proposed based upon the model for Fas receptor (panel A) and TNFR1 and their associated proteins and effector pathways. Following stimulation by Fas ligand, death-inducing proteins RIP, FADD, and FLICE are recruited to the Fas receptor via their death-domain modules in order to signal caspase activation (e.g., CPP32) and apoptosis. Activation of NSM by FAN results in the production of ceramide, which in turn activates a ceramide-activated protein kinase (CAPK). CAPK, either directly or through a second kinase, can activate NF-κB and MEKK1, which regulate nuclear events. The role of these latter pathways during apoptosis is unclear. Similar downstream effector pathways are induced by p75 and Fas receptor, including activation of JNK and NF-κB, as well as the activation of caspases. Although they are postulated, no RIP, FADD, or FLICE-like proteins with death domains have yet been identified downstream of p75.

dominant negative forms of MEKK1 or SEK1 block ceramide-mediated killing in U937 cells (Verheij et al., 1996), and withdrawal of NGF from PC12 cells or primary sympathetic neurons leads to sustained increase in JNK activity and the ensuing c-Jun phosphorylation (Estus et al., 1994; Xia et al., 1995). In fact, microinjection of c-Jun protein into SCG neurons causes spontaneous apoptosis while dominant negative forms of JNK or c-Jun (Flag∆ 169) protect these cells from apoptosis (Ham et al., 1995). Thus, p75-induced cell death might be thought to involve a ceramide>SAPK/JNK> c-Jun phosphorylation cascade that culminates in the activation of "death-specific" genes. One conceptual difficulty with this model is that apoptosis in many systems, including that mediated by TNF or FasL (Liu et al., 1996b) or that in proliferating neuroblasts (Ferrari et al., 1994), does not require new gene transcription (see Zakeri, Chapter 3, this volume). Hence, this model could be strengthened if c-Jun was shown to regulate the transcription of genes that can modulate apoptosis, such as regulators of caspases. Alternatively, JNK may phosphorylate other substrates in addition to c-Jun that mediate apoptosis. Recently, Karin and colleagues demonstrated that the TNFR1-associated death domain protein, RIP, can directly mediate apoptosis in epithelial cells (Hsu et al., 1996; Liu et al., 1996b) and using deletion mutants. These investigators were able to dissociate TNF-induced JNK activation from apoptosis. Thus expression of the C-terminal domain of RIP is sufficient to induce apoptosis but does not activate JNK. In contrast, expression of a RIP deletion mutant that lacks the death domain activates JNK but does not cause apoptosis. This study represents the first system in which the activation of JNK could be dissociated from caspase activation and the ensuing cell death, and it will then be important to determine whether this also holds true for neuronal cells.

APOPTOTIC AND ANTI-APOPTOTIC GENES IN NEURAL CELLS

ICE Proteases in Neuronal Apoptosis

Both genetic and biochemical evidence implicate ICE/CED-3 (interleukin 1β-converting enzymes) family proteases in many, if not all, physiological cell death pathways (reviewed in Nicholson, 1996; see also Cryns and Yuan, Chapter 6, this volume). This unusual class of protease superfamily (termed caspases) (Alnemri et al., 1996) shares a pentapeptide QACXG at the active site, with an absolute dependence on the cysteine for catalytic activity (Yuan et al., 1993; Nicholson et al., 1995). Once activated, all caspases cleave proteins at specific aspartate residues in the context of appropriate flanking sequences. Activated ICE, the prototype of the caspase family, is a tetrameric enzyme with heterodimeric subunits of p20 and p10 and is derived from a 45 kD proenzyme by enzymatic cleavage after four Asp-X bonds (Thornberry et al., 1992; Wilson et al., 1994). In *C. elegans,* mutation in *ced3* (the invertebrate homolog of the gene for CPP32 or caspase 3), abolishes programmed cell death, including neuronal cell death, that occurs during hemaphrodite development. These data indicate that ICE-like proteases are capable of triggering the death of numerous cell populations in this invertebrate system (Xue et al., 1996). Likewise, in cultured chick dorsal root ganglia, microinjection of *crmA* (a cowpox virus gene encoding a serpin that specifically inhibits ICE proteases) protects these neurons from apoptosis during NGF withdrawal (Gagliardini et al., 1994). More specifically, application of a cell permeable, irreversible aldehyde tetrapeptide inhibitor of ICE-like proteases (Ac-YVAD-CMK) inhibits both programmed cell death in cultured motor neurons, and motor neuron cell death *in vivo* (Milligan et al., 1995). Collectively, these data also suggest that cysteine proteases are key effectors of cell death in the vertebrate nervous system.

Despite the evidence linking caspase activation to programmed cell death and apoptosis, little is known about the regulation of these enzymes, both in terms of how they become activated after a cell receives an apoptotic stimulus (i.e., NGF withdrawal), as well as how they mediate cell death once activated. It is interesting that the pro-enzymes of caspases are present in healthy cells and do not require induction during apoptosis (Weil et al., 1996), suggesting that activation is tightly controlled by negative regulation. This issue is even more complex since at least eleven caspases have been identified and several different family members can be coexpressed in the same cell (Alnemri et al., 1996; Greidinger et al., 1996; Takahashi et al., 1996). Do different caspases perform redundant functions, or does each have a unique role in signaling cell death? Although some caspases, such as CPP32, are autocatalytic when triggered for apoptosis (Wilson et al., 1994), it is also apparent that these proteases can regulate the activity of one another. For example, CPP32 can be activated by Mch2α (Takahashi et al., 1996) or by the cytotoxic lymphocyte protease, granzyme B (Martin et al., 1996). Thus the sequential activation of a protease cascade could function to amplify the effector pathways of apoptosis, in much the same way as proteases in the coagulation cascade are activated during hemostasis. Alternatively, it is also possible that different proteases respond to different death stimuli, such as trophic factor withdrawal, p75 activation, or oxidative stress and thus a different protease would carry out the execution phase. Clearly, a thorough inventory of the proteases in a given neuron, and their mode of activation under different death stimuli, should provide insights into this question. Once activated, caspases are involved in a number of biological processes, including the destructive cleavage of cellular substrates such as poly(ADP-ribose) polymerase, lamins, topoisimerases, and actin (Fraser et al., 1996) as well as the activation of signaling pathways, most notably the sphingomyelinase pathway (Pronk et al., 1996) and the p38/JNK kinase pathway (Juo et al., 1997). Although not completely understood, the current theory postulates that the cleavage and selective degradation of one or more caspase substrates ultimately result in the catastrophic demise of the cell.

Bcl-2 Family Proteins and Neuronal Survival

In *C. elegans,* the loss of function of the ICE homolog, CED-3 results in the survival of cells that would otherwise die during development, whereas mutation of a second gene, *ced-9*, results in extensive cell death via a CED-3-dependent mechanism indicating that CED-9 is an endogenous suppressor of CED-3 function or activation (Hengartner et al., 1994). The *ced-9* gene is the functional homolog of the human proto-oncogene *bcl-2*, which encodes a 26 kD membrane-associated protein that was originally isolated from the breakpoint of the t(14;18) chromosomal translocation in follicular lymphoma (Hockenbery et al., 1990). The Bcl-2 family consists of both apoptosis-promoting proteins (e.g., Bax, Bad, Bak, Bcl-x$_s$) and apoptosis-antagonizing proteins (e.g., Bcl-2, Bcl-x$_L$, McHl, CED-9) that function, at least in part, via competing protein–protein interactions (Oltvai et al., 1994). In addition to expression in the immune system, Bcl-2 is also widely expressed throughout the developing nervous system and is retained at significant levels in the adult PNS. These findings suggest that Bcl-2 expression is not restricted to the period of neuronal development but may protect mature neurons (Merry et al., 1994; Davies, 1995). Overexpression of Bcl-2 can prevent death in some sympathetic neurons, chick embryo cranial sensory neurons, and differentiated PC12 cells deprived of NGF (Mah et al., 1993; Martinou et al., 1994; Farlie et al., 1995), and transgenic mice overexpressing the *bcl-2* gene under control of the neuron-specific enolase (NSE) promoter exhibit increased neuronal numbers in both the CNS and PNS (Farlie et al., 1995). Facial neurons from these post-natal transgenic animals are resistant to axotomy-induced cell death, suggesting that Bcl-2 can prevent both developmental and degenerative cell death (Dubois-Dauphin et al., 1994). In addition to its role in suppressing apoptosis, Bcl-2 has also been shown to promote differentiation and process formation in cultured neuronal cells (Zhang et al., 1996) and promote axonal regeneration in adult retinal ganglia cells (Chen et al., 1997). This suggests that Bcl-2 can enhance both the survival and regeneration of mature neurons in the CNS (see Gartner and Hengartner, Chapter 4; Cryns and Yuan, Chapter 6, this volume).

However, while Bcl-2 overexpression can maintain neurotrophic factor-deprived neurons that depend on NGF, BDNF, or NT3 for survival, it is ineffective in protecting neurons that are dependent upon an unrelated neurotrophic factor, CNTF (Allsopp et al., 1993; Cotman et al., Chapter 14, this volume). Thus, there are at least two intracellular mechanisms for apoptosis associated with growth factor deprivation, one Bcl-2 dependent and one Bcl-2 independent. It is also likely that other genes, including *bcl-x*, which are also expressed in the neuronal cells, have overlapping roles with *bcl-2* in regulating neuronal survival. This may indeed explain the observed normal development of the PNS and the CNS in *bcl-2* null mutant mice (Greenlund et al., 1995b). Another gene, the nuclear inhibitor of apoptosis (NAIP), has been implicated in the autosomal recessive disorder spinal muscular atrophy (SMA), as deletion of this gene in patients may lead to a failure of motor neurons to suppress apoptosis (Roy et al., 1995). While there appears to be some redundancy at the level of genes that promote neuronal survival, the *bax* null mutant mice exhibit significant increases in numbers of superior cervical ganglia and facial nuclei (Deckwerth et al., 1996). Moreover, sympathetic neurons derived from these animals are independent of NGF for survival and survive experimental axotomy in the absence of trophic factor. This suggests that neuronal apoptosis is critically dependent on death-promoting Bcl-2 family members, such as BAX.

Bcl-2 and the Mitochondria

The mechanism by which Bcl-2 suppresses apoptosis is largely unknown. While molecular ordering of the cell death pathways indicates that Bcl-2 and Bcl-x_L lie upstream of the ICE/CED-3 proteases, (Chinnaiyan et al., 1996), it is not clear how these proteins interact. Recently, however, Bcl-2 was shown to bind to a novel antiapoptotic protein BAG-1, which has sequence homology to ubiquitin (Takayama et al., 1995). Whether or not this interaction reflects a relevant connection between Bcl-2 and caspases, however, remains to be determined. In any case, clues to the potential functions of Bcl-2 emerge from immunocytochemical examination of its subcellular localization. Many Bcl-2 family members contain hydrophobic stretches of amino acids in the extreme C termini that allow insertion into membranes, including the outer mitochondrial membrane, the endoplasmic reticulum membrane, and the nuclear envelope (Givol et al., 1994; Lithgow et al., 1994). Furthermore, mutants of Bcl-2 restricted to the cytoplasm, generated by deleting the 22 amino acid C-terminus region, are rendered ineffective in suppressing apoptosis (Tanaka et al., 1993). Similarly, in the yeast *Saccharomyces cerevisiae,* cell death induced by Bax can be rescued by Bcl-2 fused to the membrane anchor of the yeast mitochondrial Mas70 protein, further suggesting that the antiapoptotic effects of Bcl-2 are linked to mitochondrial function (Greenhalf et al., 1996).

Recently, the X-ray and NMR structure of Bcl-x_L reveals a membrane insertion domain similar to that found in bacterial toxins, such as diphtheria toxin and the colicins (Muchmore et al., 1996). As such by analogy to diptheria toxin, Bcl-2 proteins may dimerize and form pH-dependent membrane pores. A direct role of Bcl-2 on mitochondrial function is suggested by the ability of Bcl-2 or Bcl-x_L to prevent the loss of mitochondrial transmembrane potential and membrane permeability that result from glutathione (GSH) depletion or chemical hypoxia induced by KCN or rotenone (Myers et al., 1995; Shimizu et al., 1996). Interesting experiments by Zamzami et al. demonstrate that intact mitochondria from apoptotic cells, or factors liberated from mitochondria treated with chemical inducers, can cause apoptosis in a cell-free system, although mitochondria from cells overexpressing Bcl-2 fail to do so (Zamzami et al., 1996). Similar results by Liu and colleagues demonstrated that cytochrome C, an essential component of the mitochondrial respiratory chain, is released from mitochondria and induces apoptosis by activating ICE/CED-3 proteases (Liu et al., 1996a; see Benites-Bribiesca, Chapter 17, this volume). In addition to critical roles in electron transport and oxidative metabolism, mitochondria are one of the principal generators of reactive oxygen species (ROS). ROS include superoxide anions ($O_2 \cdot^-$), hydrogen peroxide (H_2O_2), nitric oxide (NO), hydroxy radical (OH·), and lipid peroxides (Slater et al., 1995). Addition of H_2O_2 or lipid peroxides to neuronal cultures rapidly induces apoptosis (Whittemore et al., 1994), while addition of exogenous antioxidants, such as N-acetylcysteine (a precursor to cellular GSH) or vitamin E, to neurons prevents cell death resulting from NGF deprivation (Ferrari et al., 1995). Moreover, microinjection of copper/zinc superoxide dismutase (SOD), which catalyzes the dismutation of superoxide radicals to H_2O_2 and oxygen molecules, delays apoptosis in cultures of sympathetic neurons deprived of NGF (Greenlund et al., 1995a).

Interestingly, several studies suggest that Bcl-2 has antioxidant properties and can prevent the generation of ROS and lipid peroxides under pro-oxidant conditions (Hockenberry et al., 1993; Kane et al., 1993). Neuron cells overexpressing Bcl-2 have lower levels of oxidized glutathione (GS-SG) and the ratios of $NAD^+/NADH$ in these cells are two- to threefold less than in control cells (Ellerby et al., 1996). Collectively, these data suggest that one function of Bcl-2 may be to prevent loss of mitochondrial function and the subsequent release of toxic metabolites to the cytoplasm or other intracellular organelles, where they can exert damage or trigger the activation of caspases. However, one argument against the importance of the mitochondria in apoptosis is that Bcl-2 can protect against apoptosis in cells that lack mitochondrial DNA (Jacobson et al., 1993). Thus one can envision that while mitochondrial factors such as ROS are capable of inducing apoptosis, such factors are not obligatory for death. Indeed, other studies have shown that endoplasmic reticulum (ER) sources of Ca^{2+}, a cation that can trigger apoptosis, can be suppressed by Bcl-2 tethered to the ER membrane (Lam et al., 1994). Finally, other studies suggest that Bcl-2, when tethered to the nuclear envelope, may control the nuclear import of death effectors such as p53 and ICE proteases (Wang et al., 1996). Thus Bcl-2 may exert a multitude of biological effects and participate in multiple antioxidant signaling pathways. Clearly, one of the most exciting future areas of research will be to ascertain how Bcl-2 family members regulate the activation of caspases.

ROLE OF REACTIVE OXYGEN SPECIES (ROS) IN NEURONAL CELL DEATH

Signal Transduction by ROS

The term *oxidative stress* refers to an inbalance in cellular reduction oxidation (redox) potential such that the production of pro-oxidant molecules exceeds the antioxidant capacity of the cell, ultimately resulting in the depletion of reduced glutathione (GSH) and the progressive oxidation of macromolecules (Korsmeyer et al., 1995). However, all oxygen radicals are not deleterious to living cells, as abnormally low levels of ROS, induced by treatment of cells with high levels of antioxidants, can also induce apoptosis (Burdon et al., 1995). Studies utilizing cellular systems indicate that endogenous generation of ROS is necessary for cytokine and growth factor responses, including cell proliferation and possibly physiological cell death (Burdon et al., 1995; Sundaresan et al., 1995). Thus one must distinguish between physiological oxidant signaling and pathophysiological oxidative stress. One of the other difficulties in assessing a causative role for ROS during apoptosis arises from the fact that oxygen radicals are extremely unstable and short-lived. Therefore, most assays detect only the products generated at the late stages of oxygen damage. Furthermore, an understanding of how ROS modulate signal transduction pathways is particularly lacking. To date, there is no evidence in mammalian cells for homologs of the bacterial SoxR or OxyR operons that are capable of "sensing" ROS, and thus might be an initial step towards an antioxidant or cellular response (Hildago et al., 1994). Although ROS can modify cellular lipids, DNA, and protein, the roles of these modified macromolecules is unclear. Do specific lipid peroxides act similarly to PI-lipids to modify cellular function? Do DNA strand breaks induced by hydroxy radicals alter p53 function to trigger apoptosis? There is mounting evidence that cellular signaling proteins, including receptor tyrosine kinases and cytoplasmic kinases, phosphatases, monomeric G proteins, and transcription factors can be regulated by reversible thiol oxidation on cysteinyl residues (Burdon, 1995; Lander et al., 1996). For example, exposure of many cell types to UV light or H_2O_2 rapidly induces c-*fos* and c-*jun* gene expression, resulting in the formation of the transcription complex AP-1, which binds to AP-1 promoter sequences and regulates the transcription of target genes (Abate et al., 1990; Devary et al., 1991). Interestingly, both Fos and Jun proteins also contain redox-sensitive cysteine residues in their DNA binding domains that may regulate their actions on transcriptional activity of genes (Abate et al., 1990). Moreover, exposure of cells to UV light or H_2O_2 can activate MAP kinases and JNK, the latter of which can phosphorylate and activate c-Jun (Lo et al., 1995, 1996), and may contribute to neuronal apoptosis under certain conditions. These results suggest that ROS can modulate multiple signaling pathways leading to alterations in gene expression. The identification of redox-responsive elements in specific mammalian promoters would be a significant advance quantitatively to assess the physiological levels of ROS in tissues as well as the relationship of ROS and gene expression to conditions that result in apoptosis.

Another interesting and important ROS in neuronal signaling and pathology is nitric oxide (NO) (Schmidt et al., 1994). Initially identified as a regulator of synaptic signaling, NO has since been demonstrated to mediate a variety of cellular functions in the nervous system. NO, at high concentrations, can react with superoxide to form peroxynitrite and, as such, might play a cytotoxic role in ROS-induced apoptosis in PC12 cells and other neuronal cells (Troy et al., 1996). Consistent with this model, PC12 cell death can be induced by the down-regulation of Cu/Zn-SOD (Troy et al., 1996). This impaired SOD function results in elevated levels of ROS that can be prevented by inhibition of NO synthesis. In addition, the enhanced motor neuron loss in SOD knock-out mice after axonal injury might be related to peroxynitrite production (Reaume et al., 1996). In other studies, however, NO has been reported to prevent apoptosis, observations that might, in part, be attributable to the direct interaction of NO with regulators of cellular homeostasis (Lipton et al., 1993). Peptide and protein targets of NO whose activities are altered by nitrosothiol formation include reduced glutathione (GSH), albumin, NMDA receptor, calmodulin-stimulated adenyl cyclase and p21^{ras} (Lander et al., 1996). NO also reacts with tyrosine residues or tyrosyl radicals, may also modify enzyme activity. Interestingly, Farinelli et al. (1996) have recently reported that various NO-generating compounds can delay apoptosis in trophic factor-deprived PC12 cells and sympathetic neurons, suggesting that NO can also support survival in serum-free conditions. However, given the diversity of NO-mediated cellular responses, pharmacological agents that affect overall NO homeostasis are somewhat limiting as means of dissecting the complex actions of NO. For instance, since NGF deprivation in cultured sympathetic neurons is accompanied by an increase in ROS, application of NO-generating compounds might lead to peroxynitrite formation, and thus, the short-term survival-promoting benefit of NO on these neurons might be offset by the actions of the cytotoxic peroxynitrite molecule (Troy et al., 1996). It has also been shown that NO is cytostatic in PC12 cells and thus consumption of NO by ROS may facilitate cell cycle progression into a defective cell cycle (Peunova et al., 1995). The effects of NO generation must be considered in the larger context of ROS, and the redox state of the cell.

Role of ROS in Neurodegenerative Disease

During development, neuronal cell death clearly plays a prominent role in the sculpturing and organization of neuronal circuitry. However, neuronal loss is also a prominent feature of aging and neurodegenerative diseases (Bowling et al., 1995; Bredesen, 1995). Recent studies suggest that neuronal apoptosis may accompany the adverse conditions associated with Alzheimer's disease, Parkinson's disease, amyotrophic lateral sclerosis (ALS), Huntington's disease, and stroke-induced ischemia and aging (Bowling et al., 1995; Bredesen, 1995; and Cotman et al., Chapter 14, this volume). Although the etiologies of these diseases is clearly different, similar pathological manifestations may be associated with mitochondrial impairment and increased free-radical production, resulting in oxidative damage to macromolecules (Table 13.3). However, it is unclear if oxidative stress is a primary pathophysiological mechanism in neurodegenerative disease, or if ROS are secondary risk factors, possibly being involved as a consequence of a different primary pathological event. Genetic and biochemical studies with familial ALS (FALS) provide the most convincing evidence to support a role for ROS in disease progression

TABLE 13.3. Role of Reactive Oxygen Species in Neurodegenerative Disease

Disease	Evidence
Alzheimer's	Increased lipid peroxidation
	Increased SOD activity
	Increased catalase activity
	Increased protein oxidation
ALS	Modified SOD activity
Multiple sclerosis	Improvement by iron chelators
Parkinson's	Increased SOD activity in substantia nigra
	Decrease in ferritin
Stroke/ischemia	Free radical generation
	Protection by antioxidants
	Attenuation by elevated SOD

(Rosen et al., 1993; Wiedau-Pazos et al., 1996). ALS is a degenerative disorder of motor neurons of the cortex, brainstem, and spinal cord (Rosen et al., 1993). FALS accounts for about 10% of all clinical cases and has been associated with missense mutations in *sod*-1, the gene that encodes the Cu/Zn-su (SOD), (Sendtner et al., 1994). Two mechanisms have been proposed to account for the adverse effects of SOD1 mutations on motor neuron survival and both are proposed to involve increased oxidative stress (Orrel et al., 1995). First, most SOD1 mutants found in patients exhibit a model to significant decrease in SOD1 activity (from 30 to 70%), resulting in higher levels of ROS production (Bowling et al., 1993). Consistent with this hypothesis, down-regulation of SOD1, using antisense oligonucleotides, leads to cell death in PC12 cells, by a mechanism that is inhibitable by antioxidants or Bcl-2 (Troy et al., 1996). However, more recent studies demonstrate that SOD1 missense mutations can lead to a dominant gain of adverse function such that the mutant enzyme catalyzes oxidation of substrates by H_2O_2 or peroxynitrite at a higher rate than the wild-type enzyme (Gurney et al., 1994; Tsuda et al., 1994; Wiedau-Pazos et al., 1996; Yim et al., 1996). In fact, transgenic mice coexpressing the wild-type SOD1 allele and a transgene of mutant A4V SOD1 or G93A SOD1 (previously characterized in FALS patients), displayed a very marginal decrease in SOD1 activity, but exhibited severe motor neuron loss in the spinal cord (Gurney et al., 1994). Importantly, however, transfection of plasmids encoding the mutant G93A SOD1 protein caused apoptosis in neural cells, while wild-type protein protected against apoptosis. Collectively, these data suggest that production of ROS can contribute to neuronal apoptosis and motor neuron degeneration characteristic of FALS. ROS generation has also been implicated in the development of other pathological processes and may be a common mediator in several neurodegenerative diseases (Bondy, 1995) (Table 13.3).

SUMMARY AND FUTURE PROSPECTUS

Apoptosis is an active cellular process that is involved in both normal developmental cell death as well as in pathological conditions such as neurodegenerative disease and stroke. Under physiological conditions, neuronal survival and apoptosis are mainly controlled via the contributions of two cell surface receptors that respond to extracellular information: the Trk molecules that promote neuronal survival and differentiation, and p75 that can induce ligand-dependent apoptosis. These initial stimuli are then transmitted through intracellular pathways that are ultimately processed by two opposing sets of genes, proapoptotic genes such as ICE/CED-3 proteases, and antiapoptotic genes such as Bcl-2 and Bcl-x_L. While our discussion has emphasized the balance between Bcl-2 and ICE/CED-3 proteases in neuronal apoptosis, elucidation of the effector molecules that regulate the cross-talk between these pathways and ultimately the decision of life or death are at the beginning stages in our understanding. Some effector cascades such as the Ras > Raf > MEK and the MKK > SEK > JNK kinase pathway are shared between cell growth and cell differentiation responses, and their involvement in apoptosis must therefore depend upon the cellular context in which they are activated. Other pathways involving Bcl-2, ROS, and caspases may be specific for regulating apoptosis. By characterization of the specific regulatory pathways involved in apoptosis, future considerations will be on the development of pharmacological therapeutics that can specifically modulate neuronal apoptosis for diseases of the nervous system (see Cotman et al., Chapter 14, this volume). It is also clear that the purpose of programmed cell death and apoptosis in the nervous system will vary depending on the developmental period and on the specific system. Careful analysis of the cell death pathways in each system will be required for a complete understanding of neuronal apoptosis.

ACKNOWLEDGMENTS

This work was supported by grants from the NIH to RB and BH. We would like to thank Sanjay Chandriani and Bruce Cutler for critically reading the chapter and Moses Chao for discussing unpublished results.

REFERENCES

Abate C, Patel L, III RJR, Curran T (1990): Redox regulation of fos and jun DNA-binding activity in vivo. Science 249:1157–61.

Adam-Klages S, Adam D, Wiegmann K, Struve S, Kolanus W, Schneider-Mergener J, Kronke M (1996): FAN, a novel WD-repeat protein, couples the p55-receptor to neutral sphingomyelinase. Cell 86:937–47.

Allsopp TE, Wyatt S, Paterson HF, Davies AM (1993): The proto-oncogene bcl-2 can selectively rescue neurotrophic factor-dependent neurons from apoptosis. Cell 73:295–307.

Alnemri ES, Livingston DJ, Nicholson DW, Salvesen G, Thornberry NA, Wong WW, Yuan J (1996): Human ICE/CED-3 protease nomenclature. Cell 87:171.

Baass PC, DiGuglielmo GM, Authier F, Posner BI, Bergeron JJM (1995): Compartmentalized signal transduction by receptor tyrosine kinases. Trends Cell Biol 5:465–70.

Baker SJ, Reddy EP (1996): Transducers of life and death: TNF receptor superfamily and associated proteins. Oncogene 12:1–9.

Bar-Sagi D, Feramisco J (1985): Microinjection of the ras oncogene protein into PC12 cells induces morphological differentiation. Cell 42:841–48.

Barbacid M (1987): Ras genes. Ann Rev Biochem 56:779–827.

Barbacid M (1994): The Trk family of neurotrophin receptors. J Neurobiol 25:1386–403.

Barde Y (1989): Trophic factors and neuronal survival. Neuron 2:1525–34.

Barker PA, Shooter EM (1994): Disruption of NGF binding to the low affinity neurotrophin receptor p75LNTR reduces NGF binding to TrkA on PC12 cells. Neuron 13:203–15.

Barrett GL, Barlett PF (1994): The p75 nerve growth factor receptor mediates survival or death depending on the stage of sensory neuron development. Proc Natl Acad Sci 91:6501–5.

Boldin MP, Goncharov TM, Goltsev YV (1996): Involvement of MACH, a novel MORT1/FADD-interacting protease, in Fas/APO-1 and TNF receptor-induced cell death. Cell 85:803–15.

Bondy SC (1995): The relation of oxidative stress and hyperexcitation to neurological disease. Proc Soc Exp Biol Med 208:337–45.

Borasio GD, Wittinghofer A, Barde YA, Sendtner M, Heumann R (1989): Ras p21 protein promotes survival and fiber outgrowth of cultured embryonic neurons. Neuron 2:1087–96.

Bothwell M (1995): Functional interactions of neurotrophins and neurotrophin receptors. Ann Rev Neurosci 18:223–53.

Bothwell M (1996): p75NTR: A receptor after all. Science 272:506–7.

Bowling AC, Beal MF (1995). Bioenergetic and oxidative stress in neurodegenerative diseases. Life Sci 56:1151–71.

Bowling AC, Schulz JB, Brown J, Beal MF (1993): Superoxide dismutase activity, oxidative damage, and mitochondrial energy metabolism in familial and sporadic amyotrophic lateral sclerosis. J Neurochem 61:2322–25.

Bredesen DE (1995): Neural apoptosis. Am Neurol 38:839–51.

Burdon RH (1995): Superoxide and hydrogen peroxide in relation to mammalian cell proliferation. Free Radic Biol Med 18:775–94.

Burdon RH, Alliangana D, Gill V (1995): Hydrogen peroxide and the proliferation of BHK-21 cells. Free Radic Res 23:471–86.

Canossa M, Rovelli G, Shooter EM (1996): Transphosphorylation of the neurotrophin Trk receptors. J Biol Chem 271:5812–18.

Carpenter CL, Cantley LC (1996): Phosphoinositol kinases. Curr Opin Cell Biol 8:153–58.

Carter BD, Kaltschmidt C, Kaltschmidt B, Offenhauser N, Bohm-Matthaei R, Baeuerle PA, Barde Y-A (1996): Selective activation of NF-kB by nerve growth factor through the neurotrophin receptor p75. Science 272:542–45.

Casaccia-Bonnefil P, Carter BD, Dobrowsky RT, Chao MV (1996): Death of oligodendrocytes mediated by the interaction of nerve growth factor with its receptor p75. Nature 383:716–19.

Chao MV (1995): Ceramide: A potential second messenger in the nervous system. Mol Cell Neurosci 6:91–96.

Chao MV, Hempstead BL (1995): p75 and Trk: A two-receptor system. Trends in Neurosciences 18:321–26.

Chapman BS (1995): A region of the 75 kDa neurotrophin receptor homologous to the death domains of TNFR1 and Fas. FEBS Lett 374:215–20.

Chen DF, Schneider GE, Martinou JC, Tonegawa S (1997): Bcl-2 promotes regeneration of severed axons in mammalian CNS. Nature 385:434–39.

Chinnaiyan AM, O'Rourke K, Tewari M, Dixit VM (1995): FADD, a novel death domain-containing protein, interacts with the death domain of Fas and initiates apoptosis. Cell 81:505–12.

Chinnaiyan AM, Orth K, O'Rourke K, Duan H, Poirer GG, Dixit VM (1996): Molecular ordering of the cell death pathway: bcl-2 and bcl-x function upstream of the CED-3-like apoptotic proteases. J Biol Chem 271:4573–76.

Curtus R, Adryan KM, Stark JL, Park JS, Compton DL, Weskamp G, Huber LJ, Chao MV, Jaenish R, Lee K-F, Lindsay RM, DiStefano PS (1995): Different role of the low affinity neurotrophin receptor (p75) in retrograde axonal transport of the neurotrophins. Neuron 14:1201–11.

Cuvillier O, Pirianov G, Kleuser B, Vanek PG, Coso OA, Gutkind JS, Spiegel S (1996): Suppression of ceramide-mediated programmed cell death by sphingosine-1-phosphate. Nature 381:800–3.

Davies AM (1994): Role of neurotrophins in the developing nervous system. J Neurobiol 25:1334–48.

Davies AM (1995): The Bcl-2 family of proteins, and the regulation of neuronal survival. Trends in Neurosciences 18:355–58.

Deckwerth TL, Elliot JL, Knudson CM, Johnson EM Jr, Snider WD, Korsmeyer SJ (1996): BAX is required for neuronal cell death after trophic factor deprivation and during development. Neuron 17:401–11.

Devary Y, Gottlieb RA, Lau LF, Karin M (1991): Rapid and preferred activation of the c-jun gene during the mammalian UV response. Mol Cell Biol 11:2804–11.

DiGuglielmo GM, Baass PC, Ou W-J, Posner BI, Bergeron JJM (1994): Compartmentalization of Shc, Grb2, and mSOS, and hyperphosphorylation of Raf-1 by EGF but not insulin in liver parenchma. EMBO J 13:4269–77.

Dikic I, Schlessinger J, Lax I (1994): PC12 cells overexpressing the insulin receptor undergo insulin-dependent neuronal differentiation. Curr Biol 4:702–8.

Dobrowsky RT, Werner MH, Castellino AM, Chao MV, Hannun YA (1994): Activation of the sphingomyelin cycle through the low-affinity neurotrophin receptor. Science 265:1596–99.

Dubois-Dauphin M, Frankowski H, Tsujimoto Y, Huarte J, Marinou J-C (1994): Neonatal motoneurons overexpressing the bcl-2 protooncogene in transgenic mice are protected from axotomy-induced cell death. Proc Natl Acad Sci USA 91:3309–13.

Dudek H, Datta SR, Franke TF, Birnbaum MJ, Yao R, Cooper GM, Segal RA, Kaplan DR, Greenberg ME (1997): Regulation of neuronal survival by the serine-threonine protein kinase Akt. Science 275:661–65.

Dyand R, Hara K, Hiles I, Bax B, Gout I, Panayotou G, Fry MJ, Yonezawa K (1994): PI 3-kinase: Structural and functional properties of intersubunit interactions. EMBO J 13:511–21.

Ellerby LM, Ellerby HM, Park SM, Holleran AL, Murphy AN, Fiskum G, Kane DJ, Testa MP, Kayalar C, Bredesen DE (1996): Shift of the cellular oxidation–reduction potential in neural cells expressing Bcl-2. J Neurochem 67:1259–67.

Estus S, Faks WJ, Freeman RS, Gruda M, Bravo R, Johnson EM (1994): Altered gene expression in neurons during programmed cell death: Identification of c-jun as necessary for neuronal apoptosis. J Cell Biol 127:1717–27.

Evan G, Wyllie A, Gilbert C, Littlewood T, Land H, Brooks M, Waters C, Penn L, Hancock D (1992): Induction of apoptosis in fibroblasts by c-myc protein. Cell 69:119–28.

Eveleth DD, Bradshaw RA (1992): Nerve growth factor nonresponsive pheochromocytoma cells: Altered internalization results in signaling dysfunction. J Cell Biol 117:291–99.

Farinelli SE, Park DS, Greene LA (1996): Nitric oxide delays the death of trophic factor-deprived PC12 cells and sympathetic neurons by a cGMP-mediated mechanism. J Neurosci 6:2325–34.

Farlie PG, Dringen R, Rees SM, Kannourakis G, Bernard O (1995): Bcl-2 transgene expression can protect neurons against developmental and induced cell death. Proc Natl Acad Sci USA 92:4397–401.

Fath I, Schweighoffer F, Rey I, Multon MC, Boiziau J, Duchesne M, Tocque B (1994): Cloning of a Grb2 isoform with apoptotic properties. Science 264:971–74.

Feinstein E, Kimchi A, Wallach D, Boldin M, Varfolomeev E (1996): The death domain: A module shared by proteins with diverse cellular functions. FEBS Lett 20:342–44.

Ferrari G, Greene L (1994): Proliferative inhibition by dominant negative Ras rescues naive and neuronally differentiated PC12 cells from apoptotic death. EMBO J 13:5922–28.

Ferrari G, Yan CYI, Greene LA (1995): N-acetylcysteine (D- and L-stereoisomers) prevents apoptotic death of neuronal cells. J Neurosci 15:2857–66.

Frade JM, Rodriguez-Tebar A, Barde Y-A (1996): Induction of cell death by endogenous nerve growth factor through its p75 receptor. Nature 383:166–68.

Fraser A, Evan G (1996): A license to kill. Cell 85:781–84.

Freeman RS, Estus S, Johnson EM (1994): Analysis of cell cycle-related gene expression in postmitotic neurons: selective induction of cyclin D1 during programmed cell death. Neuron 12:343–55.

Frisch SM, Vuori K, Ruoslahti E, Chan-Hui PY (1996): Control of adhesion-dependent cell survival by focal adhesion kinase. J Cell Biol 134:793–99.

Gagliardini V, Fernandez PA, Lee RKK, Drexler HCA, Rotello RJ, Fishman MC, Yuan J (1994): Prevention of vertebrate neuronal death by the crmA gene. Science 263:826–28.

Geer Pvd, Wiley S, Lai VK-M, Olivier JP, Gish G, Stephens R, Kaplan D, Shoelson S, Pawson T (1995): A conserved amino-terminal Shc domain binds to phosphotyrosine motifs in activated receptors and phosphopeptides. Curr Biol 5:404–12.

Givol I, Tsarfaty I, Resau J, Rulong S, Silva PP, Nasioulas G, DuHadaway J, Hughes SH, Ewert DL (1994): Bcl-2 expressed using a retroviral vector is localized primarily in the nuclear membrane and the endoplasmic reticulum of chicken embryo fibroblasts. Cell Growth & Differen 5:419–29.

Glassman RH, Hempstead BL, Stainco-Coico L, Steiner MG, Hanafusa H, Birge RB (1997): v-Crk, an effector of the NGF signaling pathway, delays apoptotic cell death in neurotrophin-deprived PC12 cells. Cell Death & Differen 4:82–93.

Greene L, Kaplan DR (1995): Early events in neurotrophin signaling via TrkA and p75. Curr Opin Neurobiol 5:579–87.

Greenhalf W, Stephan C, Chaudhuri B (1996): Role of mitochondria and C-terminal membrane anchor of Bcl-2 in Bax induced growth arrest and mortality in *Saccharomyces cerevisiae*. FEBS Lett 380:169–75.

Greenlund LJS, Deckwerth TL, Johnson EM Jr (1995a): Superoxide dismutase delays neuronal apoptosis: a role for reactive oxygen species in programmed neuronal death. Neuron 14:303–15.

Greenlund LJS, Korsmeyer SJ, Johnson EM Jr (1995b): Role of BCL-2 in the survival and function of developing and mature sympathetic neurons. Neuron 15:649–61.

Greidinger EL, Miller DK, Yamin T-T, Casciola-Rosen L, Rosen A (1996): Sequential activation of three distinct ICE-like activities in Fas-ligated Jurkat cells. FEBS Lett 390:299–303.

Gurney ME, Pu H, Chiu AY, Canto MCD, Polchow CY, Alexander DD, Caliendo J, Hentati A, Kwon YW, Deng H-X, Chen W, Zhai P, Sufit RL, Siddique T (1994): Motor neuron degeneration in mice that express a human Cu,Zn-superoxide dismutase mutation. Science 264:1772–75.

Hagag N, Halegoua S, Viola M (1986): Inhibition of growth factor induced differentiation by microinjection of antibody to ras p21. Nature 319:680–82.

Ham J, Babij C, Whitfield J, Pfarr CM, Lallemand D, Yaniv M, Rubin LL (1995): A c-jun dominant negative mutant protects sympathetic neurons against programmed cell death. Neuron 14:927–39.

Hemmick LM, Perney TM, Flamm RE, Kaczmarek LK, Birnberg NC (1992): Expression of the H-ras oncogene induces potassium conductance and neuron-specific potassium channel mRNAs in the AtT20 cell line. J Neurosci 12:2007–14.

Hempstead B, Martin-Zanca D, Kaplan D, Parada L, Chao M (1991): High affinity NGF binding requires co-expression of the trk protooncogene and the low affinity NGF receptor. Nature 350:678–83.

Hempstead BL, Birge RB, Fajardo JE, Glassman R, Mahadeo O, Kraemer R, Hanafusa H (1994): Expression of the v-crk oncogene product in PC12 cells results in rapid differentiation by both nerve growth factor-dependent and epidermal growth factor-dependent pathways. Mol Cell Biol 14:1964–71.

Hengartner MO, Horvitz HR (1994): C. elegans cell survival gene ced-9 encodes a functional homolog of the mammalian proto-oncogene bcl-2. Cell 76:665–76.

Henkemeyer M, Rossi DJ, Holmyard DP, Puri MC, Mbamalu G, Harpal K, Shih TS, Jacks T, Pawson T (1995): Vascular system defects and neuronal apoptosis in mice lacking Ras GTPase-activating protein. Nature 377:695–701.

Hildago E, Demple R (1994): An iron–sulphur center essential for transcriptional activation by the redox-sensing SoxR protein. EMBO J. 13:138–46.

Hiles ID, Otsu M, Volinia S, Fry MJ, Gout I, Dyand R, Panayotou G, Ruiz LF, Thompson A, Totty NF, Hsuan JJ, Courtneidge SA, Parker PJ, Waterfield MD (1992): Phosphatidylinositol 3-kinase: Structure and expression of the 110 kD catalytic subunit. Cell 70:419–29.

Hockenberry D, Nunez G, Milliman C, Schreiber R, Korsmeyer S (1990): Bcl-2 is an inner mitochondrial membrane protein that blocks programmed cell death. Nature 348:334–36.

Hockenberry DM, Oltvai ZN, Yin X-M, Milliman CL, Korsmeyer SJ (1993): Bcl-2 functions in an antioxidant pathway to prevent apoptosis. Cell 75:241–51.

Hofmann K, Tschopp J (1995): The death domain motif found in Fas (Apo-1) and TNF receptor is present in proteins involved in apoptosis and axonal guidance. FEBS Lett 371:321–23.

Hsu H, Huang J, Shu HB, Baichwal V, Goeddel D (1996): TNF-dependent recruitment of the protein kinase RIP to the TNF receptor 1 signaling complex. Immunity 4:387–96.

Hsu H, Xiong J, Goeddel DV (1995). The TNF receptor 1-associated protein TRADD signals cell death and NF-κB activation. Cell 81:495–504.

Ibanez CF, Ebendal T, Barbany G, Murray-Rust J, Blundell TL, Persson H (1992): Disruption of the low affinity receptor-binding site in NGF allows neuronal survival and differentiation by binding to the trk gene product. Cell 69:329–41.

Itoh N, Yonehara S, Ishii A, Yonehara M, Mizushima S, Sameshima M, Hase A, Seto Y, Nagata S (1991): The polypeptide encoded by the cDNA for human cell surface antigen Fas can mediate apoptosis. Cell 66:233–43.

Jacobson MD, Burne JF, King MP, Miyashita T, Reed JC, Raff MC (1993): Bcl-2 blocks apoptosis in cells lacking mitochondrial DNA. Nature 361:365–69.

Juo P, Kuo CJ, Reynolds SE, Konz RF, Raingeaud J, Davis RJ, Biemann HP, Blenus J (1997): Fas activation of the p38 mitogen-activated protein kinase signaling pathway requires ICE/CED-3 family proteases. Mol Cell Biol 17:24–35.

Kane DJ, Sarafian TA, Anton R, Hahn H, Gralla EB, Valentine JS, Ord T, Bredesen DE (1993): Bcl-2 inhibition of neural death: decreased generation of reactive oxygen species. Science 262:1274–77.

Kaplan D, Stephens R (1994): Neurotrophin signal transduction by the Trk receptor. J Neurobiol 25:1404–17.

Khosravi-Far R, White MA, Westwick JK, Solski PA, Chrzanowska-Wodnicka M, Aelst LV, Wigler MH, Der CJ (1996): Oncogenic ras activation of Raf/mitogen-activated protein kinase-independent pathways is sufficient to cause tumorigenic transformation. Mol Cell Biol 16:3923–33.

Klein R, Parada LF, Coulier F, Barbacid M (1989): TrkB, a novel tyrosine protein kinase receptor expressed during mouse neural development. EMBO J 8:3701–9.

Klein R, S Santiago I, Smeyne RJ, Lira SA, Brambilla R, Bryant S, Zhang L, Snider WD, Barbacid M (1994): Disruption of the neurotrophin-3 receptor gene trkC eliminates Ia muscle afferents and results in abnormal movements. Nature 368:249–51.

Klein R, Smeyne RJ, Wurst W, Long LK, Auerbach BA, Joyner AL, Barbacid M (1993): Targeted disruption of the trkB neurotrophin receptor gene results in nervous system lesions and neonatal death. Cell 75:113–22.

Kordower JH, Chen E-Y, Sladek J, Mufson EJ (1994): TRK-immunoreactivity in the monkey central nervous system: forebrain. J Comp Neurol 349:20–35.

Korsching S (1993): The neurotrophin factor concept: a reexamination. J Neurosci 13:2739–48.

Korsmeyer SJ, Yin X-M, Oltvai ZN, Veis-Novack DJ, Linette GP (1995): Reactive oxygen species and the regulation of cell death by the Bcl-2 gene family. Biochim Biophys Acta 1271:63–66.

Kranenburg O, Eb AJvd, Zantema A (1996): Cyclin D1 is an essential mediator of apoptotic neuronal cell death. EMBO J 15:46–54.

Kyriakis JM, Banerjee P, Nikolakaki E, Dai T, Rubie EA, Ahmad MF, Avruch J, Woodgett JR (1994): The stress-activated protein kinase subfamily of c-jun kinases. Nature 369:156–60.

Lam M, Dubyak G, Chen L, Nunez G, Miesfeld RL, Distelhorst CW (1994): Evidence that Bcl-2 represses apoptosis by regulating endoplasmic reticulum-associated Ca^{2+} fluxes. Proc Natl Acad Sci USA 91:6569–73.

Lander HM, Milbank AJ, Tauras JM, Hajjar DP, Hempstead BL, Schwartz GD, Kraemer RT, Mirza UA, Chait BT, Burk SC, Quilliam LA (1996): Redox regulation of cell signaling. Nature 381:380–81.

Lee KF, Li E, Huber J, Landis SC, Sharpe AH, Chao MV, Jaenisch R (1992): Targeted mutation of the gene encoding low affinity NGF receptor leads to deficits in the peripheral sensory nervous system. Cell 69:737–49.

Levi-Montalcini R (1987): Nerve growth factor: Thirty five years later. Science 237:1154–64.

Lewin G, Barde Y (1996): Physiology of the neurotrophins. Ann Rev Neurosci 19:289–317.

Lindsay RM (1996): Role of neurotrophins and Trk receptors in the development and maintenance of sensory neurons: An overview. Philos Trans Roy Soc Lond [Biol] 351:365–73.

Lindsay RM, Weigand SJ, Altar CA, DiStefano PS (1994): Neurotrophic factors: From molecule to man. Trends Neurosci 17:182–90.

Lipton SA, Choi YB, Pan ZH, Lei SZ, Chen HS, Sucher NJ, Loscaizo J, Singel DJ, Stamler JS (1993): A redox-based mechanism for neuroprotective and neurodestructive effects of nitic oxide and related nitroso-compounds. Nature 364:626–32.

Lithgow T, Driel RV, Bertram JF, Strasser A (1994): The protein product of the oncogene bcl-2 is a component of the nuclear envelope, the endoplasmic reticulum, and the outer mitochondrial membrane. Cell Growth Differen 5.411–17.

Liu X, Kim CN, Yang J, Jemmerson R, Wang X (1996a): Induction of apoptotic program in cell-free extracts: Requirements for dATP and cytochrome c. Cell 86:147–57.

Liu Z-G, Hsu H, Goeddel DV, Karin M (1996b): Dissection of TNF receptor 1 effector functions: JNK activation is not linked to apoptosis while NF-κB activation prevents cell death. Cell 87:565–76.

Lo YYC, Cruz TF (1995): Involvement of reactive oxygen species in cytokine and growth factor induction of c-fos expression in chondrocytes. J Biol Chem 270:11727–30.

Lo YYC, Wong JM, Cruz TF (1996): Reactive oxygen species mediate cytokine activation of c-Jun NH2 terminal kinases. J Biol Chem 271:15703–7.

Loeb DM, Stephens RM, Copeland T, Kaplan DR, Greene LA (1994): A Trk NGF receptor point mutation affecting interaction of PLC-γ abolishes NGF-promoted peripherin induction, but not neurite outgrowth. J Biol Chem 269:8901–10.

Lowe SW, Ruley HE, Jacks T, Hous DE (1993): p53-dependent apoptosis modulates the cytotoxicity of anticancer drugs. Cell 74:957–67.

Mah S, Zhong LT, Liu Y, Roghani A, Edwards R, Bredesen D (1993): The protooncogene bcl-2 inhibits apoptosis in PC12 cells. J Neurochem 60:1183–86.

Martin SJ, Amarante-Mendes GP, Shi L, Chuang T-H, Casiano CA, O'Brien GA, Fitzgerald P, Tan EM, Bokoch GM, Greenberg AH, Green DR (1996): The cytotoxic cell protease granzyme B initiates apoptosis in a cell-free system by proteolytic processing and activation of the ICE/CED-3 family protease, CPP32, via a novel two-step mechanism. EMBO J 15:2407–16.

Martinou JC, Dubois-Dauphin M, Staple JK, Rodriguez I, Frankowski H, Missotten M, Albertini P, Talabot D, Catsicas S, Pietra C, Huarte J (1994): Overexpression of bcl-2 in transgenic mice protects neurons from naturally occurring cell death and experimental ischemia. Neuron 13:1017–30.

McGlade J, Cheng A, Pelicci G, Pelicci PG, Pawson T (1992): Shc proteins are phosphorylated and regulated by the v-Src and v-Fps protein-tyrosine kinases. Proc Natl Acad Sci USA 89:8869–73.

McMahon SB, Armanini MP, Ling LH, Phillips HS (1994): Expression and coexpression of Trk receptors in subpopulations of adult primary sensory neurons projecting to identified peripheral targets. Neuron 12:1161–71.

Meikrantz W, Gisselbrecht S, Tam SW, Schlegel R (1994): Activation of cyclin A-dependent protein kinases during apoptosis. Proc Natl Acad Sci USA 91:3754–58.

Merry DE, Veis DJ, Hickey WF, Korsmeyer SJ (1994): Bcl-2 protein expression is widespread in the developing nervous system and retained in the adult PNS. Development 120:301–11.

Mesner P, Winters T, Green S (1992): Nerve growth factor withdrawl-induced cell death in neuronal PC12 cells resembles that in sympathetic neurons. J Cell Biol 119:1669–80.

Milligan CE, Prevette D, Yaginuma H, Homma S, Cardwell C, Fritz LC, Tomaselli KJ, Oppenheim RW, Schwartz LM (1995): Peptide inhibitors of the ICE protease family arrest programmed cell death of motoneurons in vivo and in vitro. Neuron 15:385–93.

Muchmore SW, Sattler M, Liang H, Meadows RP, Harlan JE, Yoon HS, Nettesheim D, Chang BS, Thompson CB, Wong S-L, Ng S-C, Fesik SW (1996): X-Ray and NMR structure of human Bcl-XL, an inhibitor of programmed cell death. Nature 381:335–41.

Muzio M, Chinnaiyan AM, Kischkel FC, O'Rourke K, Shevchenko A, Ni J, Scaffidi C, Bretz JD, Zhang M, Gentz R, Mann M, Krammer PH, Peter ME, Dixit VM (1996): FLICE, a novel FADD-homologous ICE/CED-3-like protease, is recruited to the CD95 (Fas/APO-1) death-inducing signaling complex. Cell 85:817–27.

Myers KM, Fiskum G, Liu Y, Simmens SJ, Bredesen DE, Murphy AN (1995): Bcl-2 protects neural cells from cyanide/aglycemia-induced lipid peroxidation, mitochondrial injury, and loss of viability. J Neurochem 65:2432–40.

Nicholson DW (1996): ICE/CED3-like proteases as therapeutic targets for the control of inappropriate apoptosis. Nature Biotech 14:297–301.

Nicholson DW, Ali A, Thornberry NA, Vaillancourt JP, Ding CK, Gallant M, Gareau Y, Griffin P, Labelle M, Lazebnik YA, Munday NA, Raju SM, Smulson ME, Yamin TT, Yu VL, Miller DK (1995): Identification and inhibition of the ICE/CED-3 protease necessary for mammalian apoptosis. Nature 376:37–43.

Ninkina N, Adu J, Fischer A, Pinon LGP, Buchman VL, Davies AM (1996): Expression and function of TrkB variants in developing sensory neurons. Neuron 15: 6385–93.

Obermeier A, Bradshaw RA, Seedorf K, Choidas A, Schlessinger J, Ulrich A (1994): Neuronal differentiation signals are controlled by nerve growth factor/Trk binding sites for Shc and PLC-γ. EMBO J 13:1585–90.

Olson L (1993): NGF and the treatment of Alzheimer's disease. Exp Neurol 124:5–15.

Oltvai ZN, Korsmeyer SJ (1994): Checkpoints of dueling dimers foil death wishes. Cell 79:189–92.

Oppenheim R (1991): Cell death during development of the nervous system. Ann Rev Neurosci 14:453–501.

Orrell R, Belleroche Jd, Marklund S, Bowe F, Hallewell R (1995): A novel SOD mutant and ALS. Nature 374:504–5.

Otsu M, Hiles I, Gout I, Fry MJ, Ruiz LF, Panayotou G, Thompson A, Dhand R, Hsuan J, Totty M, Smith AD, Morgan SJ, Courtneidge SA, Parker PJ, Waterfield MD (1991): Characterization of two 85 kd proteins that associate with receptor tyrosine kinases, Middle T/pp60$^{c\text{-}src}$ complexes and PI3-kinase. Cell 65:91–104.

Park D, Rhee S (1992): Phosphorylation of Nck in response to a variety of receptors phorbol myristate acetate and cyclic AMP. Mol Cell Biol 12:5816–23.

Parker P, Miller F (1996): personal communication.

Patterson SL, Abel T, Deuel TAS, Martin KC, Rose JC, Kandel ER (1996): Recombinant BDNF rescues deficits in basal synaptic transmission and hippocampal LTP in BDNF knockout mice. Neuron 16:1137–45.

Peng X, Greene LA, Kaplan DR, Stephens RM (1995): Deletion of a conserved juxtamembrane sequence in Trk abolishes NGF-promoted neuritogenesis. Neuron 15:395–406.

Peunova N, Enikilopov G (1995): Nitric oxide triggers a switch to growth arrest during differentiation of neuronal cells. Nature 375:68–73.

Pronk GJ, Ramer K, Amiri P, Williams LT (1996): Requirement of an ICE-like protease for induction of apoptosis and ceramide generation by REAPER. Science 271:808–10.

Rabizadeh S, Oh J, Zhong L, Yang J, Bitler C, Butcher L, Bredesen D (1993): Induction of apoptosis by the low-affinity NGF receptor. Science 261:345–48.

Raffoni S, Bradshaw RA, Buxer SE (1993): The receptors for nerve growth factor and other neurotrophins. Ann Rev Biochem 62:823–50.

Reaume AG, Elliot JL, Hoffman EK, Kowell NW, Ferrante RJ, Siwek DF, Wilcox HM, Flood DG, Beal MF, Brown RH, Scott RW, Snider WD (1996): Motor neurons in Cu/Zn superoxide dismutase-deficient mice develop normally but exhibit enhanced cell death after axonal injury. Nature Genet 13:43–47.

Rosen DR, Siddique T, Patterson D, Figlewicz DA, Sapp P, Hentati A, Donaldson D, Goto J, O'Regan JP, Deng H-X, Rahmani Z, Krizus A, McKenna-Yasek D,

Cayabyab A, Gaston SM, Berger R, Tanzi RE, Halperin JJ, Herzfeldt B, Vanden Bergh, R. RVd, Hung W-Y, Bird T, Deng G, Mulder DW, Smyth C, Laing NG, Soriano E, Pericakk-Vance MA, Haines J, Rouleau GA, Gusella JS, Horvitz HR, Brown RH Jr. (1993): Mutations in Cu/Zn superoxide dismutase gene are associated with familial amyotrophic lateral sclerosis. Nature 362:59–62.

Roy N, Mahadevan MS, McLean M, Shutler G, Yaraghi Z, Farahani R, Baird S, Besner-Johnston A, Lefebvre C, Kang X, Salih M, Aubry H, Tamai K, Guan X, Ioannou P, Crawford TO, Jong PJd, Surh L, Ikeda J, Korneluk RG, MacKenzie A (1995): The gene for neuronal apoptosis inhibitory protein is partially deleted in individuals with spinal muscular atrophy. Cell 80:167–78.

Rozakis-Adcock M, McGlade J, Mbamalu G, Pelicci G, Daly R, Li W, Batzer A, Thomas S, Brugge J, Pelicci P, Schlessinger J, Pawson T (1992): Association of the Shc and Grb2 Sem5 SH2-containing proteins is implicated in activation of the Ras pathway by tyrosine kinases. Nature 360:689–92.

Rubin E (1985): Development of the rat superior cervical ganglion: ingrowth of preganglionic axons. J Neurosci 5:685–96.

Rylett RJ, Williams LR (1994): Role of neurotrophins in cholinergic-neuron function in the adult and aged CNS. Trends in Neurosciences 17:486–90.

Schall TJ, Lewis M, Koller KJ, Lee A, Rice GC, Wong GHW, Gatanaga T, Granger GA, Lentz R, Raab H, Kohr WJ, Goeddel DV (1990): Molecular cloning and expression of a receptor for human tumor necrosis factor. Cell 61:361–70.

Schlessinger J, Ullrich A (1992): Growth factor signaling by receptor tyrosine kinases. Neuron 9:383–91.

Schmidt HH, Walter U (1994): NO at work. Cell 78:919–25.

Sendtner M, Thoenen H (1994): Oxidative stress and motorneuron disease. Curr Biol 4:1036–39.

Shimizu S, Eguchi Y, Kamiike W, Waguri S, Ichiyama Y, Matsuda H, Tsujimoto Y (1996): Bcl-2 blocks loss of mitochondrial membrane potential while ICE inhibitors act at a different step during inhibition of death induced by respiratory chain inhibitors. Oncogene 13:21–29.

Slater AFG, Nobel SI, Orrenius S (1995): The role of intracellular oxidants in apoptosis. Biochim Biophys Acta 1271:59–62.

Smeyne RJ, Klein R, Schnapp A, Long LK, Bryant S, Lewin A, Lira SA, Barbacid M (1994): Severe sensory and sympathetic neuropathies in mice carrying a disrupted trk/NGF receptor gene. Nature 368:246–49.

Snider WD (1994): Functions of the neurotrophins during nervous system development: What the knockouts are teaching us. Cell 77:627–38.

Soltoff SS, Rabin L, Cantley L, D RK (1992): Nerve growth factor promotes the activation of phosphatidylinositol 3-kinase and its association with the TrkA tyrosine kinase. J Biol Chem 267:17472–77.

Stevens RM, Loeb DM, Copeland TD, Pawson T, Greene LA, Kaplan DR (1994): Trk receptors use redundant signal transduction pathways involving SHC and PLC-γ to mediate NGF responses. Neuron 12:691–705.

Sundaresan M, Yu ZX, Ferrans VJ, Irani K, Finkel T (1995): Requirement for generation of H_2O_2 for platelet derived growth factor signal transduction. Science 270:296–99.

Takahashi A, Alnemri ES, Lazebnik YA, Fernandes-Alnemri T, Litwack G, Moir RD, Goldman RD, Poirer GG, Kaufmann SH, Earnshaw WC (1996): Cleavage of lamin

A by Mch2a but not CPP32: Multiple interleukin 1β-converting enzyme-related proteases with distinct substrate recognition properties are active in apoptosis. Proc Natl Acad Sci USA 93:8395–8400.

Takami S, Getchell ML, Yamagishi M, Albers KM, Getchell TV (1995): Enhanced extrinsic innervation of nasal and oral chemosensory mucosae in keratin 14-NGF transgenic mice. Cell Tissue Res 282:481–91.

Takayama S, Sato T, Krajewski S, Kochel K, Irie S, Millan JA, Reed JC (1995): Cloning and functional analysis of BAG1: A novel Bcl-2 binding protein with anti-cell death activity. Cell 80:279–84.

Tanaka N, Ishihara M, Kitagawa M, Harada H, Kimura T, Matsuyama T, Lamphier MS, Aizawa S, Mak TW, Taniguchi T (1994): Cellular commitment to oncogene-induced transformation or apoptosis is dependent on the transcription factor IRF-1. Cell 77:829–39.

Tanaka S, Saito K, Reed J (1993): Structure–function analysis of the apoptosis-suppressing Bcl-2 oncoprotein: Substitution of a heterologous transmembrane domain restores function to truncated Bcl-2 proteins. J Biol Chem 268:10920–26.

Tepper CG, Jayadev S, Liu B, Bielawska A, Wolff R, Yonehara S, Hannun YA, Seldin MF (1995): Role for ceramide as an endogenous mediator of Fas-induced cytotoxicity. Proc Natl Acad Sci USA 92:8443–47.

Tessarollo L, Tsoulfas P, Martin-Zanca D, Gilbert DJ, Jenkins NA, Copeland NG, Parada LA (1993): TrkC, a receptor for neurotrophin 3 is widely expressed in the developing nervous system and in nonneuronal tissues. Development 118:463–475.

Thoenen H (1991): The changing scene of neurotrophic factors. Trends Neurosci 14:165–170.

Thornberry NA, Bull HG, Calaycay JR, Chapman KT, Howard AD, Kostura MJ, Miller DK, Molineaux SM, Weidner JR, Aunins J (1992): A novel heterodimeric cysteine protease is required for interleukin-1β processing in monocytes. Nature 356:768–74.

Traverse S, Gomez N, Paterson H, Marshall C, Cohen P (1995): Sustained activation of the mitogen-activated protein (MAP) kinase cascade may be required for differentiation of PC12 cells. Comparison of the effects of nerve growth factor and epidermal growth factor. Biochem J 288:351–55.

Troy CM, Derossi D, Prochiantz A, Greene LA, Shelanski ML (1996): Downregulation of Cu/Zn superoxide dismutase leads to cell death via the nitric oxide-peroxynitrite pathway. J Neurosci 16:253–61.

Tsoulfas P, Stephans RM, Kaplan DR, Parada L (1996): TrkC isoforms with inserts in the kinase domain show impaired signaling responses. J Biol Chem 271:5691–97.

Tsuda T, Munthasser S, Fraser PE, Percy ME, Rainero I, Vaula G, Pinessi L, Bergamini L, Vignocchi G, McLachlan DRC, Tatton WG, George-Hyslop PS (1994): Analysis of the functional effects of a mutation in SOD1 associated with familial amyotrophic lateral sclerosis. Neuron 13:727–36.

Venero JL, Knusel B, Beck KD, Hefti F (1994): Expression of neurotrophin and trk receptor genes in adult rats with fimbria transections: effect of intraventricular nerve growth factor and brain-derived neurotrophic factor administration. Neurosci 59:797–815.

Verheij M, Bose R, Lin XH, Yao B, Jarvis WD, Grant S, Birrer MJ, Szabo E, Zon LI, Kyriakis JM, Haimovitz-Friedman A, Fuks Z, Kolesnik RN (1996): Requirement for ceramide-initiated SAPK/JNK signaling in stress-induced apoptosis. Nature 380:75–79.

Wang H-G, Millan JA, Cox AD, Der CJ, Rapp UR, Beck T, Zha H, Reed JC (1995): R-Ras promotes apoptosis caused by growth factor deprivation via a Bcl-2 suppressible mechanism. J Cell Biol 129:1103–14.

Wang H-G, Rapp UR, Reed J (1996): Bcl-2 targets the protein kinase Raf-1 to mitochondria. Cell 87:629–38.

Weil M, Jacobson MD, Coles HSR, Davies TJ, Gardner RL, Raff KD, Raff MC (1996): Constitutive expression of the machinery for programmed cell death. J Cell Biol 133:1053–59.

Weskamp G, Reichardt LF (1991): Evidence that the biological activity of NGF is mediated through a novel subclass of high affinity receptors. Neuron 6:649–63.

Whittemore ER, Loo DT, Cotman CW (1994): Exposure to hydrogen peroxide induces cell death via apoptosis in cultured rat cortical neurons. NeuroReport 5:1485–88.

Wiedau-Pazos M, Goto JJ, Rabizadeh S, Gralla EB, Roe JA, Lee MK, Valentine JS, Bredesen DE (1996): Altered reactivity of superoxide dismutase in familial amyotrophic lateral sclerosis. Science 271:515–18.

Wiegmann K, Schutze S, Machleidt T, Witte D, Kronke M (1994): Functional dichotomy of neutral and acidic sphingomyelinases in tumor necrosis factor signaling. Cell 78:1005–15.

Wiesner DA, Dawson G (1996): Staurosporine induces programmed cell death in embryonic neurons and activation of the ceramide pathway. J Neurochcm 66:1418–25.

Wilson KP, Black J-AF, Thomson JA, Kim EE, Griffith JP, Navia MA, Murcko MA, Chambers SP, Aldape RA, Raybuck SA, Livingston DJ (1994): Structure and mechanism of interleukin-1-β-converting enzyme. Nature 370:270–75.

Xia Z, Dickens M, Raingeaud J, Davis R, Greenberg ME (1995): Opposing effects of ERK and JNK-p38 MAP kinases on apoptosis. Science 270:1326–31.

Xue D, Shaham S, Horwitz HR (1996): The *Caenorhabditis elegans* cell-death protein CED-3 is a cysteine protease with substrate specificities similar to those of the human CPP32 protease. Genes Develop 10:1073–83.

Yao R, Cooper GM (1995): Requirement for phosphatidylinositol-3 kinase in prevention of apoptosis by NGF. Science 267:2003–6.

Yim MB, Kang J-H, Yim H-S, Kwak H-S, Chock PB, Stadtman ER (1996): A gain-of-function of an amyotrophic lateral sclerosis-associated Cu,Zn-superoxide dismutase mutant: an enhancement of free radical formation due to a decrease in K_m for hydrogen peroxide. Proc Natl Acad Sci USA 93:5709–14.

Yuan J, Shaham S, Ledoux S, Ellis HM, Horvitz HR (1993): The *C. elegans* cell death gene ced-3 encodes a protein similar to mammalian interleukin-1β-converting enzyme. Cell 75:641–652.

Zamzami N, Susin SA, Marchetti P, Hirsch T, Gomez-Monterrey I, Castedo M, Kroemer G (1996): Mitochondrial control of nuclear apoptosis. J Exp Med 183:1533–44.

Van der Zee, CEEM, Ross GM, Riopelle RJ, Hagg T (1996): Survival of cholinergic forebrain neurons in developing p75-NGFR-deficient mice. Science 274:1729–32.

Zha J, Harada H, Yang E, Jockel J, Korsmeyer SJ (1996): Serine phosphorylation of death agonist BAD in response to survival factor results in binding of 14-3-3 not Bcl-xL. Cell 87:619–28.

Zhang K-Z, Westberg JA, Holtta E, Andersson L (1996): BCL2 regulates neural differentiation. Proc Natl Acad Sci USA 93:4504–8.

CHAPTER 14

CELL DEATH IN ALZHEIMER'S DISEASE

CARL W. COTMAN, DAVID H. CRIBBS, CHRISTIAN J. PIKE, and KATHRYN J. IVINS
Institute for Brain Aging and Dementia, University of California Irvine, Irvine, CA 92697-4540

INTRODUCTION

Alzheimer's disease (AD) is the fourth leading cause of death and the primary cause of dementia in the elderly. AD affects approximately 12% of all individuals over the age of 65, and this percentage increases to as high as 45% by the age of 85 (Evans et al., 1989). Typically, the presenting symptom of the disease is impaired memory, but as the disease progresses, behavioral disturbances and personality changes also occur. It is a common premise that the irreversible loss of brain function in AD is due to the disruption of synapses and the loss of neurons that make those synapses. Accordingly, identification of the mechanisms causing circuit disruption and neuronal loss is critical to understanding and interrupting the progression of AD pathology.

Recent evidence suggests that neuronal loss in AD may be caused at least in part by apoptotic mechanisms. While apoptosis is a normal process that is known to occur during the developmental elimination of excess neurons, it may be reinitiated pathologically during aging by certain stimuli, causing the loss of significant numbers of neurons and leading to dementia. Indeed, as discussed in this chapter, increasing evidence suggests that apoptosis is a mechanism that contributes to neuronal death in AD. Many apoptotic conditions appear to affect the entire cell, but because neurons project over long distances, there may be some apoptotic conditions that only affect specific parts of the cells. Accordingly, it is our hypothesis that apoptotic insults may

When Cells Die, Edited by Richard A. Lockshin, Zahra Zakeri, and Jonathan L. Tilly
ISBN 0-471-16569-7   © 1998 Wiley-Liss, Inc.

initially damage or destroy neuronal processes (neurites) by initiating local apoptotic pathways. Such events would disconnect the neural circuitry and place neurons at increased risk for apoptosis due to the loss of trophic factors and other activity-dependent processes that support neuronal functions (Trump and Berezesky, Chapter 2, this volume).

Our approach for experimentally evaluating this theory has been to use cell culture to identify possible mechanisms and markers of neuronal apoptosis, then examine *post mortem* AD brain tissues for the presence or absence of similar events. We will briefly summarize evidence demonstrating that cultured neurons are induced to undergo apoptosis when subjected to many of the conditions that develop in the AD brain, suggesting that apoptosis may be a cell death mechanism for at least some neurons in AD. Also, recent data have identified several features of AD neurons consistent with the possibility that apoptosis is an ongoing mechanism.

Because neurons are nondividing cells, the decision to enter into apoptosis may be a very regulated and protracted process that is at least temporally distinct from that occurring in dividing cells. Indeed, we propose that even when activated, the apoptotic program may be counteracted by neuronal responses to slow or reverse the program. In this chapter we will focus on AD, but our results may apply to other neurodegenerative diseases and perhaps even to conditions of acute neural injury such as head trauma and spinal cord injury. An understanding of the key initiation and modulatory sites may provide targets for intervention.

β-AMYLOID INITIATES APOPTOSIS IN CULTURED PRIMARY NEURONS

If apoptosis is a mechanism of neuritic degeneration and cell death in AD, then it may be possible to identify the inducers and molecular pathways mediating this mechanism (Beritez-Bribiesca, Chapter 17, Trump and Berezesky, Chapter 2, this volume). One such agent may be the β-amyloid peptide. β-Amyloid is a 40–42 amino acid peptide that is generated from proteolytic processing of the amyloid precursor protein (APP) and has been implicated as a causal factor in AD neurodegeneration (Selkoe et al., 1996; Yankner, 1996). In the aging and AD brain, β-amyloid accumulates in the extracellular space as small deposits and senile plaques (Fig. 14.1).

Based on the observation that neurites surrounding β-amyloid deposits show sprouting and degenerative responses, we proposed that this peptide is not metabolically inert as was initially believed, but rather possesses biological activity. We discovered that β-amyloid stimulates a transient growth of neuronal processes, and then, as it self-assembles into small aggregates and β-sheet structures, it acquires an ability to activate degenerative mechanisms (Cotman and Anderson, 1995; Cotman et al., 1996; Yankner, 1996).

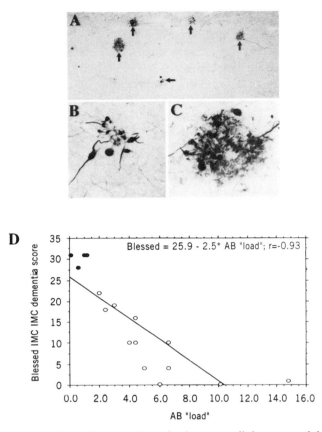

FIG. 14.1. β-Amyloid (brown) accumulates in the extracellular space of the AD brain as deposits of various sizes. (A) As the disease progresses and these deposits develop β-sheet structure, neurites (B and C) associated with them exhibit signs of degenerative and compensatory degenerative responses. (D) The accumulation of β-amyloid correlates with the decline in cognitive function (Su et al., 1996a; Cummings et al., 1996).

To determine if β-amyloid-mediated neurodegeneration *in vitro* occurs by a necrotic or apoptotic pathway, we employed a diagnostic battery of assays to examine the relevant morphological and biochemical features that differentiate these mechanisms (Loo et al., 1993; Anderson et al., 1995; Cotman and Anderson, 1995). In brief, we found that β-amyloid does indeed serve as an inducer of neuronal apoptosis (Loo et al., 1993; Watt et al., 1994), an observation since confirmed by many others (Forloni et al., 1993; Copani et al., 1995; Gschwind and Huber, 1995; Mark et al., 1995; Paradis et al., 1996). In brief, β-amyloid initially induces membrane blebbing and cell shrinkage

(Fig. 14.2). This is followed by DNA damage, and the generation of nuclear apoptotic bodies and a DNA ladder. Consistent with other neuronal apoptosis paradigms (Estus et al., 1994; Ham et al., 1995), we found that apoptosis induced by β-amyloid involves an early and sustained elevation in the immediate early gene product c-Jun within vulnerable, but not resistant, neurons (Anderson et al., 1995). Interestingly, we have observed similar increases in c-Jun levels within degenerating neurons in the AD brain (Anderson et al., 1994).

We have previously suggested that several inducers of apoptosis can generate early neuritic changes. β-Amyloid can be preassembled into small aggregates that are similar to the small deposits observed in the AD brain, which can be placed in low-density cultures of hippocampal neurons to determine the effect such deposits may have on the cells and their processes. In the presence of β-amyloid aggregates, for example, we have previously suggested that neurites are attracted to these aggregates and develop a beaded and dystrophic-like appearance (Pike et al., 1992) as illustrated in Fig. 14.3. (Note: Neurites refer to any neuronal process, whether dendritic or axonal.) The striking feature was that neurites associated with β-amyloid appeared to be affected but that those in contact with the substrate and not β-amyloid were normal in appearance. We suggested, therefore, that the neuritic changes were not a general feature of cellular degeneration but appeared to be induced by local factors. Recently, we have confirmed these observations in cultures where the cell bodies and neurites are separated such that amyloid only contacts the neurites. These data strongly suggest that neurites may be damaged by apoptotic-like mechanisms that act locally and are independent of somal damage.

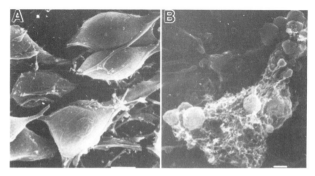

FIG. 14.2. Scanning electron micrograph showing the presence of membrane blebbing in neurons treated with β-amyloid. Control neurons show a smooth surface and do not show the presence of blebs (A), while neurons treated with β-amyloid generate blebs (B; arrows). β-Amyloid has formed a fibrous matrix on the surface of the neurons.

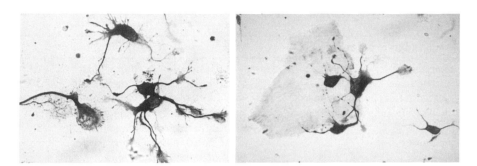

FIG. 14.3. Neurites from hippocampal neurons appear to degenerate when in contact with β-amyloid aggregates. Those neurites in contact with the normal substrate, polylysine, appear normal, suggesting that β-amyloid can induce local changes in neurites.

MANY INDUCERS OF APOPTOSIS ACCUMULATE IN THE AD BRAIN

Apoptosis can be induced in most neurons by a variety of stimuli many of which are present in the AD brain (Table 14.1). As discussed, β-amyloid can initiate apoptosis, and this inducer accumulates in proximity to neurons and neuronal processes. In parallel with characteristic AD pathology, β-amyloid induces the formation of dystrophic-like neurite morphology in cultured neurons (Fraser et al., 1994; Pike et al., 1992). Also, oxidative insults readily initiate neuronal apoptosis (Whittemore et al., 1994; Ratan et al., 1994) and oxidative damage is known to occur in the aging and AD brain (Beal, 1995; Greenlund et al., 1995; Slater et al., 1995; Smith et al., 1995). Similarly, reductions in glucose metabolism have been suggested to contribute to neurodegeneration in AD (McGeer et al., 1986; Hoyer et al., 1993; Beal et al., 1993; Goto et al., 1993; Finch and Cohen, 1997), and β-amyloid has been shown to exacerbate neurodegeneration in cultured neurons when glucose levels are reduced (Copani et al., 1991).

Excitotoxic damage initiated by glutamate and other excitatory amino acids can, under some conditions, initiate apoptosis, and many investigators have suggested that excitotoxic damage contributes to neurodegenerative diseases including AD (Choi, 1988; Bonfoco et al., 1995). Recent studies show that glutamate transport proteins may be greatly reduced in the AD brain (Masliah, 1996), which could exacerbate excitotoxic mechanisms. The profile of initiating factors strongly suggests that in the course of aging and age-related neurodegenerative disease, neurons are increasingly subjected to apoptotic inducers. In some cases, these factors can act synergistically. For example, neuronal loss is significantly potentiated when subthreshold levels of β-amyloid are combined with either excitotoxins (Koh et al., 1990; Copani et al., 1991;

TABLE 14.1. Many Inducers of Apoptosis Correspond to Conditions Present in the AD Brain

Stimuli/Inducers of Apoptosis	AD Conditions
β-Amyloid[a]	Accumulation of β-amyloid[b]
Reactive oxygen species[c]	Increased oxidative damage[d]
Elevated intracellular calcium[e]	Abnormalities in calcium homeostasis[f]
Low neurotrophic support[g]	BDNF deficiency, defect in connectivity, and retrograde transport[h]
Low energy[i]	Reduced metabolism (vascular angiopathy prevalent)[j]
Excitotoxins (e.g., glutamate)[k]	Proposed but not established[l]
4-Hydroxynoneal (HNE) oxidants[m]	Presence of lipid peroxidation products[n]
Combinations of conditions (e.g., β-amyloid, oxidation)[o]	Combinations of the above conditions increases with disease progression[p]

Genetic Risk Factors	AD Conditions
PS1, PS2[q]	Familial AD[r]
APP[s]	Familial AD[t]

[a] Cotman et al., 1996; Yankner, 1996.
[b] Cummings et al., 1996; Selkoe, 1996.
[c] Ratan et al., 1994; Greenlund et al., 1995; Whittemore et al., 1995; Slater et al., 1995.
[d] Beal, 1995; Smith et al., 1995.
[e] Johnson et al., 1992.
[f] Gibson et al., 1996a, 1996b; Shimohama et al., 1996; Mattson et al., 1997.
[g] Deckwerth and Johnson, 1993a, 1993b.
[h] Phillips et al., 1991; Murray et al., 1994.
[i] Lobner and Choi, 1996.
[j] Beal et al., 1993; Hoyer et al., 1993; Finch and Cohen, 1997.
[k] Bonfoco et al., 1995; Qin et al., 1996.
[l] Choi, 1988.
[m] Mark et al., 1997.
[n] Palmer and Burns, 1994; Lovell et al., 1995.
[o] Koh et al., 1990; Copani et al., 1991; Mattson et al., 1992; Gwag et al., 1995; Schubert and Chevion, 1995.
[p] Cotman et al., 1993.
[q] Guo et al., 1996; Wolozin et al., 1996; Deng et al., 1996.
[r] Selkoe, 1997.
[s] Yamatsuji et al., 1996a, 1996b.
[t] Mullan and Crawford, 1993; Selkoe et al., 1996.

Mattson et al., 1992; Dornan et al., 1993) or oxidative stress (Schubert and Chevion, 1995; Pike et al., submitted). Furthermore, such multicomponent insults also may be facilitated by microglia, which are found on the surface of some neurons where they may locally place neurons at risk by generating proteases and oxidative radicals (Afagh et al., 1996).

GENETIC RISK FACTORS FOR APOPTOSIS:
THE PRESENILIN GENES MAY PLACE NEURONS AT
INCREASED RISK FOR APOPTOSIS

In addition to the presence of a variety of inducers, recent evidence suggests that genetic factors may increase the risk for apoptosis in AD. Two genes, one called presenilin 1 (PS1) located on chromosome 14 and another closely related gene presenilin 2 (PS2) located on chromosome 1, appear to be primary susceptibility genes for familial AD. On the basis of our data and others, expression of the PS1 and PS2 genes appear to make neurons more vulnerable to apoptosis.

Recently, we and others have shown that both PS1 and PS2 are present in neurons in the AD brain (Cribbs et al., 1996a; Deng et al., 1996; Murphy et al., 1996; Page et al., 1996, Uchihara et al., 1996; Weber et al., 1997). We theorize that a critical factor in driving disease progression by PS mutations involves modulation of cellular viability. Discovery of this functional role resulted from attempts to isolate clones from a cDNA library that were able to inhibit apoptosis in a T-cell hybridoma (Vito et al., 1996b). Of six cDNA clones isolated, one (named ALG3) was found to correspond to the carboxy-terminal region of PS2 (Vito et al., 1996a). Overexpression of the ALG3 clone provided significant cytoprotection against a variety of apoptotic insults. More recent data suggest that the antiapoptotic activity of ALG3 reflects a dominant negative inhibition of proapoptotic PS2 activity (Vito et al., 1996b; Birge et al., Chapter 13; Benitez-Bribiesca, Chapter 17; Osborne, Chapter 8, this volume).

To gain further insight into the relationship between PS2 and apoptosis, we transfected rat PC12 cells with full-length human PS2. As we recently reported (Deng et al., 1996), overexpression of native PS2 significantly increases cellular vulnerability to the apoptotic insults staurosporine and H_2O_2, but does not appear to affect cell viability in nonstressed conditions (Fig. 14.4). These data suggest that PS2 can accelerate the onset of apoptosis by shifting the dose–response curve to lower concentrations of the inducer. In addition to the presenilins, mutations in the amyloid precursor protein are also linked to familial AD (Mullan and Crawford, 1993; Selkoe, 1996). Recent data suggest that these mutations can induce neuronal apoptosis by a G protein–linked mechanism that is inhibited by Bcl-2 (Yamatsuji et al., 1996a; Yamatsuji et al., 1996b).

Confirmation of this increased sensitivity to apoptosis has recently been reported for both wild-type and mutated forms of PS2 (Wolozin et al., 1996). A similar role in promoting apoptosis has been reported for PS1 (Guo et al., 1996). Consistent with these data, our preliminary experiments show protection by PS1 antisense oligonucleotides against staurosporine toxicity (Fig. 14.5). This antisense approach is important because it avoids potential pitfalls associated with gene overexpression by reducing endogenous levels of presenilin expression.

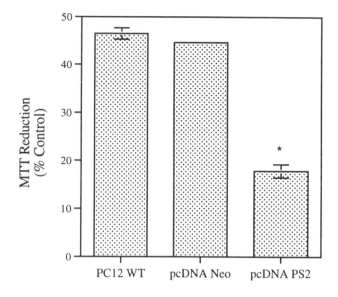

* P < 0.01 relative to matched values from WT condition.

FIG. 14.4. Cell viability in PS2 overexpressing cells is reduced in the presence of various doses of staurosporine (Deng et al., 1996).

Taken together, these data suggest that apoptosis may be a mechanism underlying neuronal loss in AD. Thus we and others have examined AD brain tissues to determine if apoptosis is a significant mechanism of cell death in the AD brain.

APOPTOSIS IN THE AD BRAIN

Apoptosis is generally defined according to strict morphological criteria. The basic features include shrinkage of the cells, membrane blebbing, chromatin condensation, and nuclear fragmentation. Nuclear fragmentation is typically accompanied by the cleavage of DNA into oligonucleosome-length fragments detectable by gel electrophoresis in many, but not all, models of apoptosis (Duke et al., 1983; Wyllie et al., 1984; Tepper and Studzinski, 1992; Zakeri et al., 1993). The process of DNA degradation produces a series of DNA fragments that have newly generated 3'-OH ends. These strand breaks can be labeled by enzymes such as terminal deoxynucleotidyl transferase (TdT).

We have recently reported that cells in the AD brain exhibit labeling for DNA strand breaks using TdT and many TdT-labeled cells exhibit an apoptotic-like morphological distribution of DNA strand breaks, including

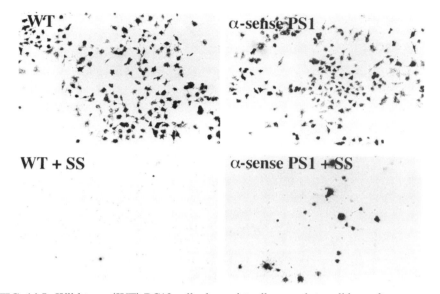

FIG. 14.5. Wild-type (WT) PC12 cells show virtually complete cell loss after exposure to 300 nM staurosporine (WT + SS); viable cells are darkly stained with MTT reagent. In contrast, WT cells pretreated with PS1 antisense oligonucleotides exhibit increased viability.

granulated and marginated patterns of intense TdT labeling, shrunken, irregular cellular shape, and the presence of apoptotic-like bodies consistent with apoptosis (Su et al., 1994; Anderson et al., 1996). We observed low levels of TdT labeling in control cases, but frequent TdT-positive nuclei in brain tissue from AD cases (Fig. 14.6). Significantly, we have observed numerous examples of cells with DNA damage in autopsy material in which AD pathology has not yet progressed sufficiently to meet the Consortium to Establish a Registry for Alzheimer's Disease (CERAD) criteria (Morris et al., 1988) for AD (Cotman and Su, 1996). The transentorhinal cortex contains scattered groups of cells that exhibit TdT labeling. The pattern of cellular damage in early cases is characteristically nonuniform, and affected cells occur in small groups that always include healthy cells among those degenerating. Consistent with our results, several investigators have reported the presence of DNA damage in AD cases (Dragunow et al., 1995; Lassmann et al., 1995; Thomas et al., 1995; Smale et al., 1995), and these data have been interpreted as supporting a role for apoptosis in AD.

Other biochemical markers of apoptosis are beginning to be recognized in AD brain tissue. These include c-Jun (Anderson et al., 1996) and possible indications of apoptotic neuritic damage. An increasing body of data suggests that synapse loss and neuritic abnormalities are an early consequence of the degenerative processes in AD (Masliah et al., 1989; Braak and Braak, 1995;

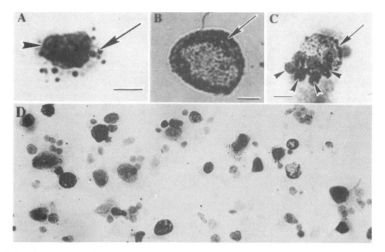

FIG. 14.6. The AD brain exhibits many neurons that express TdT labeling (D), and a portion of these display characteristics of terminal apoptosis including apoptotic bodies (A), chromatin margination (B), and the presence of microglia (C).

Su et al., 1996a), which likely occur prior to somal degradation. At present some local mechanisms of degeneration have been identified in the AD brain, such as the overexpression of the cytoskeletal protein fodrin and its breakdown products (Masliah, 1991; Sihag, 1996). Growing evidence suggests that fodrin is one of the "death substrates" in caspase-like (ICE/Ced3) protease pathways (Martin and Green, 1995; Martin et al., 1995; Vanags et al., 1996). Thus it is possible that apoptotic-type mechanisms participate in process degeneration as well as somal degeneration and accordingly it is important to define the mechanisms (Cryns and Yuan, Chapter 5, this volume).

ACTIVATION-INDUCED CELL DEATH MAY BE A COMMON THEME IN NEURONAL APOPTOSIS

If apoptosis is operative in the AD brain, as our data and others suggest, then it is critical to determine the mechanism. A common feature of many natural extracellular activators of apoptosis is that they involve interactions with membrane proteins, either disrupting a constitutive cellular survival signal required to suppress an intrinsic cell suicide program (Raff et al., 1993) or producing a novel signal that leads to apoptosis (Dellabona et al., 1990; Marrack and Kappler, 1990; Nagata and Golstein, 1995). Active generation of a death signal requires cross-linking specific membrane receptors by multivalent ligands, which initiates a signal transduction event that triggers apoptotic pathways (Lenardo, 1991; Radvanyi et al., 1993). This mechanism of cell death is called activation-induced cell death, and a number of well-characterized extracellular factors that can trigger cell degeneration by this mechanism have been identified (Nagata and Golstein, 1995; Strasser, 1995).

The most thoroughly studied model is activation of the Fas receptor, a member of the tumor necrosis factor (TNF) superfamily. When the Fas receptor is cross-linked, it activates a death cascade that includes a complex of adaptor proteins bound to the receptor including a member of the caspase family (caspase-8, also referred to as Mch5, Flice, or MACH) (Boldin et al., 1996; Muzio et al., 1996; Srinivasula et al., 1996a), which, in turn, activates downstream caspases (Birgeetal; Osborne, Chapter 8, this volume).

As noted, in the course of aging and age-related neurodegenerative diseases, select classes of molecules accumulate (e.g., β-amyloid) that may place neurons at risk for activation-induced cell death. We have suggested that the mechanism of β-amyloid-induced apoptosis involves the cross-linking of membrane receptors followed by receptor activation, which leads to aberrant signal transduction (Cotman et al., 1996; Cribbs et al., 1996b).

To test this hypothesis, we identified a ligand that (1) cross-links surface receptors and (2) can be modified so that binding is preserved but the cross-linking property is eliminated. The plant lectin Concanavalin A (Con A) has these properties and is a powerful stimulus of cell death. Like the Fas-induced model, the Con A model requires a multivalent Con A; the divalent succinylated Con A is inactive (Fig. 14.7) (Cribbs et al., 1996b), as are F_{ab} fragments of anti-Fas antibodies (Strasser, 1995). Con A–induced cell death shares many features of activation-induced cell death and β-amyloid-induced death, including the characteristics and the rates of morphological change and neuronal cell death (Cribbs et al., 1996b). Thus β-amyloid and Con A provide two insults that appear to target receptors on the plasma membrane, cause their cross-linking, and therefore may represent a Fas-like mechanism of apoptosis in neurons.

PROTOTYPICAL APOPTOSIS PATHWAYS: STAUROSPORINE VERSUS ACTIVATION OF THE FAS/APO-1 RECEPTOR

Staurosporine and activation of the Fas/Apo-1 receptor are examples of apoptosis inducers that depend on caspase (ICE/CED-3) proteases. Extensive research has been carried out on these two prototypical inducers, and the results indicate that two divergent pathways are operative in a variety of cells. For example, the cowpox virus CrmA inhibits Fas/Apo-1, TNF, and growth-factor-withdrawal-induced apoptosis (Gagliardini et al., 1994; Enari et al., 1995; Los et al., 1995; Tewari and Dixit, 1995), but CrmA does not inhibit apoptosis induced by staurosporine or DNA damaging agents (Chinnaiyan et al., 1996; Datta et al., 1996). Mch5 and Mch4 appear to be the proteases that first receive the apoptotic signals. These proteases have the ability to activate CPP32, Mch3, and all the other caspases, and also have the ability to interact with each other to generate protease amplification cycles (Srinivasula et al.,

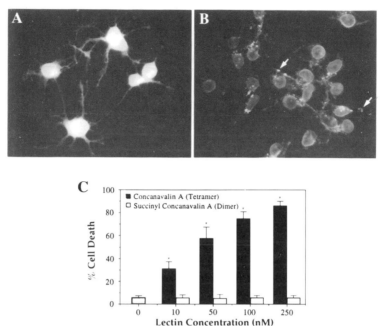

FIG. 14.7. (A) Succinyl Con A (labeled with FITC) binds, but does not cause receptor cross-linking or degeneration. (B) Con A (labeled with FITC) in its native tetrameric form induces receptor cross-linking on the neuronal surface and initiates apoptosis at concentrations as low as 10 nM (C) (Cribbs et al., 1996b).

1996b). Mch5, but not Mch4, is inhibited by CrmA, suggesting that Mch5 may mediate Fas and TNF pathways, whereas Mch4 mediates the CrmA-insensitive apoptotic pathways such as staurosporine and DNA damage pathways (Srinivasula et al., 1996b). Furthermore, it has been shown that Bcl-2 and Bcl-xl can inhibit staurosporine-induced cell death but do not attenuate Fas-induced apoptosis (Orth et al., 1996). Thus, as illustrated in Fig. 14.8, there appear to be at least two pathways that are stimulus-specific and show differential sensitivities to CrmA and the Bcl-2 family. It is likely that activation of caspases such as Mch5 and CPP32 is upstream from free radical formation because reactive oxygen species are blocked by inhibition of ICE-like protease activity (Schulz et al., 1996).

To further test the hypothesis that Con A may share features with a proposed Fas pathway, we have generated a PC-12 cell line that is transfected with Bcl-2 and have examined the susceptibility to staurosporine versus Con A. As predicted, we find that Bcl-2 protects these cells from staurosporine but not from Con A (Fig. 14.9). Thus these data suggest that Con A acts through a different mechanism and shares features with Fas receptor cross-linking. It is also suggested that β-amyloid may act in part through a similar Fas-like mechanism.

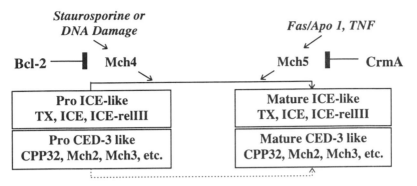

FIG. 14.8. Currently, at least two pathways can be distinguished based on their relative sensitivity to Bcl-2 versus CrmA, as well as the selective involvement of Mch4 versus Mch5.

DNA DAMAGE AND ACTIVATION VERSUS INHIBITION OF THE DEGENERATION PROGRAM

Importantly, although cells in several regions of the AD brain exhibit TdT labeling and clear morphological characteristics of apoptosis, a surprisingly large proportion (often approaching 70–80%) of nuclei in the entorhinal cortex/hippocampal formation show TdT labeling. Whereas classic apoptotic morphology is readily detected in many of these cells, others show light, agranular TdT labeling and relatively normal morphology. It is unclear whether this represents an early stage of apoptosis, necrosis, the presence of DNA damage

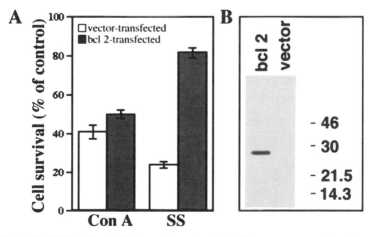

FIG. 14.9. (A) PC12 cells stably transfected with Bcl-2 are protected against staurosporine (SS) but not Con A. (B) Western blot analysis confirms high levels of Bcl-2 expression in the transfected cell line.

that is in a state of repair, or some other degenerative state. It is certain that DNA strand breaks detected with TdT reflect an increase in DNA damage in AD as compared with control brain; however, it is unlikely that all these cells are undergoing terminal cell death by either apoptosis or necrosis. It is generally believed that apoptosis is a relatively rapid process; cultured neurons completely degenerate within 1–2 days when subjected to apoptotic inducers. If all the neurons exhibiting DNA damage in early stages of AD (not uncommonly 70% in susceptible populations) followed a similar degenerative time course, then there would be far fewer neurons remaining in later stages of the disease than has been observed.

Consequently, it is essential to examine critically the implications of the large difference in TdT labeling observed between control and AD brain *in situ*. One possibility is that some of the TdT labeling in these cells represents an active process of DNA-damage/DNA-repair in a subpopulation of cells at risk. Alternatively, some TdT labeling could reflect an accumulation of cells undergoing cell death that are not effectively cleared by normal phagocytic processes. Currently, we and others have been unable to detect a DNA ladder in DNA extracted from the AD brain (Portera-Cailliau et al., 1995; Anderson et al., 1996). In the context of the large number of TdT-labeled cells present in the AD brain, this observation supports the hypothesis that some TdT labeling is the result of an ongoing process of DNA-damage/DNA-repair rather than massive and rapid terminal degeneration.

The accumulation of DNA damage as the result of deficient DNA repair has previously been suggested to have a role in AD neurodegeneration (Boerrigter et al., 1992; Mazzarello et al., 1992; Parshad et al., 1996). Additionally, this accumulation could be the result of an acceleration of DNA damage. The initiation of compensatory mechanisms in response to such events may be predicted to contribute to an active and prolonged process of DNA-damage/DNA-repair and cell death. In this case, the detection of DNA damage by TdT labeling could be used as one of the earliest markers of neuronal abnormality in AD brain, and correspondingly reflect an increased vulnerability of neuronal cells to insult. Alternatively, the apoptotic program in AD either may have been activated but then arrested or it may remain active at a low but significant level due to the chronic presence of inductive agents in the AD brain.

ANTI-APOPTOTIC MECHANISMS ARE INDUCED IN THE AD BRAIN

It is now generally recognized that various members of the Bcl-2 protein family can regulate the survival or death of cells. The loss of protective mechanisms may allow the initiation of some aspects of the cell death program such as initial DNA damage. In turn, the initiation of compensatory mechanisms in response to such events may be predicted to contribute to an active and prolonged process of DNA-damage/DNA-repair and cell death. Accordingly we have examined the AD brain to determine the status of Bcl-2. Its down-regulation would be predicted to leave neurons more susceptible to apoptosis; its up-regulation may protect neurons or arrest the program at some stage.

In human *post mortem* tissue from AD cases, immunoreactivity for Bcl-2 was examined and found to be elevated (Satou et al., 1995) (Fig. 14.10). Furthermore, there is a strong, almost one-to-one colocalization between Bcl-2 and TdT labeling in AD brain (Su et al., 1996b), suggesting that Bcl-2 is induced in neurons that exhibit DNA damage. It may be that Bcl-2 and related mechanisms arrest the program and allow cells to remain viable. Increased expression of Bcl-2 in AD and other neurodegenerative diseases has also been reported (Migheli et al., 1994; O'Barr et al., 1996). Interestingly, previous studies in human lymphocytes have shown that Bcl-2 may exert its protective effects at a stage downstream to the initiation of DNA strand cleavage (Reed, 1994). This observation is consistent with the degree of TdT labeling observed in AD brain. Bcl-2, in turn, can be regulated by other proteins, such as Bax. Current studies indicate that Bax is also increased in some neurons in AD relative to control brains (Su et al., 1997). The exact temporal order for Bcl-2 and Bax up-regulation is currently under investigation.

In accordance with the hypothesis that there may be an extended process of DNA-damage/DNA-repair and cell death in AD, the DNA repair enzyme Ref-1 is also elevated in AD neurons (Anderson et al., submitted). This increase is particularly apparent in mild to moderate AD cases. An increase in the expression of Ref-1 is particularly interesting in that this DNA repair enzyme has also been shown to regulate the DNA binding activity of the transcriptional regulator c-Jun (Abate et al., 1990), which we and others (Anderson et al., 1994; Estus et al., 1994; Ham et al., 1995) have suggested has a role in neuronal apoptosis induced by various stimuli. We also have found that the putative DNA repair enzyme GADD 45 is up-regulated in the AD brain. These proteins could be components of a dynamic and extended

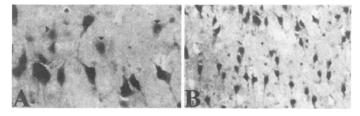

FIG. 14.10. In the AD brain many neurons show evidence of DNA damage (A). Healthy neurons show background levels of Bcl-2 staining in the cytoplasm (arrowheads). Other cells show increased TdT labeling and up-regulation of Bcl-2. Low magnification field (B) shows most of these neurons with DNA damage overexpress Bcl-2.

competition between cell death processes and compensatory protective responses in the AD brain, acting in a type of apoptosis decision cascade.

GLIAL CELLS AND VASCULAR SMOOTH MUSCLE CELLS ALSO SHOW INDICATIONS OF APOPTOSIS

In addition to neuronal loss, other cell types are vulnerable to apoptotic degeneration both *in vitro* and in the AD brain. Cell culture studies have shown that some types of microglia degenerate in the presence of β-amyloid (Korotzer et al., 1993). In AD brain, we have observed several examples of microglia and apoptotic bodies that they have apparently engulfed, or that resulted from their own degeneration (Cotman and Su, 1996). Some AD microglia also express elevated levels of Bax (Su et al., 1997). Together, these data provide evidence supporting the operation of apoptosis in the AD brain.

Similarly, cell culture data have shown that other cells in the brain degenerate when exposed to β-amyloid. For example, vascular smooth muscle cells cultured from human autopsy tissue degenerate over a several day period in the presence of β-amyloid (Davis-Salinas et al., 1995). Initially these cells express increased quantities of the amyloid precursor protein and amyloidogenic peptides, and then eventually die. The increased production of amyloid peptides may contribute to vascular amyloidosis and contribute to the loss of vascular smooth muscle and other cells in cerebral vessels. Also, vascular pericytes are extremely vulnerable to Aβ-induced degeneration (Van Nostrand et al., 1996). Consistent with these culture data, we have identified TdT-positive cells associated with blood vessels in several AD cases (Cotman and Su, 1996).

CONCLUSIONS

Apoptosis is a highly probable cell death pathway in the AD brain given the abundant presence of apoptotic inducers, compromised cellular conditions, and the response of primary neurons in cell culture to such insults and conditions. As predicted, studies on AD tissues demonstrate the presence of neurons with DNA damage, some of which display apoptotic morphology and or other diagnostic markers, such as c-Jun. Furthermore, neurons in histological sections that appear to be intact on the basis of general morphological appearance show the presence of a localized microglia response consistent with the operation of apoptotic processes. Recent genetic data have also shown that the major genes that result in familial AD, the so-called presenilin genes, may make cells more vulnerable to apoptosis. Thus it is predicted that apoptosis may be an even more significant factor in these early-onset familial cases.

Neurons are an unusual type of cell in that they send processes over great distances in the brain. This projection characteristic dictates that they will encounter multiple microenvironments in the AD brain. We suggest that apoptotic mechanisms may not only contribute to neuronal death but also to the loss of neurites, leading, in part, to a breakdown in neural circuitry. Indeed, we have shown that β-amyloid will induce the local microenvironment degeneration of neurites and that this appears to involve an apoptotic or apoptotic-like mechanism.

Neurons are also unusual in that they are nondividing and irreplaceable; thus the decisions to enter into and complete the apoptotic program are most serious. The most unexpected finding in the AD brain is the large number of neurons with DNA damage. This may be due to a delayed clearing response, arrested apoptotic program, or damage that is not immediately terminal and is being repaired. It is possible that numerous endogenous and exogenous protection mechanisms can be mobilized to protect neurons. Indeed, we have found that Bcl-2 is up-regulated in most neurons that display DNA damage. Furthermore, the DNA repair enzyme Ref-1 is similarly up-regulated in the AD brain. Thus it is possible that these protective proteins are components of a dynamic and extended decision-making competition between cell death processes and compensatory responses in the AD brain (Fig. 14.11) (Cotman and Su, 1996).

The current challenge is to derive accurate temporal relationships using autopsy tissues collected at different stages of disease progression, of different genotypes, and possibly of different dementia subtypes. This can be further defined by the study of recently developed transgenic mice and higher mammals (e.g., aged canines) that display many of the features of the AD brain. Some argue that it is not possible to establish cause and effect relationships in autopsy cases, and, thus, it is essential to rely on animal models. On the other hand, study of the presentation of the AD phenotype in the human is

Apoptosis Decision Cascade

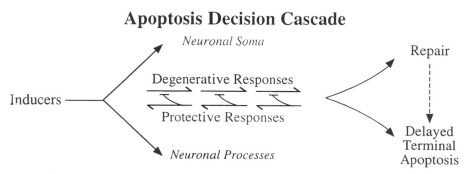

FIG. 14.11. Apoptosis in the AD brain may show competition between entry into apoptosis versus the induction of protection mechanisms, e.g., compensatory genes such as Bcl-2. This prevents neurons from undergoing terminal apoptosis.

essential if we hope to develop an accurate and comprehensive understanding of the condition and its origins. Ultimately, the requirement is a comprehensive multilevel approach, extending from cell culture to animal models to homogeneous cases of AD where the key variables are defined and controlled. In this way, we can elucidate the degenerative mechanisms and develop rational and optimal intervention strategies.

REFERENCES

Abate C, Patel L, Rauscher III FJ, Curran T (1990): Redox regulation of fos and jun DNA-binding activity *in vitro*. Science 249:1157–61.

Afagh A, Cummings BJ, Cribbs DH, Cotman CW, Tenner AJ (1996): Localization and cell association of C1q in Alzheimer's disease brain. Exp Neurol 138:22–32.

Anderson AJ, Cummings BJ, Cotman CW (1994): Increased immunoreactivity for Jun- and Fos-related proteins in Alzheimer's disease: Association with pathology. Exp Neurol 125:286–95.

Anderson AJ, Pike CJ, Cotman CW (1995): Differential induction of immediate early gene proteins in cultured neurons by β-amyloid (Aβ): Association of c-jun with Aβ-induced apoptosis. J Neurochem 65:1487–98.

Anderson AJ, Su JH, Cotman CW (1996): DNA damage and apoptosis in Alzheimer's disease: Colocalization with c-Jun immunoreactivity, relationship to brain area, and effect of postmortem delay. J Neurosci 16:1710–19.

Anderson AJ, Su JH, Cotman CW (submitted): Increase in immunoreactivity for the DNA repair enzyme Ref-1 in Alzheimer's disease brain.

Ankarcrona M, Dypbukt JM, Bonfoco E, Zhivotovsky B, Orrenius S, Lipton SA, Nicotera P (1995): Glutamate-induced neuronal death: A succession of necrosis or apoptosis depending on mitochondrial function. Neuron 15:961–73.

Beal MF, Hyman BT, Koroshetz W (1993): Do deficits in mitochondrial energy metabolism underlie the pathology of neurodegenerative diseases? TINS 16:178–84.

Beal MF (1995): Aging, energy, and oxidative stress in neurodegenerative diseases. Ann Neurol 38:357–66.

Boerrigter ME, Wei JY, Vijg J (1992): DNA repair and Alzheimer's disease. J Gerontol 47:B177–84.

Boldin MP, Goncharov TM, Goltsev YV, Wallach D (1996): Involvement of MACH, a novel MORT1/FADD-interacting protease, in Fas/APO-1- and TNF receptor-induced cell death. Cell 85:803–15.

Bonfoco E, Krainc D, Ankarcrona M, Nicotera P, Lipton SA (1995): Apoptosis and necrosis: Two distinct events induced, respectively, by mild and intense insults with N-methyl-D-aspartate or nitric oxide/superoxide in cortical cell cultures. Proc Natl Acad Sci USA 92:7162–66.

Braak H, Braak E (1995): Staging of Alzheimer's disease-related neurofibrillary changes. Neurobiol Aging 16:271–78; discussion 8–84.

Chinnaiyan AM, Orth K, O'Rourke K, Duan H, Poirier GG, Dixit VM (1996): Molecular ordering of the cell death pathway. Bcl-2 and Bcl-xL function upstream of the CED-3-like apoptotic proteases. J Biol Chem 271:4573–76.

Choi DW (1988): Glutamate neurotoxicity and diseases of the nervous system. Neuron 1:623–34.

Copani A, Koh JY, Cotman CW (1991): β-Amyloid increases neuronal susceptibility to injury by glucose deprivation [see comments]. NeuroReport 2:763–65.

Copani A, Bruno V, Battaglia G, Leanza G, Pellitteri R, Russo A, Stanzani S, Nicoletti F (1995): Activation of metabotropic glutamate receptors protects cultured neurons against apoptosis induced by β-amyloid peptide. Mol Pharmacol 47:890–97.

Cotman CW, Cummings BJ, Pike CJ (1993): Molecular cascades in adaptive versus pathological plasticity. In: A. Goria (ed): Neuroregeneration. New York: Raven Press, pp 217–40.

Cotman CW, Anderson AJ (1995): A potential role for apoptosis in neurodegeneration and Alzheimer's disease. Mol Neurobiol 10:19–45.

Cotman CW, Su JH (1996): Mechanisms of neuronal death in Alzheimer's disease. Brain Pathol 6:493–506.

Cotman CW, Cribbs DH, Anderson AJ (1996): The β-amyloid model of Alzheimer's disease In Wasco W, Tanzi RE (eds): Molecular Mechanisms of Dementia Totowa, N.J.: Humana Press, Inc., pp 73–90.

Cribbs DH, Chen LS, Bende SM, LaFerla FM (1996a): Widespread neuronal expression of the presenilin-1 early-onset Alzheimer's disease gene in the murine brain. Am J Pathol 148:1797–806.

Cribbs DH, Kreng VM, Anderson AJ, Cotman CW (1996b): Crosslinking of concanavalin A receptors on cortical neurons induces programmed cell death. Neurosci 75:173–85.

Cummings BJ, Pike CJ, Shankle R, Cotman CW (1996): β-Amyloid deposition and other measures of neuropathology predict cognitive status in Alzheimer's disease. Neurobiol Aging 17:921–33.

Datta R, Banach D, Kojima H, Talanian RV, Alnemri ES, Wong WW, Kufe DW (1996): Activation of the CPP32 protease in apoptosis induced by 1-β-D-arabinofuranosylcytosine and other DNA-damaging agents. Blood 88:1936–43.

Davis-Salinas J, Saporito-Irwin SM, Cotman CW, Van Nostrand WE (1995): Amyloid β-protein induces its own production in cultured degenerating cerebrovascular smooth muscle cells. J Neurochem 65, 931–34.

Deckwerth TL, Johnson Jr EM (1993a): Temporal analysis of events associated with programmed cell death (apoptosis) of sympathetic neurons deprived of nerve growth factor. J Cell Biol 123:1207–22.

Deckwerth TL, Johnson Jr EM (1993b): Neurotrophic factor deprivation-induced death. Ann NY Acad Sci 679:121–31.

Dellabona P, Peccoud J, Kappler J, Marrack P, Benoist C, Mathis D (1990): Superantigens interact with MHC Class II molecules outside of the antigen groove. Cell 62:1115–21.

Deng G, Pike CJ, Cotman CW (1996): Alzheimer associated presenilin-2 confers increased sensitivity to apoptosis in PC12 cells. FEBS Lett 397:50–54.

Dornan WA, Kang DE, McCampbell A, Kang EE (1993): Bilateral injections of βA(25–35)+IBO into the hippocampus disrupts acquisition of spatial learning in the rat. NeuroReport 5:165–68.

Dragunow M, Faull RLM, Lawlor P, Beilharz EJ, Singleton K, Walker EB, Mee E (1995): *In situ* evidence for DNA fragmentation in Huntington's disease striatum and Alzheimer's disease temporal lobes. NeuroReport 6:1053–57.

Duke RC, Chervenak R, Cohen JJ (1983): Endogenous endonuclease-induced DNA fragmentation: An early event in cell-mediated cytolysis. Proc Natl Acad Sci USA 80:6361–65.

Enari M, Hug H, Nagata S (1995): Involvement of an ICE-like protease in Fas-mediated apoptosis. Nature 375:78–81.

Estus S, Zaks W, Freeman RS, Gruda M, Bravo R, Johnson Jr EM (1994): Altered gene expression in neurons during programmed cell death: Identification of c-jun as necessary for neuronal apoptosis. J Cell Biol 127:1717–27.

Evans DA, Funkenstein HH, Albert MS, Scher Pa, Cook NR, Chown MJ, Hebert LE, Hennekens CH, Taylor JO (1989): Prevalence of Alzheimer's disease in a community population of older persons. JAMA 262:2551–65.

Finch CE, Cohen DM (1997): Aging, metabolism, and Alzheimer disease: Review and hypotheses. Exp Neurol 143:82–102.

Forloni G, Chiesa R, Smiroldo S, Verga L, Salmona M, Tagliavini F, Angeretti N (1993): Apoptosis mediated neurotoxicity induced by chronic application of β-amyloid fragment 25–35. NeuroReport 4:523–26.

Fraser PE, McLachlan DR, Surewicz WK, Mizzen CA, Snow AD, Nguyen JT, Kirschner DA (1994): Conformation and fibrillogenesis of Alzheimer Aβ-peptides with selected substitution of charged residues. J Mol Biol 244:64–73.

Gagliardini V, Fernandez PA, Lee RK, Drexler HC, Rotello RJ, Fishman MC, Yuan J (1994): Prevention of vertebrate neuronal death by the crmA gene [see comments] [published erratum appears in Science 1994 Jun 3:264(5164):1388]. Science 263:826–28.

Gibson G, Martins R, Blass J, Gandy S (1996a): Altered oxidation and signal transduction systems in fibroblasts from Alzheimer patients. Life Sci 59:477–89.

Gibson GE, Zhang H, Toral-Barza L, Szolosi S, Tofel-Grehl B (1996b): Calcium stores in cultured fibroblasts and their changes with Alzheimer's disease. Biochim Biophys Acta 1316:71–77.

Goto I, Taniwaki T, Hosokawa S, Otsuka M, Ichiya Y, Ichimiya A (1993): Positron emission tomographic (PET) studies in dementia. J Neurol Sci 114:1–6.

Greenlund LJ, Deckwerth TL, Johnson Jr EM (1995): Superoxide dismutase delays neuronal apoptosis: A role for reactive oxygen species in programmed neuronal death. Neuron 14:303–15.

Gschwind M, Huber G (1995): Apoptotic cell death induced by β-amyloid 1–42 peptide is cell type dependent. J Neurochem 65:292–300.

Guo Q, Furukawa K, Sopher BL, Pham DG, Xie J, Robinson N, Martin GM, Mattson MP (1996): Alzheimer's PS-1 mutation perturbs calcium homeostasis and sensitizes PC12 cells to death induced by amyloid β-peptide. NeuroReport 8:379–83.

Gwag BJ, Lobner D, Koh JY, Wie MB, Choi DW (1995): Blockade of glutamate receptors unmasks neuronal apoptosis after oxygen-glucose deprivation *in vitro*. Neurosci 68:615–19.

Ham J, Babij C, Whitfield J, Pfarr CM, Lallemand D, Yaniv M, Rubin LL (1995): A c-Jun dominant negative mutant protects sympathetic neurons against programmed cell death. Neuron 14:927–39.

Hoyer S (1993): Abnormalities in brain glucose utilization and its impact on cellular and molecular mechanisms in sporadic dementia of Alzheimer type. Ann NY Acad Sci 695:77–80.

Johnson Jr EM, Koike T, Franklin J (1992): A "calcium set point hypothesis" of neuronal dependence on neurotrophic factor. Exp Neurol 115:163–66.

Koh JY, Yang LL, Cotman CW (1990): β-Amyloid protein increases the vulnerability of cultured cortical neurons to excitotoxic damage. Brain Res 533:315–20.

Korotzer A, Pike C, Cotman C (1993): β-Amyloid peptides induce degeneration of cultured rat microglia. Brain Res 624:121–25.

Lassmann H, Bancher C, Breitschopf H, Wegiel J, Bobinski M, Jellinger K, Wisniewski H (1995): Cell death in Alzheimer's disease evaluated by DNA fragmentation *in situ*. Acta Neuropathol (Berl) 89:35–41.

Lenardo MJ (1991): Interleukin-2 programs mouse α β T lymphocytes for apoptosis. Nature 353:858–61.

Lobner D, Choi DW (1996): Preincubation with protein synthesis inhibitors protects cortical neurons against oxygen-glucose deprivation-induced death. Neurosci 72:335–41.

Loo DT, Copani AG, Pike CJ, Whittemore ER, Walencewicz AJ, Cotman CW (1993): Apoptosis is induced by β-amyloid in cultured central nervous system neurons. Proc Natl Acad Sci USA 90:7951–55.

Los M, Van de Craen M, Penning LC, Schenk H, Westendorp M, Baeuerle PA, Droge W, Krammer PH, Fiers W, Schulze-Osthoff K (1995): Requirement of an ICE/ CED-3 protease for Fas/APO-1-mediated apoptosis. Nature 375:81–83.

Lovell MA, Ehmann WD, Butler SM, Markesbery WR (1995): Elevated thiobarbituric acid-reactive substances and antioxidant enzyme activity in the brain in Alzheimer's disease. Neurology 45:1594–601.

Mark RJ, Hensley K, Butterfield DA, Mattson MP (1995): Amyloid β-peptide impairs ion-motive ATPase activities: Evidence for a role in loss of neuronal Ca^{2+} homeostasis and cell death. J Neurosci 15:6239–49.

Mark RJ, Lovell MA, Markesbery WR, Uchida K, Mattson MP (1997): A role for 4-hydroxynonenal, an aldehydic product of lipid peroxidation in disruption of ion homeostasis and neuronal death induced by amyloid β-peptide. J Neurochem 68:255–64.

Marrack P, Kappler J (1990): The staphylococcal enterotoxins and their relatives [published erratum appears in Science 248(4959):1066]. Science 248:705–11.

Martin SJ, Green DR (1995): Protease activation during apoptosis: death by a thousand cuts? Cell 82:349–52.

Martin SJ, O'Brien GA, Nishioka WK, McGahon AJ, Mahboubi A, Saido TC, Green DR (1995): Proteolysis of fodrin (non-erythroid spectrin) during apoptosis. J Biol Chem 270:6425–28.

Masliah E, Terry RD, DeTeresa RM, Hansen LA (1989): Immunohistochemical quantification of the synapse-related protein synaptophysin in Alzheimer's disease. Neurosci Lett 103:234–39.

Masliah E, Hansen L, Mallory T, Albright T, Terry RD (1991): Abnormal brain spectrin immunoreactivity in sprouting neurons in Alzheimer's disease. Neurosci Lett 129:1–5.

Masliah E, Alford M, DeTeresa R, Mallory M, Hansen L (1996): Deficient glutamate transport is associated with neurodegeneration in Alzheimer's disease. Ann Neurol 40:759–66.

Mattson MP, Cheng B, Davis D, Bryant K, Lieberburg I, Rydel RE (1992): β-Amyloid peptides destabilize calcium homeostasis and render hormonal cortical neurons vulnerable to excitotoxicity. J Neurosci 12:376–89.

Mattson MP, Furukawa F, Bruce AJ, Mark RJ, Blanc E (1997): Calcium homeostasis and free radical metabolism as convergence points in the pathophysiology of dementia. In Wasco W, Tanzi RE (eds.): Molecular Mechanisms of Dementia, Totowa, N.J.: Humana Press, Inc., pp 103–43.

Mazzarello P, Poloni M, Spadari S, Focher F (1992): DNA repair mechanisms in neurological diseases: Facts and hypotheses. J Neurol Sci 112:4–14.

McGeer PL, Kamo H, Harrop R, Li DK, Tuokko H, McGeer EG, Adam MJ, Ammann W, Beattie BL, Calne DB et al. (1986): Positron emission tomography in patients with clinically diagnosed Alzheimer's disease. Can Med Assoc J 134:597–607.

Migheli A, Cavalla P, Piva R, Giordana MT, Schiffer D (1994): bcl-2 protein expression in aged brain and neurodegenerative diseases. NeuroReport 5:1906–8.

Morris J, Mohs R, Rogers H, Fillenbaum G, Heyman A (1988): Consortium to establish a registry for Alzheimer's disease (CERAD) clinical and neuropsychological assessment of Alzheimer's disease. Psychopharmacol Bull 24:641–44.

Mullan M, Crawford F (1993): Genetic and molecular advances in Alzheimer's disease. TINS 16:398–403.

Murphy GM Jr, Forno LS, Ellis WG, Nochlin D, Levy-Lahad E, Poorkaj P, Bird TD, Jiang Z, Cordell B (1996): Antibodies to presenilin proteins detect neurofibrillary tangles in Alzheimer's disease. Am J Pathol 149:1839–46.

Murray KD, Gall CM, Jones EG, Isackson PJ (1994): Differential regulation of brain-derived neurotrophic factor and type II calcium/calmodulin-dependent protein kinase messenger RNA expression in Alzheimer's disease. Neurosci 60:37–48.

Muzio M, Chinnaiyan AM, Kischkel FC, O'Rourke K, Shevchenko A, Ni J, Scaffidi C, Bretz JD, Zhang M, Gentz R, Mann M, Krammer PH, Peter ME, Dixit VM (1996): FLICE, a novel FADD-homologous ICE/CED-3-like protease, is recruited to the CD95 (Fas/APO-1) death-inducing signaling complex. Cell 85:817–27.

Nagata S, Golstein P (1995): The Fas death factor. Science 267:1449–56.

O'Barr S, Schultz J, Rogers J (1996): Expression of the protooncogene bcl-2 in Alzheimer's disease brain. Neurobiol Aging 17:131–36.

Olanow CW (1993): A radical hypothesis for neurodegeneration. TINS 16:439–44

Orth K, Chinnaiyan AM, Garg M, Froelich CJ, Dixit VM (1996): The CED-3/ICE-like protease Mch2 is activated during apoptosis and cleaves the death substrate lamin A. J Biol Chem 271:16443–64.

Page K, Hollister R, Tanzi RE, Hyman BT (1996): *In situ* hybridization analysis of presenilin/mRNA in Alzheimer's disease and in lesioned rat brain. Proc Natl Acad Sci USA 93:14020–24.

Palmer AM, Burns MA (1994): Selective increase in lipid peroxidation in the inferior temporal cortex in Alzheimer's disease. Brain Res 645:338–42.

Paradis E, Douillard H, Koutroumanis M, Goodyer C, LeBlanc A (1996): Amyloid β peptide of Alzheimer's disease downregulates Bcl-2 and upregulates bax expression in human neurons. J Neurosci 16:7533–39.

Parshad RP, Sanford KK, Price FM, Melnick LK, Nee LE, Schapiro MB, Tarone RE, Robbins JH (1996): Fluorescent light-induced chromatid breaks distinguish Alzheimer disease cells from normal cells in tissue culture. Proc Natl Acad Sci USA 93:5146–50.

Phillips HS, Hains JM, Armanini M, Laramee GR, Johnson SA, Winslow JW (1991): BDNF mRNA is decreased in the hippocampus of individuals with Alzheimer's disease. Neuron 7:695–702.

Pike CJ, Cummings BJ, Cotman CW (1992): β-Amyloid induces neuritic dystrophy *in vitro:* Similarities with Alzheimer pathology. NeuroReport 3:769–72.

Pike CJ, Ramezan-Arab N, Cotman CW: β-Amyloid toxicity in vitro: Evidence of oxidative stress but not protection by antioxidants. J Neurochem 69:1601–1611.

Portera-Cailliau C, Herdeen JC, Price DL, Koliatsos VE (1995): Evidence for apoptotic cell death in Huntington disease and excitotoxic animal models. J Neurosci 15:3775–87.

Qin ZH, Wang Y, Chase TN (1996): Stimulation of N-methyl-D-aspartate receptors induces apoptosis in rat brain. Brain Res 725:166–176.

Radvanyi LG, Mills GB, Miller RG (1993): Religation of the T cell receptor after primary activation of mature T cells inhibits proliferation and induces apoptotic cell death. J Immunol 150:5704–15.

Raff MC, Barres BA, Burne JF, Coles HS, Ishizaki Y, Jacobson MD (1993): Programmed cell death and the control of cell survival: lessons from the nervous system. Science 262:695–700.

Ratan RR, Murphy TH, Baraban JM (1994): Oxidative stress induces apoptosis in embryonic cortical neurons. J Neurochem 62:376–79.

Reed JC (1994): Bcl-2 and the regulation of programmed cell death. J Cell Biol 24:1–6.

Satou T, Cummings BJ, Cotman CW (1995): Immunoreactivity for Bcl-2 protein within neurons in the Alzheimer's disease brain increases with disease severity. Brain Res 697:35–43.

Schubert D, Chevion M (1995): The role of iron in β amyloid toxicity. Biochem Biophys Res Commun 216:702–7.

Schulz JB, Weller M, Klockgether T (1996): Potassium deprivation-induced apoptosis of cerebellar granule neurons: A sequential requirement for new mRNA and protein synthesis, ICE-like protease activity, and reactive oxygen species. J Neurosci 16:4696–706.

Selkoe DJ (1996): Amyloid β-protein and the genetics of Alzheimer's disease. J Biol Chem 271:18295–98.

Selkoe DJ, Yamazaki T, Citron M, Podlisny MB, Koo EH, Teplow DB, Haass C (1996): The role of APP processing and trafficking pathways in the formation of amyloid β-protein. Ann NY Acad Sci 777:57–64.

Selkoe DJ (1997): Alzheimer's disease: genotypes, phenotypes, and treatments. Science 275:630–31.

Shimohama S, Chachin M, Taniguchi T, Hidaka H, Kimura J (1996): Changes of neurocalcin, a calcium-binding protein, in the brain of patients with Alzheimer's disease. Brain Res 716:233–36.

Sihag RK, Cataldo AM (1996): Brain β-spectrin is a component of senile plaques in Alzheimer's disease. Brain Res 743:249–57.

Slater AF, Nobel CS, Orrenius S (1995): The role of intracellular oxidants in apoptosis. Biochim Biophys Acta 1271:59–62.

Smale G, Nichols NR, Brady DR, Finch CE, Horton Jr WE (1995): Evidence for apoptotic cell death in Alzheimer's disease. Exp Neurol 133:225–30.

Smith MA, Sayre LM, Monnier VM, Perry G (1995): Radical AGEing in Alzheimer's disease. Trends Neurosci 18:172–76.

Srinivasula SM, Ahmad M, Fernandes-Alnemri T, Litwack G, Alnemri ES (1996a): Molecular ordering of the Fas-apoptotic pathway: The Fas/APO-1 protease Mch5 is a CrmA-inhibitable protease that activates multiple Ced-3/ICE-like cysteine proteases. Proc Natl Acad Sci USA 93:14486–91.

Srinivasula SM, Fernandes-Alnemri T, Zangrilli J, Robertson N, Armstrong RC, Wang L, Trapani JA, Tomaselli KJ, Litwack G, Alnemri ES (1996b): The Ced-3/interleukin 1β converting enzyme-like homolog Mch6 and the lamin-cleaving enzyme Mch2α are substrates for the apoptotic mediator CPP32. J Biol Chem 271:27099–106.

Strasser A (1995): Apoptosis. Death of a T cell [news; comment]. Nature 373:385–86.

Su JH, Anderson AJ, Cummings BJ, Cotman CW (1994): Immunohistochemical evidence for apoptosis in Alzheimer's disease. NeuroReport 5:2529–33.

Su JH, Cummings BJ, Cotman CW (1996a): Plaque biogenesis in brain aging and Alzheimer's disease. I. Progressive changes in phosphorylation states of paired helical filaments and neurofilaments. Brain Res 739:79–87.

Su JH, Satou T, Anderson AJ, Cotman CW (1996b): Up-regulation of Bcl-2 is associated with neuronal DNA damage in Alzheimer's disease. NeuroReport 7:437–40.

Su JH, Deng G, Cotman CW (1997): Bax protein expression is increased in Alzheimer's brain: Correlations with DNA damage, Bcl-2 expression, and brain pathology. J Neuropathol Exp Neurol 56:86–93.

Tepper CG, Studzinski GP (1992): Teniposide induces nuclear but not mitochondrial DNA degradation. Cancer Res 52:3384–90.

Tewari M, Dixit VM (1995): Fas- and tumor necrosis factor-induced apoptosis is inhibited by the poxvirus crmA gene product. J Biol Chem 270:3255–60.

Thomas LB, Gates DJ, Richfield EK, O'Brian TF, Schweitzer JB, Steindler DA (1995): DNA end labeling (TUNEL) in Huntington's disease and other neuropathological conditions. Exp Neurol 133:265–72.

Uchihara T, el Hachimi HK, Duyckaerts C, Foncin JF, Fraser PE, Levesque L, St George-Hyslop PH, Hauw JJ (1996): Widespread immunoreactivity of presenilin in neurons of normal and Alzheimer's disease brains: Double-labeling immunohistochemical study. Acta Neuropathol (Berl) 92:325–30.

Van Nostrand WE, Davis-Salinas J, Saporito-Irwin SM (1996): Amyloid β-protein induces the cerebrovascular cellular pathology of Alzheimer's disease and related disorders. Ann NY Acad Sci 777:297–302.

Vanags DM, Porn-Ares MI, Coppola S, Burgess DH, Orrenius S (1996): Protease involvement in fodrin cleavage and phosphatidylserine exposure in apoptosis. J Biol Chem 271:31075–85.

Vito P, Lacana E, D'Adamio L (1996a): Interfering with apoptosis: Ca^{2+}-binding protein ALG-2 and Alzheimer's disease gene ALG-3. Science 271:521–25.

Vito P, Wolozin B, Ganjei JK, Iwasaki K, Lacana E, D'Adamio L (1996b): Requirement of the familial Alzheimer's disease gene PS2 for apoptosis. Opposing effect of ALG-3. J Biol Chem 271:31025–28.

Watt JA, Pike CJ, Walencewicz-Wasserman AJ, Cotman CW (1994): Ultrastructural analysis of β-amyloid-induced apoptosis in cultural hippocampal neurons. Brain Res 661:147–56.

Weber LL, Leissering MA, Yang AJ, Glabe CG, Cribbs DH, LaFerla FM (1997): Presenilin-1 immunoreactivity is localized intracellularly in Alzheimer's disease brain, but not detected in amyloid plaques. Exp Neurol 143:37–44.

Whittemore ER, Loo DT, Cotman CW (1994): Exposure to hydrogen peroxide induces cell death via apoptosis in cultured rat cortical neurons. NeuroReport 5:1585–88.

Wolozin B, Iwasaki K, Vito P, Ganjei JK, Lacana E, Sunderland T, Zhao B, Kusiak JW, Wasco W, D'Adamio L (1996): Participation of presenilin 2 in apoptosis: Enhanced basal activity conferred by an Alzheimer mutation. Science 274:1710–13.

Wyllie AH, Morris RG, Smith AL, Dunlop D (1984). Chromatin cleavage in apoptosis: Association with condensed chromatin morphology and dependence on macromolecular synthesis. J Pathol 142:67–77.

Yamatsuji T, Matsui T, Okamoto T, Komatsuzaki K, Takeda S, Fukumoto H, Iwatsubo T, Suzuki N, Asami-Odaka A, Ireland S, Kinane TB, Giambarella U, Nishimoto I (1996a): G protein-mediated neuronal DNA fragmentation induced by familial Alzheimer's disease-associated mutants of APP. Science 272:1349–52.

Yamatsuji T, Okamoto T, Takeda S, Murayama Y, Tanaka N, Nishimoto I (1996b): Expression of V642 APP mutant causes cellular apoptosis as Alzheimer trait-linked phenotype. EMBO J 15:498–509.

Yankner BA (1996): Mechanisms of neuronal degeneration in Alzheimer's disease. Neuron 16:921–32.

Zakeri ZF, Quaglino D, Latham T, Lockshin RA (1993): Delayed internucleosomal DNA fragmentation in programmed cell death. FASEB J 7:470–78.

CHAPTER 15

REGULATION OF APOPTOSIS BY CELLULAR INTERACTIONS WITH THE EXTRACELLULAR MATRIX

JAMES G. TIDBALL and DOUGLAS E. ALBRECHT

Department of Physiological Science, University of California, Los Angeles, CA 90095-1527

INTRODUCTION

The relationship between the extracellular matrix (ECM) and cells that adhere to it can play important regulatory roles in many basic cellular processes by influencing enzyme activity (e.g., Burridge et al., 1992; Kornberg et al., 1992) and phospholipid metabolism (e.g., McNamee et al., 1993), and modifying transcriptional and translational activities of the cell (e.g., Benecke et al., 1978). In aggregate, these regulatory influences can control key events in the life of a cell, such as cell motility, proliferation, growth, and differentiation (e.g., Adams and Watts, 1993). Early investigations of the influence of the ECM on basic cellular functions showed that the effects of ECM contact on cell function could be immediate. For example, early investigations by Folkman and Moscona (1978) and Ben-Ze'ev and colleagues (1980) found that moments after a suspended cell contacts a substratum there was a tremendous increase in protein and RNA synthesis in the cell, showing that signals generated by cell–substratum contact are rapidly transduced to changes in gene expression. Thus the formation of cell–ECM contacts can have a rapid anabolic effect on cells, and can provide an important regulatory step in basic cellular processes (Cotman et al., Chapter 14, this volume).

Disruptions of existing cellular interactions with the ECM, from experimental, pathological, or normal physiological changes, were also shown in early studies to be closely linked to changes in the functional capacity of cells. In an especially provocative, early investigation, Chen et al. (1976) demonstrated

When Cells Die, Edited by Richard A. Lockshin, Zahra Zakeri, and Jonathan L. Tilly
ISBN 0-471-16569-7 © 1998 Wiley-Liss, Inc.

that cells experiencing transformation in which there was loss of normal growth control, also displayed a disruption in normal cell–substratum interaction. The transformed cells also showed a decrease in production of a protein called at that time "the large, external transformation sensitive (LETS) protein," which is now known as the ECM protein fibronectin. This investigation was one of the first to identify a relationship between pathological cell growth and changes in cell–substratum contact accompanied by defects in synthesis of ECM components, suggesting these events may be interdependent.

These early investigations showed that cell–substratum contact is closely linked to cell growth, in at least some cell types, and that the relationship between substratum contact and cell growth and differentiation is positive for nonpathological cells. Those general findings laid the groundwork for the hypothesis that there could also be a positive relationship between loss of substratum contact and cell death in nonpathological cells, particularly by apoptosis. Because modifications in cell–substratum interactions can result in changes in gene expression and protein synthesis, apoptosis could feasibly result from either the turning off of an inhibitor of cell death whose function or expression was positively regulated by substratum contact, or the turning on of a promoter of cell death whose function or expression was negatively regulated by substratum contact.

THE BIOLOGICAL SIGNIFICANCE OF REGULATING CELL DEATH BY CELL–ECM INTERACTIONS

Cells can be generally separated into two groups, according to the nature of their interactions with the substratum. Epithelial and endothelial cells form cell layers that are organized into structured tissues that can form distinct organs, while mesenchymal cells tend to be isolated and migratory. Typically, epithelial and endothelial cells form persistent associations with well-structured connective tissue components, especially with basement membranes (Fig. 15.1). Basement membranes are specialized ECM structures that are composed of three separate connective tissue layers. These layers, the lamina lucida, lamina densa, and lamina reticularis are highly enriched in the glycoproteins laminin, type IV collagen, and fibronectin, respectively. Other structural molecules involved in cell–substratum adhesion, such as vitronectin, tenascin, and heparan sulfate proteoglycan, are also present (Laurie et al., Mayne and Sanderson, 1985; Timpl and Brown, 1996). Basement membranes typically form a continuous layer that provides an attachment substratum at the basal surface of epithelial and endothelial cells. However, in muscle fibers, basement membranes provide a sheath that surrounds the entire cell. Basement membrane proteins, when used as substrate for cell culture *in vitro,* can promote cell attachment and spreading, and stimulate proliferation or differentiation (e.g., Ocalan et al., 1988; von der Mark and Ocalan, 1989). These basement membrane–mediated effects on cell function are attributable,

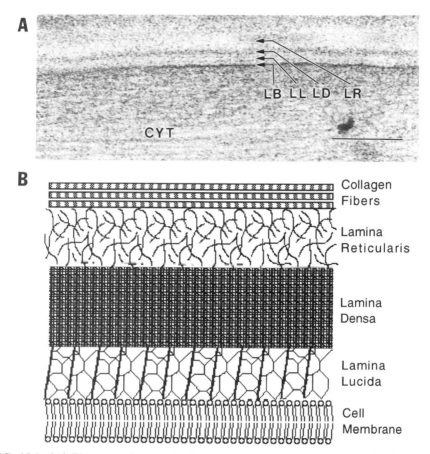

FIG. 15.1. (A) Electron micrograph of the basement membrane of a skeletal muscle fiber. LB: lipid bilayer, LL: lamina lucida, LD: lamina densa, LR: lamina reticularis, and CYT: cytoplasm of the muscle cell. Scale bar is 250 nm. (B) Schematic diagram showing the layers of the basement membrane.

at least in part, to activation and expression of specific genes associated with cell growth and differentiation (Vachon et al., 1996).

Epithelial cells in the basal layers of the skin lie in close contact with the underlying basement membrane, and are a proliferative population of cells. A natural consequence of continuous cell proliferation in the basal layers is that the number of cells soon exceeds the space available for attachment to the supporting basement membrane. This resulting separation of some epithelial cells from the basement membrane deprives the detached cells of matrix-derived signals, and apoptosis occurs (Adams and Watts, 1993). This form of apoptosis, which results from loss of normal cell–substratum contact, has been called *anoikis,* meaning homelessness (Frisch and Francis, 1994). Anoikis of skin epithelial cells provides a valuable regulatory mechanism for controlling skin thickness by providing a mechanism for removal of skin cells as new cells are added through mitosis. An important finding that indicates the possible mechanism by which apoptosis is regulated in skin epithelial cells was the observation that bcl-2, an inhibitor of apoptosis (Vaux et al., 1988; McDonnell et al., 1989; Nunez et al., 1990), is expressed by cells in the basal layer of the skin epithelium, but not in more superficial layers (Hockenberry et al., 1991). This indicates that the loss of cell–substratum contact causes a cell rescue signal to be turned off, rather than a proapoptosis signal to be turned on. Thus apoptosis may be the default mode for fully differentiated epithelial cells of the skin, which is actively prevented by continued expression of a survival factor, bcl-2, that is regulated by cell interaction with the basement membrane (Birge et al., Chapter 13, this volume).

Apoptosis resulting from disruption in cell–substratum interactions also provides an important regulatory mechanism for mammary gland involution following weaning. The alveolar epithelium of the mammary gland, which is the milk-producing cell layer, proliferates during pregnancy, but the constituent cells remain in contact with the underlying basement membrane throughout proliferation. Following weaning and at the onset of mammary gland involution, the alveolar epithelium is dramatically reduced by apoptosis (Walker et al., 1989; Strange et al., 1992). However, unlike the skin epithelium, these cells remain in contact with the basement membrane at the onset of apoptosis (Walker et al., 1989). Despite the presence of cell–substratum contact at the onset of alveolar cell apoptosis, current evidence indicates that perturbation of cell–substratum attachment also underlies the initiation of apoptosis of these cells. Four days post weaning in mouse mammary glands, apoptosis is apparent both in histological assays for changes in nuclear morphology an in DNA extracts that show internucleosomal cleavage (Strange et al., 1992), although persistent contacts between the epithelial cells and the morphologically discernible basement membrane are present at this stage. However, three to four days following weaning, the mouse mammary epithelium increases expression of stromelysin and plasminogen activator (Talhouk et al., 1992; Strange et al., 1992), which are capable of causing proteolysis of collagens, fibronectin, and other ECM proteins. At this same time, there is a decrease in the expression of tissue inhibitor of metalloproteinases (TIMP), which inhibits the protease activities of stromelysin and other proteases that cleave ECM proteins (Strange et al., 1992). These findings indicate that the proteolysis of the basement membrane occurs prior to apoptosis of mammary epithelial cells and before morphologically detectible changes in basement membrane structure or cell–basement membrane interactions occur. A role for ECM proteolysis in apoptosis of mammary epithelial cells is also supported by observations that involuting glands receiving applications of exogenous TIMP displayed a reduced rate of alveolar regression (Talhouk et al., 1992). Thus perturbation of cell–ECM interactions also underlies the occurrence of apoptosis in mammary epithelial cells, although the process is not anoikis, as defined by morphological criteria (Cryns and Yuan, Chapter 6; Sikorska and Walker, Chapter 7, this volume).

This possible role of basement membrane proteolysis in serving to activate apoptosis in the involuting mammary gland has also been supported by observations *in vitro*. Primary cultures of mammary cells from pregnant mice (Pullan et al., 1996) and CID-9 mammary epithelial cells (Boudreau et al., 1995) became apoptotic when cultured on non-basement-membrane substrata, such as plastic or type I collagen, even though the cells adhered to these substrata. However, growth on Engelbreth–Holm–Swarm (EHS) tumor matrix, which contains basement membrane proteins such as laminin and type IV collagen, decreased the occurrence of apoptosis *in vitro*. Furthermore, time-lapse video of individual cells showed that morphological changes indicating apoptosis appeared prior to cell detachment, confirming that apoptosis in these cells was not anoikis, although it may have been ECM regulated (Pullan et al., 1996). ECM regulation of apoptosis in mammary epithelial cells also differed importantly from that occurring in the skin epithelium in that there was no apparent relationship between bcl-2 expression in mammary epithelium and apoptosis (Pullan et al., 1996) although experimentally induced overexpression of bcl-2 did reduce apoptosis of these cells *in vivo*. However, bax, a promoter of apoptosis, was found to be up-regulated at the time of apoptosis (Pullan et al., 1996). Thus, while loss of normal cell–substratum interactions may turn off cell survival factors in the skin epithelium to initiate apoptosis, in the mammary epithelium apoptosis appears to be initiated by turning on proapoptotic factors when specific cell–substratum interactions are perturbed.

ECM RECEPTORS AND THEIR ROLE IN REGULATING APOPTOSIS

Integrins, which comprise a large, well-conserved family of transmembrane receptors for ECM proteins, have been shown to play important roles in the transduction of information across the cell membrane (Hynes, 1992). Ligand binding by integrin heterodimers can rapidly stimulate intracellular responses including changes in cytosolic pH, intracellular calcium concentrations, phospholipid metabolism, and kinase activation. These basic responses activate tremendously complex cascades of regulatory events involving the more than 20 signaling molecules located at integrin clusters (Miyamoto et al., 1995). Activation of these signals can result in changes in cell shape, motility, proliferation, differentiation, and, as shown in more recent investigations, they can also regulate cell viability by controlling or initiating apoptosis.

Integrins are formed by the noncovalent association of α and β chains, both of which exist as transmembrane polypeptides that can interact with the cytoskeleton via their cytosolic domains and bind ECM molecules or other cell surface receptors via their extracellular domains (Hynes 1992). Specificity for their extracellular ligands is conferred by the combination of α and β chains that form the integrin heterodimer (Table 15.1). The relationship between loss of cell–substratum adhesion and the initiation of apoptosis suggested the possibility that integrin binding could have an inhibitory effect on apoptosis. Numerous investigations have now shown this to be an important event in the regulation of apoptosis (Table 15.2), although the relationships between integrin binding and the occurrence of apoptosis vary between cell types, and the mechanisms by which integrin binding is transduced to survival or proapoptotic signals have not been well characterized (Birge et al., Chapter 13, this volume).

Blood vessel endothelial cells, which normally reside *in vivo* as flattened cells that are tightly attached to the basement membrane, have provided a valuable *in vitro* and *in vivo* model for examining the role of integrins in regulating apoptosis. These endothelial cells (EC) soon become apoptotic *in vitro* if they are not cultured on substrata that resemble their *in vivo* substrata. This substratum-related rescue was shown to be an integrin-dependent event by Meredith and colleagues (1993), who demonstrated that EC could be rescued from apoptosis if cultured on substrata coated with anti-integrin $\beta1$, but not on substrata coated with anti-class I histocompatibility antigen (HA)

TABLE 15.1. Specificity of Integrins for Extracellular Ligands

Integrins	Ligands
$\alpha1\beta1$	collagen, laminin
$\alpha2\beta1$	collagen, laminin, tenascin, integrin $\alpha3\beta1$, fibronectin
$\alpha3\beta1$	collagen, fibronectin, laminin, integrin $\alpha2\beta1$, integrin $\alpha3\beta1$, epiligrin, entactin
$\alpha4\beta1$	fibronectin, invasin, VCAM-1, ICAM-2
$\alpha5\beta1$	fibronectin, collagen
$\alpha6\beta1$	laminin
$\alpha7\beta1$	laminin
$\alpha8\beta1$	tenascin, fibronectin, vitronectin
$\alpha9\beta1$	tenascin
$\alpha v\beta1$	fibronectin, RGD, vitronectin, collagen, osteopontin
$\alpha L\beta2$	ICAMs
$\alpha m\beta2$	ICAMs
$\alpha x\beta2$	fibrinogen
$\alpha II\beta3$	collagen, disintegrin, fibronectin, RGD, vitronectin
$\alpha v\beta3$	tenascin, disintegrin, collagen, fibronectin, RGD, vitronectin, osteopontin
$\alpha6\beta4$	laminin
$\alpha v\beta5$	vitronectin, RGD
$\alpha v\beta6$	fibronectin, tenascin
$\alpha4\beta7$	fibronectin, VCAMs, mucosal ACAM-1
$\alpha E\beta7$	E-cadherin
$\alpha v\beta8$	vitronectin

TABLE 15.2. The Effects of Integrin–Ligand Binding on Apoptosis.
+, Integrin–Ligand Binding Causes Apoptosis in that Cell Type.
−, Integrin–Ligand Binding Prevents Apoptosis in that Cell Type.

Integrin Type	Influence on Apoptosis	Cell Type
$\alpha 5\beta 1$	−	Chinese hamster ovary cells (CHO)[1]
$\alpha ?\beta 1$	−	Mammary epithelial cells[2]
$\alpha ?\beta 1$	−	Human umbilical vein endothelial cells (HUVECs)[3]
$\alpha v\beta 3$	−	Endothelial cells (angiogenesis)[4]
$\alpha 6\beta 4$	+	Epithelial cells (rectal carcinoma, RKO)[5]
$\alpha v\beta ?$	−	Colon carcinoma cells (LIM 1863)[6]

1. Zhang et al., 1995.
2. Howlett et al., 1995.
3. Meredith et al., 1993.
4. Brooks et al., 1994.
5. Clarke et al., 1995.
6. Bates et al., 1994.

or with anti-vascular cell adhesion molecule-1 (V-CAM), even though the EC expressed HA and V-CAM on their surfaces and would bind to surfaces coated with their antibodies. Subsequent investigations showed that rescue of EC from apoptosis also depended directly or indirectly on the quantity of integrin molecules bound by ligands. For example, EC cells grown on surfaces coated with low concentrations of fibronectin or vitronectin would attach weakly to the substratum, but still become apoptotic, whereas substrata coated with higher concentrations of the ligands would inhibit apoptosis (Re et al., 1994).

Current findings indicate that integrin-mediated binding to the ECM also protects EC cells from apoptosis *in vivo,* and that this may play a significant physiological role. Vasculature can be rapidly formed and remodeled during development, growth, wound healing, and in response to exercise. Although growth of new vasculature *in vivo* is partially attributable to the effects of soluble factors, such as platelet-derived growth factor (PDGF) and fibroblast growth factor (FGF) (Schollmann et al., 1992; Kraiss et al., 1993; Kobayashi et al., 1994), EC survival during capillary growth is also dependent on integrin binding. During angiogenesis of the chick chorioallantoic membrane, EC express $\alpha v\beta 3$ integrin (the vitronectin receptor), and angiogenesis can be blocked *in situ* by the application of anti-$\alpha v\beta 3$ integrin to the membrane (Brooks et al., 1994). Angiogenesis can also be blocked *in vivo* by a single intravenous injection of anti-$\alpha v\beta 3$ integrin or injection with a peptide that competes with $\alpha v\beta 3$ binding. Interestingly, this block of angiogenesis by blocking $\alpha v\beta 3$ integrin–ligand binding results from apoptosis of only the proliferating EC cells; nonproliferative EC cells not involved in angiogenesis are not affected by the treatment.

In at least several systems, the integrin-mediated rescue from apoptosis is attributable to kinase activation. This was initially suggested by Meredith and colleagues (1993), who observed that not only would integrin binding rescue EC from apoptosis, but suspended EC cells that would normally become apoptotic were protected by treatment with tyrosine phosphatase inhibitors, and that the protective effect would be reversed by treatment with the tyrosine kinase inhibitor, herbimycin A. (See also Birge et al., Chapter 13, and Cotman et al., Chapter 14, this volume.) FAK (focal adhesion kinase) is a prominent, but not exclusive, possibility for a kinase that may mediate this rescue. FAK is an integrin-binding tyrosine kinase that is activated by cell adhesion to substrata or by integrin binding to substrata coated with fibronectin, type IV collagen, or laminin (Kornberg et al., 1992). Furthermore, cells treated with soluble fibronectin do not display FAK activation, which is similar to observations that substratum-bound integrin ligands can rescue cells from apoptosis, although the soluble ligands have no rescue effect. Although there are similarities between the conditions that activate FAK and those that induce integrin-mediated rescue from apoptosis, the role of FAK in regulating apoptosis is not clear. FAK activation has been shown recently to be insufficient to prevent apoptosis in at least some cell types. Zhang and colleagues (1995) found that CHO cells expressing $\alpha5\beta1$ integrin and grown on fibronectin experienced FAK activation and did not become apoptotic; however, CHO cells expressing $\alpha v\beta1$ integrin, which is also a fibronectin receptor, likewise bound to fibronectin and experienced FAK activation, but they then became apoptotic. Alternatively, FAK activation has been implicated as a negative regulator of apoptosis in other cell types. For example, epithelial cell lines that express constitutively activated FAK were protected from apoptosis that normally resulted from loss of substratum attachments in epithelial cells (Frisch et al., 1996). In addition, recent findings (Crouch et al., 1996) showed that FAK proteolysis is an early event in c-myc induced apoptosis of fibroblasts that precedes other indicators of apoptotic change. Thus FAK may also be required to protect against apoptosis in these cells. More detailed information concerning intracellular signaling events that regulate apoptosis are discussed elsewhere in this volume.

Integrin-mediated binding of cells to the ECM is generally, but not always, an event that protects against apoptosis. In some pathological cell types, integrin-mediated binding can induce apoptosis. Growth of tumor cells *in vivo* and *in vitro* frequently correlates with the occurrence of defects in integrin-mediated binding. For example, tumorigenicity of many colon carcinoma cells corresponds with their loss of normal integrin-mediated attachments to substrata. RKO colon carcinoma cells lack the ability to express $\beta4$ integrin, although they do express $\beta1$ (Clarke et al., 1995). This enables RKO cells to adhere to laminin via $\alpha6\beta1$, a relatively weak association, but not $\alpha6\beta4$, which is stronger (Clarke et al., 1995). Transfection of these cells with full-length $\beta4$, so that $\alpha6\beta4$ was expressed, resulted in apoptosis. Interestingly, transfection with truncated $\beta4$ that did not express the cytosolic domain did not induce apoptosis in these cells, indicating that the apoptosis-promoting signal depends on β4 interaction with cytoskeletal proteins or intracellular regulatory proteins.

Increased expression of integrin isoforms that are not normally expressed also occurs in some transformed cell types. For example, neoplasia of melanoma cells is associated with increased expression of αvβ3 integrin (Cheresh, 1991; Albeda et al., 1991; Felding-Habermann et al., 1992), which causes an increase in tumorigenicity. Expression of this integrin heterodimer figures in the pathological role of these cells, because it allows these cells to bind proteolytic fragments of type I collagen. Normally, the only members of the β1 family of integrin heterodimers that can bind type I collagen are α1, α2, and α3β1 integrins. However, when collagen is proteolyzed or denatured, cryptic sites that contain the tripeptide arginine–glycine–aspartate (RGD) are exposed, which is a ligand for αvβ3 integrin. If melanoma cells that do not express αvβ3 are grown in a matrix of denatured type I collagen, they rapidly die of apoptosis (Montgomery et al., 1994). However, if the same melanoma cells are transfected with αv integrin cDNA, they are rescued from apoptosis, although the rescue can be prevented by treatment with anti-αv integrin (Montgomery et al., 1994). The scenario proposed by Montgomery and colleagues to interpret these findings is that melanoma cells first attach to collagen via α2β1 integrin, and then as part of their invasive behavior, they degrade type I collagen and expose cryptic RGD sites. Further cell adhesion occurs via αvβ3, which promotes tumor cell survival, proliferation, and tumorigenicity.

All currently known examples in which ECM regulation of apoptosis occurs involve integrins as receptors. However, this absence of evidence for a role of other receptor–ECM interactions in regulating apoptosis should not be taken as evidence that they do not exist or are not important; rather, it more likely reflects the broad importance of integrins in basic biological functions, and the relatively intense focus of current cell biological interest in integrin-mediated functions. Other nonintegrin receptor–ligand interactions that induce or protect against apoptosis have been characterized, but they involve intercellular associations or receptor binding by soluble factors, rather than ECM. A possible exception to this is found in dystrophin-deficient muscle, which experiences increased rates of apoptosis *in vivo* (Tidball et al., 1995; Matsuda et al., 1995) and *in vitro* (Smith et al., 1995) that may be attributable to defects in cell–ECM interactions. Dystrophin is a membrane-associated cytoskeletal protein that binds actin thin filaments and also a transmembrane complex of proteins (reviewed by Ervasti and Campbell, 1993). The transmembrane, dystrophin-associated proteins (DAP) then bind basement membrane laminin via their extracellular domain. Thus this dystrophin complex is analogous to integrin and its associated proteins in forming a transmembrane link between cytosolic and extracellular structural proteins. Although not yet proved, it is hypothesized that the increase in apoptosis of dystrophic muscle cells, at least *in vitro,* may be attributable to defective cell–ECM association. Further support of this hypothesis comes from recent evidence showing that muscle fibers that fail to express the muscle specific form of laminin, known as merosin, undergo apoptosis *in vitro* (Vachon et al., 1996). Whether apoptosis resulting from merosin deficiency is attributable to the absence of merosin–DAP interaction or merosin–integrin interaction, or perhaps some other unidentified factor, is unknown.

ECM BINDING, CELL SHAPE, CELL DEFORMATION, AND THE REGULATION OF APOPTOSIS

Through the studies cited previously, there is ample evidence that the binding of integrins and perhaps other ECM receptors to their specific ligands can regulate apoptosis. However, simple receptor–ligand binding is not adequate to protect cells from apoptosis. For example, in the studies by Re et al. (1994), growth of EC cells on surfaces coated with integrin ligands at concentrations adequate to allow integrin binding and clustering at contact sites was inadequate to prevent apoptosis. However, growth of the cells on substrata coated with ligand at concentrations that were adequate to cause cell spreading were found to protect against apoptosis, indicating that some variable associated with cell shape contributed to the protective role. Cell shape or a cell shape associated factor is not solely responsible for the protective effect. This, too, is evident in previously cited studies, which showed that mammary epithelial cells initiate apoptosis while they remain spread on their substratum. Thus the protective effect of cell–ECM interactions relies on the combined effects of both appropriate integrin–ligand interactions and cell shape. However, at least one additional layer of regulatory complexity exists in that soluble substances, such as select growth factors present in serum, must also be present or apoptosis will occur in cells that have attached and spread on substrate. This growth factor dependence for survival may also be attributed to cell shape or some variable associated with cell shape change. For example, cells that were allowed to attach and spread on substrate *in vitro* and then deprived of serum in the culture media were shown to begin to round up and detach from the substratum within minutes, followed by eventual apoptosis (Kulkarni and McCulloch, 1994).

The complex relationship between apoptosis and the regulatory roles of cell–ECM binding, cell shape, and soluble factors is conceptually, and probably mechanistically, related to how these factors and events regulate cell growth. Cell adhesion to substrata is commonly a requirement for cell growth in normal differentiated cells (Stoker et al., 1968), although simple adhesion alone is not adequate to stimulate growth. Folkman and Moscona (1978) showed that cell spreading following adhesion was an important positive regulator of growth. The relationship of cell spreading to growth may be attributable, at least to some degree, to an increased availability of receptors for growth factors on the spread cell than on the rounded cell (Tucker et al., 1981). While in some cases this increased availability may reflect the increase in surface area of cells that occurs during spreading, for at least some cells the regulatory effect is more subtle and not understood. For example, although PDGF (platelet-derived growth factor) binding by PDGF receptor bearing cells stimulates cell proliferation and inhibits apoptosis (Albrecht and Tidball, 1996), changes in cell shape can affect cell responsiveness to PDGF stimulation that is not explicable by changes in the number of PDGF receptors at the cell surface. This has been well demonstrated in studies utilizing fibroblasts grown in collagen gels to which they attached and spread (Lin and Grinell, 1993; Tingstrom et al., 1992). When the gels were allowed to retract, so that the attached cells were less spread, the cells became refractory to PDGF stimulation, although there was no significant reduction in PDGF receptor concentration at the cell surface. There is also evidence that the expression and activation of integrins and growth factor receptors may be functionally related. Increases in cell–ECM association can increase a cell's ability to be stimulated by growth factors that protect against apoptosis, and that increased activation of growth factor receptors can increase expression of integrins and increase cell spreading. For example, PDGF stimulates $\alpha2\beta1$ integrin synthesis (Ahlen and Rubin, 1994), which promotes cell adhesion and spreading on type I collagen, and in reciprocation, attachment, and spreading of cells on type I collagen, fibronectin or anti-integrin-coated substrata stimulate tyrosine phosphorylation of PDGF-B receptors (Sundberg and Rubin, 1996).

Cell adhesion to the ECM also has a profound influence on the mechanical environment of the cells, and mechanical factors may also play a role in regulating apoptosis. For example, mechanically straining fetal rat lung cells increases tyrosine kinase activity in the cells, as well as generating increased concentrations of diacylglycerol and inositol triphosphate (Liu et al., 1996). Although not specifically examined in the rat lung cell model, these changes in kinase activity and modifications in concentration of second messenger have been shown to have important regulatory influences on apoptosis in many other cell types. In a model using cardiac myocytes, the relationship between mechanical loads, generation of signaling molecules, and the occurrence of apoptosis was examined more explicitly and showed that myocytes that were strained beyond their physiological range experienced a tremendous increase in the rate of apoptosis that was attributable to an increase in nitric oxide synthesis by the cells (Cheng et al., 1995).

IDENTIFICATION OF CELL–ECM INTERACTIONS AS A REGULATOR OF APOPTOSIS *IN VIVO*

Despite the intuitive appeal of cell–ECM interactions acting as a regulator of apoptosis for cells *in vivo*, it is difficult to demonstrate conclusively that apoptosis is playing a significant regulatory role in most *in vivo* systems. These limitations equally hinder investigations of the role of cell–ECM interactions in regulating apoptosis as investigations of other studies of apoptosis *in vivo*.

A particular limitation to convincingly demonstrating apoptosis as a regulatory mechanism *in vivo* is that it is a rapid process that leaves only subtle traces. Analysis of tissue *in vivo* for the presence of apoptosis generally relies on: (1) the morphological characteristics, especially the electron microscopic assessment for changes in chromatin organization; (2) histochemical analysis, especially the use of terminal deoxynucleotidyl transferase–mediated end labeling of apoptotic DNA fragments using a microscopically discernible label (the TUNEL technique; Gavrieli et al., 1992); or (3) biochemical analysis, in which DNA is extracted and fragments separated by size to test for internucleosomal cleavage of chromosomes that indicates apoptosis ("DNA ladders"). An ideal experimental approach would employ each assay to substantiate the occurrence of apoptosis in the tissue. However, the observation of an apoptotic nucleus by electron microscopy or TUNEL labeling is not adequate to prove that apoptosis plays a significant regulatory role in the tissue. Similarly, the absence of evidence of DNA ladders in gels is not adequate to prove that apoptosis is not functionally important in tissue because of limits of sensitivity of the technique (Trump and Berezesky, Chapter 2; Zakeri, Chapter 3; Benitez-Bribiesca, Chapter 17, this volume).

Each of the most useful assays for apoptosis *in vivo* is limited either by its sensitivity or by sample size, because apoptosis is detectable for only a brief period in many systems, and apoptosis is not always synchronized in populations of cells. For example, Gavrieli and his colleagues (1992) determined that dexamethasone-stimulated thymocytes are only detectably apoptotic by TUNEL labeling for a period of 1 to 3 hours following the onset of apoptosis. Based on that determination, one can calculate that an entire tissue could die via apoptosis in the period of 20 days if only 0.4% of the nuclei in the tissue were detectably apoptotic by TUNEL labeling at any given time. This calculation relies on the assumptions that there was a 2-hour period of detectability for each apoptotic event, that apoptosis occurred at a constant rate over the 20 day period, and that no mitosis occurred during the period. This low frequency of appearance of apoptotic nuclei in *in vivo* systems, even when apoptosis is an important regulatory event, renders the phenomenon easy to overlook. For example, if nuclei made up 5% of the total volume of the tissue, the nuclei were typically 4 μm in diameter, and the tissue was examined in 10 μm sections that measured 0.5 by 0.5 cm, only 4 apoptotic nuclei would be TUNEL labeled in the typical section. Although this may seem a puny contribution to the entire population of nuclei, this small component can reflect important events in the regulation of the tissue through apoptosis.

Electron microscopic assays for apoptosis *in vivo* is daunting because of the tremendous sampling that must occur to establish whether apoptosis is occurring at a rate that is large enough to play a significant regulatory role in the tissue. In the sample tissue discussed, the volume of each section sampled is approximately 500 times greater than the typical section for electron microscopy; thus an apoptotic nucleus would appear in fewer than 1 out of

100 random sections. Obtaining a sufficient sample size by electron microscopy to establish the actual frequency of apoptosis would be a huge task.

Analysis of apoptosis by assaying for internucleosomal cleavage of DNA, which produces "DNA ladders" in electrophoretic separation, circumvents sampling limitations because the DNA from the entire tissue can be sampled at one time. However, this assay is limited by sensitivity, because the great majority of DNA in the tissue will be nonapoptotic. Sensitivity can be improved by 3' end labeling extracted DNA with $[\alpha^{32}P]$-ddATP using TdT (Tilly and Hsueh, 1993), which increases detection of DNA fragments. However, we have found when using 3' end labeling extracted DNA that the limit of detection of apoptotic DNA occurs between 0.5 and 1.5% of the total nuclei experiencing apoptosis at any given time (Tidball et al., 1995). Thus the example of an apoptotic system described in which 0.4% of the nuclei are apoptotic at any instant would be undetected in DNA ladders.

Overall, TUNEL labeling provides the best currently available assay to test the possibility that perturbations of cell–ECM interactions influence the occurrence of apoptosis *in vivo*, in that it is adequately sensitive, it is quantifiable, and it permits identification of the specific cell and the cell's location at the time of apoptosis. Thus testable hypotheses can be framed, such as: Does the frequency of occurrence of apoptosis increase when a specific cell type loses contact with the ECM during a specific event? These correlative, *in vivo* observations can then give rise to more mechanistic studies, in which specific components of the cell–ECM association are perturbed either *in vitro* or *in vivo*, to test their role in regulating apoptosis.

SUMMARY

The importance of cell–ECM interactions in regulating so many vital functions in cells provided the foundation for studies that have now shown these interactions to be equally important in regulating apoptosis. Currently, the only well-characterized cell–ECM interactions that regulate apoptosis are integrin mediated; however, this may reflect the extensive current research interest in integrin-mediated signaling, rather than the absence of other important cell–ECM interactions in regulating apoptosis. Our current knowledge provides the most extensive evidence for a role for cell–ECM interactions in regulating apoptosis *in vivo* in pathological cells, or in tissues that are undergoing a rapid elimination of large numbers of cells as a result of a prominent event in differentiation. This should not be taken as evidence that other important systems in which cell–ECM interactions regulate apoptosis do not also exist. Instead, it may reflect current technical limitations to identifying less flagrant instances of apoptosis *in vivo*. In view of the ancient origins of apoptosis as a regulatory process, it seems likely that as we broaden our studies of systems in which cell–ECM associations regulate apoptosis, we will find great diversity and breadth to the mechanisms and systems in which cell–ECM interactions can regulate this process.

REFERENCES

Adams J, Watt F (1993): Regulation of development and differentiation by the extracellular matrix. Development 117:1183–1198.

Ahlen K, Rubin K (1994): Platelet-derived growth factor-BB stimulates synthesis of the integrin alpha 2-subunit in human diploid fibroblasts. Exp Cell Res 215:347–353.

Albelda SM, Mette SA, Elder DE, Stewart R, Damjanovich L, Herlyn M, Buck CA (1990): Integrin distribution in malignant melanoma: association of the beta 3 subunit with tumor progression. Cancer Res. 50:6757–6764.

Albrecht DE, Tidball JG (1996): PDGF and basement membrane proteins rescue muscle cells from apoptosis. Basic Appl Myology 6:297.

Bates RC, Buret A, van Helden DF, Horton MA, Burns GF (1994): Apoptosis induced by inhibition of intercellular contact. J Cell Biol 125:403–415.

Benecke BJ, Ben-Ze'ev A, Penman S (1978): The control of mRNA production, translation and turnover in suspended and reattached anchorage-dependent fibroblasts. Cell 14:931–939.

Ben Ze'ev A, Farmer SR, Penman S (1980): Protein synthesis requires cell surface contact while nuclear events respond to cell shape in anchorage dependent fibroblasts. Cell 21:365–372.

Boudreau N, Sympson CJ, Werb Z, Bissell MJ (1995): Suppression of ICE and apoptosis in mammary epithelial cells by extracellular matrix. Science 267:891–893.

Brooks PC, Montgomery AMP, Rosenfeld M, Reisfeld RA, Hu T, Klier G, Cheresh DA (1994): Integrin $\alpha v \beta 3$ antagonists promote tumor regression by inducing apoptosis of angiogenic blood vessels. Cell 79:1157–1164.

Burridge K, Turner CE, Romer LH (1992): Tyrosine phosphorylation of paxillin and pp125FAK accompanies cell adhesion to extracellular matrix: a role in cytoskeletal assembly. J Cell Biol 119:893–903.

Cheresh DA (1991): Structure, function and biological properties of integrin alpha v beta 3 on human melanoma cells. Cancer Metastasis Rev 10:3–10.

Chen LB, Gallimore PH, McDougall JK (1976): Correlation between tumor induction and the large external transformation sensitive protein on the cell surface. Proc Natl Acad Sci USA 73:3570–3574.

Cheng W, Li B, Kajstura J, Li P, Wolin MS, Sonnenblick EH, Hintze TH, Olivetti G, Anversa P (1995): Stretch-induced programmed myocyte cell death. J Clin Invest 96:2247–2259.

Clarke AS, Lotz MM, Chao C, Mercurio AM (1995): Activation of the p21 pathway of growth arrest and apoptosis by the $\beta 4$ integrin cytoplasmic domain. J Biol Chem 270:22673–22676.

Crouch DH, Fincham VJ, Frame MC (1996): Targeted proteolysis of the focal adhesion kinase pp125FAK during c-myc-induced apoptosis is suppressed by integrin signaling. Oncogene 12:2689–2696.

Ervasti, JM, Campbell KP (1993): Dystrophin and the membrane skeleton. Curr Opin Cell Biol 5:82–87.

Felding-Habermann B, Mueller BM, Romerdahl CA, Cheresh DA (1992): Involvement of integrin alpha v gene expression in human melanoma tumorigenicity. J Clin Invest 89:2018–2022.

Folkman J, Moscona A (1978): Role of cell shape in growth control Nature 273:345–349.

Frisch SM, Francis H (1994): Disruption of epithelial cell-matrix interactions induces apoptosis. J Cell Biol 124:619–626.

Frisch SM, Vuori K, Ruoslahti E, Chan-Hui P-Y (1996): Control of adhesion-dependent cell survival by focal adhesion kinase. J Cell Biol 134:793–799.

Gavrieli Y, Sherman Y, Ben-Sasson SA (1992): Identification of programmed cell death in situ via specific labeling of nuclear DNA fragmentation. J Cell Biol 119:493–501.

Hockenberry D, Zutter M, Hickey W, Nahm M, Korsmeyer SJ (1991): Bcl-2 protein is topographically restricted in tissues characterized by apoptotic cell death. Proc Natl Acad Sci USA 88:6961–6965.

Howlett AR, Bailey N, Damsky C, Petersen OW, Bissel MJ (1995): Cellular growth and survival are mediated by β1 integrins in normal human breast epithelium but not in breast carcinoma. J Cell Sci 108:1945–1957.

Hynes, RO (1992): Integrins: versatility, modulation and signaling in cell adhesion. Cell 69:11–25.

Kobayashi S, Nishimura J, Kanaide H (1994): Cytosolic Ca2+ transients are not required for platelet-derived growth factor to induce cell cycle progression of vascular smooth muscle cells in primary culture. Actions of tyrosine kinase. J Biol Chem 269:9011–9018.

Kornberg LJ, Earp HS, Parsons JT, Schaller M, Juliano RL (1992): Cell adhesion or integrin clustering increases phosphorylation of a focal adhesion-associated tyrosine kinase. J Biol Chem 267:23439–23442.

Kraiss LW, Raines E, Wilcox JN, Seifert RA, Barrett TB, Kirkman TR, Hart CE, Bowen-Pope DF, Ross R, Clowes AW (1993): Regional expression of the platelet-derived growth factor and its receptor in a primate graft model of vessel wall assembly. J Clin Invest 92:338–348.

Kulkarni GV, McCulloch CAG (1994): Serum deprivation induces apoptotic cell death in a subset of Balb/c 3T3 fibroblasts. J Cell Sci 107:1169–1179.

Laurie GW, Leblond CP, Martin GR (1982): Localization of type IV collagen, laminin, heparan sulfate proteoglycan, and fibronectin to the basal lamina of basement membranes. J Cell Biol 95:340–344.

Lin Y, Grinnell F (1993): Decreased level of PDGF-stimulated receptor autophosphorylation by fibroblasts in mechanically relaxed collagen matrices. J Cell Biol 122:663–672.

Liu M, Qin Y, Liu J, Tanswell AK, Post M (1996): Mechanical strain induces pp60src activation and translocation to cytoskeleton in fetal rat lung cells. J Biol Chem 271:7066–7071.

Matsuda R, Nishikawa A, Tanaka H (1995): Visualization of dystrophic muscle fibers in mdx mouse by vital staining with evans blue: evidence of apoptosis in dystrophin-deficient muscle. J Biochem 118:959–964.

Mayne R, Sanderson RD (1985): The extracellular matrix of skeletal muscle. Collagen Rel Res 5:449–468.

McDonnell TJ, Deane N, Platt FM, Nunez G, Jaegaer U, McKearn JP, Korsmeyer SJ (1989): Bcl-2-immunoglobulin transgenic mice demonstrate extended B cell survival and follicular lymphoproliferation. Cell 57:79–88.

McNamee, HP, Ingber DE, Schwartz MA (1993): Adhesion to fibronectin stimulates inositol lipid synthesis and enhances PDGF-induced inositol lipid breakdown. J Cell Biol 121:673–678.

Meredith JE, Fazeli B, Schwartz MA (1993): The extracellular matrix as a cell survival factor. Mol Biol Cell 4:953–961.

Miyamoto S, Termoto H, Coso OA, Gutkind JS, Burbelo PD, Akiyama SK, Yamada KM (1995): Integrins can collaborate with growth factors for phosphorylation of receptor tyrosine kinases and MAP kinase activation: roles of integrin aggregation and occupancy of receptors. J Cell Biol 131:791–805.

Montgomery, AMP, Reisfeld RA, Cheresh DA (1994): Integrin alpha v beta 3 rescues melanoma cells from apoptosis in three-dimensional dermal collagen. Proc Natl Acad Sci USA 91:8856–8860.

Nunez G, London L, Hockenbery D, Alexander M, McKearn JP, Korsmeyer SJ (1990): Deregulated bcl-2 gene expression selectively prolongs survival of growth factor-deprived hemopoietic cell lines. J Immunol 144:3602–3610.

Ocalan M, Goodman SL, Kuhl U, Hauschka SD, von der Mark K (1988): Laminin alters cell shape and stimulates motility and proliferation of murine skeletal myoblasts. Develop Biol 125:158–167.

Pullan S, Wilson J, Metcalfe A, Edwards GM, Goberdhan N, Tilly J, Hickman JA, Dive C, Streuli CH (1996): Requirement of basement membrane for the suppression of programmed cell death in mammary epithelium. J Cell Sci 109:631–642.

Re F, Zanetti A, Sirone M, Polentarutti N, Lanfrancone L, Dejana E, Colotta F (1994): Inhibition of anchorage-dependent cell spreading triggers apoptosis in cultured human endothelial cells. J Cell Biol 127:537–546.

Schollmann C, Grugel R, Tatje D, Hoppe J, Folkman J, Marme D, Weich HA (1992): Basic fibroblast growth factor modulates the mitogenic potency of the platelet-derived growth factor (PDGF) isoforms by specific upregulation of the PDGF alpha receptor in vascular smooth muscle cells. J Biol Chem 267:18032–18039.

Smith J, Fowkes G, Schofield PN (1995): Programmed cell death in dystrophic (mdx) muscle is inhibited by IGF-II. Cell Death Diff 2:243–251.

Stoker MGP, O'Neill C, Berryman S, Waxman V (1968): Anchorage and growth regulation in normal and virus-transformed cells. Int J Cancer 3:683–693.

Strange R, Li F, Saurer S, Burkhardt A, Friis RR (1992): Apoptotic cell death and tissue remodeling during mouse mammary gland involution. Development 115:49–58.

Sundberg C, Rubin K (1996): Stimulation of $\beta 1$ integrins on fibroblasts induces PDGF independent tyrosine phosphorylation of PDGF β-receptors. J Cell Biol 132:741–752.

Talhouk RS, Bissel MJ, Werb Z (1992): Coordinated expression of extracellular matrix-degrading proteinases and their inhibitors regulates mammary epithelial function during involution. J Cell Biol 118:1271–1282.

Tidball JG, Albrecht DE, Lokensgard B, Spencer MJ (1995): Apoptosis precedes necrosis of dystrophin-deficient muscle. J Cell Sci 108:2197–2204.

Tilly JL, Hseuh AJW (1993): Microscale autoradiographic method for the qualitative and quantitative analysis of apoptotic DNA fragmentation. J Cell Physiol 154:519–526.

Timpl R, Brown JC (1996): Supramolecular assembly of basement membranes. BioEssays 18:123–132.

Tingstrom A, Heldin C-H, Rubin K (1992): Regulation of fibroblast-mediated collagen gel contraction by platelet-derived growth factor, interleukin-1 alpha and transforming growth factor-beta-1. J Cell Sci 102:315–322.

Tucker RW, Butterfield CE, Folkman J (1981): Interaction of serum and cell spreading affects the growth of neoplastic and non-neoplastic fibroblasts. J Supramol Struc Cell Biochem 15:29–40.

Vachon PH, Loechel F, Xu H, Wewer UM, Engvall E (1996): Merosin and laminin in myogenesis; specific requirement for merosin in myotube stability and survival. J Cell Biol 134:1483–1497.

Vaux DL, Cory S, Adams JM (1988): Bcl-2 gene promotes haemopoietic cell survival and cooperates with c-myc to immortalize pre-B cells. Nature 335:440–442.

von der Mark K, Ocalan M (1989): Antagonistic effects of laminin and fibronectin on the expression of the myogenic phenotype. Differentiation 40:150–157.

Walker NI, Bennett RE, Kerr JFR (1989): Cell death by apoptosis during involution of the lactating breast in mice and rats. Amer J Anat 185:19–32.

Zhang Z, Vuori K, Reed JC, Ruoslahti E (1995): The $\alpha 5 \beta 1$ integrin supports survival of cells on fibronectin and up-regulates Bcl-2 expression. Proc Natl Acad Sci USA 92:6161–6165.

PART V

THE CLINICAL RELEVANCE
OF APOPTOSIS

CHAPTER 16

CELL DEATH AND SPECIES PROPAGATION: MOLECULAR AND GENETIC ASPECTS OF APOPTOSIS IN THE VERTEBRATE FEMALE GONAD

JONATHAN L. TILLY

Department of Obstetrics, Gynecology and Reproductive Biology, Harvard Medical School, and The Vincent Center for Reproductive Biology, Massachusetts General Hospital, VBK137E—GYN, 55 Fruit Street, Boston, Massachusetts 02114

INTRODUCTION

Survival and propagation of the species is arguably the most fundamental driving force in the function of all multicellular organisms. In most vertebrates, from the moment primordial germ cells (PGCs) are formed in the early embryo, cellular survival mechanisms must be put in place to ensure that these cells successfully migrate from the allantois along the hindgut mesentery to a specialized fetal tissue, referred to as the genital ridge, that ultimately develops into the adult gonad (reviewed in Peters, 1970; Byskov, 1986). It is believed that local endocrine or paracrine signals, possibly secreted by the genital ridges, "guide" the PGCs via chemotaxis to the developing indifferent gonads (Godin et al., 1990, 1991). Furthermore, the survival of PGCs during migration is thought to be conveyed, at least in part, via the actions of stem cell growth factor (SCF; also referred to as *steel* factor/SF, multipotent growth factor/MGF or kit-ligand/KL) produced by the somatic cells along the migratory route (Matsui et al., 1990), although it appears unlikely that SCF also functions as the chemotactic factor (Godin et al., 1990, 1991). Nevertheless, it is believed that any PGCs straying too far from this rich, but localized, source of survival factors during migration are rapidly lost via apoptosis (see

When Cells Die, Edited by Richard A. Lockshin, Zahra Zakeri, and Jonathan L. Tilly
ISBN 0-471-16569-7 © 1998 Wiley-Liss, Inc.

the following and Tidsall and Albrecht, Chapter 15; Osborne, Chapter 8, this volume). However, a relatively large number of PGCs (in the human, this number is approximately 1000) successfully complete this migration and colonize the genital ridge to initiate successive mitotic divisions (reviewed in Nicosia, 1983; Byskov, 1986). This process allows for rapid clonal expansion of the germline prior to meiotic cell division that generates the haploid gametes in each sex. Importantly, apoptosis impacts on future reproductive potential not only at the level of PGC migration, since this process of controlled cell elimination is also a prominent feature of mitotic and meiotic cell divisions in the germline during gonadal development and function (Coucouvanis et al., 1993; Pesce and De Felici, 1994). In fact, it has been estimated that one-half to two-thirds of male germ cells produced during spermatogenesis normally undergo apoptosis (Roosen-Runge, 1964; Huckins, 1978; Allan et al., 1987; Miething, 1992). Similarly, over two-thirds of the female germ cell pool is lost via apoptosis by the time of birth (Block, 1953; Beaumont and Mandl, 1961; Borum, 1961; Pinkerton et al., 1961; Baker, 1963; Forabosco et al., 1991), and only 5–6% of the potential germ cell stockpile in women remains by the time of puberty (reviewed in Tilly and Ratts, 1996). Consequently, characterization of the regulatory pathways responsible for directing germ cell fate during fetal and post-natal life has become one of the most important research areas in reproductive biology. However, since the majority of experimental data concerning physiological cell death in the gonads have been derived from studies of the ovary, this chapter will not specifically cover apoptosis in the context of spermatogenesis and testicular function (Zakeri, Chapter 3, this volume). If interested, the reader is referred to a recent review of this topic (Dunkel et al., 1997), as well as to several pertinent publications (Russell and Clermont, 1977; Allan et al., 1987; Miething, 1992; Tapanainen et al., 1993; Sinha Hikim et al., 1995).

The occurrence of physiological cell death in the ovary is by no means a new concept, as the first observations in this regard were made in 1885 by Walther Flemming, who described a process of "chromatolysis" as being responsible for the degeneration of inner epithelial (granulosa) cells in rabbit ovarian follicles. Morphological features of these cells included cellular and nuclear condensation followed by fragmentation (Flemming, 1885), key events now known to be hallmarks of the physiological cell death process defined as apoptosis by Kerr and colleagues almost nine decades later (Kerr et al., 1972). With more sophisticated techniques available in recent years for the assessment of apoptosis, considerable progress has been made in understanding the molecular basis of cellular demise in the ovary (reviewed in Tilly, 1996; Tilly et al., 1997). At present, the most well-studied episode of apoptosis in the female gonad concerns the loss of follicles by the degenerative process of atresia. It is now well-established that apoptosis is the primary mechanism responsible for the initiation and progression of atresia, and in most species studied this process is first observed in the somatic granulosa cells that support and nourish the oocyte until its release from the follicle at ovulation (reviewed in Tilly, 1996). By comparison, efforts to understand the role and regulation of apoptosis in the female germ cell have been minimal. The attrition of oogonia and oocytes during the perinatal period is no less important than post-natal follicular atresia to the overall function of the ovary later in life. In fact, since two-thirds of a woman's potential germ cell stockpile is lost as a direct consequence of apoptosis of oogonia and oocytes prior to birth, the timing of reproductive senescence (e.g., exhaustion of the follicle pool and the menopause) is probably more affected by the magnitude of this germ cell attrition as opposed to the loss of the majority of remaining germ cells via follicular atresia following birth (Tilly and Ratts, 1996). Another key issue surrounding the fate of the germ cell lineage concerns pathological destruction of oocytes, and the ensuing acceleration of reproductive senescence in females. Therefore, the final section of this chapter will briefly cover, among other things, the emerging role of apoptosis and cell death–associated signaling events in primary reproductive failure resulting from exposure of the ovary to chemotherapeutic drugs or environmental/industrial toxicants (Benitez-Bribiesca, Chapter 17, this volume).

ENDOCRINE REGULATION OF APOPTOSIS IN THE OVARY

General Concepts

As would be expected of a major endocrine organ, the control of apoptosis in the ovary by systemic and locally derived (e.g., ovarian) hormones is extremely complex (reviewed in Tilly, 1996). Essentially all of the cell death observed in the female gonad during reproductive life is driven by the menstrual/estrous cycle-related changes in hormonal secretion patterns occurring within the pituitary and the ovary. Much of the hypophyseal–gonadal interplay is critically important for two events: (1) release of a healthy, mature ovum at ovulation for potential fertilization; and (2) establishment of an optimum uterine environment capable of supporting blastocyst implantation and fetal gestation following ovulation if fertilization occurs. Since this "cross-talk" between organs of the female reproductive tract represents such an intriguing example of how controlled cell death in one tissue directly modulates apoptosis in a second tissue, a few comments will be made here about this concept to underscore the significance and complexity of the endocrine control of physiological cell death in the female reproductive system.

Each menstrual cycle in the human female is characterized initially by the gonadotropin-dependent growth of a cohort of rapidly maturing follicles during the follicular phase of the cycle (reviewed in Knobil, 1980; Hodgen, 1982; Gougeon, 1996). By as yet incompletely described mechanisms, one follicle is selected for continued development to the preovulatory stage (e.g., the dominant follicle), whereas the remaining or subordinate follicles of this cohort begin the degenerative process of atresia (reviewed in Hodgen, 1982), characterized by granulosa cell apoptosis (reviewed in Tilly, 1996). During this phase of the cycle, increased secretion of estradiol-17β from the ovary, most notably from granulosa cells of the dominant follicle once selected, leads to a substantial thickening of the uterine endometrium due to the potent effects of estrogens acting either directly or indirectly via local growth factors on uterine glandular epithelial cell proliferation (Murphy and Gahary, 1990; Nelson et al., 1992). This estrogen-dependent communication between the ovary and the uterus is the first event of many that must be set in motion for successful pregnancy to ensue if fertilization occurs.

Once ovulation is triggered by the surge of luteinizing hormone (LH) from the pituitary gland, the remaining cells of the ruptured dominant follicle rapidly undergo the process of luteinization to form the corpus luteum (reviewed in Niswender and Nett, 1988; Patton and Stouffer, 1991; Smith et al., 1994). This event demarcates the beginning of the luteal phase of the cycle, and a switch from an estrogen-dominant to progesterone-dominant endocrine environment in the body (reviewed in Knobil, 1980; Niswender and Nett, 1988). The primary function of the corpus luteum is the secretion of substantial levels of progesterone, and to a lesser degree estradiol-17β, required for continued growth of the uterine lining and differentiation of the glandular epithelial cells into secretory cells (reviewed in Niswender and Nett, 1988). Both of these events are critically important for successful embryo implantation (reviewed in de Ziegler and Bouchard, 1993; Edwards, 1995). If fertilization of the ovum does not take place, the corpus luteum possesses a finite lifespan such that regression of this ovarian structure will be initiated and is accomplished via the induction of apoptotic cell death (Juengel et al., 1993; Dharmarajan et al., 1994; Rueda et al., 1995a,b; Shikone et al., 1996). The process of luteolysis yields a rapid loss of ovarian steroid support to the uterus, and this in turn leads to an induction of apoptosis in the uterine endometrial cells at the end of the secretory phase (Hopwood and Levison, 1975; Kokawa et al., 1996; Tao et al., 1997), followed by menses.

However, if fertilization occurs in the human female, the developing trophoblastic cells synthesize and secrete large amounts of a luteal cell survival factor critical for signaling the corpus luteum that a conceptus is present in the reproductive tract and that apoptosis should not be initiated (see the following, and Patton and Stouffer, 1991). The continued production of progesterone and estrogen by the "rescued" corpus luteum of pregnancy then likely ensures the maintenance of the endometrium by, among other things, inhibiting activation of apoptosis in the glandular epithelial cells (Terada et al., 1989; Rotello et al., 1992). Without this dynamic communication between organs of the female reproductive tract, a process controlled in large part by the precise endocrine regulation of survival or death in various cell types at specific points in the cycle, the endometrium would be quickly shed leading to pregnancy loss. Keeping these general comments in mind, the remainder of this chapter will focus specifically on the occurrence, regulation, and significance of physiological cell death in the ovary.

Endocrine Control of Ovarian Somatic Cell Death

The majority of available research data concerning the hormonal regulation of apoptosis in the ovary have been derived from studies of granulosa cells or isolated follicles in the context of follicular atresia. In this regard, the function of the post-natal ovary, as with the testis, is primarily controlled by the gonadotropins, follicle-stimulating hormone (FSH) and LH, secreted from the pituitary gland. In response to changes in serum levels of FSH and LH during the menstrual or estrous cycle, the ovary produces a number of steroids and bioactive peptides (i.e., activin, inhibin) that serve as positive and negative feedback regulators of pituitary secretory function, either directly via actions on gonadotrophs or indirectly by affecting hypothalamic release of gonadotropin-releasing hormone that in turn stimulates hypophyseal gonadotropin secretion (reviewed in Knobil, 1980; Hodgen, 1982). In addition to pronounced effects of gonadotropins on follicular development and granulosa cell differentiation (Richards, 1980, 1994; Hirshfield, 1991), FSH and LH serve important roles as primary survival factors for ovarian follicles. Evidence to support this observation has been derived from both *in vivo* and *in vitro* approaches, and in all studies reported thus far a gonadotropin-rich environment prevents apoptosis in granulosa cells and the ensuing onset of follicular atresia (Tilly et al., 1992, 1995a; Chun et al., 1994; Tilly and Tilly, 1995).

In the rodent ovary, the primary experimental system used to study the occurrence and regulation of ovarian cell fate, there apparently exists a complex stromal cell–epithelial cell interaction that mediates many of the actions of gonadotropins, including the anti-apoptotic effects, within the follicle (reviewed in Skinner and Parrot, 1994; Tilly, 1996; Tidball and Albrecht, Chapter 15, this volume). As is the case with the endocrine regulation of apoptosis in a number of major organ systems, the mesenchymal-derived cells of the follicle (theca-interstitial cells) are the primary source of several growth factors thought to be important for survival of the adjacent epithelial-derived cells (granulosa cells). In addition, an important role for insulin-like growth factor-I (IGF-I) as a "master coordinator" of the actions of gonadotropins within the follicle has been reported (Chun et al., 1994), and there also clearly exists an important function for locally produced IGF-binding proteins in titering the bioactivity of IGF-I (Holly et al., 1990; Chun et al., 1994; Flaws et al., 1995a; Peng et al., 1996). It should be mentioned that innervation of the follicle may also be important in the process of follicular selection (granulosa cell survival) versus atresia (granulosa cell death), as evidenced by the ability of vasoactive intestinal peptide, an intrafollicular neuropeptide, to mimic the actions of gonadotropins in promoting granulosa cell survival *in vitro* in diverse species (Flaws et al., 1995a).

Interestingly, in the case of the small percentage of follicles that, for as yet unknown reasons, possess the inherent capacity to resist atresia and survive to the preovulatory stage of development, the fate of the somatic cells within the follicle post-ovulation apparently remains apoptotic death. As discussed earlier, the remaining follicular tissue, following release of the ovum into the reproductive tract for potential fertilization, quickly transforms into the "progesterone-producing factory," termed the corpus luteum. If pregnancy does not occur, the corpus luteum initiates its own program of physical destruction (luteolysis) at the end of each menstrual/estrous cycle that involves the induction of apoptosis (Juengel et al., 1993; Dharmarajan et al., 1994; Rueda et al., 1995a,b; Shikone et al., 1996). The endocrine regulation of this instance of ovarian cell death is not well described, although it is known that human chorionic gonadotropin (hCG) produced by the trophoblast can act as a luteal cell survival factor (Dharmarajan et al., 1994). As such, in those species that produce CG during conceptus development and fetal gestation, the "rescue" of the corpus luteum of pregnancy likely involves the anti-apoptotic actions of this gonadotropin in luteal cells. Considering that luteal-derived progesterone is the primary factor responsible for ensuring uterine endometrial integrity during blastocyst implantation and gestation, there is an obligate necessity for corpus luteum survival in the maintenance of early pregnancy in the human (reviewed in Patton and Stouffer, 1991). It is important to stress, however, that the process of luteolysis is complex and is probably regulated by a plethora of endocrine factors in addition to (or in some species, a lieu of) CG, including cytokines, prostaglandins, and steroids (reviewed in Niswender and Nett, 1988; Pate, 1994; Zakeri, Chapter 3, this volume).

Hormonal Regulation of Apoptosis in Germ Cells

In contrast to ovarian somatic cells, comparatively little is known of the endocrine signaling events involved in modulating the fate of germ cells in the developing ovary during fetal life. It is now well accepted that fetal germ cell attrition *in vivo* and *in vitro* occurs as a direct consequence of apoptosis. Using *in vitro* approaches, roles for SCF (Dolci et al., 1991; Pesce et al., 1993), leukemia inhibitory factor (LIF; Pesce et al., 1993), and interleukin-4 (Cook et al., 1996) as germ cell survival factors have been established. Additionally, mice that are homozygous null for SCF or its receptor (c-*kit*) exhibit gonadal dysgenesis and infertility (Mintz and Russell, 1957; Besmer et al., 1993), providing *in vivo* data to support a fundamental role for the SCF/c-*kit* system as a critical signaling pathway in normal oogenesis and follicle endowment. By comparison, "gene knockout" of LIF apparently does not result in impaired gonadal development nor fertility problems in mice of either sex, albeit uterine defects related to implantation failure were noted in female mice (Stewart et al., 1992). Thus, although LIF can promote the survival of PGCs *in vitro,* the significance of these findings to the *in vivo* regulatory events involved in fetal oogenesis remains to be elucidated. It should also be mentioned that retinoic acid has recently been identified as a potent stimulator of primordial germ cell proliferation (Koshimizu et al., 1995), although the actions of this vitamin A metabolite in directly modulating germ cell apoptosis have not yet been characterized.

INTRACELLULAR EVENTS INVOLVED IN APOPTOSIS IN THE OVARY

General Comments

Due to the complexity and species specificity that exist in the hormonal regulation of apoptosis in the ovary, data provided by analysis of the endocrine control of the various paradigms of ovarian cell death in animal models may or may not hold true in the human, depending of course upon the hormone(s) in question. For example, the anti-apoptotic role of gonadotropins in ovarian follicles is, in all likelihood, conserved across species in light of the universal dependence of the gonads on these endocrine factors for normal development, function, and homeostasis. However, the reported action of intrafollicular-derived transforming growth factor-α as a paracrine mediator of gonadotropin-promoted follicular survival in the rat (Tilly et al., 1992; Tilly, 1996) may represent an example of a species-specific mechanism for apoptosis regulation in the ovary. As such, redirecting research efforts towards evaluation of the intracellular effectors of physiological cell death may provide data of greater importance to understanding the basis of apoptosis in the ovary in all vertebrate species. This proposal is supported by a solid foundation of research data derived from molecular and genetic analyses of apoptosis in extragonadal tissues and tumor cell lines (reviewed in Korsmeyer, 1995; Wyllie, 1995; Patel et al., 1996; Yang and Korsmeyer, 1996). As will be discussed, these investigations have revealed the existence of a number of genes and intracellular signals conserved through evolution that appear to function as key mediators of cellular survival or death in many tissues, including the ovary.

The *bcl-2* Gene Family

Among the increasing number of cell death effectors that have been identified to date, members of the *bcl-2* (B-cell lymphoma/leukemia-2 protein) gene family have emerged as principal players in the cascade of events that activate or inhibit apoptosis (reviewed in Korsmeyer, 1995; Yang and Korsmeyer, 1996). Members of this gene family, now numbering at least ten and defined by conserved functional domains, can be segregated into two groups: proteins that provide resistance to apoptosis (i.e., BCL-2, BCL-X_{long}, BAG-1, MCL-1, A1) and proteins that accelerate or induce cell death (i.e., BAX, BCL-X_{short}, BAK, BIK, BAD). Through hetero- and homodimeric interactions, proteins encoded by various members of the *bcl-2* gene family are thought to be primary determinants of the susceptibility of a given cell to an apoptotic stimulus (Yang and Korsmeyer, 1996). Although the mechanisms by which BCL-2 and related proteins exert their effects within cells remain to be fully characterized, a recent report has identified an intriguing association between BCL-2 and mitochondrial function (Zamzami et al., 1996). Prior to this series of observations, a fair amount of indirect evidence supporting mitochondrial involvement in apoptosis was provided by investigations of the role of cellular metabolism and the ensuing generation of reactive oxygen species in various paradigms of apoptosis (reviewed in Buttke and Sandstrom, 1994; Briehl and Baker, 1996). In this regard, BCL-2 has been shown to prevent cell death induced by accumulation of oxygen free radicals (Hockenberry et al., 1993; Kane et al., 1993), and this function was attributed to its intracellular localization within the mitochondrial membrane (Hockenberry et al., 1990; Krajewski et al., 1993). Several pieces of new information further implicate mitochondria as direct contributors in the final commitment to initiate apoptosis via activation of the mitochondrial membrane permeability transition (Zamzami et al., 1996; Marchetti et al., 1996). Although the details of this system are beyond the scope of this chapter, it is clear that mitochondrial permeability transition is a crucial, and possibly final, event in the induction of apoptosis (Henkart and Grinstein, 1996; Birge et al., Chapter 13; Trump and Berezesky, Chapter 2, this volume). Furthermore, a direct link between the anti-apoptotic actions of BCL-2 (Zamzami et al., 1996), and more recently the pro-apoptotic actions of BAX (Xiang et al., 1996), and the mitochondrial permeability transition has been established.

In the ovary, many of the apoptosis regulatory proteins encoded by *bcl-2* gene family members have been identified (Tilly et al., 1995a, 1996a; Johnson et al., 1996; Kugu et al., 1997; Rueda et al., 1997), and recent evidence strongly implies a fundamental role for BCL-2 and related proteins in ovarian cell death. For example, increased expression of the death-susceptibility factor, BAX, is positively correlated with apoptosis in granulosa cells (Tilly et al., 1995; Kugu et al., 1998) and luteal cells (Rueda et al., 1997). Moreover, mice with a targeted disruption in the *bax* gene, and thus not able to express functional BAX protein, show a number of phenotypic abnormalities in the ovary, including apparent defects in the normal induction of apoptosis in granulosa cells during atresia (Knudson et al., 1995) and a surfeit oocyte-containing primordial follicles (Knudson et al., 1996). By comparison, mice lacking functional BCL-2 protein possess reduced numbers of primordial follicles relative to their wild-type (e.g., BCL-2-intact) sister littermates (Ratts et al., 1995). These data, coupled with more recent studies documenting resistance of BAX-deficient mouse oocytes to apoptosis induced by exposure to potent ovotoxicants (Perez et al., 1997a,b; reviewed in Tilly et al., 1997), collectively support a critical role for these proteins in ovarian germ cell endowment and depletion under normal and pathophysiological conditions. Whether or not BCL-2 and related family members act in ovarian cells via the mitochondrial permeability transition is unknown, and thus future investigations are certainly needed to directly assess the role, if any, of this newly identified signaling event in ovarian cell death. Nevertheless, understanding the role of *bcl-2* gene family members in normal ovarian function has permitted recent breakthroughs in elucidation of the molecular and genetic events responsible for pathological oocyte destruction, such as that caused by chemotherapy (Perez et al., 1997b; Tilly et al., 1997). The clinical significance of these observations to the maintenance of fertility in young women treated for cancer will be addressed in the final section of this chapter (Benitez-Bribiesca, Chapter 17, this volume).

The p53 Tumor Suppressor Protein

Another central component of the death signaling pathway in many cells, particularly in tumor cells exposed to chemo- or radiotherapy, is the anti-oncogenic protein, p53. This transcription factor represents an interesting example of a bifunctional protein involved in the regulation of both cell proliferation and cell death (reviewed in Ko and Prives, 1996). In the context of apoptosis, the mechanisms by which p53 leads to activation of cellular suicide are ill-defined; however, it is known that p53 can interact with other cell death regulatory factors at the level of transcriptional regulation of their cognate genes. For instance, a p53-response element has been identified in the promoter of the *bax* death-susceptibility gene, yielding higher levels of *bax* expression when this response element in bound and activated (Miyashita and Reed, 1995). Futhermore, the *bcl-2* gene promoter contains a repressor element that binds p53, leading to a suppression of gene expression (Miyashita et al., 1994). In keeping with the reported role of p53 as a pro-apoptotic factor, the combined actions of p53 on the *bax* (increase) and *bcl-2* (decrease) genes would lead to a situation inside the cell favoring greater BAX bioactivity and hence increased death potential (Oltvai et al., 1993; Oltvai and Korsmeyer, 1994; Yang and Korsmeyer, 1996).

In the ovary, nuclear accumulation of p53 protein has been identified in granulosa cells of ovarian follicles destined for atresia (Tilly et al., 1995b), consistent with a role for p53 in promoting apoptosis via enhanced *bax* expression within this population of cells (Tilly et al., 1995a). In addition, follicles exposed to a gonadotropin-rich environment *in vivo,* and thus devoid of apoptosis, do not contain any detectable p53 (Tilly et al., 1995b). Importantly, gene transfer of p53 in granulosa cells has confirmed the functional role of this protein in apoptosis induction in this cell type (Keren-Tal et al., 1995). At present, the stimulus for nuclear translocation of p53 in granulosa cells during follicular atresia is unknown. However, in other cell systems, a primary trigger for p53 stabilization and translocation is DNA damage (Kastan et al., 1991), such as that caused by accumulation of reactive oxygen species (Yu, 1994). Although a potential cause–effect relationship between oxidative stress and p53 translocation in granulosa cells remains to be established, it is known that anti-oxidant enzymes and oxygen free radical scavengers can mimic the ability of FSH to suppress apoptosis in granulosa cells of ovarian follicles (Tilly and Tilly, 1995; Birge et al., Chapter 13). These data, taken with recent evidence for the existence of a FSH- and IGF-I-sensitive ascorbic acid pump in granulosa cells (Behrman et al., 1996) and the reported induction of anti-oxidant gene expression by gonadotropins in the ovary (Tilly and Tilly, 1995), reinforce the view that oxidative stress may be one component of the atresia process. This hypothesis is further supported by a number of observations implicating the unopposed generation of oxygen free radicals as an important component of the death pathway in many cell lineages (reviewed in Buttke and Sandstrom, 1994; Briehl and Baker, 1996), including cells of the corpus luteum during luteolysis (Rueda et al., 1995a).

The *CASP* Gene Family

A third category of conserved cell death regulators is composed of a cohort of intracellular proteases referred to as caspases (cysteine aspartic acid–specific proteases; Alnemri et al., 1996; Cryns and Yuan, Chapter 6; Sikorska and Walker, Chapter 7, this volume) or CED-3/interleukin-1β-converting enzyme (ICE) homologs (reviewed in Martin and Green, 1995; Kumar and Lavin, 1996; Patel et al., 1996). Members of this gene family in vertebrates number at least eleven, and are characterized by two distinct features: a pentapeptide domain (QXCRG) that functions as the active site of the enzyme, and specificity for cleavage of substrate proteins at an aspartic acid residue (reviewed in Kumar and Lavin, 1996; Patel et al., 1996). A number of recent studies have defined a fundamental role for caspases in apoptosis via their ability to cleave a wide spectrum of cellular homeostatic proteins, including cytoskeletal elements, proteins that comprise the nuclear scaffold, and enzymes responsible for DNA repair and mRNA processing (reviewed in Patel et al., 1996; see also Casciola-Rosen et al., 1996). Moreover, caspases generally exist as inactive proenzymes within cells, and recent data suggest that there exists a discrete ordering of these enzymes so that activation of one leads to activation of others through direct catalysis (Enari et al., 1996; Fernandes-Alnemri et al., 1996; Liu et al., 1996a). Importantly, the requirement of these enzymes for cleavage at aspartic acid residues has permitted the design of highly specific inhibitors of caspases, tools that have become fundamental to assessing the role of these enzymes in cellular demise (Nicholson et al., 1995; Cain et al., 1996; Jacobson et al., 1996; reviewed in Patel et al., 1996).

Expression of several members of the *CASP* gene family in the rat (Flaws et al., 1995b), avian (Johnson et al., 1997), and human (Kugu et al., 1998) ovary has been reported, and the levels of mRNA encoding at least two of these pro-apoptosis enzymes are reduced in the rat ovary, *in vivo,* by exogenous gonadotropin treatment (Flaws et al., 1995b). Interestingly, expression of caspase-1 (ICE), the prototypical member of the caspase family in vertebrates first identified to trigger apoptosis by *in vitro* overexpression (Miura et al., 1993), is extremely low in the ovary and is not regulated by gonadotropins (Flaws et al., 1995b). Although arguably caspase-1 could be primarily controlled at the level of enzymatic activity, as opposed to *de novo* gene expression, this does not appear to be the case, since caspase-1 activity is below detectable limits in extracts prepared from either healthy or atretic antral follicles of the rat ovary (Flaws et al., 1995b). Finally, it is important to note that one of the primary products of caspase-1 action, the generation of active IL-1β from the pro-form of the cytokine, has been implicated as an important contributor to apoptosis activation in some cell lineages (Friedlander et al., 1996). In this regard, the role of IL-1β in the ovary remains controversial, as various reports indicate a cytotoxic effect (Hurwitz et al., 1992), an absence of an effect (Flaws et al., 1995b), or an anti-apoptotic effect (Chun et al., 1995) of this cytokine in ovarian cells or follicles *in vitro*. The reasons for the discrepancy in these findings may prove difficult to resolve, but it should be noted that "gene knockout" of caspase-1 in mice apparently produces defects in apoptosis only in select immune cell lineages (Kuida et al., 1995; Li et al., 1995). Furthermore, the absence of functional IL-1β receptors in genetically manipulated mice does not yield abnormalities in normal follicular development or ovulation rates (Abbondanzo et al., 1996), phenotypes that would be predicted if IL-1β were critically important for follicular survival.

Recent studies to inhibit activity of caspases in ovarian cells are more supportive of a role for these enzymes in apoptosis induction in granulosa cells, although some perplexing questions were raised by initial investigations along these lines. Specifically, a complete suppression of internucleosomal DNA cleavage in granulosa cells of rat antral follicles incubated *in vitro* without tropic hormone support was achieved by treating follicles with a putative inhibitor of caspases, sodium aurothiomalate (SAM; Flaws et al., 1995b). Although these data would argue that caspases are critical for the execution of cell death in this model, morphological features of apoptosis in granulosa cells (cellular condensation, nuclear pyknosis, fragmentation into apoptotic bodies) and atresia of the follicles (disorganization of the granulosa cell layer) were maintained in SAM-treated follicles (Flaws et al., 1995b). Subsequent studies have shown that higher-order DNA fragmentation, detectable by pulsed-field but not conventional agarose gel electrophoresis, is widespread in granulosa cells of follicles treated with SAM (Trbovich et al., 1998; Sikorska and Walker, Chapter 7, this volume). These data collectively indicate that, as with other cell systems, higher-order DNA cleavage of rosettes and loops without internucleosomal fragmentation is sufficient for apoptosis to proceed in granulosa cells. In fact, additional studies have demonstrated that SAM is a potent inhibitor of nucleases (Trbovich et al., 1998). Collectively, these findings suggest that extreme caution should be used in the interpretation of experimental data derived from the use of this compound in assessing the role of caspases in apoptosis, particularly when DNA oligonucleosomes are used as the sole criterion for identification of cell death (Kaipia et al., 1996).

More convincing evidence of a functional role for caspases in granulosa cell apoptosis has been derived from the use of specific peptide inhibitors directed against the active site of these enzymes. These studies have revealed that the occurrence of granulosa cell death during atresia of mouse follicles induced *in vitro* by hormone deprivation can be markedly attenuated by peptide inhibitors of caspases (Maravei et al., 1997). Additionally, proteolysis of the cytoskeletal element fodrin, a known target for caspase-3 (CPP32) action during cell death induction (Martin et al., 1995, 1996; Maravei et al., 1997), has been identified in mouse ovarian follicles during atresia (Maravei et al., 1997). Although more work is clearly needed, these data provide clear evidence of a fundamental role for members of the *CASP* gene family in ovarian cell apoptosis (Cryns and Yuan, Chapter 6; Trump and Berezesky, Chapter 2; Tidhall and Albrecht, Chapter 15; Zakeri, Chapter 3, this volume).

Early Signaling Pathways

Before concluding this section, a few points will be made concerning the identity of early or immediate response signals that may be involved in the subsequent activation of the downstream effectors of apoptosis discussed earlier. One of the most well studied of these early signals is the lipid second messenger, ceramide (reviewed in Kolesnick et al., 1994; Hannun, 1996). The production of ceramide in cells occurs via multiple pathways (reviewed in Hannun, 1996), and there is evidence that cell type specificity exists in the mechanisms responsible for generation of this second messenger in response to the same stimulus (Bose et al., 1995; Jaffrézou et al., 1996) (Zakeri, Chapter 3, this volume). Although ceramide can exert many effects within cells, a primary function of ceramide is the induction of apoptotic death (Obeid et al., 1993; reviewed in Kolesnick et al., 1994; Hannun, 1996). In addition to a number of endocrine factors capable of eliciting ceramide generation following hormone–receptor interaction (such as tumor necrosis factor-α/TNF-α and Fas-ligand; reviewed in Kolesnick et al., 1994), recent investigations have also supported the involvement of ceramide in apoptosis in tumor cells exposed to chemotherapeutic drugs (Bose et al., 1995; Jaffrézou et al., 1996). Interestingly, regardless of the stimulus for ceramide generation (e.g., physiologic versus pharmacologic), the downstream events activated by ceramide that lead to cell death appear to involve several enzymes collectively referred to as stress-activated protein kinases (SAPK; Verheij et al., 1996).

In the ovary, little is known about the significance of the ceramide or SAPK signaling pathways in cell death. Several recent studies have shown that *in vitro* treatment of either avian granulosa cells (Witty et al., 1996) or rat ovarian follicles (Kaipia et al., 1996; Tilly et al., 1996; Martimbeau and Tilly, 1997) with a membrane-permeable ceramide analog is capable of inducing apoptosis. However, all the published studies to date concerning ceramide and cell death in the ovary have failed to include an analysis of changes in endogenous ceramide generation following experimental manipulation. As such, future studies are needed to address this issue, particularly in the context of assessing if various endocrine factors, such as TNF-α, actually lead to cell death in the ovary via increased ceramide generation (Kaipia et al., 1996), as opposed to the multitude of other signaling molecules also coupled to the TNF-α receptor (Liu et al., 1996b; Zakeri, Chapter 3, this volume). Both components of another ligand–receptor pathway coupled to ceramide generation, namely, Fas-ligand and Fas, have recently been implicated in ovarian cell turnover. Expression of Fas in granulosa cells and Fas-ligand in germ cells of rat ovarian follicles has been demonstrated (Hakuno et al., 1996). Moreover, in the human ovary increased immunohistochemical staining for Fas was noted in granulosa cells of atretic follicles (Kondo et al., 1996), consistent with a role for this signaling pathway in apoptosis activation during follicle degeneration. Although the precise cellular source of Fas-ligand in this regard remains to be defined, observations from the rat ovary suggest that germ cell–somatic cell cross-talk may be at least partly involved (Hakuno et al., 1996). This hypothesis, as well as the potential involvement of ceramide in Fas-ligand-promoted intracellular signaling in the ovary, awaits further testing. However, it should be noted that Fas activation has been shown to trigger apoptosis in human granulosa–luteal cells maintained *in vitro* (Quirk et al., 1995).

APOPTOSIS AND THE OVARY: IMPLICATIONS AND FUTURE DIRECTIONS

Examination of the available literature suggests that a greater understanding of the occurrence and regulation of apoptosis in the ovary will have profound clinical significance. For example, many instances of premature pregnancy loss may be the result of inappropriate activation of apoptosis in the corpus luteum of pregnancy (e.g., luteal phase defect). Since progesterone produced by the corpus luteum is required for uterine endometrial integrity during the first eight weeks of gestation in the human, a luteal phase defect involving inappropriate activation of apoptosis in this ovarian structure would have dire consequences for early pregnancy maintenance. A second application may be revealed by the development of methods to slow oocyte depletion from the ovary, and thus prevent the natural menopause as well as premature ovarian failure of unknown etiology. Alternatively, activation of controlled cellular deletion in the female gonad could be used to develop new contraceptives, capable of sterilization (induction of germ cell apoptosis), monthly fertility loss (induction of granulosa cell apoptosis and atresia of the maturing follicles), or post-conception pregnancy termination (induction of apoptosis in the corpus luteum).

Finally, knowledge of the regulatory mechanisms involved in ovarian cell death may have a tremendous and long-overdue impact on understanding, and possibly overcoming, pathological conditions leading to oocyte depletion. Unfortunately, research efforts in this area have just been initiated, and thus very little published information exists on the topic. Preliminary investigations have indicated that mouse oocytes exposed *in vitro* to either chemotherapeutic drugs or a class of environmental toxicants referred to as polycyclic aromatic hydrocarbons exhibit typical morphological (cellular condensation, budding, fragmentation) and biochemical (DNA fragmentation) features of apoptosis (Perez et al., 1997a,b; Tilly et al., 1997). Furthermore, using molecular genetic approaches and *in vitro* manipulations, several intriguing observations have now implicated ceramide, BAX, and caspases as integral components of the machinery required for oocyte apoptosis following exposure to chemotherapeutic drugs and other ovotoxicants (Perez et al., 1997a,b). Thus knowledge gained by such investigations may allow the development of new strategies, such as gene therapy, to prevent pathological oocyte loss. Although the clinical application of these and other approaches may not be realized in the near future, it should be emphasized that this information is a critical foundation for such goals. Certainly, without the basic knowledge of what must be overcome when combating the destruction of oocytes in women treated for cancer or exposed to other noxious ovotoxicants, it can be concluded that progress towards this goal would remain essentially nonexistent.

ACKNOWLEDGMENTS

The work conducted by the author and discussed herein was supported by grants from the NIH (R01-AG12279, R55-HD31188, R01-HD34226, R01-ES06999, Office of Research on Women's Health), the American Federation For Aging Research, and the Vincent Memorial Research Fund. The Massachusetts General Hospital Fund for Medical Discovery and The Lalor Foundation are also acknowledged for their support of Postdoctoral Research Fellows in the author's laboratory during the course of these studies.

REFERENCES

Abbondanzo SJ, Cullinan EB, McIntyre K, Labow MA, Stewart CL (1996): Reproduction in mice lacking a functional type 1 IL-1 receptor. Endocrinology 137:3598–601.

Allan DJ, Harmon BV, Kerr JFR (1987): Cell death in spermatogenesis. In Potten CS (ed): Perspectives on Mammalian Cell Death. Oxford (UK):Oxford University Press, pp 229–58.

Alnemri ES, Livingston DJ, Nicholson DW, Salvesen G, Thornberry NA, Wong WW, Yuan J (1996): Human ICE/CED-3 protease nomenclature. Cell 87:171.

Baker TG (1963): A quantitative and cytological study of germ cells in human ovaries. Proc R Soc Lond (Biol) 158:417–33.

Beaumont HM, Mandl AM (1961): A quantitative and cytological study of oogonia and oocytes in the foetal and neonatal rat. Proc R Soc Lond (Biol) 155:557–79.

Behrman HR, Preston SL, Aten RF, Rinuldo P, Zreik TG (1996): Hormone induction of ascorbic acid transport in immature granulosa cells. Endocrinology 137:4316–21.

Besmer P, Manova K, Duttlinger R, Huang EJ, Packer A, Gyssler C, Bachvarova RF (1993): The kit-ligand (steel factor) and its receptor c-kit/W: Pleiotropic roles in gametogenesis and melanogenesis. Development (Supplement) 125–37.

Block E (1953): A quantitative morphological investigation of the follicular system of newborn female infants. Acta Anat 17:201–6.

Borum K (1961): Oogenesis in the mouse: A study of the meiotic prophase. Exp Cell Res 24:495–507.

Bose R, Verheij M, Haimovitz-Friedman A, Scotto K, Fuks Z, Kolesnick RN (1995): Ceramide synthase mediates daunorubicin-induced apoptosis: An alternative mechanism for generating death signals. Cell 82:405–14.

Briehl MM, Baker AF (1996): Modulation of the antioxidant defence as a factor in apoptosis. Cell Death Differ 3:63–70.

Buttke TM, Sandstrom PA (1994): Oxidative stress as a mediator of apoptosis. Immunol Today 15:7–10.

Byskov AG (1986): Differentiation of the mammalian embryonic gonad. Physiol Rev 66:71–117.

Cain K, Inayat-Hussain SH, Couet C, Cohen GM (1996): A cleavage-site-directed inhibitor of interleukin-1β-converting enzyme-like proteases inhibits apoptosis in primary cultures of rat hepatocytes. Biochem J 314:27–32.

Casciola-Rosen L, Nicholson DW, Chong T, Rowan KR, Thornberry NA, Miller DK, Rosen A (1996): Apopain/CPP32 cleaves proteins that are essential for cellular repair: A fundamental principle of apoptotic death. J Exp Med 183:1957–64.

Chun S-Y, Billig H, Tilly JL, Furuta I, Tsafriri A, Hsueh AJW (1994): Gonadotropin suppression of apoptosis in cultured preovulatory follicles: Mediatory role of endogenous insulin like growth factor I. Endocrinology 135:1845–53.

Chun S-Y, Eisenhauer KM, Kubo M, Hsueh AJW (1995): Interleukin-1β suppresses apoptosis in rat ovarian follicles by increasing nitric oxide production. Endocrinology 136:3120–27.

Cooke JE, Heasman J, Wylie CC (1996): The role of interleukin-4 in the regulation of mouse primordial germ cell numbers. Develop Biol 174:14–21.

Coucouvanis EC, Sherwood SW, Carswell-Crumpton C, Spack EG, Jones PP (1993): Evidence that the mechanism of prenatal germ cell death in the mouse is apoptosis. Exp Cell Res 209:238–47.

de Ziegler, Bouchard P (1993): Understanding endometrial physiology and menstrual disorders in the 1990s. Curr Opin Obstet Gynecol 5:378–88.

Dharmarajan AM, Goodman SB, Tilly KI, Tilly JL (1994): Apoptosis during functional corpus luteum regression: Evidence of a role for chorionic gonadotropin in promoting luteal cell survival Endocrine J (Endocrine) 2:295–303.

Dolci S, Williams DE, Ernst MK, Resnick JL, Brannan CI, Lock LF, Lyman SD, Boswell S, Donovan PJ (1991): Requirement for mast cell growth factor for primordial germ cell survival in culture. Nature 352:809–11.

Dunkel L, Hirvonen V, Erkkilä K (1997): Clinical aspects of male germ cell apoptosis during testis development and spermatogenesis. Cell Death Differ 4:171–79.

Edwards RG (1995): Physiological and molecular aspects of human implantation. Hum Reprod 10 (Supplement 2):1–13.

Enari M, Talanian RV, Wong W, Nagata S (1996): Sequential activation of ICE-like and CPP32-like proteases during Fas-mediated apoptosis. Nature 380:723–26.

Fernandes-Alnemri, T, Armstrong RC, Krebs J, Srinivasula SM, Wang L, Bullrich F, Fritz LC, Trapani JA, Tomaselli KJ, Litwack G, Alnemri ES (1996): *In vitro* activation of CPP32 and Mch3 by Mch4, a novel human apoptotic cysteine protease containing two FADD-like domains. Proc Natl Acad Sci USA 93:7464–69.

Flaws JA, DeSanti A, Tilly KI, Javid RO, Kugu K, Johnson AL, Hirshfield AN, Tilly JL (1995a): Vasoactive intestinal peptide-mediated suppression of apoptosis in the ovary: Potential mechanisms of action and evidence of a conserved anti-atretogenic role through evolution. Endocrinology 136:4351–59.

Flaws JA, DeSanti A, Kugu K, Trbovich AM, Tilly KI, DeSanti A, Hirshfield AN, Tilly JL (1995b): Interleukin-1β-converting enzyme-related proteases (IRPs) and mammalian cell death: Dissociation of IRP-induced oligonucleosomal endonuclease activity from morphological apoptosis in granulosa cells of the ovarian follicle. Endocrinology 136:5042–53.

Flemming W (1885): Über die bildung von richtungsfiguren in säugethiereiern beim untergang Graff'scher follikel. Archiv vur Anatomie und Entwickelungsgeschichte (Archives fur Anatomie und Physiologie) 221–44.

Forabosco A, Sforza C, De Pol A, Vizzotto L, Marzona L, Ferrario VF (1991): Morphometric study of the human neonatal ovary. Anat Rec 231:201–8.

Friedlander RM, Gagliardini V, Rotello RJ, Yuan J (1996): Functional role of interleukin-1β (IL-1β) in IL-1β-converting enzyme-mediated apoptosis. J Exp Med 184:717–24..

Godin I, Wylie C, Heasman J (1990): Genital ridges exert long-range effects on mouse primordial germ cell numbers and direction of migration in culture. Development 108:357–63.

Godin I, Deed R, Cooke J, Zsebo K, Dexter M, Wylie CC (1991): Effects of the *steel* gene product on mouse primordial germ cells in culture. Nature 352:807–9.

Gougeon A (1996): Regulation of ovarian follicular development in primates: Facts and hypotheses. Endocr Rev 17:121–55.

Hakuno N, Koji T, Yano T, Kobayashi N, Tsutsumi O, Taketani Y, Nakane P (1996): Fas/APO-1/CD95 system as a mediator of granulosa cell apoptosis in ovarian follicle atresia. Endocrinology 137:1938–48.

Hannun YA (1996): Functions of ceramide in coordinating cellular responses to stress. Science 274:1855–59.

Henkart PA, Grinstein S (1996): Apoptosis: Mitochondria resurrected? J Exp Med 183:1293–95.

Hirshfield AN (1991): Development of follicles in the mammalian ovary. Int Rev Cytol 124:43–101.

Hockenberry D, Nuñez G, Milliman C, Schreiber RD, Korsmeyer SJ (1990): Bcl-2 is an inner mitochondrial membrane protein that blocks programmed cell death. Nature 348:334–36.

Hockenberry DM, Oltvai ZN, Yin X-M, Milliman CL, Korsmeyer SJ (1993): Bcl-2 functions in an antioxidant pathway to prevent apoptosis. Cell 75:241–51.

Hodgen GD (1982): The dominant ovarian follicle. Fertil Steril 38:281–300.

Holly JMP, Eden JA, Alaghband-Zadeh J (1990): Insulin-like growth factor binding protein profiles in follicular fluid from normal dominant and cohort follicles, polycystic and multicystic ovaries. Clin Endocrinol 33:53–64.

Hopwood D, Levison DA (1975): Atrophy and apoptosis in the cyclical human endometrium. J Pathol 119:159–66.

Huckins C (1978): The morphology and kinetics of spermatogonial degeneration in normal adult rats: An analysis using a simplified classification of the germinal epithelium. Anat Rec 190:905–26.

Hurwitz A, Hernandez ER, Payne DW, Dharmarajan AM, Adashi EY (1992): Interleukin-1 is both morphogenic and cytotoxic to cultured rat ovarian cells: Obligatory role for heterologous, contact-independent cell–cell interaction. Endocrinology 131:1643–49.

Jacobson MD, Weil M, Raff MC (1996): Role of Ced-3/ICE-family proteases in staurosporine-induced programmed cell death. J Cell Biol 133:1041–51.

Jaffrézou J-P, Levade T, Bettaïeb A, Andrieu N, Bezombes C, Maestre N, Vemeersch S, Rousse A, Laurent G (1996): Daunorubicin-induced apoptosis: Triggering of ceramide generation through sphingomyelin hydrolysis. EMBO J 15:2417–24.

Johnson AL, Bridgham JT, Bergeron L, Yuan J (1997): Characterization of the avian *Ich-1* cDNA and expression of *Ich-1* $_{long}$ mRNA in the hen ovary. Gene 192:227–33.

Johnson AL, Bridgham JT, Witty JP, Tilly JL (1996): Susceptibility of avian ovarian granulosa cells to apoptosis is dependent upon state of follicle development and is related to endogenous levels of *bcl-x*$_{long}$ gene expression. Endocrinology 137:2059–66.

Juengel JL, Garverick HA, Johnson AL, Youngquist RS, Smith MF (1993): Apoptosis during luteal regression in cattle. Endocrinology 132:249–54.

Kaipia A, Chun S-Y, Eisenhauer KM, Hsueh AJW (1996): Tumor necrosis factor-α and its second messenger, ceramide, stimulate apoptosis in cultured ovarian follicles. Endocrinology 137:4864–70.

Kane DJ, Sarafian TA, Anton R, Hahn H, Gralla EB, Valentine JS, Ord T, Bredesen DE (1993): Bcl-2 inhibition of neural death: Decreased generation of reactive oxygen species. Science 262:1274–77.

Kastan MB, Onyekwere O, Sidransky D, Vogelstein B, Craig RW (1991): Participation of p53 protein in the cellular response to DNA damage. Cancer Res 51:6304–11.

Keren-Tal I, Suh B-S, Dantes A, Lindner S, Oren M, Amsterdam A (1995): Involvement of p53 expression in cAMP-mediated apoptsis in immortalized granulosa cells. Exp Cell Res 218:283–95.

Kerr JFR, Wyllie AH, Currie AR (1972): Apoptosis: A basic biological phenomenon with wide-ranging implications in tissue kinetics. Br J Cancer 26:239–57.

Knobil E (1980): The neuroendocrine control of the menstrual cycle. Rec Prog Horm Res 36:53–88.

Knudson CM, Tung KSK, Tourtellotte WG, Brown GAJ, Korsmeyer SJ (1995): Bax-deficient mice with lymphoid hyperplasia and male germ cell death. Science 270:96–99.

Knudson CM, Tung KSK, Flaws JA, Brown GAJ, Tilly JL, Korsmeyer SJ (1996): Oocyte survival but spermatocyte death in Bax-deficient mice. Chicago, IL: Proceedings of the International Symposium on Cell Death in Reproductive Physiology, p 33.

Ko LJ, Prives C (1996): p53: Puzzle and paradigm. Genes Dev 10:1054–72.

Kokawa K, Shikone T, Nakano R (1996): Apoptosis in the human uterine endometrium during the menstrual cycle. J Clin Endocrinol Metab 81:4144–47.

Kolesnick RN, Haimovitz-Friedman A, Fuks Z (1994): The sphingomyelin signal transduction pathway mediates apoptosis for tumor necrosis factor, Fas, and ionizing radiation. Biochem Cell Biol 72:471–74.

Kondo H, Maruo T, Peng X, Mochizuki M (1996): Immunological evidence for the expression of the Fas antigen in the infant and adult human ovary during follicular regression and atresia. J Clin Endocrinol Metab 81:2702–10.

Korsmeyer SJ (1995): Regulators of cell death. Trends Genet 11:101–105.

Koshimizu U, Watanabe M, Nakatsuji N (1995): Retinoic acid is a potent growth activator of mouse primordial germ cells *in vitro*. Develop Biol 168:683–85.

Krajewski S, Tanaka S, Takayama S, Schibler MJ, Fenton W, Reed JC (1993): Investigation of the subcellular distribution of the Bcl-2 onocprotein: Residence in the nuclear envelope, endoplasmic reticulum, and outer mitochondrial membrane. Cancer Res 53:4701–14.

Kugu K, Ratts VS, Piquette GN, Tilly KI, Tao X-J, Martimbeau S, Aberdeen GW, Krajewski S, Reed JC, Pepe GJ, Albrecht ED, Tilly JL (1998): Analysis of apoptosis and expression of *bcl-2* gene family members in the human and baboon ovary. Cell Death Differ 5 (in press).

Kuida K, Lippke JA, Ku G, Harding MW, Livingston DJ, Su MS-S, Flavell RA (1995): Altered cytokine export and apoptosis in mice deficient in interleukin-1β-converting enzyme. Science 267:2000–3.

Kumar S, Lavin MF (1996): The ICE family of cysteine proteases as effectors of cell death. Cell Death Differ 3:255–67.

Li P, Allen H, Benerjee S, Franklin S, Herzog L, Johnston C, McDowell J, Paskind M, Rodman L, Salfeld J, Towne E, Tracey D, Wardwell S, Wei F-Y, Wong W, Kamen R, Seshadri T (1995): Mice deficient in IL-1β-converting enzyme are defective in production of mature-IL-1β and resistant to endotoxic shock. Cell 80:401–11.

Liu X, Kim N, Pohl J, Wang X (1996a): Purification and characterization of an interleukin-1β-converting enzyme family protease that activates cysteine protease P32 (CPP32). J Biol Chem 271:13371–76.

Liu Z-g, Hsu H, Goeddel DV, Karin M (1996): Dissection of TNF receptor 1 effector functions: JNK activation is not linked to apoptosis while NF-κB activation prevents cell death. Cell 87:565–76.

Luciano AM, Pappalardo A, Ray C, Peluso JJ (1994): Epidermal growth factor inhibits large granulosa cell apoptosis by stimulating progesterone synthesis and regulating the distribution of intracellular free calcium. Biol Reprod 51:646–54.

Maravei DV, Trbovich AM, Perez GI, Tilly KI, Banach D, Talanian RV, Wong WW, Tilly JL (1997): Cleavage of cytoskeletal proteins by caspases during ovarian cell death: Evidence that cell-free systems do not always mimic apoptotic events in intact cells. Cell Death Differ 4 (in press).

Marchetti P, Castedo M, Susin SA, Zamzami N, Hirsch T, Macho A, Haeffner A, Hirsch F, Geuskens M, Kroemer G (1996): Mitochondrial permeability transition is a central coordinating event of apoptosis. J Exp Med 184:1155–1160.

Martimbeau S, Tilly JL (1997): Physiological cell death in endocrine dependent tissues: An ovarian perspective. Clin Endocrinol 46:241–54.

Martin SJ, Green DR (1995): Protease activation during apoptosis: Death by a thousand cuts? Cell 82:349–52.

Martin SJ, O'Brien GA, Nishioka WK, McGahon AJ, Mahboubi A, Saido TC, Green DR (1995): Proteolysis of fodrin (non-erythroid spectrin) during apoptosis. J Biol Chem 270:6425–28.

Martin SJ, Amarante-Mendes GP, Shi L, Chuang T-H, Casiano CA, O'Brien GA, Fitzgerald P, Tan EM, Bokoch GM, Greenberg AH, Green DR (1996): The cytotoxic cell protease granzyme-B initiates apoptosis in a cell-free system by proteolytic processing and activation of the ICE/CED-3 family protease, CPP32, via a novel two-step mechanism. EMBO J 15:2407–16.

Matsui Y, Zsebo KM, Hogan BL (1990): Embryonic expression of a haematopoietic growth factor encoded by the *Sl* locus and the ligand for *c-kit*. Nature 347:667–69.

Miething A (1992): Germ-cell death during prespermatogenesis in the testis of the golden hamster. Cell Tissue Res 267:583–90.

Mintz B, Russel ES (1957): Gene-induced embryological modifications of primordial germ cells in the mouse. J Exp Zool 134:207–30.

Miura M, Zhu H, Rotello R, Hartweig EA, Yuan J (1993): Induction of apoptosis in fibroblasts by IL-1β-converting enzyme, a mammalian homolog of the *C. elegans* cell death gene *ced-3*. Cell 75:653–60.

Miyashita T, Reed JC (1995): Tumor suppressor p53 is a direct transcriptional activator of the human *bax* gene. Cell 80:293–99.

Miyashita T, Harigai M, Hanada M, Reed JC (1994): Identification of a p53-dependent negative response element in the *bcl-2* gene. Cancer Res 54:3131–35.

Murphy LJ, Gharary A (1990): Uterine insulin-like growth factor I: Regulation of expression and its role in estrogen-induced uterine proliferation. Endocr Rev 11·443–53.

Nelson GK, Takahashi T, Lee DC, Luetteke NC, Mclachlan JA (1992): Transforming growth factor-α is a potential mediator of estrogen action in the mouse uterus. Endcrinology 131:1657–64.

Nicholson DW, Ali A, Thornberry NA, Vaillancourt JP, Ding CK, Gallant M, Gareau Y, Griffin PR, Labelle M, Lazebnik Y, Munday NA, Raiu SM, Smulson ME, Yamin T-T, Yu VL, Miller DK (1995): Identification and inhibition of the ICE/CED-3 protease necessary for mammalian apoptosis. Nature 376:37–43.

Nicosia SV (1983): Morphological changes in the human ovary throughout life. In Serra GB (ed): The Ovary, New York: Raven Press pp 57–81.

Niswender GD, Nett TM (1988): The corpus luteum and its control. In Knobil E (ed): The Physiology of Reproduction. New York: Raven Press, pp 489–525.

Obeid LM, Linardic CM, Karolak LA, Hunnun YA (1993): Programmed cell death induced by ceramide. Science 259:1769–71.

Oltvai ZN, Korsmeyer SJ (1994): Checkpoints of dueling dimers foil death wishes. Cell 79:189–92.

Oltvai ZN, Milliman CL, Korsmeyer SJ (1994): Bcl-2 heterodimerizes *in vivo* with a conserved homolog, Bax, that accelerates programmed cell death. Cell 74:609–19.

Pate JL (1994): Cellular components involved in luteolysis. J Anim Sci 72:1884–90.

Patel T, Gores GJ, Kaufmann SH (1996): The role of proteases during apoptosis. FASEB J 10:587–97.

Patton PE, Stouffer RL (1991): Current understanding of the corpus luteum in women and nonhuman primates. Clin Obstet Gynecol 34:127–43.

Peng X, Maruo T, Samoto T, Mochizuki M (1996): Comparison of immunocytologic localization of insulin-like growth factor binding protein-4 in normal and polycystic ovary syndrome human ovaries. Endocrine J 43:269–78.

Perez GI, Knudson CM, Brown GAJ, Korsmeyer SJ, Tilly JL (1997a): Resistance of BAX-deficient mouse oocytes to apoptosis induced by 7,12-dimethylbenz(a)anthracene (DMBA) *in vitro*. Toxicologist 36 (supplement):250.

Perez GI, Knudson CM, Leykin L, Korsmeyer SJ, Tilly JL (1997b): Apoptosis-associated signaling pathways are required for chemotherapy-mediated female germ cell destruction. Nature Med 3: (in press).

Pesce M, De Felici M (1994): Apoptosis in mouse primordial germ cells: A study by transmission and scanning electron microscope. Anat Embryol 189:435–40.

Pesce M, Farrace MG, Piacentini M, Dolci S, De Felici M (1994): Stem cell factor and leukemia inhibitory factor promote primordial germ cell survival by suppressing programmed cell death (apoptosis). Development 118:1089–94.

Peters H (1970): Migration of gonocytes into the mammalian gonad and their differentiation. Proc R Soc Lond (Biol) 259:91–101.

Pinkerton JHM, McKay DG, Adams EC, Hertig TA (1961): Development of the human ovary: A study using histochemical techniques. Obstet Gynecol 18:152–81.

Quirk SM, Cowan RG, Joshi SG, Henrikson KP (1995): Fas antigen-mediated apoptosis in human granulosa-luteal cells. Biol Reprod 52:279–87.

Ratts VS, Flaws JA, Kolp R, Sorenson CM, Tilly JL (1995): Ablation of bcl-2 gene expression decreases the numbers of oocytes and primordial follicles established in the post-natal female mouse gonad. Endocrinology 136:3665–68.

Richards JS (1980): Maturation of ovarian follicles: Actions and interactions of pituitary and ovarian hormones on follicular cell differentiation. Physiol Rev 60: 51–89.

Richards JS (1994): Hormonal control of gene expression in the ovary. Endocr Rev 15:725–51.

Roosen-Runge EC (1964): The degeneration of pre-spermatogonial germ cells in the rat after birth. Anat Rec 148:328.

Rotello RJ, Lieberman RC, Lepoff RB, Gerschenson LE (1992): Characterization of uterine epithelium apoptotic death kinetics and regulation by progesterone and RU 486. Am J Pathol 140:449–56.

Rueda BR, Tilly KI, Hansen TR, Hoyer PB, Tilly JL (1995a): Expression of superoxide dismutase, catalase and glutathione peroxidase in the bovine corpus luteum: Evidence supporting a role for oxidative stress in luteolysis. Endocrine 3:227–32.

Rueda BR, Wegner JA, Marion SL, Wahlen DD, Hoyer PB (1995b): Internucleosomal DNA fragmentation in ovine luteal tissue associated with luteolysis: in vivo and in vitro analyses. Biol Reprod 52:305–12.

Rueda BR, Tilly KI, Botros I, Jolly PD, Hansen TR, Hoyer PB, Tilly JL (1997): Increased bax and interleukin-1β-converting enzyme (Ice) messenger RNA levels coincide with apoptosis in the bovine corpus luteum during structural regression. Biol Reprod 56:186–93.

Russell LD, Clermont Y (1977): Degeneration of germ cells in normal, hypophysectomized and hormone-treated hypophysectomized rats. Anat Rec 187:347–66.

Shikone Y, Yamoto M, Kokawa K, Yamashita K, Nishimori K, Nakano R (1996): Apoptosis of human copora lutea during cyclic luteal regression and early pregnancy. J Clin Endocrinol Metab 81:2376–80.

Sinha Hikim AP, Wang C, Leung A, Swerdloff R (1995): Involvement of apoptosis in the induction of germ cell degeneration in adult rats after gonadotropin-releasing hormone agonist treatment. Endocrinology 136:2770–75.

Skinner MK, Parrot JA (1994): Growth factor-mediated cell–cell interactions in the ovary. In Findlay JK (ed): Cellular and Molecular Mechanisms in Female Reproduction. New York: Academic Press, pp 67–79.

Smith MF, McIntush EW, Smith GW (1994): Mechanisms associated with corpus luteum development. J Animal Sci 72:1857–72.

Stewart CL, Kaspar P, Brunet LJ, Bhatt H, Gadi I, Kontgen F, Abbondanzo SJ (1992): Blastocyst implantation depends on maternal expression of leukaemia inhibitory factor. Nature 359:76–79.

Tao X-J, Tilly KI, Maravei DV, Shifren JL, Krajewski S, Reed JC, Tilly JL, Isaacson KB (1997): Differential expression of members of the *bcl-2* gene family in proliferative and secretory human endometrium: Glandular epithelial cell apoptosis is associated with increased expression of *bax*. J Clin Endocrinol Metab 82:2738–46.

Tapanainen J, Tilly JL, Vihko KK, Hsueh AJW (1993): Hormonal control of apoptotic cell death in the testis: Gonadotropins and androgens as testicular cell survival factors. Mol Endocrinol 7:643–50.

Terada N, Yamamoto R, Takada T, Miyake T, Terakawa N, Wakimoto H, Taniguchi H, Li W, Kitamura Y, Matsumoto K (1989): Inhibitory effect of progesterone on cell death of mouse uterine epithelium. J Steroid Biochem 33:1091–96.

Tilly JL (1996): Apoptosis and ovarian function. Rev Reprod 1:162–72.

Tilly JL, Perez GI: Mechanisms and genes of physiological cell death: A new direction for toxicological risk assessments? In Sipes IG, McQueen CA, Gandolfi AJ (eds): Comprehesive Toxicology, Vol. 10. Oxford: Elsevier, pp 379–85.

Tilly JL, Ratts VS (1996): Biological and clinical importance of ovarian cell death. Contemp Obstet Gynecol 41:59–86.

Tilly JL, Tilly KI (1995): Inhibitors of oxidative stress mimic the ability of follicle-stimulating hormone to suppress apoptosis in cultured rat ovarian follicles. Endocrinology 136:242–52.

Tilly JL, Billig H, Kowalski KI, Hsueh AJW (1992): Epidermal growth factor and basic fibroblast growth factor suppress the spontaneous onset of apoptosis in cultured rat ovarian granulosa cells and follicles by a tyrosine kinase-dependent mechanism. Mol Endocrinol 6:1942–50.

Tilly JL, Tilly KI, Kenton ML, Johnson AL (1995a): Expression of members of the *bcl-2* gene family in the immature rat ovary: Equine chorionic gonadotropin-mediated inhibition of apoptosis is associated with decreased *bax* and constitutive *bcl-2* and *bcl-x*$_{\text{long}}$ messenger ribonucleic acid levels. Endocrinology 136:232–41.

Tilly KI, Banerjee S, Banerjee PP, Tilly JL (1995b): Expression of the p53 and Wilm's tumor suppressor genes in the rat ovary: Gonadotropin repression *in vivo* and immunohistochemical localization of nuclear p53 protein to apoptotic granulosa cells of atretic follicles. Endocrinology 136:1394–402.

Tilly KI, Perez GI, Tilly JL (1996): Ceramide-induced apoptosis in ovarian follicles: Kinetic analysis and modulation by potassium chloride efflux. Mol Biol Cell 7 (Supplement): 194a.

Tilly JL, Tilly KI, Perez GI (1997): The genes of cell death and cellular susceptibility to apoptosis in the ovary: A hypothesis. Cell Death Differ 4:180–87.

Trbovich AM, Hughes Jr FM, Perez GI, Kugu K, Tilly KI, Cidlowski JA, Tilly JL (1998): High and low molecular weight DNA cleavage in ovarian granulosa cells:

Characterization and protease modulation in intact cells and cell-free nuclear auto-digestion assays. Cell Death Differ 5 (in press).

Verheij M, Bose R, Lin XH, Yao B, Jarvis WD, Grant S, Birrer MJ, Szabo E, Zon LI, Kyriakis JM, Haimovitz-Friedman A, Fuks Z, Kolesnick RN (1996): Requirement for ceramide-initiated SAPK/JNK signalling in stress-induced apoptosis. Nature 380:75–79.

Witty JP, Bridgham JT, Johnson AL (1996): Induction of apoptotic cell death in hen granulosa cells by ceramide. Endocrinology 137:5269–77.

Wyllie AH (1995): The genetic regulation of apoptosis. Curr Opin Genet Develop 5:97–104.

Xiang J, Chao DT, Korsmeyer SJ (1996): BAX-induced cell death may not require interleukin-1β-converting enzyme-like proteases. Proc Natl Acad Sci USA 93:14559–63.

Yang E, Korsmeyer SJ (1996): Molecular thanatopsis: A discourse on the BCL-2 family and cell death. Blood 88:386–401.

Yu BP (1994): Cellular defenses against damage from reactive oxygen species. Physiol Rev 74:139–62.

Zamzami N, Susin SA, Marchetti P, Hirsch T, Gómez-Montgomery I, Castedo M, Kroemer G (1996): Mitochondrial control of nuclear apoptosis. J Exp Med 183:1533–44.

CHAPTER 17

ASSESSMENT OF APOPTOSIS IN TUMOR GROWTH: IMPORTANCE IN CLINICAL ONCOLOGY AND CANCER THERAPY

LUIS BENÍTEZ-BRIBIESCA
"Leon Weiss" Professor Oncology, University of Mexico (UNAM); Chief, Oncological Research Unit, Oncology Hospital, National Medical Center, IMSS, Mexico, D.F. Mexico

INTRODUCTION

Malignant tumors are generally regarded as diseases of growth and differentiation. Although there are more than 200 different types of malignancies, they all share common biological features that have prompted the metaphor of one disease with hundreds of faces.

Perhaps the most apparent common denominator in all cancers is their accelerated growth, which can be witnessed both clinically and experimentally. The rate of tumor growth is a question of great importance both for the biological study of malignant progression and for practical purposes in the treatment of cancer. The study of the kinetics of cell proliferation in tumors has produced important information, particularly in relation to the derangement of cell cycling. This information contributed to the development of anticancer treatments aimed at suppressing cell proliferation (Shackney et al., 1978).

Until recently, the increased mitotic rate and the lack of differentiation were thought to be the two main factors for explaining the accelerated growth of malignancies. During the past decade, however, it became apparent that deregulation of programmed cell death played an important role in this phenomenon. It has been shown that, in most malignant tumors, apoptosis is inhibited and that the increase in cell numbers during the development of

When Cells Die, Edited by Richard A. Lockshin, Zahra Zakeri, and Jonathan L. Tilly
ISBN 0-471-16569-7 © 1998 Wiley-Liss, Inc.

neoplasias is due not only to an increased mitotic rate, but also to the diminished rate of cell deletion. This latter occurs physiologically in all tissues for maintaining cell balance in all organs (Kerr et al., 1994).

The current excitement over research in this area has been driven even more by discoveries demonstrating that physiological cell elimination is regulated by a genetic program that involves oncogenes and antioncogenes (Steller, 1995). Many of these promoting and suppressor DNA sequences were known to play a central role in carcinogenesis before their regulatory actions in apoptosis were learned. Thus scientists agree that understanding the biochemical and molecular pathways that control apoptosis is of paramount importance to the cancer problem (Thompson, 1995).

The study of programmed cell death in malignancies has allowed a better understanding of tissue homeostasis; it has completed the multistep carcinogenesis scheme; it has offered other insights into the mechanisms of chemoresistance and chemotherapy; and above all, it has opened new routes for understanding the biology of tumors and the development of novel types of therapies (Clark, 1991; Meyn, 1994). In this chapter we will review how deranged control of apoptosis helps to explain many so far unclear facts about immortalization, cell renewal kinetics, and carcinogenesis and how *the assessment of cell death kinetics applies to the biology of tumors including clinical diagnosis and oncologic therapeutics.*

MULTISTEP CARCINOGENESIS AND PROGRAMMED CELL DEATH

With the discovery of chemical carcinogens it became possible to produce tumors experimentally in animals. With these experimental models it was also possible to state that carcinogenesis occurs in at least two separate steps, initiation and progression (Fig. 17.1). The first is a rapid process that alters the cellular DNA through a mutation that induces a transformation that, in turn, renders the cell immortal. The latter is a slow, progressive change that promotes cell proliferation and later invasion and metastasis (Ryser, 1971).

Further studies have complicated this relatively simple scheme, and it is now fairly well proven that other intermediate stages occur in carcinogenesis to generate a clinical malignancy. Molecular genetic studies carried on in a heritable type of colon carcinoma have demonstrated that a series of genetic and chromosomal abnormalities accumulate progressively in the colonic mucosa of susceptible individuals to generate a full-blown invasive adenocarcinoma (Service, 1994). It is likely that most cancers develop in a similar multistep process, since genomic rearrangements suggestive of acquired DNA instability are a feature of malignant tumors. Two types of instability are recognized in the biology of the tumors: The first are clones of aneuploid cells with reduplication of chromosomes, and the second are diploid tumors with instability at microsatellite loci (Reichmann et al., 1981).

PROGRAMMED CELL DEATH
IN
MULTISTEP CARCINOGENESIS

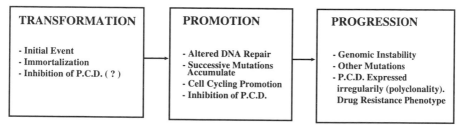

FIG. 17.1. Diagrammatic representation of the simplified orthodox scheme of multistep carcinogenesis and the role played by altered programmed cell death (PCD) in each step.

The first change induced by a chemical, physical, or biological initiator is a mutation that is not repaired and thus is inherited by the cell progeny. For many years it was known that cells thus affected were rendered immortal, but it was unclear why this happened. Although immortalization is not synonymous with carcinogenesis, it is a prerequisite for the further development of a malignant phenotype. Most cell lines from tumors are immortal, meaning that they have lost their capacity to die after a certain species-specific number of cell duplications ("Hayflick limit"; in humans about 60 doublings in culture). Soon after the discovery of the genetic control of programmed cell death, it became clear that the derangement of this genetic apparatus could be responsible for the lack of response, by malignant cells, to death signals, since many of the proto-oncogenes and suppressor genes altered during carcinogenesis also play a key role in the regulation of cell death (Kerr et al., 1994; Strasser and Vaux, 1996).

Analysis of multistep carcinogenesis at the molecular level indicates, therefore, that the process of neoplastic evolution is more complicated. The colon cancer model shows that multiple genetic changes must occur after the initiated cells undergo clonal expansion (Vogelstein et al., 1988). These changes include mutations, chromosomal rearrangements, gene amplification, and chromosomal gains and losses. It has been estimated that somewhere between three and seven independent random genetic events are typically required to turn a normal cell into a cancer cell and probably even more to invade and metastasize (Baker et al., 1989; Hollstein et al., 1991).

These genetic derangements are usually expressed as increased activity of some oncogenes and deletions or mutations of suppressor genes leading to increased cell proliferation and altered differentiation. These two abnormalities had been regarded as sufficient for explaining tumor growth, but the inhibition of normal cell deletion that regulates tissue homeostasis offers an additional parameter to explain the uncontrolled growth of malignant neoplasias. The genetic abnormalities that occur during multistep carcinogenesis have been shown to affect apoptosis. *Therefore, inhibition of cell death must also play an important role in tumor growth, along with stimulation of the mitotic cycle and with suppression of cell differentiation* (Fig. 17.1; Clark, 1991; Harrington et al., 1994).

Cell Death: A Major Factor in Loss of Tissue Homeostasis in Cancer

The balance between cell production and cell loss in adult tissues that guarantees a constant cell number in tissues and organs in mature adult organisms is termed *tissue homeostasis.* When this balance is lost, either atrophy or hyperplasia results (Meyn, 1994; Thompson, 1995).

The molecular mechanisms that provide the fine control of tissue homeostasis are just beginning to be understood. They include myriad signals, receptors, and second messengers under genetic and epigenetic control. For many years uncontrolled growth of cells was considered to be the result of defects in genes controlling either cell proliferation or cell differentiation. There is now substantial evidence that the rate of cell death can be equally important in maintaining tissue homeostasis. Cancer growth represents the most obvious abnormality of tissue equilibrium towards a progressive gain in cell numbers that cannot be explained by only the continuous stimulation of cell division in the malignant clones.

Although it is generally accepted that malignancies show a high mitotic index, there are many examples in which it is very difficult to find a mitotic figure in numerous tissue sections, such as in carcinoma of the prostate. But the study of the doubling time and length of the cell cycle in malignant tumors has rendered surprising results. In human tumors the number of replicating cells can be even lower than that of their tissues of origin, and the duration of the cell cycle is usually longer (Nowell, 1976, 1990). *It is thus clear that increased cell replication is not the only explanation for the continuous gain of cells. The other side of the coin, the inhibition of the normal rate of cell loss, seems to be another major factor in this process.*

In different tissues, cells seem to coexist under a social control and require survival signals from other cells in order to suppress the built-in program for cell death. A cell with a heritable mutation affecting cell death, or stimulating the production of autocrine survival factors, grows in an antisocial deregulated way. The relationship between cell replication and cell death will define the rate of tumor growth (Clark, 1991; Kerr et al., 1994). Most cancers arise from a single cell and therefore are monoclonal in origin; however, during tumor progression and due to the great genetic instability, several clones develop. Fully developed malignancies show polyclonality, which is expressed in different ways (Nowell, 1976). Despite the fact that probably since initiation the apoptotic mechanisms are altered, later on some tumor cell clones retain the capacity to respond to death signals; in fact, some malignant tumors, such as testicular germ cell neoplasias, are composed mainly of cells highly responsive to inducers of cell death such as chemotherapeutic agents (Huddart et al., 1995). Apoptosis occurs in most malignancies at different rates, and it has been shown that chemotherapeutic agents and radiotherapy are capable of inducing cell death both *in vitro* and *in vivo,* but also that most neoplasias develop a resistant phenotype in the most advanced stages (Huschtscha et al., 1996).

Although suppression of cell death appears to be strongly selected during carcinogenesis, it is intriguing that it never appears to involve deletion of the entire apoptotic mechanism. Cells that lose key apoptotic genes such as *p53* or that have a high level of expression of the antiapoptotic gene *bcl-2* retain the ability to undergo apoptosis (see the following). They just raise their threshold (Harrington et al., 1994), rendering oncological treatments ineffective.

In terms of cell replication, tumors contain four populations of neoplastic cells: (1) cycling cells, (2) cells that can be recruited into cycling, (3) cells unable to divide because they are partially differentiated, and (4) dying or apoptotic cells (Steel, 1967, 1977). For many years therapeutic strategies have been directed against populations (1) and (2) but recently there have been considerable efforts to design novel therapies to stimulate cell death. Hence assessment of apoptosis in malignant tumors is of paramount importance both for experimental and for clinical oncologists (Fisher, 1994).

CELL REPLICATION, DIFFERENTIATION AND APOPTOSIS IN GRADING MALIGNANT NEOPLASIAS

Cell Death: Important but Difficult to Assess in Neoplasia

Until recently, only cell replication and differentiation were considered as key indicators for histological grading of malignancies. It has been demonstrated that the measurement of the replication rate and the degree of differentiation bear some relationship to the natural history of cancers and that in individual patients assessment of those two parameters could indicate the possible rate of growth and the invasive potential of the tumor. Most malignancies are routinely classified according to their degree of differentiation and recently also to their replication capacity and ploidy—hence the terms well, poorly, or undifferentiated; diploid, aneuploid, hyperploid, and high and low mitotic index. In general, well-differentiated, diploid, low-mitotic-index neoplasias have a better clinical course and prognosis (Shackney et al., 1978; Nowell, 1990; Majno and Joris, 1996).

Apoptosis seems now to play a very important role in tumor growth and in modulating resistance to chemotherapeutic agents and radiation therapy. Therefore, it appears mandatory to assess apoptosis as another key factor to grade tumors and help determine tumor growth rate and progression. Recent studies have shown that the quantification of apoptosis in malignant tumors might be indicative of the type of response to different forms of therapy.

While there are a number of well-proven methods for assessing cell replication and cell differentiation that offer precise quantitative data in experimental models, the estimation of apoptosis in human tumors still remains elusive. There are two main reasons for this difference. The first is that cell replication can be traced very efficiently, from purely morphological grounds to specific markers such as DNA labeling, the proliferation-associated nuclear antigen Ki-67, cyclin-dependent kinases, DNA ploidy, and flow cytometric determinations of the various cycle cell stages M, G_1-S-G_2 (Meyer et al., 1977; Baak et al., 1985). Furthermore, cell differentiation can also be easily assessed in routine histology and with the aid of a number of cell markers. Using immunocytochemical techniques to identify proteins such as keratins, myosins, hormone receptors, hormone proteins, cytoskeleton molecules, and surface immunoglobulins, it is quite simple to define clearly the rate of cell replication and the degree of differentiation in any given tumor, both in biopsies and in cell lines (Silvestrini et al., 1979; Rosai, 1981). The second reason is that estimation of cell death kinetics is still quite problematic, mainly because apoptosis is a comparatively fast phenomenon and leaves inconspicuous traces, as opposed to mitotic figures. Perhaps the greatest difficulty in studying programmed cell death in turnover is twofold: that there are many different signals that trigger the death program, and that these signals vary in different tissues and circumstances (Schwartzman et al., 1993; Kane, 1995). In cancer cells, the death pathway is altered in many different ways according to the tissue of origin and the degree of differentiation (Stauton and Gaffney, 1995).

Most techniques currently in use for detecting apoptosis in tumors rely on the changes that take place in the late stages of the apoptotic phenomenon, mainly the progressive hydrolysis of DNA, which seems to be the last cellular change before the cell disintegrates. Until now, however, there is no known universal marker of cell death that can detect a dying cell prior to DNA breakage, although two candidates have been proposed for cells committed to die: ceramide synthesis and activation of ICE-like proteases (Pronk et al., 1996).

Potential for the Evaluation of Cell Death as a Prognostic Indicator in Human Malignant Neoplasias

DNA-Based Techniques. For many years the cell replication rate has been regarded as the most reliable indicator of tumor growth and aggressiveness. Many methods have been applied to quantify the degree of proliferative activity, from the most simple, which count mitotic figures in tissue sections, to the most sophisticated, which estimate the S fraction by flow cytometry or by means of DNA labeling.

Several studies have established the prognostic value of determining mitotic activity in solid tumors, such as breast cancer. Although apparently straightforward, there have been many variations for establishing strict protocols for accurate calculation of the mitotic index (MI). Unfortunately, discrepancies among laboratories and different observers in establishing an MI are common and can be accounted for in several ways. First, there are differences in the areas studied due to different microscopic lenses when counts are referred to in a series of microscopic fields studied. Second, mitotic figures may not be indicative of malignant cells alone if abundant fibroblasts (desmoplastic reaction) or lymphoid populations infiltrate the tumor; and third, sections vary in cellularity and size. Recently, Gaffney et al. (1996) standardized an MI based on direct microscopic observations of mitotic figures in ten microscopic fields, and the average cell number was determined by counts of population density in three of those fields. These authors found a striking correlation of MI with ploidy, S-phase fraction, hormone receptor status, and disease progression. The combination of MI with other risk factors such as tumor size and lymph node metastasis adds up to the prognosis prediction (Clark et al., 1989). Nevertheless, mitotic activity is the best single predictor of prognosis and has the advantage of being inexpensive and fairly simple, and therefore can be applied in most pathology laboratories. A similar approach has been attempted to establish a morphological apoptotic index counting apoptotic cells and apoptotic bodies, but results are not yet encouraging.

The more sophisticated techniques for studying cell replication include 3H-thymidine uptake, bromodeoxy-uridine labeling, and DNA flow cytometric assays. The thymidine labeling index is a measure of the proportion of cells in a population engaged in replication of nuclear DNA during a brief period of exposure to the radioactive label. As such it is a measure of the rate of replication or potential growth rate of a tumor. However, the rate of replication does not necessarily predict the volume growth rate of the tumor, because loss of cells due to migration, necrosis, and apoptosis may account for more than 50% of the new cells produced (Steel, 1967, 1977). This technique is time-consuming and expensive and is not readily available in most pathology laboratories.

Recently, flow cytometric DNA analysis has offered a refined and valuable technique for studying cell replication in malignant cells. Histograms obtained from flow cytometric analysis result from counts of thousands of cells and thus have a high statistical reliability. Several parameters, such as G_0/G_1-phase, S-phase, and G_2/M-phase cell cycle fractions, DNA index and the number and size of aneuploid cell clones can be determined from the DNA histogram (Darzynkiewcz and Melamed, 1993). Several studies have validated the usefulness of flow cytometric DNA analysis in both solid and hematological malignancies. The estimation of ploidy and S-phase fractions in epithelial cancers, such as breast ductal carcinoma, are important predictors of prognosis and bear a strong correlation with cell differentiation and hormone receptor status (Dressler et al., 1988; Kallioniemi et al., 1988). Despite the high cost of flow cytometers, these instruments are becoming increasingly available in most well-equipped pathology laboratories, and the techniques are relatively simple.

However, cell replication is only part of the story of tumor growth. Cell death is the other side of the coin, and it must be considered that the growth rate pattern of malignancy is the result of both. High replication and low cell death rates ensure a rapid tumor growth, while low replication with high cell death rates would indicate slow growth or even regression (i.e., after chemo- or radiotherapy). It thus seems mandatory to explore and quantify cell death in cancers parallel to the quantitation of cell replication properly to establish prognostic factors including response to oncological treatment. Due to their versatility, flow cytometric techniques have also been applied to quantify apoptotic cell fractions in tumors but with much less definition than that of DNA analysis (Darzynkiewcz et al., 1992).

Attempts have been made to establish an apoptotic index in spontaneous human tumors, as well as in experimental neoplasias. Also, the induction of apoptosis in different malignancies after chemotherapy or radiotherapy has been studied *in vitro* and *in vivo*. Using a morphological approach, similar to that used for the calculation of MI, Staunton and Gaffney (1995) established an apoptotic index (AI) in a variety of tumors. They found that every tumor type has a different apoptotic index and that this index correlates with the MI, so that in most cases with low mitotic index there was also a low AI. Most benign tumors had a very low AI, although in colonic adenomas a more extensive apoptosis than in the corresponding adenocarcinomas was found. In contrast, high AI did not correlate with high MI in most tumors.

A number of investigators using different techniques have studied the occurrence of apoptosis in various tumors. Aihara et al. (1994) found that the frequency of apoptotic bodies correlates positively with the Gleason grading in prostate cancer. The role of apoptosis in modulation of growth of different colonic adenomas, and the inhibition of apoptosis in the development of colorectal adenocarcinoma, have been studied by Arai and Kino (1995) and Bedi et al. (1995), respectively, and in gastric cancer by Kasagi et al. (1994). In malignant non-Hodgkin's lymphomas a correlation of AI and MI was found by Leoncini et al. (1993). Recently, Tatebe et al. (1996) studied the AI and MI in advanced colorectal carcinomas and found a direct correlation of cell death and cell proliferation but no correlation with expression of p53. These indices were higher in lymph node and liver metastasis, suggesting that apoptosis might reflect not only cell loss but also the proliferative activity in colorectal malignancies.

In summary, most advanced tumors show varying degrees of spontaneous apoptosis. Each histological type exhibits a different apoptotic index, and the more aggressive and metastasizing, the higher the AI. But since apoptotic and proliferative indices have a high correlation in some cancers, it is likely that cell death kinetics also reflect the proliferative activity of tumors (Tatebe et al., 1996). It seems that cell replication and cell deletion kinetics in malignant neoplasias are tightly coupled phenomena.

Currently Available and Needed Techniques to Estimate Apoptosis

There are a number of methods for identifying apoptotic cells, but few have practical applications for studying human tumors in patients before and after anticancer treatment. In fact, there is increasing interest in searching for a simple and effective means of specific identification and quantification of cell proliferation and cell deletion that can be used routinely in oncological patients (Cotter and Martin, 1996). Most methods currently in use rely chiefly on two phenomena: the condensation of nuclear chromatin and DNA fragmentation. The most simple of all is the study of the morphological pattern in routine tissue sections whereby apoptotic cells are identified by nuclear condensation, and cytoplasmic fragmentation into apoptotic bodies that are frequently found already engulfed by macrophages. But apoptotic cells are not easily distinguishable by light microscopy from other elements with condensed chromatin or hyperchromatic nuclei such as cells in telophase or small infiltrating lymphocytes (Hockenberry, 1995; Majno and Joris, 1995). Nevertheless, this is the method that has been used to establish the apoptotic index in human tumors on purely histological grounds.

In contrast to cell replication, which can be studied with a variety of well-defined markers as discussed previously, the study of apoptosis relies almost exclusively on two parameters: chromatin condensation and DNA internucleosomal breakage. These changes can be studied by a variety of techniques but pose several problems. First, both chromatin condensation and DNA internucleosomal hydrolysis are very late events in PCD. Second, these changes might not be apparent in some cell types, and third, the rapid elimination of late stage apoptotic cells in tissues and circulating blood cells makes it difficult to estimate properly the number of dying cells (Collins et al., 1992).

The cardinal feature of apoptotic cells is DNA fragmentation, although it does not always evolve according to the scheme of internucleosomal breakage (Zakeri et al., 1993). It is believed that the linker regions between nucleosomes are the targets of endonuclease attack, resulting in fragments of 180–200 base pair (bp) and multiples of this unit length. In DNA extracted from a cell population, this type of cleavage can be assessed by the appearance of a ladder of bands on a conventional agarose gel electrophoresis. This type of analysis is essentially qualitative, but is the most common biochemical method used for the detection of apoptosis and is often considered the hallmark of apoptosis. Apart from being solely a qualitative method, the ladder is produced only when a considerable number of the cell population under study is in the late apoptotic stages (Benítez-Bribiesca, 1996). If only few cells are undergoing DNA cleavage, the laddering of DNA is obscured, especially if random DNA degradation is also taking place due to necrotic cell death, as frequently occurs in large human tumors. Another drawback is that apoptosis can occur in the absence of internucleosomal fragmentation and that larger kilobase pair (kbp)-sized DNA fragments as well as single-strand DNA breaks are also generated during apoptosis (Cohen, 1992; Brown et al., 1993; Peitsch et al., 1993). This method is of very limited value when studying cell death in human malignancies.

The recent application of *in situ* end labeling of DNA at sites of cleavage has also proved to be a useful tool in quantifying the percentage of apoptosis in a cell population and in identifying individual apoptotic cells. This technique can be applied to intact fixed tissues to demonstrate the presence of strand breaks. The technique relies on the use of exogenous enzymes, such as DNA polymerase I and terminal deoxynucleotidyl transferase (TdT) to incorporate labeled nucleotides into the 3′ hydroxyl termini of DNA strand breaks. In general these techniques are termed *in situ* end labeling (ISEL) and more specifically *in situ* nick translation (ISNT) or TdT-mediated dUTP nick end labeling or TUNEL (Gavrieli et al., 1992; Gold et al., 1993; Fehsel et al., 1994; Negoescu et al., 1996). With these methods there are still a good number of false negative and some false positive stained cells for reasons that are still unclear (Grasl-Kraupp et al., 1995). This method can be also used in flow cytometric analysis, but it is quite expensive for routine use in clinical oncology.

A number of flow cytometric techniques to discriminate live from dead cells have been designed. Changes in cell size as well as chromatin and DNA structure provide the basis for these methods. An apoptotic index can be calculated, but as Darzinkiewicz states, regardless of the assay used, the mode of cell death should be positively identified by morphological microscopic examination, and should be the deciding factor in situations where ambiguity arises (Darzynkiewicz et al., 1992; Sun et al., 1994; Darzykiewicz and Li, 1995). Until now, flow cytometric methods have not gained general acceptance for estimation of cell death in human tumors, mainly because they are complicated and require large numbers of cells.

A potentially useful method to explore DNA strand breaks is single-cell gel electrophoresis or "comet assay." This method, originally designed to study single-strand DNA oxidative damage, has gained general acceptance for assessing DNA mutagenic damage. Recently, the possibility of adapting this technique to detect double-strand DNA breaks has allowed its application for studying apoptotic cell death. The method is quite simple and inexpensive and can be rendered quantitative with the aid of image analysis. The possibility of estimating the degree of both random and nonrandom DNA degradation, as well as the presence of single and double strand breaks, makes this procedure a potentially useful tool for the study of apoptosis in cancer patients. The sensitivity and specificity are very high when used in different cell lines subjected to various types of DNA damage including apoptosis, but its application to the study of malignant neoplasias awaits further confirmation (Olive et al., 1990, 1993; Olive and Banáth, 1995).

The general denominator of these techniques is that they rely mainly on the final stages of DNA degradation during apoptosis, which is usually a late phenomenon or sometimes does not take place at all in PCD. Furthermore, due to the rapidity of apoptosis that is usually completed in around 3 h the "time window" for most of these methods makes it difficult to have a realistic estimation of the degree of apoptosis taking place in a malignant tumor (Bursch et al., 1985, 1990). There is, therefore, an urgent need to find early markers that can also detect the cells committed to die before they reach the late stage of DNA degradation (Fig. 17.2).

Despite the disparity of techniques and the different tumors and clinical stages used to study cell death in human cancers, some general conclusions can be drawn:

1. The study of cell death kinetics is at least as important as cell proliferation kinetics in human tumors.
2. Currently available methods to assess cell death kinetics in malignant human neoplasias are restricted to the end-stage changes in chromatin and DNA of apoptotic cells.

Programmed Cell Death Diagram and Site of Applicable Techniques

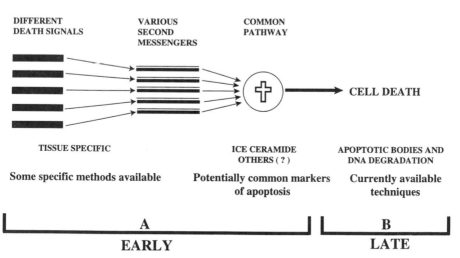

FIG. 17.2. Simplified diagram of programmed cell death. It is arbitrarily divided into early (A) and late (B) stages. Signals capable of triggering apoptosis are multiple (both endogenous and exogenous) and seem to be tissue and cell specific. Second messengers also vary according to the triggering signal, but most if not all converge to a common pathway when cells are irreversibly committed to die. (ICE and ceramide are candidates for this final pathway. See text). The late stages begin when DNA starts to be cleaved and cells show morphological changes such as chromatin condensation and apoptotic bodies. Most available techniques to assess apoptosis in malignant tumors detect only stage B.

3. Other methods that identify cell populations committed to die before DNA changes occur are needed.

4. Genetic markers of apoptotic-resistant and apoptotic-susceptible cell populations might help to understand the role of cell death in tumor progression.

5. *In vitro* techniques to detect chemo- and radiation-susceptible cell populations in leukemias and some solid tumors might be helpful to monitor response to oncological treatments.

GENETIC DEREGULATION OF APOPTOSIS IN MALIGNANT TUMORS

Genes Involved in Growth, Cell Cycle, or Unknown Function Whose Mutations Correlate with Malignancy

In previous chapters of this book the genetic control of programmed cell death has been explained in detail. In this section only the genetic abnormalities leading to alteration of programmed cell death in malignant neoplasias will be discussed.

The mutations known to occur in the process of carcinogenesis, from initiation through promotion and progression, affect many of the genes known to regulate programmed cell death. Genes that regulate apoptosis can be grouped into two broad categories: (1) repressors, whose expression can prevent or delay cell death, and (2) inducers, whose expression can directly induce cell death or increase susceptibility of cells to death-inducing stimuli (Kerr et al., 1994; Strasser and Vaux, 1996).

Many malignant cells show changes in expression, mutation, or deletion of oncogenes and antioncogenes, such as *bcl-2, p53, myc,* or *ras,* altering the genetic program of cell elemination. Others show changes in the APO-Fas ligand and TNF receptors, and still others might show changes in the activity of proteases of the ICE family, or in the ceramide–sphingomyelin pathway (Cory, 1994; Harrington, 1994; Strasser and Vaux, 1996).

Tumor Suppressor Gene p53: Cell Cycle Arrest, Apoptosis, or Both?

It is now firmly established that mutations of the p53 gene are the most commonly observed genetic lesions in spontaneously arising tumors. More than 50% of all human cancers bear some p53 anomaly and p53 mutations also characterize the germline mutations that occur in certain inherited human cancers such as the Li–Fraumeni Syndrome (Hollstein, 1991).

The key role of p53 alterations in carcinogenesis has been confirmed by a number of animal studies. In transgenic mice containing multiple copies of a mutant p53 transgene a variety of tumors develop in 30% of the animals in a few months. Homozygous mice with both germline p53 alleles inactivated develop normally but eventually all die of cancer by the age of 10 months (Donehower, 1994).

This suppressor gene seems to be a central player in mediating cell cycle arrest following DNA damage. Kastan et al. have demonstrated that cells treated with various DNA damaging agents undergo G1 arrest accompanied by a dramatic increase in wild-type p53 expression. In contrast, tumor or normal cells lacking p53 fail to arrest in G1 following irradiation (Kastan et al., 1992).

It has been postulated that p53 mediates G1 arrest to allow the cell sufficient time to repair DNA damage so that genetic lesions would not be propagated if the cells enter the S phase prematurely. In this sense, p53 should have an important role in maintaining genetic stability, and its loss would lead to increased genetic abnormalities and consequently to increased susceptibility to malignant transformation (Fig. 17.3).

But p53 seems also to play a very important role in apoptosis. Presumably, when DNA damage produced by ionizing radiation or other DNA damaging agents accumulates and becomes irreparable, p53 triggers the cell death program to prevent propagation of the genetic damage. Deletions or mutations of p53 that occur in a great number of cancers render affected cells incapable of undergoing apoptosis. Thus in cells that have lost p53 a failure to undergo cell cycle arrest, as a consequence of DNA damage, allows unrestricted entry

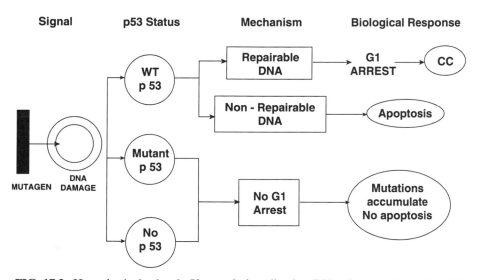

FIG. 17.3. Hypothetical role of p53 protein in cells after DNA damage. Normally p53 arrests cell cycle in G1 following DNA damage; if it can be repaired, cell cycle (CC) resumes. If it does not, the programmed cell death mechanism takes place. When p53 is mutated or deleted, no DNA repair occurs after DNA damage, and accumulation of mutations, genetic instability, and apoptosis inhibition follows, leading to malignant transformation.

into S phase before repair occurs, increasing the accumulation of mutations and leading to enhanced survival due to inhibition of apoptosis (Kastan, 1991; Yonish-Roauch et al., 1991; Shaw et al., 1992). There are still some paradoxes and unresolved questions in this field. It is not clear how DNA damage stimulates expression of p53, how this gene regulates the activation of programmed cell death, or why it is that wild-type p53 induces just G1 cell cycle arrest in some cancer cells while in others it induces apoptosis.

The Inhibitor Gene *bcl-2* a Potential Prognosticator?

The paradigm inhibitor of apoptosis is the bcl-2 gene overexpressed in human follicular lymphoma when translocated to an immunoglobulin locus (Korsmeyer, 1992).

The biological function of the 26-kD protein coded by that gene is to sustain the survival of interleukin-3-dependent cells in the absence of the cytokine, allowing them to enter a quiescent G_0 state. This protein is implicated in lymphocyte homeostasis, since its down-regulation is associated with deletion of unwanted lymphocytes, and conversely, up-regulation ensures the survival of the selected repertoire. Furthermore, bcl-2 overexpression enhances cell survival in the face of diverse insults such as genotoxic drugs, ionizing radiation, and glucocorticoids (Sentman et al., 1991; Reed, 1994).

The role of this antiapoptosis gene bcl-2 in the genesis of lymphoma and other cancers such as androgen-independent prostatic cancer is still unclear. In bcl-2 transgenic mice B-lymphoid tumors appear but only after a long latency period, suggesting the need for additional mutations. It has been found that constitutive myc overexpression can cooperate with bcl-2 to promote lymphomagenesis (Cory et al., 1994), and presumably other oncogene cooperations such as ras, are needed for complete malignant transformation (Fernandez Sarabia and Bischoff, 1993).

Tumors expressing bcl-2, however, can be regarded as chemo- or radioresistant and thus with a poor clinical prognosis. The possibility of using determination of Bcl-2 as an indirect marker of cell death resistance is of interest and has been used already in some hemato-immunological neoplasias as well as in a few solid cancers.

Myc, a Putative Proapoptotic Protein

Myc can promote apoptosis as well as proliferation, and this double activity is probably coupled with survival signals. A mechanism that couples proliferation and apoptosis could provide a biological safety net. But in malignancy a mutation up-regulating a survival gene like bcl-2 would inhibit the myc apoptotic pathway and enhance the proliferative activity. This in fact has been observed in bcl-2/myc transgenic mice that develop lymphoma (Cory, 1994).

The proto-oncogene c-myc seems to cooperate with the tumor suppressor gene p53. In quiescent mouse fibroblasts expressing wild-type p53 protein, activation of c-myc induces both apoptosis and cell cycle reentry. In contrast, in quiescent p53 null fibroblasts, activation of c-myc induces cell cycle reentry but no apoptosis (Hermeking and Eick, 1994). These observations show that p53 mediates apoptosis as a safeguard mechanism to prevent cell proliferation induced by oncogene activation.

The Fas/APO-1–Mediated Apoptosis and Its Loss in Tumors

Discovery of a cell surface protein belonging to the tumor necrosis factor (TNF)/nerve growth factor (NGF)/receptor superfamily and its further characterization, along with its specific ligand as a death factor, has prompted a number of studies in malignant neoplasias (Owen-Schaub, 1994).

It has been well established that the gain of function of growth-promoting factors, such as oncogene products, or the loss of function of growth inhibitory and differentiation factors, contribute to the development and progression of malignant disease. The loss of function of death factors regulating tissue homeostasis contributes equally to tumor development. Along this paradigm Fas/APO-1 can be viewed as a tumor suppressor gene, owing to its positive role in the induction of apoptosis, whereas a nonfunctional Fas/APO-1 could be considered as an enhancer of oncogenesis as a consequence of its inability to trigger the apoptotic cell death pathway (Nagata and Golstein, 1995). It is now clear that Fas/APO-1 is unevenly expressed in many tumor cell lines of both hematologic and nonhematologic origin, while others fail to express it at all (Laucat, et al., 1995). Along this simple scheme it was possible to assume that those malignant neoplasms expressing Fas/APO-1 could be more susceptible to spontaneous or induced apoptosis and therefore be more amenable to immunological therapy with anti-Fas antibodies with a better clinical outcome (Owen-Schaub, 1994). Although treatment with anti-Fas Ab has been successfully exploited in mice bearing human hematopoietic tumors, a number of different studies have shown that in most human malignancies the Fas/APO-1 pathway is frequently nonfunctional. This explains why even in Fas/APO-1 positive cells the anti-Fas antibody engagement may not uniformly elicit cell death. Even more, some cells might show growth stimulation after anti-Fas treatment, due probably to the regulation of constitutively produced cellular proteins such as Bcl-2, Bax, Bcl-X, and superoxide dismutase.

The use of Fas/APO-1 system as a marker for cell death susceptibility is thus so far quite limited and should await further research. The same can be said for the use of anti-Fas antibodies as effectors of the Fas/APO-1 death pathway in human tumors, although this is a very promising line of investigation (Cheng, 1994; Nagata and Golstein, 1995).

On the other hand, the protein that binds to Fas receptor, known as Fas ligand (Fas L), is a key molecule in normal immune development, homeostasis, modulation, and function, and it also plays a significant role in the mechanism of immune escape of malignant tumors. Fas L has been found to be constitutively expressed in some NK lymphomas and T cell–type lymphocyte leukemias and recently in colon cancer cells lines. In the latter a novel mechanism of immune escape has been described (O'Connell et al., 1996) whereby the colon cancer cells resist Fas-mediated T cell cytotoxicity but express functional Fas L, which triggers the death signal in infiltrating activated T cells. This Fas L–mediated cancer cell counterattack could correlate with tumor aggressive and metastasizing potential of certain solid tumors like colon cancer.

Signaling and Effectors May Be Detected As Indicators of Cell Death

ICE-Like Proteases and Ceramide Generation. Several members of a new class of cysteine protease genes have been discovered recently as regulators of programmed cell death, mainly ICE (interleukin-1β converting enzyme), Nedd 2, CPP32, Ich-2, and mch-2. All of these genes encode cysteine proteinases that are initially translated as inactive precursor polypeptides. Only a handful of substrates for these cell death proteases are known (Thornberry et al., 1992; Fernandes et al., 1995a,b).

ICE liberates bioactive interleukin-1β from the inactive precursor molecule, but that certainly is not the signal that triggers apoptosis. It is not known what initiates the proteolytic cascade of cysteine proteases and which proteins are their specific substrates. What is clear is that almost all mammalian cells express several death proteases, even when they are not undergoing apoptosis. However, the involvement of these proteases in cell death has been established by the use of specific inhibitors and with studies in ICE-mutated transgenic mice in which induction of apoptosis is inhibited. It has also been demonstrated that Bcl-2 and similar proteins can inhibit the action of the cysteine proteases, either by blocking their activation or by preventing them from reaching their targets (Bump et al., 1995). Although some cells undergo PCD without ICE activation, in most human malignant cells induction of apoptosis is mediated by ICE-like proteases. Kondo et al. found that malignant glioma cells show an overexpression of ICE gene after treatment with cisplatin and that mice transient transfection causes apoptosis in these cells (Kondo et al., 1995). Treatment with specific cysteine proteases inhibitors or with bcl-2 and related genes suppressed cisplatin-induced apoptosis in these cells. However, ICE is not always required for cell death induced by radiation, corticosteroids, or growth factor deprivation. It therefore appears that different apoptosis-triggering signals may activate distinct cysteine proteinases (Vaux et al., 1988).

Another enzyme that has been implicated in cell death is sphingomyelinase, which hydrolyzes sphingomyelin from phosphocholine and ceramide. Ceramide functions as a second messenger in the induction of apoptosis by TNF-β, Fas ligand, and radiation, and it has been shown that production of this sphingomyelin derivative requires functional ICE-like protease activity (Pronk et al., 1996).

Sphingosine, another product of sphingolipid breakdown derived from ceramide, has been shown to have inhibitory effects on *in vitro* as well as *in vivo* tumor cell growth and to induce apoptotic DNA fragmentation during PMA-induced differentiation in HL-60 promyelocytic leukemia cells (Ohta et al., 1995).

These studies, taken together, suggest that ICE-like activity and sphingolipid breakdown products may be good indicators for monitoring the apoptotic potential in human tumors. The significance of these potential markers in tumor growth and sensitivity to anticancer treatments remains to be established.

Other Factors and the Common Pathway of Apoptosis

No simple factor can explain the deregulation of apoptosis in malignant cells. Aside from bcl-2, p53, and Fas/APO ligand, other factors such as ICE, TNF, sphingomyelinase, and ceramide production certainly play a role in the derangement of the cell death program in neoplastic cells and are able to alter the response to anticancer treatments. Nevertheless, it seems likely that the great number of signals, second messengers, and genetic products that become altered during carcinogenesis to render the malignant cell resistant to apoptosis could act in a particular domain of that program to block the final pathway of cell death (Cory et al., 1994).

Since Bcl-2 can inhibit apoptosis induced by such a wide variety of cytotoxic insults, it seems likely that several independent routes to cell destruction converge in a final common pathway regulated by Bcl-2. There is some evidence that this final common pathway could be at the proICE-like cysteine protease level, because Bcl-2 can inhibit apoptosis in fibroblasts overexpressing ICE (Miura et al., 1993). Paradoxically Bcl-2 does not block apoptosis triggered by certain stimuli such as Fas or TNF receptor (Vanhaesebroeck et al., 1993). This apparent bcl-2 resistance could be explained if a critical effector molecule acts downstream from Bcl-2 or because the receptor induces antagonists. Since Bcl-2 also fails to block killing by cytotoxic T cells, the granzymes released into the target cell may cleave critical ICE targets, thereby escaping Bcl-2 inhibition (Sentman et al., 1991). The relevant substrates for the apoptosis-inducing proteases remain to be determined, but there is some indication that ICE-like cysteine proteases contribute to the generation of ceramide from sphingomyelin (Pronk et al., 1996). Whether Bcl-2-ICE regulated activity is truly a common final pathway for cell death remains to be established, but the possibility that these molecules could be universal markers certainly looks very promising and could provide a simple and effective way of studying apoptosis during carcinogenesis and particularly during oncological treatment.

CURRENT ASSESSMENT OF CELL DEATH IN MALIGNANT NEOPLASIAS

Cancers have many common features but also show great differences according to the tissue of origin and the degree of differentiation. Moreover, in fully developed tumors there is a significant cell heterogeneity due to the intrinsic generation of a number of clones with different genotypes and phenotypes, despite the fact that the dominant cell type is usually maintained during tumor development. Furthermore, it has become evident during the past few years that the type and rate of cell death also varies in different therapeutic regimens. It would therefore be appropriate to analzye the apoptotic phenomenon separately in various types of malignant neoplasias, as examples of the complexity and variability of the alteration of cell death program that takes place in clinical malignancies.

Apoptosis in Prostatic Adenocarcinoma

Prostatic cancer, one of the most frequent malignancies in adult men, is usually treated with surgical or chemical castration when it has grown beyond glandular confinement. Although in most cases there is an initial dramatic response, essentially all patients eventually relapse into a hormonal and chemoresistant state. This universal relapse of invasive prostate cancer after androgen ablation is explained by the heterogeneous presence of androgen-independent and androgen-dependent malignant cells in the primary tumor. After castration, only the androgen-dependent cells are eliminated, leaving the androgen-independent clones that are unresponsive to further therapy, including some types of chemotherapy (Isaacs, 1989; Furuya and Isaacs, 1994).

Involution of the normal prostate by androgen ablation induces programmed cell death in androgen-dependent glandular epithelial cells (Kyprianou and Isaacs, 1988), as explained in previous chapters. In human prostatic adenocarcinoma, androgen withdrawal induces apoptosis equally in androgen-dependent cells. It has been demonstrated that 1 day after castration, the number of cancer cells entering cell cycle declines to one-third of the initial values, and at 48 hours the proliferative activity is less than 1%, while the number of apoptotic cells increases progressively. This demonstrates that programmed cell death in androgen-dependent malignant prostatic cells can occur without recruitment of these cells into the proliferative cell cycle (Martikainen et al., 1991). On the contrary, programmed cell death is not induced in androgen-independent clones due to a defect in the initiation step. These cells retain the cell death program and can be stimulated to undergo apoptosis by increasing the intracellular free Ca^{2+} by means of calcium ionophores. After 6 to 12 hours of ionomycin treatment cells arrest in G_0, and 24 to 48 h later, cells begin DNA fragmentation into nucleosomal-sized pieces. This demonstrates that androgen ablation is incapable of inducing sustained elevation of intracellular Ca^{2+} in androgen-independent cells (Martikainen et al., 1991; Furuya and Isaacs, 1994).

Furthermore, bcl-2, the apoptosis inhibitory oncogene, is overexpressed during progression of prostatic cancer from an androgen-dependent to an androgen-independent state, meaning that expression of this oncogene is critical for the development of the androgen-independent phenotype in the progression of prostate cancer following hormonal ablation (Colombel et al.,

1993). It follows that estimation of apoptosis rate in this malignancy is indicative of the androgen-dependent or -independent stages and the type of response to different treatments.

Leukemias and Lymphomas. The bcl-2 Connection

It is probably in hematological and lymphoid malignancies that the study of apoptosis has provided the most convincing information of its application and usefulness in clinical oncology. Although many genetic and biochemical mechanisms are known to be deregulated in leukemias and lymphomas, the altered expression of the bcl-2 oncogene seems to be a common denominator. The bcl-2 gene was identified as a transcriptional element associated with the t(14;18) translocation commonly found in human follicular lymphoma and mapped to chromosome 18q 21 (Cleary and Sklar, 1985). Deregulation of this proto-oncogene leads to extended viability of cells by blocking apoptotic cell death and renders them susceptible to accumulate additional mutations, which can result in transformation. Although in normal adult human tissues expression of bcl-2 protein occurs mainly in B lymphocytes, it is also found in hematopoietic precursor cells in all lineages, in hormonally responsive epithelia in the epidermis, in intestinal mucosa, and in central nervous system neurons (Hockenbery et al., 1991). These observations suggest that bcl-2 expression might be critical in maintaining tissue homeostasis and could be involved in transformation and progression of other malignancies different from the t(14,18) in follicular lymphomas. In fact, high levels of bcl-2 protein have been demonstrated in acute and chronic lymphocytic leukemias, myelogenous leukemias, and intermediate high-grade lymphomas that lack the specific translocation of the bcl-2 gene to the immunoglobulin heavy-chain locus, suggesting that other potential mechanisms may be involved in bcl-2 gene deregulation (Hsu et al., 1994). Recently bcl-2 overexpression has been found in other tumors, such as prostate cancer, adenocarcinoma of the colon and rectum, gastric epithelial dysplasia, adenocarcinoma of the breast, neuroblastomas, Ewing sarcomas, retinoblastomas, and neurofibromas, suggesting its role in deregulation of cell death in malignancies other than leukemias and lymphomas (Willingham and Bhalla, 1994; Lauwers et al., 1994).

These data indicate that bcl-2 overexpression can occur either as a primary or a secondary event in carcinogenesis but that apoptosis inhibition and the resulting increased cell survival might be indicative of an aggressive course of the disease. For instance, in chronic lymphocytic leukemia there is an expansion of a subpopulation of CD5+ B cells that are functionally inactive and long-lived. In these cells a high-level expression of bcl-2, up to 31 times higher than that occurring in normal mononuclear circulating cells, has been observed. In acute promyelocytic leukemia NB4 cell lines, bcl-2 is constitutively expressed in association with bax, bcl-XL, p53, and c-myc. When these cells are treated with As2 03, a novel therapeutic agent for promyelocytic leukemia, a down-regulation of bcl-2 protein occurs, and the malignant cells enter the apoptotic pathway of cell death (Guo Qiang et al., 1996). These observations suggest that in cells with bcl-2 overexpression its product acts only as a brake on the apoptosis machinery without destroying it. Furthermore, it has been shown that short-term culture of chronic lymphocytic leukemia (CLL) cells generally leads to cell destruction via the apoptotic pathway. There is, however, a marked heterogeneity of apoptotic response in patients with CLL. Both spontaneous and drug-induced apoptosis are significantly greater in lymphocytes obtained from patients with early-stage, untreated CLL; in contrast, the lowest levels are observed in patients with heavily pretreated refractory disease (Robertson et al., 1994). These data indicate that the varying propensity of CLL cells to undergo apoptosis *in vitro* correlates with the clinical features and that a relatively apoptosis-resistant phenotype develops with disease progression. Although classic bcl-2 translocation or molecular evidence of bcl-2 rearrangements is lacking in CLL, bcl-2 expression is common. Bcl-2 immunolocalization can be helpful in distinguishing between malignant lymphoma and reactive follicular hyperplasia. The spectrum of hematopoietic neoplasms in which Bcl-2 may confer survival advantage extends beyond the t(14;18)-bearing tumors (Pezzella et al., 1990; Korsmeyer, 1992).

Testicular Tumors

Another example of the great variation of apoptotic potential in tumors is that of germ cell testicular tumors. While for most types of malignancies the response rate and long-term survival following chemotherapy are quite disappointing, mainly due to increased drug resistance, testicular germ cell tumors are the most chemosensitive malignant neoplasias even in advanced clincical stages (Walker et al., 1986; Morrow et al., 1993). High chemosensitivity has been observed as well in other selected groups of tumors such as some pediatric sarcomas and lymphomas. The reason for this special sensitivity remains obscure, and studies in cell cycle kinetics, tumor microenvironment, and MDR genotypes have thus far failed to explain this phenomenon.

It is likely that cells from these groups of tumors are particularly "sensitive" to induction of apoptosis as opposed to "resistant" cells of the mostly chemoresistant cancers. In fact, some cell lines from lymphomas, leukemias, and germ cell tumors can be easily induced to enter the death pathway of apoptosis in cell cultures. Germ cell tumors of the testis in general seem to be composed of these type of "sensitive" or "primed" cells for programmed cell death, and this is why they respond quite well to relatively mild chemo- and radiotherapy (Huddart et al., 1995).

So far there is no marker to distinguish "sensitive" from "insensitive" malignant cells in histological sections. In cell culture, however, a challenge with chemotherapeutic agents or mild radiation could distinguish between the two populations, and with the aid of any quantitative method an apoptotic index could be calculated. This assay could be of great value for clinicians to establish prognosis and predict response to oncological treatments.

INDUCTION OF APOPTOSIS IN THE TREATMENT OF CANCER

Medical treatment of malignant neoplasias can be grouped into three main types: chemotherapy, radiation, and adjuvant therapy. Either alone or in combination these agents were originally aimed to destroy proliferating cells by interfering directly or indirectly with the cell cycle (Stockdale, 1987; Morrow et al., 1993). It has become apparent in the past few years that these therapeutic strategies, aside from their damaging action on proliferating cells, are also capable of inducing apoptosis, both in normal and in malignant cells (Fig. 17.4; Fisher, 1994). But it is also known that malignant tumors respond differently to these therapeutic regimes and that eventually all advanced tumors become refractory to their cytotoxic activity (Huschtscha, 1996).

During treatment a chemosensitive or radiosensitive tumor can evolve into a full refractory state to these treatments. Recently, the capacity of malignant cells to resist the effects of chemotherapy, radiation, or some hormonal treatments has been related to their ability to escape the induction of apoptosis. It follows that measurement and estimation of apoptotic potential in tumors could be of considerable help in predicting the outcome of therapeutic regimes in cancer patients (Dive et al., 1992).

The way in which anticancer drugs induce apoptosis is largely unknown. Apoptosis occurs in many types of cancer cells exposed to a variety of chemotherapeutic drugs, including antimetabolites, deoxynucleotide synthesis inhibitors, DNA topoisomerase inhibitors, alkylating agents, and drugs that interfere with microtubules. Also, hormones, such as glucocorticoids, widely used in some oncological treatment regimes, induce apoptosis in many normal and

INDUCTION OF APOPTOSIS
IN ONCOLOGICAL TREATMENT

● Radiation	● Hormone withdrawal
● Chemotherapy	● Ab vs. APO-1-Fas
● Mild Hyperthermia	● Cytotoxic Lymphocytes

FIG. 17.4. All known oncological treatments are capable of inducing apoptosis in sensitive cells. Chemo- and radio-resistant phenotypes arise frequently due to inhibition of cell death program (see text for explanation).

malignant cell lines, and hormonal withdrawal, mainly of testosterone, can also induce apoptosis in androgen-dependent prostate cancer clones.

A major obstacle encountered in cancer therapy is multidrug resistance (MDR). This acquired-resistance phenotype that emerges in many tumors after they are challenged with a variety of chemotherapeutic agents has been traced to the expression of two genes: mdr-1 and the mdr-2, that encode for the p170 glycoprotein, which acts as an ATP-dependent efflux pump expelling the drug from the cell interior (Morrow et al., 1993). Therefore, the cancer cells expressing these genes become resistant and continue to progess. Recent studies, however, support the idea that some genes that inhibit apoptosis when overexpressed can also confer the MDR phenotype, mainly the bcl-2 and p53 (Miyashita and Reed, 1993). This has been proven experimentally in both human and murine hematolymphoid cell lines that become resistant to apoptosis induced by a number of antineoplastic compounds after transfection with bcl-2. Expression of this gene does not protect from DNA damage since the number of strand breaks induced by camptothecin in normal and bcl-2 transfected cells (Hsu et al., 1994) is similar, suggesting that bcl-2 must function downstream to block apoptosis normally induced by the accumulation of DNA strand breaks. There is, however, considerable variation of the cell death inhibitory capacity of bcl-2 in different cell lines. Also the p53 status correlates well with efficacy of cancer therapy (Lowe et al., 1994). Recent reports suggest that determination of bcl-2 protein levels directly in leukemic cell samples from patients may be predictive of the outcome to treatment. For instance, high levels of bcl-2 protein correlate with lower rates of complete remission and shorter survival in patients with acute myeloid leukemia.

Ionizing radiation is commonly used in the treatment of cancer. Depending on the dose, radiation can induce either apoptosis or necrosis. At small or moderate doses, the main effect of ionizing radiation is the induction of apoptosis mainly in rapidly proliferating normal tissues or malignant neoplasias. There have been few studies of apoptosis in irradiated tumors, but it is clear that the extent of apoptosis induced by radiotherapy varies enormously from one malignancy to another. There is, however, good evidence that there may be a strong correlation between the magnitude of the apoptotic response and radio-curability.

The way in which ionizing radiation triggers apoptosis in normal and neoplastic cells is not completely understood, but it is probably a consequence of the damage inflicted on DNA. The cell has very delicate sensors to monitor the integrity of the genome, the most active being the p53 gene. When DNA is damaged, the p53 product accumulates through a post-translational stabilization mechanism and arrests the cell cycle at G_1 to allow extra time for DNA repair. If repair fails, p53 triggers deletion of the abnormal cell via the apoptotic pathway. Thymocytes lacking p53 are resistant to radiation, and those of mice transgenic for bcl-2 also exhibit resistance to gamma-induced cell death. On the other hand, increased levels of normal p53 in certain tumor-derived cell lines are known to induce apoptosis. It is therefore reasonable to speculate that in resistance to radiation-induced cell death, several genes are deregulated; bcl-2 overexpression and p53 mutation might be overlapping events (Donehower, 1994; Wyllie et al., 1994).

A further complication in studying human cancer and its response to therapeutic agents is the acquired genetic defects taking place during tumor progression. These mutations can alter genes involved in the cell death program and generate apoptosis-resistant clones. Lowe et al. have demonstrated the generation of p53 *de novo* mutations in tumors derived from p53$^{+/+}$ cells that had either relapsed or displayed acquired resistance to therapy. Initially responsive tumors become resistant to cancer therapy upon tumor relapse due to a clonal selection by the therapeutic agents (Lowe et al., 1994). Apoptosis-sensitive clones can be effectively eliminated by chemo- or radiotherapy, but resistant clones remain and expand in later stages. Also, within a certain tumor type considerable variations exist in regard to chemosensitivity. In Wilms' tumor, 85% of patients survive after appropriate treatments (chemotherapy and surgery), but the most aggressive and anaplastic form is usually resistant to treatment, and patients harboring this variety have a very poor prognosis. Bardeesy et al. found that p53 mutations in this renal tumor correlated with strong anaplasia and a substantial decrease in spontaneous apoptosis compared to nonanaplastic tumors that do not show p53 mutations (Bardeesy et al., 1994; Lowe et al., 1994). This suggests that the presence of a wild type of p53 could be indicative of a better prognosis and that p53 mutated types characterize the more aggressive behavior and a poor response to antioncological treatments. In fact, elevated (mutated) p53 expression has been described in a number of human tumors, and some have established a direct correlation with prognostic indicators such as estrogen receptor status, while others fail to find any predictive value of clinical outcome (Scott et al., 1991; Xia et al., 1995; Kang et al., 1996).

All these data point to the possibility of assessing sensitivity or resistance to different oncological treatments by means of an accurate study of apoptosis and through the expression of the various genes that control this cell death mechanism. The problem lies in the great variation of the rate of cell deletion that occurs in tumors and the different degree of expression of genes that regulate apoptosis in malignancies of various cellular origins. For instance, the ability of bcl-2 to inhibit cell death induced by chemotherapeutic agents varies in different cell lines. For example, bcl-2 transfection to fibroblasts and lymphoid cells make them resistant to etoposide, but the same is not true when bcl-2 is transfected to a lung cancer–derived cell line. Furthermore, the lack of bcl-2 expression in non-Hodgkin's lymphomas appears to represent an adverse prognostic factor (Hsu et al., 1994).

Nevertheless, these investigations have stimulated the idea that induction of cell death in malignant tumors could enhance the sensitivity to current anticancer regimes, and in the not too distant future new therapeutic strategies could be developed. Targeted induction of apoptosis would potentially be less toxic than antimitotics, and certainly would be a different and possible synergistic modality.

The primary importance of the apoptosis concept in the treatment of cancer lies in its being a regulated phenomenon subject to stimulation and inhibition. Although little is known about how antioncological therapeutic agents affect its initiation, it seems reasonable to suggest that apoptosis plays a central role. It can be anticipated that through a better understanding of the molecular regulation of apoptosis, new and more sensitive methods can be developed, allowing a more accurate assessment of this phenomenon. We now recognize that the deregulation of PCD is an important component of most malignancies and that susceptibility for undergoing cell death may be regulated by the products of oncogenes and tumor suppressor genes. This has led to increasing interest in exploiting the induction of cell death, specifically in malignant cell populations, as a potential area for the design of new anticancer strategies.

REFERENCES

Aihara M, Truong LD, Dunn JK, Wheeler TM, Scardino PT, Thompson TC (1994). Frequency of apoptotic bodies positively correlates with Gleason grade in prostate cancer. Hum Pathol 25:797–801.

Aria T, Kino I (1995): Role of apoptosis in modulation of the growth of human colorectal tubular and villous adenomas. J Pathol 176:37–44.

Baak JPA, Van Dop H, Kurver PHJ, Hermans J (1985): The Value of morphometry to classic prognosticators in breast cancer. Cancer 56:374–82.

Baker SJ, Fearon ER, Nigro J, et al. (1989): Chromosome 17 deletions and p53 gene mutations in colorectal carcinomas. Science 244:217–21.

Bardeesy N, Falkoff D, Petruzzi MJ, Nowak N, Zabel B, Adam M, Aguiar MC, Grundy P, Shows T (1994): Anaplastic Wilm's tumour, a subtype displaying poor prognosis, harbours p53 mutations. Nat Genet 7:91–97.

Bedi A, Pasricha PJ, Akhtar AJ, Barber JP, Bedi GC, Giardiello FM, Zehnbauer BA, Hamilton SR, Jones RJ (1995): Inhibition of apoptosis during development of colorectal cancer. Cancer Res 55:1811–16.

Benítez-Bribiesca L (1996): Imágenes de la apoptosis, Gac Méd Méx. 132:641.

Brown DG, Sun XM, Cohen GM (1993): Desamethasone-induced apoptosis involves cleavage of DNA to large fragments prior to internucleosomal fragmentation. J Biol Chem 268:3037–39.

Bump NJ, Hackett M, Hugunin M, Seshagiri S, Brady K, Chen P, Ferenz C, et al. (1995): Inhibition of ICE family proteases by baculovirus antiapoptotic protein p35. Science 269:1885–88.

Brusch W, Paffe S, Putz B, Barthel G, Shulte-Hermann R (1990): Determination of the length of the histological stages of apoptosis in normal liver and in altered hepatic foci of rats. Carcinogenesis 11:847–53.

Bursch W, Taper HS, Lauer B, Schulte-Hermann R (1985): Quantitative histological and histochemical studies on the occurrence and stages of controlled cell death (apoptosis) during regression of rat liver hyperplasia. Virchows Arch [Cell Pathol] 50:153–66.

Cheng J, Zhou T, Liu CH, Shapiro PJ, Braver JM, Kiefer CM, Barr JP, Mountz DJ (1994): Protection from Fas-mediated apoptosis by a soluble form of the Fas molecule. Science 263:1759–62.

Clark GM, Dressler LG, Owens MA, Pounds G, Oldaker T, McGuire WL (1989): Prediction of relapse or survival in patients with node-negative breast cancer by DNA flow cytometry. N Engl J Med 320:627–33.

Clark WH (1991): Tumour progression and the nature of cancer. Br J Cancer 64:631–44.

Cleary ML, Sklar J (1985): Nucleotide sequence of a t(14;18) chromosomal breakpoint in follicular lymphoma and demonstration of a breakpoint cluster region near a transcriptionally active locus on chromosome 18. Proc Natl Acad Sci USA 82:7439–43.

Cohen GM, Sun XM, Snowden RT, Dinsdale D, Skilleter DN (1992): Key morphological features of apoptosis may occur in the absence of internucleosomal DNA fragmentation. Biochem J 286:331–34.

Collins RJ, Harmon BV, Gobé GC, Kerr JFR (1992): Internucleosomal DNA cleavage should not be the sole criterion for identifying apoptosis. Int J Radiat Biol 61:451–53.

Colombel M, Symmans F, Gil S, O'Toole KM, Chopin D, Benson M, Olsson CA, Korsmeyer S, Buttyan R (1993): Detection of the apoptosis-suppressing oncoprotein bcl-2 in hormone-refractory human prostate cancers. Am J Pathol 143:390–400.

Cory S, Strasser A, Jacks T, Corcoran LM, Metz T, Harris AW, Adams JM (1994): Enhanced cell survival and tumorigenesis. In The Molecular Genetics of Cancer. Cold Spring Harbor Symposia on Quantitative Biology. Vol. LIX. Cold Spring Harbor Laboratory Press. pp 365–75.

Cotter TG, Martin SJ (1996): Techniques in apoptosis. A user's guide. London: Protland Press Ltd.

Darzynkiewicz Z, Bruno S, Del Bino G, Gorczyca W, Hotz MA, Lassota P, Traganos F (1992): Features of apoptotic cells measured by flow cytometry. Cytometry 13:795–808.

Darzynkiewicz Z, Li X. Labelling DNA strand breaks with BrdUTP (1995): Detection of apoptosis and cell proliferation. Cell Prolif 28:571–79.

Darzynkiewicz Z, Melamed MR (1993): Flow cytometry and sorting. An historical perspective. Cytometry Res 3:1–9.

Dive C, Evans CA, Whetton AD (1992): Induction of apoptosis new targets for cancer chemotherapy. Semin Cancer Biol 3:417–27.

Donehower LA (1994): Tumor suppressor gene p53 and apoptosis. Cancer Bull 46:161–66.

Dressler LG, Seamer LC, Owens MA, Clark GM, McGuire WL (1988): DNA flow cytometry and prognostic factors in 1331 frozen breast cancer specimens. Cancer 61:420–27.

Fehsel K, Kröncke KD, Kolb H, Kolb-Bachofen V (1994): In situ nick-translation detects focal apoptosis in thymuses of glucocorticoid- and lipopolysaccharide-treated mice. J Histochem Cytochem 42:613–19.

Fernandes-Alnemri T, Litwack G, Alnemri ES (1995a): Mch2, a new member of the apoptotic Ced-3/ice cysteine protease gene family. Cancer Res 55:2737–42.

Fernandes-Alnemri T, Takahashi A, Armstrong R, Krebs J, Fritz L, Tomaselli KJ, et al. (1995b): Mch3, a novel human apoptotic cysteine protease highly related to CPP32[1]. Cancer Res 55:6045–52.

Fernandez-Sarabia MJ, Bischoff JR (1993): Bcl-2 associates with the ras-related protein R-ras p23. Nature 366:274–75.

Fisher DE (1994): Apoptosis in cancer therapy: Crossing the threshold. Cell 78:539.

Furuya Y, Isaacs JT (1994): Proliferation-independent programmed cell death as a therapeutic target for prostate cancer. Cancer Bull 46:173–79.

Gaffney EV, Venz-Williamson TL, Hutchinson G, Biggs PJ, Nelson KM (1996): Relationship of standardized mitotic indices to other prognostic factors in breast cancer. Arch Pathol Lab Med 120:473–77.

Gavrieli Y, Sherman Y, Ben-Sasson SA (1992): Identification of programmed cell death in situ via specific labeling of nuclear DNA fragmentation. J Cell Biol 119:493–501.

Gold R, Schmied M, Rothe G, Zischler H, Brietschopf H, Wekerle H, Lassmann H (1993): Detection of DNA fragmentation in apoptosis: Application of in situ nick translation to cell culture systems and tissue sections. J Histochem Cytochem 41:1023–30.

Grasl-Kraupp B, Ruttkay-Nedecky B, Koudelka H, Bukowska K, Brusch W, Schulte-Hermann R (1995): In situ detection of fragmented DNA (TUNEL assay) fails to discriminate among apoptosis, necrosis and autolytic cell death: A cautionary note. Hepatology 21:1465–68.

Guo-Qiang CH, Jun Z, Xue-Geng S, Jian-Hua N, Hao-Jie Z, Gui-Ying S, Xiao-Long J, et al. (1996): In vitro studies on cellular and molecular mechanisms of arsenic trioxide (As_2O_3) in the treatment of acute promyelocytic leukemia: As_2O_3 induces NB_4 cell apoptosis with downregulation of Bcl-2 expression and modulation of PML-RAR/PML proteins. Blood 88:1052–61.

Harrington EA, Fanidi A, Evan GI (1994): Oncogenes and cell death. Curr Opin Genetics Development 4:120–29.

Hermeking H, Eick D (1994): Mediation of c-Myc-induced apoptosis by p53. Science 265:2091–93.

Hockenbery D (1995): Defining apoptosis. Amer J Pathol 146:16–19.

Hockenbery DM, Zutter M, Hickey W, Nahm M, Korsmeyer ST (1991): BCL-2 protein is topographically restricted in tissues characterized by apoptotic cell death. Proc Natl Acad Sci 88:6961–65.

Hollstein M, Sidransky D, Vogelstein B, Harris CC (1991): p53 mutations in human cancers. Science 253:49–53.

Hsu B, Marin MC, Brisbay S, McConnell K (1994): Expression of bcl-2 gene confers multidrug resistance to chemotherapy-induced cell death. Cancer Bull 46:125–29.

Huddart RA, Titley J, Robertson D, Williams GT, Horwich A, Cooper CS (1995): Programmed cell death in response to chemotherapeutic agents in human germ cell tumour lines. Eur J Cancer 31A:739–46.

Huschtscha LI, Bartier WA, Andersson Ross CE, Tattersall MHN (1996): Characteristics of cancer cell death after exposure of cytotoxic drugs in vitro. Brit J Cancer 73:54–60.

Isaacs JT (1989): Relationship between tumor size and curability of prostatic cancer by combined chemo–hormonal therapy in rats. Cancer Res 49:6290–94.

Kallioniemi OP, Blanco G, Alavaikko M, Hietanen T, Mattila J, Lauslahti K, Lehtinen M, Koivula T (1988): Improving the prognostic value of DNA flow cytometry in breast cancer by combining DNA index and S-phase fraction. Cancer 62:2183–90.

Kane AB (1995): Redefining cell death. Amer J Pathol 146:1–2.

Kang Y, Cortina R, Perry RR (1996): Role of c-myc in tamoxifen-induced apoptosis in estrogen-independent breast cancer cells. J Nat Cancer Inst 88:279–90.

Kasagi N, Gomyo Y, Shirai H, Tsujitani S, Ito H (1994): Apoptotic cell death in human gastric carcinoma: Analysis by terminal-deoxynucleotidyl-transferase-mediated dUTP-biotin nick end labeling. Jap J Cancer Res 85:939–45.

Kastan MB, Onyekwere O, Sidransky D et al. (1991): Participation of p53 protein in the cellular response to DNA damage. Cancer Res 51:6304–11.

Kastan MB, Zhan Q, El-Deiry WS (1992): A mammalian cell cycle checkpoint pathway utilizing p53 and GADD45 is defective in ataxia-telangiectasia. Cell 71:587–97.

Kerr JFR, Winterford CM, Harmon BV (1994): Apoptosis. Its significance in cancer and cancer therapy. Cancer 73:2013–26.

Kondo S, Barna BP, Morimura T, Takeuchi J, Yuan J, Akbasak A, Barnett GH (1995): Interleukin-1β-converting enzyme mediates cisplatin-induced apoptosis in malignant glioma cells. Cancer Res 55:6166–71.

Korsmeyer SJ (1992): Bcl-2 initiates a new category of oncogenes: Regulators of cell death. Blood 80:879–86.

Kyprianou N, Isaacs JT (1988): Activation of programmed cell death in the rat ventral prostate following castration. Endocrinology 122:552–62.

Laucat FR, Le Deist F, Hivroz C, Roberts IAG, Debatin KM, Fischer A, De Villartay JP (1995): Mutations in Fas associated with human lymphoproliferative syndrome and autoimmunity. Science 268:1347–49.

Lauwers GY, Scott GV, Hendricks J (1994): Immunohistochemical evidence of aberrant bcl-2 protein expression in gastric epithelial dysplasia. Cancer 83:2900–4.

Leoncini L, Vecchio MTD, Megha T, Barbini P, Galieni SP, Sabattini E, Gherlinzoni F, Tosi P, Kraft R, Cottier H (1993): Correlations between apoptotic and proliferative indices in malignant non-Hodgkin's lymphomas. Amer J Pathol 142:755–63.

Lowe SW, Bodis S, McClatchey A, Remington L, Ruley HE, Fisher DE, Housman DE, Jacks T (1994): p53 status and the efficacy of cancer therapy in vivo. Science 266:807–10.

Majno G, Joris Y (1996): Cells, tissues and disease. Principles of general pathology. 1st. Ed. Blackwell Science USA, Chap. 26, pp 727–78.

Majno G, Joris Y (1995): Apoptosis, oncosis, and necrosis. An overview of cell death. Amer J Pathol 146:3–15.

Martikainen P, Kyprianou N, Tucker RW, Isaacs JT (1991): Programmed death of nonproliferating androgen-independent prostatic cancer cells. Cancer Res 51:4693–701.

Meyer JS, Rao R, Stevens SC, White WL (1977): Low incidence of estrogen receptor in breast carcinomas with rapid rates of cellular replication. Cancer 40:2290–98.

Meyn R, Milas L, Stephens C (1994): Programmed cell death in normal development and disease. Cancer Bull 46:120–24.

Miura M, Zhu H, Rotello R, Hartweig EA, Yuan J (1993): Induction of apoptosis in fibroblast by IL-1β-converting enzyme, a mammalian homolog of the C. elegans cell death gene ced-3. Cell 75:653.

Miyashita T, Reed JC (1993): Bcl-2 oncoprotein blocks chemotherapy-induced apoptosis in a human leukemia cell line. Blood 81:151–57.

Morrow CS, Cowan KH, Devita VT, Hellman S, Rosenberg SA (eds.) (1993): Mechanisms of antineoplastic drug resistance. In Cancer Principles and Practice of Oncology, 4th Edition, Philadelphia: J.B. Lippincott, pp 340–48.

Nagata S, Golstein P (1995): The Fas death factor. Science 267:1449–56.

Negoescu A, Lorimier P, Labat-Moleur F, Frouet Ch, Robert C, Guillermet Ch, Brambilla Ch, Brambilla E (1996): In situ apoptotic cell labeling by the TUNEL method: Improvement and evaluation on cell preparations. J Histochem Cytochem 44:959–68.

Nowell PC (1976): The clonal evolution of tumor cell populations. Science 194:23–28.

Nowell PC (1990): Cytogenetics of tumor progression. Cancer 65:2172–77.

O'Connell J, Sulivan GC, Collins LJ, Shanahan F (1996): The fas counterattack: Fas-mediated T cell killing by colon cancer cells expressing Fas ligand. J Exp Med 184:1075–82.

Ohta L, Sweency EA, Masamune A, Yatomi Y, Hakomori Sen-itiroh (1995): Induction of apoptosis by sphingosine in human leukemic HL-60 cells: A possible endogenous modulator of apoptotic DNA fragmentation occurring during phorbol ester-induced differentiation. Cancer Res 55:691–97.

Olive PL, Banáth JP, Durand RE (1990): Heterogeneity in radiation-induced DNA damage and repair in tumor and normal cells measured using the "comet" assay. Radiat Res 122:86–94.

Olive PL, Banáth JP (1995): Sizing highly fragmented DNA in individual apoptotic cells using the comet assay and a DNA crosslinking agent. Exper Cell Res 221:19–26.

Olive PL, Frazer G, Banáth JP (1993): Radiation-induced apoptosis measured in TK6 human B lymphoblast cells using the comet assay. Radiat Res 136:130–36.

Owen-Schaub L (1994): Fas/APO-1: A cell surface protein mediating apoptosis. Cancer Bull 46:141–45.

Peitsch MC, Müller Ch, Tschopp J (1993): DNA fragmentation during apoptosis is caused by frequent single-strand cuts. Nucl Acids Res 21:4206–9.

Pezzella F, Tse AGD, Cordell JL, Pulford KAF, Gatter KC, Mason DY (1990): Expression of the bcl-2 oncogene protein is not specific for the 14;18 chromosomal translocation. Amer J Pathol 137:225–32.

Pronk GJ, Ramer K, Amiri P, Williams LT (1996): Requirement of an ICE-like protease for induction of apoptosis and ceramide generation by REAPER. Science 271:808–10.

Reed JC (1994) Bcl-2 and the regulation of programmed cell death. J Cell Biol 124:1.

Reichmann A, Martin P, Levin B (1981): Chromosome banding patterns in human large bowel cancer. Int J Cancer 28:431.

Robertson LE, Huang P, Keating M, Plunkett W (1994): Apoptosis in chronic lymphocytic leukemia. Cancer Bull 46:130–35.

Rosai J (1981): Ackerman's Surgical Pathology. Sixth ed. The C.V. Mosby Co.

Ryser HJ (1971): Chemical carcinogenesis. N Engl J Med 285:721–34.

Schwartzman RA, Cidlowski JA (1993): Apoptosis: The biochemistry and molecular biology of programmed cell death. Endocrine Rev 14:133–50.

Scott N, Sagar P, Stewart J, Blair GE, Dixon MF, Quirke P (1991): p53 in colorectal cancer: Clinicopathological correlation and prognostic significance. Br J Cancer 63:317–19.

Sentman CL, Shutter JR, Hockenbery D, Kanagawa O, Korsmeyer SJ (1991): bcl-2 inhibits multiple forms of apoptosis but not negative selection in thymocytes. Cell 67:879.

Service RF (1994): Stalking the start of colon cancer. Science 263:1559.

Shackney SE, McCormack GW, Cuchural GJ (1978): Growth rate patterns of solid tumors and their relation to responsiveness to therapy. Ann Intern Med 89:107–21.

Shaw P, Bovey R, Tardy S, Sahli R, Sordat B, Costa J (1992): Induction of apoptosis by wild-type p53 in a human colon tumor-derived cell line. Proc Natl Acad Sci 89:4495–98.

Silvestrini R, Daidone MG, Di Fonzo G (1979): Relationship between proliferative activity and estrogen receptors in breast cancer. Cancer 44:665–70.

Statunton MJ, Gaffney EF (1995): Tumor type is a determinant of susceptibility to apoptosis. Am J Clin Pathol 103:300–7.

Steel GG (1967): Cell loss as a factor in the growth rate of human tumours. Eur J Cancer 3:381–87.

Steel GG (1977): Growth kinetics of tumours. Oxford: Clarendon Press.

Steller H (1995): Mechanisms and genes of cellular suicide. Science 267:1445–62.

Stockdale FE (1987): Cancer growth and chemotherapy. In Rubenstein E, Federman DD, (eds). Scientific American Medicine. Oncology. New York: Scientific American, 1–13.

Strasser A, Vaux DL (1996): The molecular biology of apoptosis. Proc Natl Acad Sci 93:2239–44.

Sun XM, Snowden RT, Dinsdale D, Ormerod MG, Cohen GM (1994): Changes in nuclear chromatin precede internucleosomal DNA cleavage in the induction of apoptosis by etoposide. Biochem Pharmacol 47:187–95.

Tatebe S, Ishida M, Kasagi N, Tsujitani S, Kaibara N, Ito H (1996): Apoptosis occurs more frequently in metastatic foci than in primary lesions of human colorectal carcinomas: Analysis by terminal-deoxynucleotidyl-transferase-mediated dUTP-biotin nick end labeling. Int J Cancer 65:173–77.

Thompson CB (1995): Apoptosis in the pathogenesis and treatment of disease. Science 267:1456–62.

Thornberry NA, Bull HG, Calaycay JR, Chapman KT, Howard AD, Kostura MJ, Miller DK, et al. (1992): A novel heterodimeric cysteine protease is required for interleukin-1β processing in monocytes. Nature 356:768–74.

Vanhaesebroeck B, Reed JC, de Valck D, Grooten J, Miyashita T, Tanaka S, Beyaert R, van Roy R, Fiers W (1993): Effect of bcl-2 proto-oncogene expression on cellular sensitivity to tumor necrosis factor-mediated cytotoxicity. Oncogene 8:1075.

Vaux DL, Cory S, Adams JM (1988): Bcl-2 gene promotes haematopoietic cell survival and cooperates with c-myc to immortalize pre-B cells. Nature 335:440.

Vogelstein B, Fearon ER, Hamilton SR, et al. (1988): Genetic alterations during colorectal-tumor development. N Engl J Med 319:525–32.

Walker MC, Parris CN, Masters JRW (1986): Differential sensitivity of human testicular and human bladder cell lines to chemotherapeutic agents. J Natl Cancer Inst 79:213–16.

Willingham MC, Bhalla K (1994): Transient mitotic phase localization of Bcl-2 onco-protein in human carcinoma cells and its possible role in prevention of apoptosis. J Histochem Cytochem 42:441–50.

Wyllie AH, Carder PJ, Clarke AR, Cripps KJ, Gledhili S, Greaves MF, et al. (1994): Apoptosis in carcinogenesis: The role of p53 in the molecular genetics of cancer. Cold Spring Harbor Simposia on Quantitative Biology. Vol. LIX. pp 403–9.

Xia F, Wang X, Wang YH, Tsang NM, Yandell DW, Kelsey KT, Liber HL (1995): Altered p53 status correlates with differences in sensitivity to radiation-induced mutation and apoptosis in two closely related human lymphoblast lines. Cancer Res 55:12–15.

Yonish-Rouach E, Resnitzky D, Lotem J, Sachs L, Kimchi A, Oren M (1991): Wild-type p53 induces apoptosis of myeloid leukaemic cells that is inhibited by interleukin-6. Nature 352:345–47.

Zakeri ZF, Quaglino D, Latham T, Lockshin RA (1993): Delayed internucleosomal DNA fragmentation in programmed cell death. FASEB J 7:470–78.

INDEX